INTERNATIONAL FUEL GAS CODE®

COMMENTARY

2003

First Printing: May 2004

ISBN # 1-58001-132-2

COPYRIGHT © 2004
by
INTERNATIONAL CODE COUNCIL, INC.

ALL RIGHTS RESERVED. This 2003 International Fuel Gas Code® Commentary is a copyrighted work owned by the International Code Council, Inc. Without advance written permission from the copyright owner, no part of this book may be reproduced, distributed, or transmitted in any form or by any means, including, without limitation, electronic, optical or mechanical means (by way of example and not limitation, photocopying, or recording by or in an information storage retrieval system). For information on permission to copy material exceeding fair use, please contact: Publications, 4051 West Flossmoor Road, Country Club Hills, IL 60478-5795 (Phone 800-214-4321).

Trademarks: "International Code Council," the "International Code Council" logo and the "International Fuel Gas Code" are trademarks of the International Code Council, Inc.

Material designated IFGS
by
AMERICAN GAS ASSOCIATION
400 N. Capitol Street, N.W. • Washington, DC 20001
(202) 824-7000

Copyright © American Gas Association, 2002. All rights reserved.

PRINTED IN THE U.S.A.

PREFACE

The principal purpose of the Commentary is to provide a basic volume of knowledge and facts relating to building construction as it pertains to the regulations set forth in the 2003 *International Fuel Gas Code®*. The person who is serious about effectively designing, constructing and regulating buildings and structures will find the Commentary to be a reliable data source and reference to almost all fuel gas components of the built environment

As a follow-up to the *International Fuel Gas Code*, we offer a companion document, the *International Fuel Gas Code Commentary*. The basic appeal of the Commentary is that it provides in a small package and at reasonable cost thorough coverage of many issues likely to be dealt with when using the *International Fuel Gas Code*. It then supplements that coverage with historical and technical background. Reference lists, information sources and bibliographies are also included.

Throughout all of this, strenuous effort has been put into keeping the vast quantity of material accessible and its method of presentation useful. With a comprehensive yet concise summary of each section, the Commentary provides a convenient reference for regulations applicable to the construction of buildings and structures. In the chapters that follow, discussions focus on the full meaning and implications of the code text. Guidelines suggest the most effective method of application and the consequences of not adhering to the code text. Illustrations are provided to aid understanding; they do not necessarily illustrate the only methods of achieving code compliance.

The format of the Commentary includes the full text of each section, table and figure in the code, followed immediately by the commentary applicable to that text. At the time of printing, the Commentary reflects the most up-to-date text of the 2003 *International Fuel Gas Code*. The format of the Commentary includes the full text of each section, table, and figure in the code, followed immediately by the commentary for that text. Each chapter's narrative includes a section on "General Comments" and "Purpose," and each section's narrative usually includes a discussion about why the requirement commands the conditions set forth. Code text is reproduced as it appears in the *International Fuel Gas Code*, and commentary is indented below the code text, beginning with the symbol ❖. Code figures and tables are reproduced as they appear in the *International Fuel Gas Code*. Commentary figures and tables are identified in the text by the word "Commentary" (as in "see commentary Figure 704.3"), and each has a full border.

The *International Fuel Gas Code* is segregated by section numbers into two categories: code and standard. Code sections are identified as IFGC; standard sections are identified as IFGS.

Commentary is to be used in conjunction with the *International Fuel Gas Code* and not as a substitute for the code. **The Commentary is advisory only;** the code official alone possesses the authority and responsibility for interpreting the code.

Comments and recommendations are encouraged, for through your input, we can improve future editions. Please direct your comments to the Codes and Standards Development Department at the Chicago District Office.

TABLE OF CONTENTS

CHAPTER 1	ADMINISTRATION	1
CHAPTER 2	DEFINITIONS	19
CHAPTER 3	GENERAL REGULATIONS	47
CHAPTER 4	GAS PIPING INSTALLATIONS	97
CHAPTER 5	CHIMNEYS AND VENTS	185
CHAPTER 6	SPECIFIC APPLIANCES	281
CHAPTER 7	GASEOUS HYDROGEN SYSTEMS	319
CHAPTER 8	REFERENCED STANDARDS	333
APPENDIX A	SIZING AND CAPACITIES OF GAS PIPING	339
APPENDIX B	SIZING OF VENTING SYSTEMS SERVING APPLIANCES EQUIPPED WITH DRAFT HOODS, CATEGORY I APPLIANCES, AND APPLIANCES LISTED FOR USE WITH TYPE B VENTS (IFGS)	349
APPENDIX C	EXIT TERMINALS OF MECHANICAL DRAFT AND DIRECT-VENT VENTING SYSTEMS (IFGS)	359
APPENDIX D	RECOMMENDED PROCEDURE FOR SAFETY INSPECTION OF AN EXISTING APPLIANCE INSTALLATION (IFGS)	361
INDEX		363

CHAPTER 1
ADMINISTRATION

General Comments

The law of building regulation is grounded on the police power of the state. In terms of how it is used, this is the power of the state to legislate for the general welfare of its citizens. This power enables passage of such laws as a fuel gas code. It is from the police power delegated by the state legislature that local governments are able to enact building regulations. If the state legislature has limited this power in any way, the municipality may not exceed these limitations. Although the municipality may not further delegate its police power (e.g., by delegating the burden of determining code compliance to the building owner, contractor or architect), it may turn over the administration of building regulations to a municipal official, such as a code official, if he or she is given sufficient criteria to clearly establish the basis for decisions as to whether or not a proposed building, including its fuel gas systems, conforms to the code.

Chapter 1 is largely concerned with maintaining "due process of law" in enforcing the performance criteria contained in the code. Only through careful observation of the administrative provisions can the code official reasonably hope to demonstrate that "equal protection under the law" has been established. Although it is generally assumed that the administrative and enforcement sections of a code are geared toward the code official, this is not entirely true. The provisions also establish the rights and privileges of the design professional, contractor and building owner. The position of the code official is merely to review the proposed and completed work and determine whether a fuel gas installation conforms to the code requirements. The design professional is responsible for the design of a safe, sanitary fuel gas system. The contractor is responsible for installing the system in strict accordance with the plans.

During the course of the construction of a fuel gas system, the code official reviews the activity to verify that the spirit and intent of the law are being met and that the fuel gas system provides adequate protection of public health. As a public servant, the code official enforces the code in an unbiased, proper manner. Every individual is guaranteed equal enforcement of the code. Furthermore, design professionals, contractors and building owners have the right of due process for any requirement in the code.

Purpose

A fuel gas code, like any other code, is intended to be adopted as a legally enforceable document to safeguard health, safety, property and public welfare. A fuel gas code cannot be effective without adequate provisions for its administration and enforcement. The official charged with the administration and enforcement of fuel gas regulations has a great responsibility, and with this responsibility goes authority. No matter how detailed the fuel gas code may be, the code official must, to some extent, exercise judgment in determining compliance. The code official has the responsibility for establishing that the homes in which the citizens of the community reside and the buildings in which they work are designed and constructed to be reasonably free from hazards associated with the presence and use of fuel gas appliances, appurtenances, fixtures and systems. The code intends to establish a minimum acceptable level of safety.

SECTION 101 (IFGC)
GENERAL

101.1 Title. These regulations shall be known as the *Fuel Gas Code* of [NAME OF JURISDICTION], hereinafter referred to as "this code."

❖ This section identifies the adopted regulations by insertion of the name of the adopting jurisdiction into the code.

101.2 Scope. This code shall apply to the installation of fuel-gas piping systems, fuel-gas utilization equipment and related accessories in accordance with Sections 101.2.1 through 101.2.5.

Exceptions:

1. Detached one- and two-family dwellings and multiple single-family dwellings (townhouses) not more than three stories high with separate means of egress and their accessory structures shall comply with the *International Residential Code*.

2. As an alternative to the provisions of this code, fuel-gas piping systems, fuel-gas utilization equipment and related accessories in existing buildings that are undergoing repairs, alterations, changes in occupancy or construction of additions shall be permitted to comply with the provisions of the *International Existing Building Code*.

❖ This section describes the types of fuel gas systems to which the code is intended to apply and specifically lists those systems to which the code does not apply. The applicability of the code spans the initial design of fuel gas systems, through the installation and construction phases, and into the maintenance of operating systems. Exception 1 defers to the *International Residen-*

tial Code® (IRC®) for the regulation of installations in buildings that fall within the scope of that code. Chapter 24 of the IRC covers fuel gas systems and is a duplication of the applicable *International Fuel Gas Code®* (IFGC®) text. The provisions of Chapter 24 of the IRC and the IFGC are identical. Exception 2 refers to the *International Existing Building Code®* (IEBC®) as an option to the provisions of the IFGC with regard to existing buildings (see IEBC, Section 503.3.1).

101.2.1 Gaseous hydrogen systems. Gaseous hydrogen systems shall be regulated by Chapter 7.

❖ See general comments for Chapter 7.

101.2.2 Piping systems. These regulations cover piping systems for natural gas with an operating pressure of 125 pounds per square inch gauge (psig) (862 kPa gauge) or less, and for LP-gas with an operating pressure of 20 psig (140 kPa gauge) or less, except as provided in Section 402.6.1. Coverage shall extend from the point of delivery to the outlet of the equipment shutoff valves. Piping systems requirements shall include design, materials, components, fabrication, assembly, installation, testing, inspection, operation and maintenance.

❖ The code does not limit the operating pressure of systems, but rather limits the code's coverage of piping systems to those having pressures less than or equal to the stated pressures. Consistent with the definition, piping systems begin at the point of delivery and end at the outlet of the appliance shutoff valves (see Section 101.2.3 and definition of "Piping systems").

101.2.3 Gas utilization equipment. Requirements for gas utilization equipment and related accessories shall include installation, combustion and ventilation air and venting and connections to piping systems.

❖ The piping and connectors between appliance shutoff valves and the appliance served are covered by the code, although outside the scope of the definition of "piping systems."

101.2.4 Systems and equipment outside the scope. This code shall not apply to the following:

1. Portable LP-gas equipment of all types that is not connected to a fixed fuel piping system.
2. Installation of farm equipment such as brooders, dehydrators, dryers and irrigation equipment.
3. Raw material (feedstock) applications except for piping to special atmosphere generators.
4. Oxygen-fuel gas cutting and welding systems.
5. Industrial gas applications using gases such as acetylene and acetylenic compounds, hydrogen, ammonia, carbon monoxide, oxygen and nitrogen.
6. Petroleum refineries, pipeline compressor or pumping stations, loading terminals, compounding plants, refinery tank farms and natural gas processing plants.
7. Integrated chemical plants or portions of such plants where flammable or combustible liquids or gases are produced by, or used in, chemical reactions.
8. LP-gas installations at utility gas plants.
9. Liquefied natural gas (LNG) installations.
10. Fuel gas piping in power and atomic energy plants.
11. Proprietary items of equipment, apparatus or instruments such as gas-generating sets, compressors and calorimeters.
12. LP-gas equipment for vaporization, gas mixing and gas manufacturing.
13. Temporary LP-gas piping for buildings under construction or renovation that is not to become part of the permanent piping system.
14. Installation of LP-gas systems for railroad switch heating.
15. Installation of hydrogen gas, LP-gas and compressed natural gas (CNG) systems on vehicles.
16. Except as provided in Section 401.1.1, gas piping, meters, gas pressure regulators and other appurtenances used by the serving gas supplier in the distribution of gas, other than undiluted LP-gas.
17. Building design and construction, except as specified herein.
18. Piping systems for mixtures of gas and air within the flammable range with an operating pressure greater than 10 psig (69 kPa gauge).
19. Portable fuel cell appliances that are neither connected to a fixed piping system nor interconnected to a power grid.

❖ This section lists the specific installations and equipment that the code does not intend to regulate. Item 19 relates to Chapter 7 and addresses portable fuel cell appliances as defined in Chapter 2.

101.2.5 Other fuels. The requirements for the design, installation, maintenance, alteration and inspection of mechanical systems operating with fuels other than fuel gas shall be regulated by the *International Mechanical Code*.

❖ This section simply defers the coverage of all equipment other than gas-fired equipment to the *International Mechanical Code®* (IMC®). The IRC also regulates the installation of residential equipment that is not gas fired.

101.3 Appendices. Provisions in the appendices shall not apply unless specifically adopted.

❖ This section certifies that the appendices are not part of the code unless specifically included in the adopting ordinance of the jurisdiction. Otherwise, the appendices are not intended to be enforceable.

101.4 Intent. The purpose of this code is to provide minimum standards to safeguard life or limb, health, property and public welfare by regulating and controlling the design, construction, installation, quality of materials, location, operation and maintenance or use of fuel gas systems.

❖ The intent of the code is to set forth requirements that establish the minimum acceptable level to safeguard life or limb, health, property and public welfare. The intent becomes important in the application of such sections as Sections 102, 104.2, 105.2 and 108, as well as any enforcement-oriented interpretive action or judgement. Like any code, the written text is subject to interpretation. Interpretations should not be affected by economics or the potential impact on any party. The only consideration should be protection of the public health, safety and welfare.

101.5 Severability. If a section, subsection, sentence, clause or phrase of this code is, for any reason, held to be unconstitutional, such decision shall not affect the validity of the remaining portions of this code.

❖ Once the code is adopted, only a court can set aside any provisions of the code. This is essential to safeguard the application of the code text if a provision of the code is declared illegal or unconstitutional. This section would preserve the legislative action that put the legal provisions in place.

SECTION 102 (IFGC)
APPLICABILITY

102.1 General. The provisions of this code shall apply to all matters affecting or relating to structures and premises, as set forth in Section 101. Where, in a specific case, different sections of this code specify different materials, methods of construction or other requirements, the most restrictive shall govern.

❖ The scope of the code as described in Section 101 is referenced in this section. The most restrictive code requirement is to apply where different requirements may be specified in the code for a specific installation. The code is designed to regulate new construction and new work and is not intended to be applied retroactively to existing buildings except where existing fuel-gas-related systems are specifically addressed in this section and Section 108.

102.2 Existing installations. Except as otherwise provided for in this chapter, a provision in this code shall not require the removal, alteration or abandonment of, nor prevent the continued utilization and maintenance of, existing installations lawfully in existence at the time of the adoption of this code.

❖ Existing installations are generally considered to be "grandfathered" with code adoption if the system meets a minimum level of safety. Frequently the criteria for this level are the regulations (or code) under which the existing building was originally constructed. If there are no previous code criteria to apply, the code official is to apply those provisions of the code that are reasonably applicable to existing buildings. A specific level of safety is dictated by provisions dealing with hazard abatement in existing buildings and maintenance provisions as contained in this code, the *International Property Maintenance Code®* (IPMC®), and the *International Fire Code®* (IFC®).

[EB] 102.2.1 Existing buildings. Additions, alterations, renovations or repairs related to building or structural issues shall be regulated by the *International Building Code*.

❖ This section states that the IBC regulates construction related to building or structural issues.

102.3 Maintenance. Installations, both existing and new, and parts thereof shall be maintained in proper operating condition in accordance with the original design and in a safe condition. Devices or safeguards which are required by this code shall be maintained in compliance with the code edition under which they were installed. The owner or the owner's designated agent shall be responsible for maintenance of installations. To determine compliance with this provision, the code official shall have the authority to require an installation to be reinspected.

❖ All fuel gas systems and equipment are subject to deterioration resulting from aging, wear, accumulation of dirt and debris, corrosion and other factors. Maintenance is necessary to keep these systems and equipment in proper operating condition. Required safety devices and controls must be maintained to continue providing the protection that they afford. Existing equipment and systems could be equipped with safety devices or other measures that were necessary because of the nature of the equipment, and such safeguards may have been required by a code that predates the current code. All such safeguards required by previous or present codes must be maintained for the life of the equipment or system served by those safeguards.

Maintenance of fuel gas systems as prescribed in this section is the responsibility of the property owner. The owner may authorize another party to be responsible for the property, in which case that party is responsible for the maintenance of the fuel gas systems involved.

The reinspection authority of the code official is needed to assure compliance with the maintenance requirements in this section.

[EB] 102.4 Additions, alterations or repairs. Additions, alterations, renovations or repairs to installations shall conform to that required for new installations without requiring the existing installation to comply with all of the requirements of this code. Additions, alterations or repairs shall not cause an existing installation to become unsafe, hazardous or overloaded.

Minor additions, alterations, renovations and repairs to existing installations shall meet the provisions for new construction, unless such work is done in the same manner and arrangement as was in the existing system, is not hazardous and is approved.

❖ Simply stated, new work must comply with the current requirements for new work. Any alteration or addition to an existing system involves some extent of new work, and such new work is subject to the requirements of the code. Additions or alterations can place additional loads or different demands on an existing system, and such loads or demands could necessitate changing all or part of the existing system. Additions and alterations must not cause an existing system to be any less in compliance with the code than it was before the changes.

[EB] 102.5 Change in occupancy. It shall be unlawful to make a change in the occupancy of a structure which will subject the structure to the special provisions of this code applicable to the new occupancy without approval. The code official shall certify that such structure meets the intent of the provisions of law governing building construction for the proposed new occupancy and that such change of occupancy does not result in any hazard to the public health, safety or welfare.

❖ When a building undergoes a change of occupancy, the fuel gas systems must be evaluated to determine what effect the change of occupancy has on them. For example, if a mercantile building is converted to a restaurant, additional fuel gas piping system capacity and modifications may be required. If an existing system serves an occupancy that is different from the occupancy it served when the code went into effect, the fuel gas system must comply with the applicable code requirements for a system serving the newer occupancy. Depending on the nature of the previous occupancy, changing a building's occupancy classification could result in a change to the fuel gas system.

[EB] 102.6 Historic buildings. The provisions of this code relating to the construction, alteration, repair, enlargement, restoration, relocation or moving of buildings or structures shall not be mandatory for existing buildings or structures identified and classified by the state or local jurisdiction as historic buildings when such buildings or structures are judged by the code official to be safe and in the public interest of health, safety and welfare regarding any proposed construction, alteration, repair, enlargement, restoration, relocation or moving of buildings.

❖ This section gives the code official the widest possible flexibility in enforcing the code when the building in question has historic value. This flexibility, however, is not without conditions. The most important criterion for application of this section is that the building must be specifically classified as being of historic significance by a qualified party or agency. Usually a state or local authority does this after considerable scrutiny of the historical value of the building. The agencies with this authority typically exist at the state or local government level.

102.7 Moved buildings. Except as determined by Section 102.2, installations that are a part of buildings or structures moved into or within the jurisdiction shall comply with the provisions of this code for new installations.

❖ Buildings that have been relocated are subject to the requirements of the code as if they were new construction. Placing a building where one did not previously exist is the same as constructing a new building. This section requires that the existing fuel gas systems be altered to the extent necessary to bring them into compliance with the provisions of the code applicable to new construction or that the existing fuel gas system comply with Section 102.2.

102.8 Referenced codes and standards. The codes and standards referenced in this code shall be those that are listed in Chapter 8 and such codes and standards shall be considered part of the requirements of this code to the prescribed extent of each such reference. Where differences occur between provisions of this code and the referenced standards, the provisions of this code shall apply.

Exception: Where enforcement of a code provision would violate the conditions of the listing of the equipment or appliance, the conditions of the listing and the manufacturer's installation instructions shall apply.

❖ A referenced standard or portion of one is an enforceable extension of the code as if the content of the standard were included in the body of the code.

The use and application of referenced standards are limited to those portions of the standards that are specifically identified. It is the intention of the code to be in harmony with the referenced standards. In the unlikely event that conflicts occur, the code text governs. The exception recognizes the extremely unlikely but possible occurrence of the code requiring or allowing something less restrictive or stringent than the product listing or manufacturer's instructions. If the code conflicts with or deviates from the conditions of the listing, this may or may not mean that the code violated the listing. For example, the listing for an appliance might allow a particular application of an appliance that is expressly prohibited by the code. In this case, the code has not violated the listing, but instead has simply limited the application allowed by the listing. The intent is for the highest level of safety to prevail.

102.9 Requirements not covered by code. Requirements necessary for the strength, stability or proper operation of an existing or proposed installation, or for the public safety, health and general welfare, not specifically covered by this code, shall be determined by the code official.

❖ New technology and unusual installations will sometimes result in a situation or circumstance not specifically covered by the code. This section of the code gives the code official the authority to decide whether and how the code can be used to cover the new situation. Clearly such a section is needed, and the code official's reasonable application of the section is necessary. The purpose of the section, however, is not to impose requirements that may be preferred when the code provides alternative methods or is not silent on the circumstances. Additionally, the section can be used to implement the general performance-oriented language of the code to specific enforcement situations.

SECTION 103 (IFGC)
DEPARTMENT OF INSPECTION

103.1 General. The Department of Inspection is hereby created and the executive official in charge thereof shall be known as the code official.

❖ The executive official in charge of inspections is named the "code official" by this section. In actuality, the person who is in charge of the department may hold a different title such as building commissioner, plumbing inspector, mechanical inspector, or construction official. For the purpose of the code, that person is referred to as the "code official."

103.2 Appointment. The code official shall be appointed by the chief appointing authority of the jurisdiction; and the code official shall not be removed from office except for cause and after full opportunity to be heard on specific and relevant charges by and before the appointing authority.

❖ This section establishes the code official as an appointed position and gives the circumstances under which the official can be removed from office.

103.3 Deputies. In accordance with the prescribed procedures of this jurisdiction and with the concurrence of the appointing authority, the code official shall have the authority to appoint a deputy code official, other related technical officers, inspectors and other employees.

❖ This section gives the code official the authority to appoint other individuals to assist with the administration and enforcement of the code. These individuals have the authority and responsibility as designated by the code official.

103.4 Liability. The code official, officer or employee charged with the enforcement of this code, while acting for the jurisdiction, shall not thereby be rendered liable personally, and is hereby relieved from all personal liability for any damage accruing to persons or property as a result of an act required or permitted in the discharge of official duties.

Any suit instituted against any officer or employee because of an act performed by that officer or employee in the lawful discharge of duties and under the provisions of this code shall be defended by the legal representative of the jurisdiction until the final termination of the proceedings. The code official or any subordinate shall not be liable for costs in an action, suit or proceeding that is instituted in pursuance of the provisions of this code; and any officer of the Department of Inspection, acting in good faith and without malice, shall be free from liability for acts performed under any of its provisions or by reason of any act or omission in the performance of official duties in connection therewith.

❖ Any suit instituted against any officer or employee because of an act performed by that officer or employee in the lawful discharge of duties and under the provisions of this code must be defended by the legal representative of the jurisdiction until the final termination of the proceedings. The code official or any subordinate is not liable for costs in an action, suit or proceeding that is instituted in pursuance of the provisions of this code; and any officer of the department of mechanical inspection who is acting in good faith and without malice is free from liability for acts performed under any of its provisions or by reason of any act or omission in the performance of official duties.

SECTION 104 (IFGC)
DUTIES AND POWERS OF THE CODE OFFICIAL

104.1 General. The code official shall enforce the provisions of this code and shall act on any question relative to the installation, alteration, repair, maintenance or operation of systems, except as otherwise specifically provided for by statutory requirements or as provided for in Sections 104.2 through 104.8.

❖ The duty of the code official is to enforce the code. Because the code official must also act on all questions related to this responsibility except as specifically exempted by law or elsewhere in the code, the code official is the "authority having jurisdiction" for all matters relating to the code and its enforcement. It is the duty of the code official to interpret the code and to determine compliance with the code. Code compliance will not always be easy to determine and will require the judgement and expertise of the code official. When the code is silent with respect to a particular mechanical installation or lack thereof, the code official is obligated to secure the intent of the code by using the best possible judgement in acting on the matter.

104.2 Rule-making authority. The code official shall have authority as necessary in the interest of public health, safety and general welfare to adopt and promulgate rules and regulations; interpret and implement the provisions of this code; secure the intent thereof and designate requirements applicable because of local climatic or other conditions. Such rules shall not have the effect of waiving structural or fire performance requirements specifically provided for in this code, or of violating accepted engineering methods involving public safety.

❖ The code official has the administrative authority to make rules and regulations that serve to interpret or supplement the provisions of the code as long as those rules conform to the intent of the code. Most importantly, those rules are not intended to set aside or waive any code provisions, but are intended to implement code compliance. Additionally, the rules are not to conflict with accepted engineering practice or reduce the level of safety prescribed by the code. The most frequent use of this authority is making the administrative rules and procedures for the department's efficient and effective operation. An example of the content of administrative rules is the type of work that must be inspected and the notice that must be provided to the department for such inspections. Certain conditions may exist or arise that are geographically dependent, and the department may establish rules related to the climate, soils or other local environmental conditions.

104.3 Applications and permits. The code official shall receive applications and issue permits for installations and alterations under the scope of this code, inspect the premises for which such permits have been issued and enforce compliance with the provisions of this code.

❖ The code enforcement process normally begins with an application for a permit. The code official is responsible for processing the applications and issuing permits for the installation, replacement, addition to or modification of fuel gas systems in accordance with the code.

104.4 Inspections. The code official shall make all of the required inspections, or shall accept reports of inspection by approved agencies or individuals. All reports of such inspections shall be in writing and shall be certified by a responsible officer of such approved agency or by the responsible individual. The code official is authorized to engage such expert opinion as deemed necessary to report upon unusual technical issues that arise, subject to the approval of the appointing authority.

❖ The code official must make inspections as necessary to determine compliance with the code or to accept written reports of inspections by an approved agency. The inspection of the work in progress or accomplished is another significant element in determining code compliance. Although a department might not have the resources to inspect every aspect of all work, the required inspections are those that are dictated by administrative rules and procedures based on many factors, including available inspection resources. In order to expand the available resources for inspection purposes, the code official may approve an inspection agency that, in the code official's opinion, possesses the proper qualifications to perform the inspections. When unusual, extraordinary or complex technical issues arise relative to either fuel gas installations or the safety of an existing fuel gas system, the code official has the authority to seek the opinion and advice of experts. A technical report from an expert requested by the code official can be used to assist the code official in the approval process.

104.5 Right of entry. Whenever it is necessary to make an inspection to enforce the provisions of this code, or whenever the code official has reasonable cause to believe that there exists in a building or upon any premises any conditions or violations of this code that make the building or premises unsafe, dangerous or hazardous, the code official shall have the authority to enter the building or premises at all reasonable times to inspect or to perform the duties imposed upon the code official by this code. If such building or premises is occupied, the code official shall present credentials to the occupant and request entry. If such building or premises is unoccupied, the code official shall first make a reasonable effort to locate the owner or other person having charge or control of the building or premises and request entry. If entry is refused, the code official has recourse to every remedy provided by law to secure entry.

When the code official has first obtained a proper inspection warrant or other remedy provided by law to secure entry, an owner or occupant or person having charge, care or control of the building or premises shall not fail or neglect, after proper request is made as herein provided, to promptly permit entry therein by the code official for the purpose of inspection and examination pursuant to this code.

❖ The first part of this section establishes the right of the code official to enter premises to conduct the permit inspections required by Section 107. Permit application forms typically include a statement in the certification signed by the applicant (who is the owner or the owner's agent) granting the code official the authority to enter areas covered by the permit in order to enforce code provisions related to the permit. The right to enter other structures or premises is more limited. First, to protect the right of privacy, the owner or occupant must grant the code official permission before an interior inspection of the property can be conducted. Permission is not required for inspections that can be accomplished from within the public right-of-way. Second, the owner or occupant may deny access. Unless the inspector has reasonable cause to believe that a violation of the code exists, access may be unattainable. Third, code officials must present proper identification (see commentary, Section 104.6) and request admittance during reasonable hours—usually the normal business hours of the establishment. Fourth, inspections must be aimed at securing or determining compliance with the provisions and intent of the regulations that are specifically within the established scope of the code official's authority.

Searches to gather information for the enforcement of other codes, ordinances or regulations are considered unreasonable and are prohibited by the Fourth Amendment to the U.S. Constitution. "Reasonable cause" in the context of this section must be distinguished from "probable cause," which is required to gain access to property in criminal cases. The burden of proof establishing reasonable cause may vary among jurisdictions. Usually, an inspector must show that the property is subject to inspection under the provisions of the code; that the interests of the public health, safety and welfare outweigh the individual's right to maintain privacy; and that such an inspection is required solely to determine compliance with the provisions of the code.

Many jurisdictions do not recognize the concept of an administrative warrant and may require the code official to prove probable cause in order to gain access. This burden of proof often requires the code official to state in advance why access is needed (usually access is restricted to gathering evidence for seeking an indictment or making an arrest); what specific items or information is sought; the information's relevance to the case against the subject; how knowledge of the relevance of the information or items sought was obtained; and how the evidence sought will be used. In all such cases, the right to privacy must always be weighed against the right of the code official to conduct an inspection to verify that the public health, safety and welfare are not in jeopardy. Such important and complex constitutional issues should be discussed with the jurisdiction's legal counsel. Jurisdictions should establish procedures for securing the necessary court orders when an inspection is deemed necessary following a refusal.

The last paragraph in this section requires the owner or occupant to permit entry for inspection if a proper inspection warrant or other documentation required by law has been obtained.

104.6 Identification. The code official shall carry proper identification when inspecting structures or premises in the performance of duties under this code.

❖ This section requires the code official (including all authorized inspection personnel) to carry identification when conducting the duties of the position. The identification removes any question of the purpose and authority of the inspector.

104.7 Notices and orders. The code official shall issue all necessary notices or orders to ensure compliance with this code.

❖ An important element of code enforcement is the necessary advisement of deficiencies and correction, which is accomplished through notices and orders. The code official is required to issue orders to abate illegal or unsafe conditions. Sections 108.7, 108.7.1, 108.7.2 and 108.7.3 contain additional information for these notices.

104.8 Department records. The code official shall keep official records of applications received, permits and certificates issued, fees collected, reports of inspections, and notices and orders issued. Such records shall be retained in the official records as long as the building or structure to which such records relate remains in existence, unless otherwise provided for by other regulations.

❖ Good business practice requires the code official to keep official records pertaining to permit applications, permits, fees collected, inspections, notices and orders issued. This documentation is a valuable source of information if questions arise "throughout the life of the building and its occupancy." The code does not require that the construction documents be kept after the project is complete, but other regulations or the administrative rules may provide for retaining the construction documents for a minimum period.

SECTION 105 (IFGC)
APPROVAL

105.1 Modifications. Whenever there are practical difficulties involved in carrying out the provisions of this code, the code official shall have the authority to grant modifications for individual cases, provided the code official shall first find that special individual reason makes the strict letter of this code impractical and that such modification is in compliance with the intent and purpose of this code and does not lessen health, life and fire safety requirements. The details of action granting modifications shall be recorded and entered in the files of the Department of Inspection.

❖ The code official may amend or make exceptions to the code as needed where strict compliance with the code is impractical. Only the code official has the authority to grant modifications. Consideration of a particular difficulty is to be based on the application of the owner and a demonstration that the intent of the code is satisfied. This section is not intended to permit setting a code provision aside or ignoring a provision; rather, it is intended to provide for the acceptance of equivalent protection. Such modifications do not, however, extend to actions that are necessary to correct violations of the code. In other words, a code violation or the expense of correcting a code violation cannot constitute a practical difficulty.

Details of modification action must be filed for retrieval of substantiation of the reasons for the modification. Comprehensive written records are an essential part of an effective administrative system. Unless clearly written records of the considerations and documentation used in the modification process are created and maintained, subsequent enforcement action cannot be supported.

105.2 Alternative materials, methods and equipment. The provisions of this code are not intended to prevent the installation of any material or to prohibit any method of construction not specifically prescribed by this code, provided that any such alternative has been approved. An alternative material or method of construction shall be approved where the code official finds that the proposed design is satisfactory and complies with the intent of the provisions of this code, and that the material, method or work offered is, for the purpose intended, at least the equivalent of that prescribed in this code in quality, strength, effectiveness, fire resistance, durability and safety.

❖ The code is not intended to inhibit innovative ideas or technological advances. A comprehensive regulatory document such as a fuel gas code cannot envision and then address all future innovations in the industry. As a result, a performance code must be applicable to and provide a basis for the approval of an increasing number of newly developed, innovative materials, systems and methods for which no code text or referenced standards yet exist. The fact that a material, product or method of construction is not addressed in the code is not an indication that the material, product or method is intended to be prohibited. The code official is expected to apply sound technical judgement in accepting materials, systems or methods that, while not anticipated by the drafters of the current code text, can be demonstrated to offer equivalent performance. By virtue of its text, the code regulates new and innovative construction practices while addressing the relative safety of building occupants. The code official is responsible for determining whether a requested alternative provides a level of protection of the public health, safety and welfare as required by the code.

105.3 Required testing. Whenever there is insufficient evidence of compliance with the provisions of this code, evidence that a material or method does not conform to the requirements of this code, or in order to substantiate claims for alternative materials or methods, the code official shall have the authority to require tests as evidence of compliance to be made at no expense to the jurisdiction.

❖ Sufficient technical data, test reports and documentation must be submitted for evaluation by the code official to provide the basis on which the code official can make a decision regarding an alternative material or type of equipment. If evidence satisfactory to the code official proves that the alternative equipment, material or construction method is equivalent to that required by the code, the code official is obligated to approve it. Any such approval cannot have the effect of waiving any requirements of the code. The burden of proof of equivalence lies with the applicant who proposes the use of alternative equipment, materials or methods.

105.3.1 Test methods. Test methods shall be as specified in this code or by other recognized test standards. In the absence of recognized and accepted test methods, the code official shall approve the testing procedures.

❖ The code official must require the submission of any appropriate information and data to assist in the determination of equivalency. This information must be submitted before a permit can be issued. The type of information required includes test data in accordance with the referenced standards, evidence of compliance with the referenced standard specifications and design calculations. A research report issued by an authoritative agency is particularly useful in providing the code official with the technical basis for evaluation and approval of new and innovative materials and components. The use of authoritative research reports can greatly assist the code official by reducing the time-consuming engineering analysis necessary to review materials and products. Failure to substantiate a request for the use of an alternative is a valid reason for the code official to deny a request.

105.3.2 Testing agency. All tests shall be performed by an approved agency.

❖ The code official must approve the testing agency, and the agency should have technical expertise, test equipment and quality assurance to properly conduct and report the necessary testing.

105.3.3 Test reports. Reports of tests shall be retained by the code official for the period required for retention of public records.

❖ Test reports are to be retained as required by public record laws. The attorney of the jurisdiction could be asked for the specific time period in applicable laws of the localities.

105.4 Material and equipment reuse. Materials, equipment and devices shall not be reused unless such elements have been reconditioned, tested and placed in good and proper working condition, and approved.

❖ The code criteria for materials, equipment and devices have changed over the years. Evaluation of testing and materials technology has permitted the development of new criteria that the old materials may not satisfy. As a result, used materials must be evaluated in the same manner as new materials. Used (previously installed) equipment must be equivalent to that required by the code if it is to be used again in a new installation.

SECTION 106 (IFGC)
PERMITS

106.1 When required. An owner, authorized agent or contractor who desires to erect, install, enlarge, alter, repair, remove, convert or replace an installation regulated by this code, or to cause such work to be done, shall first make application to the code official and obtain the required permit for the work.

Exception: Where equipment replacements and repairs are required to be performed in an emergency situation, the permit application shall be submitted within the next working business day of the Department of Inspection.

❖ In general, a permit is required for all activities that are regulated by the code, and these activities cannot begin until the permit is issued.

A fuel gas permit is required for the installation, replacement, alteration or modification of fuel gas systems and components that are in the scope of applicability of the code. Replacement of an existing piece of equipment or related piping is treated no differently than a new installation in new building construction. The purpose of a permit is to cause the work to be inspected to determine compliance with the intent of the code.

The exception provides for prompt permit applications for situations where equipment and appliance replacements and repairs are done to address an emergency. This action enables the department having jurisdiction to promptly inspect the work.

106.2 Permits not required. Permits shall not be required for the following:

1. Any portable heating appliance.
2. Replacement of any minor component of equipment that does not alter approval of such equipment or make such equipment unsafe.

Exemption from the permit requirements of this code shall not be deemed to grant authorization for work to be done in violation of the provisions of this code or of other laws or ordinances of this jurisdiction.

❖ The installations exempt from the requirement for a permit are very limited as evidenced by items 1 and 2. Item 1 pertains to appliances and equipment that are used temporarily and that are not designed for permanent installation. Examples of portable heating appliances and equipment include space heaters and construction site heaters.

Item 2 applies to the replacement of minor equipment and appliance components. A permit would be required if the component replacement could potentially affect either the safety of the equipment or the conditions of approval of the equipment. For example, replacement of a defective control with a control of the same type and specifications as the original factory-supplied control would not require a permit. Replacement of a burner assembly with a burner assembly having a different input capacity or one that is designed to burn a different fuel would require a permit.

106.3 Application for permit. Each application for a permit, with the required fee, shall be filed with the code official on a form furnished for that purpose and shall contain a general description of the proposed work and its location. The application shall be signed by the owner or an authorized agent. The permit application shall indicate the proposed occupancy of all parts of the building and of that portion of the site or lot, if any, not covered by the building or structure and shall contain such other information required by the code official.

❖ This section states that the building owner or an authorized agent of the owner is the only person who can apply for a building permit. An owner's authorized agent could be anyone who is given written permission to act in the owner's interest for the purpose of obtaining a permit, such as an architect, engineer, contractor or tenant. Permit forms will generally have enough space for a very brief description of the work to be accomplished, which is sufficient for small jobs. For larger projects, the description will be contained in construction documents.

106.3.1 Construction documents. Construction documents, engineering calculations, diagrams and other data shall be submitted in two or more sets with each application for a permit. The code official shall require construction documents, computations and specifications to be prepared and designed by a registered design professional when required by state law. Construction documents shall be drawn to scale and shall be of sufficient clarity to indicate the location, nature and extent of the work proposed and show in detail that the work conforms to the provisions of this code. Construction documents for buildings more than two stories in height shall indicate where penetrations will be made for installations and shall indicate the materials and methods for maintaining required structural safety, fire-resistance rating and fireblocking.

Exception: The code official shall have the authority to waive the submission of construction documents, calculations or other data if the nature of the work applied for is such that reviewing of construction documents is not necessary to determine compliance with this code.

❖ When the work is of a "minor nature," either in scope or needed description, the code official may use judgement in determining the need for a detailed description of the work. An example of minor work that may not involve a detailed description is the replacement of an existing piece of equipment in a mechanical system or the replacement or repair of a defective portion of a piping system.

These provisions are intended to reflect the minimum scope of information needed to determine code compliance. A statement on the construction documents, such as "All fuel gas systems must comply with the 2003 edition of the *International Fuel Gas Code®*," is not an acceptable substitute for showing the required information.

This section also requires the code official to determine that state professional registration laws are complied with as they apply to the preparation of construction documents.

106.4 Permit issuance. The application, construction documents and other data filed by an applicant for a permit shall be reviewed by the code official. If the code official finds that the proposed work conforms to the requirements of this code and all laws and ordinances applicable thereto, and that the fees specified in Section 106.5 have been paid, a permit shall be issued to the applicant.

❖ This section requires the code official to review submittals for a permit for compliance with the code and verify that the project will be carried out in accordance with other applicable laws. This may involve interagency communication and cooperation so that all laws are being obeyed. Once the code official finds this to be so, a permit may be issued upon payment of the required fees.

106.4.1 Approved construction documents. When the code official issues the permit where construction documents are required, the construction documents shall be endorsed in writing and stamped "APPROVED." Such approved construction documents shall not be changed, modified or altered without authorization from the code official. Work shall be done in accordance with the approved construction documents.

The code official shall have the authority to issue a permit for the construction of part of an installation before the construction documents for the entire installation have been submitted or approved, provided adequate information and detailed statements have been filed complying with all pertinent requirements of this code. The holder of such permit shall proceed at his or her own risk without assurance that the permit for the entire installation will be granted.

❖ Construction documents that reflect compliance with code requirements form an integral part of the permit process. Successful completion of the work depends on these documents. This section requires the code official to stamp the complying construction documents as being "APPROVED" and fix the status of the document in time. Once approved, the documents may not be revised without the express authorization of the code official. This maintains the code-compliance level of the documents.

106.4.2 Validity. The issuance of a permit or approval of construction documents shall not be construed to be a permit for, or an approval of, any violation of any of the provisions of this code or of other ordinances of the jurisdiction. A permit presuming to give authority to violate or cancel the provisions of this code shall be invalid.

The issuance of a permit based upon construction documents and other data shall not prevent the code official from thereafter requiring the correction of errors in said construction documents and other data or from preventing building operations from being carried on thereunder when in violation of this code or of other ordinances of this jurisdiction.

❖ This powerful code section states the fundamental premise that the permit is only "a license to proceed with the work." It is not a license to "violate, cancel or set aside any provisions of the code." This statement is important because despite any errors in the approval process, the permit applicant is responsible for code compliance.

106.4.3 Expiration. Every permit issued by the code official under the provisions of this code shall expire by limitation and become null and void if the work authorized by such permit is not commenced within 180 days from the date of such permit, or is suspended or abandoned at any time after the work is commenced for a period of 180 days. Before such work recommences, a new permit shall be first obtained and the fee, therefor, shall be one-half the amount required for a new permit for such work, provided no changes have been or will be made in the original construction documents for such work, and further that such suspension or abandonment has not exceeded one year.

❖ The permit becomes invalid in two distinct situations, but both are based on a 6-month period. The first situation is when no work has started 6 months from issuance of the permit. The second situation is when there is no continuation of authorized work for 6 months. The person who was issued the permit should be notified in writing that the permit is invalid and what steps must be taken to restart the work.

This section also gives the administrative authority a means of offsetting the costs associated with the administration of expired, reissued permits by charging a nominal fee for permit reissuance. If, however, the nature or scope of the work to be resumed is different from that indicated on the original permit, the permit process essentially starts from scratch, and full fees are charged. The same procedure would also apply if the work has not commenced within 1 year of the date of permit issuance or if work has been suspended for a year or more.

106.4.4 Extensions. A permittee holding an unexpired permit shall have the right to apply for an extension of the time within which he or she will commence work under that permit when work is unable to be commenced within the time required by this section for good and satisfactory reasons. The code official shall extend the time for action by the permittee for a period not exceeding 180 days if there is reasonable cause. A permit shall not be extended more than once. The fee for an extension shall be one-half the amount required for a new permit for such work.

❖ Although it is customary for a project to begin immediately following issuance of a permit, there may be occasions when an unforeseen delay may occur. This section affords the permit holder an opportunity to apply for and receive a single, 180-day extension within which to begin a project under a still-valid permit (less than 180 days old). The applicant must, however, submit to the code official an adequate explanation for the delay in starting a project, which could include such things as the need to obtain approvals or permits for the project from other agencies having jurisdiction. This section requires the code official to determine what constitutes "good and satisfactory" reasons for any delay and further allows the jurisdiction to offset its administrative costs for extending the permit by charging one-half the permit fee for the extension.

106.4.5 Suspension or revocation of permit. The code official shall revoke a permit or approval issued under the provisions of this code in case of any false statement or misrepresentation of fact in the application or on the construction documents upon which the permit or approval was based.

❖ A permit is in reality a license to proceed with the work. The code official, however, must revoke all permits shown to be based, all or in part, on any false statement or misrepresentation of fact. An applicant may subsequently reapply for a permit with the appropriate corrections or modifications made to the application and construction documents.

106.4.6 Retention of construction documents. One set of construction documents shall be retained by the code official until final approval of the work covered therein. One set of approved construction documents shall be returned to the applicant, and said set shall be kept on the site of the building or work at all times during which the work authorized thereby is in progress.

❖ Once the code official has stamped or endorsed as approved the construction documents on which the permit is based (see commentary, Section 106.4.1), one set of approved construction documents must be kept on the construction site to serve as the basis for all subsequent inspections. To avoid confusion, the construction documents on the site must be precisely the documents that were approved and stamped because inspections are to be performed based on the approved documents. Additionally, the contractor cannot determine compliance with the approved construction documents unless those documents are readily available. Unless the approved construction documents are available, the inspection should be postponed and work on the project halted.

ADMINISTRATION

106.5 Fees. A permit shall not be issued until the fees prescribed in Section 106.5.2 have been paid, nor shall an amendment to a permit be released until the additional fee, if any, due to an increase of the installation, has been paid.

❖ All fees are to be paid prior to permit issuance. This requirement facilitates payment and also establishes that the permit applicant intends to proceed with the work.

106.5.1 Work commencing before permit issuance. Any person who commences work on an installation before obtaining the necessary permits shall be subject to 100 percent of the usual permit fee in addition to the required permit fees.

❖ This section is intended to serve as a deterrent to proceeding with work on a fuel gas system without a permit (except as provided in Section 106.2). As a punitive measure, it doubles the cost of the permit fee to be charged. This section does not, however, intend to penalize a contractor called upon to do emergency work after hours if he or she promptly notifies the code official the next business day, obtains the requisite permit for the work done and has the required inspections performed.

106.5.2 Fee schedule. The fees for work shall be as indicated in the following schedule.

[JURISDICTION TO INSERT APPROPRIATE SCHEDULE]

❖ A published fee schedule must be established for plans examination, permits and inspections. Ideally, the department should generate revenues that cover operating costs and expenses. The permit fee schedule is an integral part of this process.

106.5.3 Fee refunds. The code official shall authorize the refunding of fees as follows.

1. The full amount of any fee paid hereunder which was erroneously paid or collected.
2. Not more than [SPECIFY PERCENTAGE] percent of the permit fee paid when no work has been done under a permit issued in accordance with this code.
3. Not more than [SPECIFY PERCENTAGE] percent of the plan review fee paid when an application for a permit for which a plan review fee has been paid is withdrawn or canceled before any plan review effort has been expended.

The code official shall not authorize the refunding of any fee paid, except upon written application filed by the original permittee not later than 180 days after the date of fee payment.

SECTION 107 (IFGC)
INSPECTIONS AND TESTING

107.1 Required inspections and testing. The code official, upon notification from the permit holder or the permit holder's agent, shall make the following inspections and other such inspections as necessary, and shall either release that portion of the construction or notify the permit holder or the permit holder's agent of violations that are required to be corrected. The holder of the permit shall be responsible for scheduling such inspections.

1. Underground inspection shall be made after trenches or ditches are excavated and bedded, piping is installed and before backfill is put in place. When excavated soil contains rocks, broken concrete, frozen chunks and other rubble that would damage or break the piping or cause corrosive action, clean backfill shall be on the job site.
2. Rough-in inspection shall be made after the roof, framing, fireblocking and bracing are in place and components to be concealed are complete, and prior to the installation of wall or ceiling membranes.
3. Final inspection shall be made upon completion of the installation.

The requirements of this section shall not be considered to prohibit the operation of any heating equipment installed to replace existing heating equipment serving an occupied portion of a structure in the event a request for inspection of such heating equipment has been filed with the department not more than 48 hours after replacement work is completed, and before any portion of such equipment is concealed by any permanent portion of the structure.

❖ Inspections are necessary to determine that an installation conforms to all code requirements. Because the majority of a fuel gas piping system could be hidden within the building enclosure, periodic inspections are necessary before portions of the system are concealed. The code official is required to determine that fuel gas systems and equipment are installed in accordance with the approved construction documents and the applicable code requirements. Inspections that are necessary to provide such verification must be conducted. Generally, the administrative rules of a department may list the interim inspections judged to be required. Construction that occurs in steps or phases may necessitate multiple inspections; therefore, an exact number of required inspections cannot be specified. Where violations are noted and corrections are required, reinspections may be necessary. As time permits, frequent inspections of some job sites, especially where the work is complex, can be beneficial if the inspector detects code-compliance problems or potential problems before they develop or become more difficult to correct. The contractor, builder, owner or other authorized party is responsible for arranging for the required inspections and coordinating inspections to prevent work from being concealed prior to being inspected.

1. Inspection of underground piping is especially important because once it is covered, it is the most challenging part of a fuel gas system in which to detect a leak. If repairs are necessary, underground repairs are proportionally more expensive because of the need for heavy equipment and the more labor-intensive nature of working below ground. To reduce possible damage to pipe from rubble, rocks and other rough materials, excava-

tions must be bedded and backfilled with clean fill materials spread and tamped to provide adequate support and protection for piping.

2. A rough-in inspection is an inspection of all parts of the fuel gas system that will eventually be concealed in the building structure. The inspection must be made before any of the system is closed up or hidden from view. To gain approval, the fuel gas systems must pass the required rough-in tests.

3. A rough-in inspection may be completed all at one time or as a series of inspections. This is administratively determined by the local inspections department and is typically dependent on the size of the job.

4. A final inspection may be done as a series of inspections or all at one time, similar to a rough-in inspection. A final inspection is required prior to the approval of mechanical work and fuel gas installations. For the construction of a new building, final approval is required prior to the issuance of the certificate of occupancy as specified in the building code. To verify that all previously issued correction orders have been complied with and to determine whether subsequent violations exist, a final inspection must be made. Violations observed during the final inspection must be noted, and the permit holder must be advised of them.

The final inspection is made after the completion of the work or installation. Typically, the final inspection is an inspection of all that was installed after the rough-in inspection and not concealed in the building construction. Subsequent reinspections are necessary if the final inspection has generated a notice of violation.

The last paragraph of this section is emergency related and provides for prompt operation of replacement heating equipment, allowing the occupied areas of a facility to be heated as soon as the new heating equipment is installed. Such installations are subject to all requirements of the code.

107.1.1 Approved inspection agencies. The code official shall accept reports of approved agencies, provided that such agencies satisfy the requirements as to qualifications and reliability.

❖ As an alternative to conducting the inspection, the code official can accept inspections and reports by approved inspection agencies. Appropriate criteria on which to base approval of an inspection agency include competence, objectivity, certifications and experience.

107.1.2 Evaluation and follow-up inspection services. Prior to the approval of a prefabricated construction assembly having concealed work and the issuance of a permit, the code official shall require the submittal of an evaluation report on each prefabricated construction assembly, indicating the complete details of the installation, including a description of the system and its components, the basis upon which the system is being evaluated, test results and similar information and other data as necessary for the code official to determine conformance to this code.

❖ As an alternative to the physical inspection by the code official in the plant or location where prefabricated components are fabricated (such as modular homes and prefabricated structures), the code official has the option of accepting an evaluation report from an approved agency detailing such inspections. These evaluation reports can serve as the basis from which the code official will determine code compliance.

107.1.2.1 Evaluation service. The code official shall designate the evaluation service of an approved agency as the evaluation agency, and review such agency's evaluation report for adequacy and conformance to this code.

❖ The code official must review all submitted reports for conformity to the applicable code requirements. If in the judgement of the code official the submitted reports are acceptable, the code official should document the basis for the approval.

107.1.2.2 Follow-up inspection. Except where ready access is provided to installations, service equipment and accessories for complete inspection at the site without disassembly or dismantling, the code official shall conduct the in-plant inspections as frequently as necessary to ensure conformance to the approved evaluation report or shall designate an independent, approved inspection agency to conduct such inspections. The inspection agency shall furnish the code official with the follow-up inspection manual and a report of inspections upon request, and the installation shall have an identifying label permanently affixed to the system indicating that factory inspections have been performed.

❖ The owner must provide special inspections of fabricated assemblies at the fabrication plant. The code official or an approved inspection agency must conduct periodic in-plant inspections to assure conformance to the approved evaluation report described in Section 107.1.2. Such inspections are required if the fuel gas systems can be inspected completely at the job site.

107.1.2.3 Test and inspection records. Required test and inspection records shall be available to the code official at all times during the fabrication of the installation and the erection of the building; or such records as the code official designates shall be filed.

❖ All testing and inspection records related to a fabricated assembly must be filed with the code official to maintain a complete and legal record of the assembly and erection of the building.

107.2 Testing. Installations shall be tested as required in this code and in accordance with Sections 107.2.1 through 107.2.3. Tests shall be made by the permit holder and observed by the code official.

❖ Testing of fuel gas systems is required where testing is specified in the technical chapters of the code. See Section 406 for specific requirements and methods of testing gas-piping systems.

107.2.1 New, altered, extended or repaired installations. New installations and parts of existing installations, which have been altered, extended, renovated or repaired, shall be tested as prescribed herein to disclose leaks and defects.

❖ Testing is necessary to make sure that the system is free from leaks and other defects. To the extent specified in the technical chapters of the code, testing is also required for portions of existing systems that have been altered, extended, renovated or repaired.

107.2.2 Apparatus, instruments, material and labor for tests. Apparatus, instruments, material and labor required for testing an installation or part thereof shall be furnished by the permit holder.

❖ The permit holder is responsible for performing tests as well as for supplying all of the labor and apparatus necessary to conduct the tests. The code official observes but never performs the test.

107.2.3 Reinspection and testing. Where any work or installation does not pass an initial test or inspection, the necessary corrections shall be made so as to achieve compliance with this code. The work or installation shall then be resubmitted to the code official for inspection and testing.

❖ If a system or portion of a system does not pass the initial test or inspection, violations must be corrected and the system must be reinspected.
   To encourage code compliance and to cover the expense of the code official's time, many code enforcement jurisdictions charge fees for reinspections that are required subsequent to the first inspection.

107.3 Approval. After the prescribed tests and inspections indicate that the work complies in all respects with this code, a notice of approval shall be issued by the code official.

❖ After the code official has performed the required inspections and observed any required equipment and system tests (or has received written reports of the results of such tests), he or she must determine whether the installation or work is in compliance with all applicable sections of the code. The code official must issue a written notice of approval if the subject work or installation is in apparent compliance with the code. This notice is given to the permit holder, and a copy of the notice is retained on file by the code official.

107.4 Temporary connection. The code official shall have the authority to allow the temporary connection of an installation to the sources of energy for the purpose of testing the installation or for use under a temporary certificate of occupancy.

❖ Typical procedure for a local jurisdiction is to withhold the issuance of the certificate of occupancy until approvals have been received from each code official responsible for inspection of the structure. The code official is permitted to issue a temporary authorization to make connections to the public utility system prior to the completion of all work. The certification is intended to acknowledge that, because of seasonal limitations, time constraints or the need for testing or partial operation of equipment, some building systems may be connected even though the building is not suitable for final occupancy. The intent of this section is that a request for temporary occupancy or the connection and use of mechanical equipment or systems should be granted when the requesting permit holder has demonstrated to the code official's satisfaction that the public health, safety and welfare will not be endangered. The code official should view the issuance of a "temporary authorization or certificate of occupancy" as substantial an act as the issuance of the final certificate. Indeed, the issuance of a temporary certificate of occupancy offers a greater potential for conflict because once the building or structure is occupied, it is very difficult to remove the occupants through legal means.

SECTION 108 (IFGC)
VIOLATIONS

108.1 Unlawful acts. It shall be unlawful for a person, firm or corporation to erect, construct, alter, repair, remove, demolish or utilize an installation, or cause same to be done, in conflict with or in violation of any of the provisions of this code.

❖ Violations of the code are unlawful. This is the basis for all citations and correction notices related to violations of the code.

108.2 Notice of violation. The code official shall serve a notice of violation or order to the person responsible for the erection, installation, alteration, extension, repair, removal or demolition of work in violation of the provisions of this code, or in violation of a detail statement or the approved construction documents thereunder, or in violation of a permit or certificate issued under the provisions of this code. Such order shall direct the discontinuance of the illegal action or condition and the abatement of the violation.

❖ The code official must notify the person responsible for the erection or use of a building found to be in violation of the code. The section that is allegedly being violated must be cited so that the responsible party can respond to the notice.

108.3 Prosecution of violation. If the notice of violation is not complied with promptly, the code official shall request the legal counsel of the jurisdiction to institute the appropriate proceeding at law or in equity to restrain, correct or abate such violation, or to require the removal or termination of the unlawful occupancy of the structure in violation of the provisions of this code or of the order or direction made pursuant thereto.

❖ The code official must pursue legal means to correct the violation through the use of legal counsel of the jurisdiction. This is not optional.

Any extensions of time so that the violations may be corrected voluntarily must be for a reasonable, bona fide cause, or the code official may be subject to criticism for "arbitrary and capricious" actions. In general, it is better to have a standard time limitation for correction of violations. Departures from this standard must be for a clear and reasonable purpose, usually stated in writing by the violator.

108.4 Violation penalties. Persons who shall violate a provision of this code, fail to comply with any of the requirements thereof or erect, install, alter or repair work in violation of the approved construction documents or directive of the code official, or of a permit or certificate issued under the provisions of this code, shall be guilty of a [SPECIFY OFFENSE], punishable by a fine of not more than [AMOUNT] dollars or by imprisonment not exceeding [NUMBER OF DAYS], or both such fine and imprisonment. Each day that a violation continues after due notice has been served shall be deemed a separate offense.

❖ This section prescribes a standard fine or other penalty as deemed appropriate by the jurisdiction. Additionally, this section identifies a principle that "each day that a violation continues . . . shall be deemed a separate offense" for the purpose of applying the prescribed penalty in order to facilitate the prompt resolution.

108.5 Stop work orders. Upon notice from the code official that work is being done contrary to the provisions of this code or in a dangerous or unsafe manner, such work shall immediately cease. Such notice shall be in writing and shall be given to the owner of the property, the owner's agent, or the person doing the work. The notice shall state the conditions under which work is authorized to resume. Where an emergency exists, the code official shall not be required to give a written notice prior to stopping the work. Any person who shall continue any work on the system after having been served with a stop work order, except such work as that person is directed to perform to remove a violation or unsafe condition, shall be liable for a fine of not less than [AMOUNT] dollars or more than [AMOUNT] dollars.

❖ Upon receipt of a violation notice from the code official, the builders must immediately cease construction activities identified in the notice, except as expressly permitted to correct the violation. A stop work order can result in inconvenience and monetary loss to the affected parties; therefore, justification must be evident and judgement must be exercised before a stop work order is issued. A stop work order can prevent a violation from becoming worse and more difficult or expensive to correct.

A stop work order may be issued where work is proceeding without a permit to perform the work. Hazardous conditions could develop when the code official is unaware of the nature of the work and a permit for the work has not been issued.

The issuance of a stop work order on a fuel gas system may result from work done by the contractor that affects a building component that is not fuel-gas related. For example, if a contractor cuts a structural element to install piping, the structure may be weakened enough to cause a partial or complete structural failure. As determined by the adopting jurisdiction, a penalty may be assessed for failure to comply with this section, and the dollar amount is to be inserted in the blanks shown.

108.6 Abatement of violation. The imposition of the penalties herein prescribed shall not preclude the legal officer of the jurisdiction from instituting appropriate action to prevent unlawful construction, restrain, correct or abate a violation, prevent illegal occupancy of a building, structure or premises, or stop an illegal act, conduct, business or utilization of the installations on or about any premises.

❖ Despite the assessment of a penalty in the form of a fine or imprisonment against a violator, the violation itself must still be corrected. Failure to make the necessary corrections will result in the violator being subject to additional penalties as described in the preceding section.

108.7 Unsafe installations. An installation that is unsafe, constitutes a fire or health hazard, or is otherwise dangerous to human life, as regulated by this code, is hereby declared an unsafe installation. Use of an installation regulated by this code constituting a hazard to health, safety or welfare by reason of inadequate maintenance, dilapidation, fire hazard, disaster, damage or abandonment is hereby declared an unsafe use. Such unsafe installations are hereby declared to be a public nuisance and shall be abated by repair, rehabilitation, demolition or removal.

❖ Unsafe conditions include those that constitute a health hazard, fire hazard, explosion hazard, shock hazard, asphyxiation hazard, physical injury hazard or are otherwise dangerous to human life and property.

In the course of performing duties, the code official may identify a hazardous condition that must be declared in violation of the code and, therefore, must be abated.

108.7.1 Authority to condemn installations. Whenever the code official determines that any installation, or portion thereof, regulated by this code has become hazardous to life, health or property, he or she shall order in writing that such installations either be removed or restored to a safe condition. A time limit for compliance with such order shall be specified in the written notice. A person shall not use or maintain a defective installation after receiving such notice.

When such installation is to be disconnected, written notice as prescribed in Section 108.2 shall be given. In cases of immediate danger to life or property, such disconnection shall be made immediately without such notice.

❖ When a system is determined to be unsafe, the code official must notify the owner or agent of the building as the first step in correcting the difficulty. The notice is to describe the repairs and improvements necessary to correct the deficiency or require the unsafe equipment or system to be removed or replaced. Such notices must specify a time frame in which the corrective actions must occur. Additionally, the notice should require the immediate response of the owner or agent. If the owner or agent is not available, public notice of the declaration should suffice for complying with this section. The code official may also determine that disconnecting the system is necessary to correct an unsafe condition and must give written notice to that effect (see commentary, Section 108.2), unless immediate disconnection is essential for public health and safety reasons (see commentary, Section 108.7.2).

108.7.2 Authority to disconnect service utilities. The code official shall have the authority to require disconnection of utility service to the building, structure or system regulated by the technical codes in case of emergency where necessary to eliminate an immediate hazard to life or property. The code official shall notify the serving utility, and wherever possible, the owner and occupant of the building, structure or service system of the decision to disconnect prior to taking such action. If not notified prior to disconnection, the owner or occupant of the building, structure or service system shall be notified in writing, as soon as practicable thereafter.

❖ Disconnecting the utility service is the most radical method of hazard abatement available to the code official and should be reserved for cases in which all other lesser remedies have proven ineffective. Such an action must be preceded by written notice to the owner and any occupants of the building being ordered to disconnect. Disconnection must be accomplished within the time frame established by the code official in the written notification to disconnect. When the hazard to the public health and welfare is so imminent as to mandate immediate disconnection, the code official has the authority and even the obligation to cause disconnection without notice.

108.7.3 Connection after order to disconnect. A person shall not make energy source connections to installations regulated by this code which have been disconnected or ordered to be disconnected by the code official, or the use of which has been ordered to be discontinued by the code official until the code official authorizes the reconnection and use of such installations.

When an installation is maintained in violation of this code, and in violation of a notice issued pursuant to the provisions of this section, the code official shall institute appropriate action to prevent, restrain, correct or abate the violation.

❖ When any fuel gas system is maintained in violation of the code and in violation of any notice issued pursuant to the provisions of this section, the code official is to institute appropriate action to prevent, restrain, correct or abate the violation.

Once the reason for discontinuation of use or disconnection of the fuel gas system no longer exists, only the code official may authorize resumption of use or reconnection of the system after it is demonstrated to the code official's satisfaction that all repairs or other work are in compliance with applicable sections of the code. This section also requires the code official to take action to abate code violations (see commentary, Section 108.2).

SECTION 109 (IFGC)
MEANS OF APPEAL

109.1 Application for appeal. A person shall have the right to appeal a decision of the code official to the board of appeals. An application for appeal shall be based on a claim that the true intent of this code or the rules legally adopted thereunder have been incorrectly interpreted, the provisions of this code do not fully apply or an equally good or better form of construction is proposed. The application shall be filed on a form obtained from the code official within 20 days after the notice was served.

❖ This section holds that any aggrieved party with a material interest in the decision of the code official may appeal that decision before a board of appeals. This provides a forum other than the court of jurisdiction in which to review the code official's actions.

This section allows any person to appeal a decision of the code official. In practice, this section has been interpreted to permit appeals only by those aggrieved parties with a material or definitive interest in the decision of the code official. An aggrieved party may not appeal a code requirement per se. The intent of the appeal process is not to waive or set aside a code requirement; rather, it is intended to provide a means of reviewing a code official's decision on an interpretation or application of the code or to review the equivalency of protection to the code requirements.

109.2 Membership of board. The board of appeals shall consist of five members appointed by the chief appointing authority as follows: one for five years; one for four years; one for three years; one for two years and one for one year. Thereafter, each new member shall serve for five years or until a successor has been appointed.

❖ The board of appeals is to consist of five members appointed on a rotating basis by the "chief appointing authority"—typically, the mayor or city manager. This method of appointment allows for a smooth transition of board of appeals members, thus ensuring continuity of action over the years.

109.2.1 Qualifications. The board of appeals shall consist of five individuals, one from each of the following professions or disciplines.

1. Registered design professional who is a registered architect; or a builder or superintendent of building construction with at least 10 years' experience, five of which shall have been in responsible charge of work.

2. Registered design professional with structural engineering or architectural experience.
3. Registered design professional with fuel gas and plumbing engineering experience; or a fuel gas contractor with at least 10 years' experience, five of which shall have been in responsible charge of work.
4. Registered design professional with electrical engineering experience; or an electrical contractor with at least 10 years' experience, five of which shall have been in responsible charge of work.
5. Registered design professional with fire protection engineering experience; or a fire protection contractor with at least 10 years' experience, five of which shall have been in responsible charge of work.

❖ The board of appeals consists of five persons with the qualifications and experience indicated in this section. One must be a registered design professional (see Item 2) with structural or architectural experience. The others must be registered design professionals, construction superintendents or contractors with experience in the various areas of building construction. These requirements are important in that technical people rule on technical matters. The board of appeals is not the place for policy or political deliberations. The intent is that these matters be decided purely on their technical merits, with due regard for state-of-the-art construction technology.

109.2.2 Alternate members. The chief appointing authority shall appoint two alternate members who shall be called by the board chairman to hear appeals during the absence or disqualification of a member. Alternate members shall possess the qualifications required for board membership and shall be appointed for five years, or until a successor has been appointed.

❖ This section authorizes the chief appointing authority to appoint two alternate members who are to be available if the principal members of the board are absent or disqualified. Alternate members must possess the same qualifications as the principal members and are appointed for a term of 5 years, or until such time that a successor is appointed.

109.2.3 Chairman. The board shall annually select one of its members to serve as chairman.

❖ It is customary to determine chairmanship annually so that a regular opportunity is available to evaluate and either reappoint the current chairman or appoint a new one.

109.2.4 Disqualification of member. A member shall not hear an appeal in which that member has a personal, professional or financial interest.

❖ Members must disqualify themselves regarding any appeal in which they have a personal, professional or financial interest.

109.2.5 Secretary. The chief administrative officer shall designate a qualified clerk to serve as secretary to the board. The secretary shall file a detailed record of all proceedings in the office of the chief administrative officer.

❖ The chief administrative officer is to designate a qualified clerk to serve as secretary to the board. The secretary is required to file a detailed record of all proceedings in the office of the chief administrative officer.

109.2.6 Compensation of members. Compensation of members shall be determined by law.

❖ Members of the board of appeals need not be compensated unless required by the local municipality or jurisdiction.

109.3 Notice of meeting. The board shall meet upon notice from the chairman, within 10 days of the filing of an appeal, or at stated periodic meetings.

❖ The board must meet within 10 days of the filing of an appeal or at regularly scheduled meetings.

109.4 Open hearing. All hearings before the board shall be open to the public. The appellant, the appellant's representative, the code official and any person whose interests are affected shall be given an opportunity to be heard.

❖ Hearings before the board must be open to the public. The person who filed the appeal, his or her representative, the code official and any person whose interests are affected must be heard.

109.4.1 Procedure. The board shall adopt and make available to the public through the secretary procedures under which a hearing will be conducted. The procedures shall not require compliance with strict rules of evidence, but shall mandate that only relevant information be received.

❖ The board is required to establish and make available to the public written procedures detailing how hearings are to be conducted. Additionally, this section provides that although strict rules of evidence are not applicable, the information presented must be deemed relevant.

109.5 Postponed hearing. When five members are not present to hear an appeal, either the appellant or the appellant's representative shall have the right to request a postponement of the hearing.

❖ When all five members of the board are not present, either the person making the appeal or his representative may request a postponement of the hearing.

109.6 Board decision. The board shall modify or reverse the decision of the code official by a concurring vote of three members.

❖ A concurring vote of three members of the board is needed to modify or reverse the decision of the code official.

109.6.1 Resolution. The decision of the board shall be by resolution. Certified copies shall be furnished to the appellant and to the code official.

❖ A formal decision in the form of a resolution is required as an official record. Copies of this resolution are to be furnished to both the person making the appeal and the code official. The code official is bound by the action of the board of appeals, unless it is the opinion of the code official that the board of appeals has acted improperly. In such cases, corporate council may seek relief through the court having jurisdiction.

109.6.2 Administration. The code official shall take immediate action in accordance with the decision of the board.

❖ To avoid any undue delay in the progress of construction, the code official is required to act quickly on the board's decision. This action may be to enforce the decision or to seek legislative relief if the board's action can be demonstrated to be inappropriate.

109.7 Court review. Any person, whether or not a previous party to the appeal, shall have the right to apply to the appropriate court for a writ of certiorari to correct errors of law. Application for review shall be made in the manner and time required by law following the filing of the decision in the office of the chief administrative officer.

❖ This section allows any person to request a review by the court of jurisdiction if that person believes errors of law have occurred. Review must be applied for after the decision of the board is filed with the chief administrative officer. This helps to establish the observance of due process for all concerned.

Bibliography

The following resource materials are referenced in this chapter or are relevant to the subject matter addressed in this chapter.

Legal Aspects of Code Administration. Country Club Hills, IL: Building Officials and Code Administrators International, Inc., 1996; Whittier, CA: International Conference of Building Officials, 1996; Birmingham, AL: Southern Building Code Congress International, 1996.

Readings in Code Administration, Volume 1: History/Philosophy/Law. Country Club Hills, IL: Building Officials and Code Administrators International, Inc., 1974.

Rhyne, Charles S. *Survey of the Law and Building Codes.* The American Institute of Architects and the National Association of Home Builders.

CHAPTER 2
DEFINITIONS

General Comments

The words or terms defined in this chapter are deemed to be of prime importance in either specifying the subject matter of code provisions or in giving meaning to certain terms used throughout the code for administrative or enforcement purposes. The code user should be familiar with the terms found in this chapter, because the definitions are essential to the correct interpretation of the code and because the user might not be aware that a particular term found in the text is defined.

Purpose

Codes, by their very nature, are technical documents. Literally every word, term and punctuation mark can add to or change the meaning of the intended result. This is especially important with a performance code where the desired result often takes on more importance than the specific words.

Furthermore, the code, with its broad scope of applicability, includes terms used in a variety of construction disciplines. These terms can often have multiple meanings depending on the context or discipline being used at the time.

For these reasons, a consensus must be reached on the specific meaning of terms contained in the code. Chapter 2 performs this function by stating clearly what specific terms mean for the purpose of the code.

SECTION 201 (IFGC)
GENERAL

201.1 Scope. Unless otherwise expressly stated, the following words and terms shall, for the purposes of this code and standard, have the meanings indicated in this chapter.

❖ In the application of the code, the terms used have the meanings given in Chapter 2.

201.2 Interchangeability. Words used in the present tense include the future; words in the masculine gender include the feminine and neuter; the singular number includes the plural and the plural, the singular.

❖ Although the definitions contained in Chapter 2 are to be taken literally, gender and tense are considered to be interchangeable; thus, any grammatical inconsistencies within the code text will not hinder the understanding or enforcement of the requirements.

201.3 Terms defined in other codes. Where terms are not defined in this code and are defined in the ICC *Electrical Code*, *International Building Code*, *International Fire Code*, *International Mechanical Code* or *International Plumbing Code*, such terms shall have meanings ascribed to them as in those codes.

❖ When a word or term appears in the code that is not defined in this chapter, other references may be used to find its definition. These include the *International Building Code* (IBC®), the *ICC Electrical Code*® (ICC EC™), the *International Fire Code* (IFC®), the *International Mechanical Code* (IMC®), or the *International Plumbing Code*® (IPC®). These codes contain additional definitions (some parallel and duplicative) that may be used in the enforcement of this code or in the enforcement of the other codes by reference.

201.4 Terms not defined. Where terms are not defined through the methods authorized by this section, such terms shall have ordinarily accepted meanings such as the context implies.

❖ Another resource for defining words or terms not defined in this chapter or in other codes is their "ordinarily accepted meanings." The intent of this statement is that a dictionary definition may suffice if the definition is in context.

Construction terms used throughout the code might not be defined in Chapter 2 or in a dictionary. In such cases, one would first turn to the definitions contained in the referenced standards (see Chapter 7) and then to textbooks on the subject in question.

SECTION 202 (IFGC)
GENERAL DEFINITIONS

❖ This section contains definitions of terms that are associated with the subject matter of this chapter. These terms are not exclusively related to this chapter but are applicable everywhere the term is used in the code. Definitions of terms are necessary for the understanding and application of the code requirements.

ACCESS (TO). That which enables a device, appliance or equipment to be reached by ready access or by a means that first requires the removal or movement of a panel, door or similar obstruction (see also "Ready access").

❖ Access to equipment and appliances is necessary to facilitate inspection, observation, maintenance, adjustment, repair or replacement. Access to equipment means that the equipment can be physically reached without someone having to remove a permanent portion of the structure. It is acceptable, for example, to install equipment in an interstitial space that would require removal of lay-in suspended ceiling panels to gain access. Equipment would not be considered as being provided with access if it were necessary to remove or open any portion of a structure other than panels, doors, covers or similar obstructions intended to be removed or opened [also see the definition of "Ready access (to)"].

Access can be described as the capability of being reached or approached for the purpose of inspection, observation, maintenance, adjustment, repair or replacement. Achieving access may first require the removal or opening of a panel, door or similar obstruction and may require overcoming an obstacle such as elevation.

AIR CONDITIONER, GAS-FIRED. A gas-burning, automatically operated appliance for supplying cooled and/or dehumidified air or chilled liquid.

❖ Gas-fired air conditioners are not specifically referred to in the code but would include systems fueled with natural or propane gas including absorption type units and reciprocating units powered by internal combustion engines.

AIR CONDITIONING. The treatment of air so as to control simultaneously the temperature, humidity, cleanness and distribution of the air to meet the requirements of a conditioned space.

❖ Air conditioning is commonly referred to only in the context of cooling and dehumidifying air; however, the definition also indicates a much broader scope. In essence, the process of providing ventilation air to a space constitutes air conditioning because the introduction of any ventilation air is an attempt to control the indoor environment.

AIR, EXHAUST. Air being removed from any space or piece of equipment and conveyed directly to the atmosphere by means of openings or ducts.

❖ Exhaust air may be from a space or a piece of equipment. Exhaust air systems are terminated outside the building, in some cases after the exhaust air has been treated to remove any harmful emissions. Exhaust air is not recirculated.

AIR-HANDLING UNIT. A blower or fan used for the purpose of distributing supply air to a room, space or area.

❖ In addition to blowers, air-handling units may contain heat exchangers, filters and means to control air volume.

AIR, MAKEUP. Air that is provided to replace air being exhausted.

❖ Makeup air is not to be confused with combustion air. Makeup air replaces the air being exhausted through such systems as bathroom and toilet exhausts, kitchen exhaust hoods, hazardous exhaust systems and clothes dryer exhaust systems. Refer to Sections 610 and 613.5 for specific requirements for makeup air for direct-fired makeup air heaters and clothes dryers, respectively. Exhaust systems cannot function at design capacity without adequate volumes of makeup air to replace the air being exhausted.

ALTERATION. A change in a system that involves an extension, addition or change to the arrangement, type or purpose of the original installation.

❖ An alteration is any modification or change made to an existing installation. For example, increasing the size of piping for a portion of the system to accommodate different appliances would be an alteration.

ANODELESS RISER. A transition assembly in which plastic piping is installed and terminated above ground outside of a building.

❖ As the name implies, these riser assemblies accomplish corrosion protection of the steel riser by means other than cathodic protection involving a sacrificial anode. Some anodeless risers allow the termination of plastic piping aboveground by encasing the piping in a steel conduit equipped with a plastic-to-steel transition fitting [see commentary Figure 202(1)].

APPLIANCE (EQUIPMENT). Any apparatus or equipment that utilizes gas as a fuel or raw material to produce light, heat, power, refrigeration or air conditioning.

❖ An appliance is a manufactured component or assembly of components that converts one source of energy into a different form of energy to serve a specific purpose. The term "appliance" generally refers to residential- and commercial-type utilization equipment that is manufactured in standardized sizes or types. The term "appliance" is generally not associated with industrial-type equipment. For the application of the IFGC code provisions, the terms "appliance" and "equipment" are interchangeable.

Examples of appliances regulated by this code include furnaces, boilers, water heaters, room heaters, decorative gas log sets, cooking equipment, clothes dryers, pool, spa and hot tub heaters, unit heaters, ovens and similar gas-fired equipment.

APPLIANCE, FAN-ASSISTED COMBUSTION. An appliance equipped with an integral mechanical means to either draw or force products of combustion through the combustion chamber or heat exchanger.

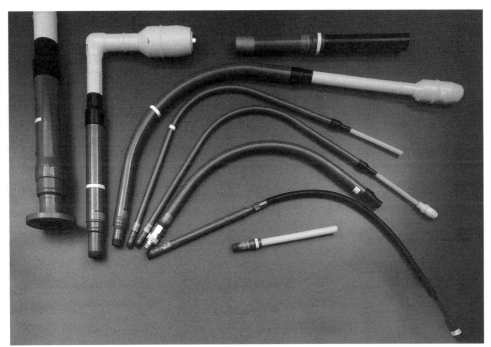

**Figure 202(1)
ANODELESS RISERS**

❖ Fan-assisted appliances are a specific type of Category I appliance, typically furnaces and boilers. They are not to be confused with Category II, III or IV appliances; power-vented appliances; or appliances served by exhausters. As the name implies, these appliances use a fan (blower) to "assist" the combustion process by helping the flue gases to overcome the internal flow resistance of the appliance heat exchanger. Some fan-assisted appliances are field-convertible to Category III appliances.

APPLIANCE, AUTOMATICALLY CONTROLLED. Appliances equipped with an automatic burner ignition and safety shutoff device and other automatic devices which accomplish complete turn-on and shutoff of the gas to the main burner or burners, and graduate the gas supply to the burner or burners, but do not affect complete shutoff of the gas.

❖ With respect to the *International Fuel Gas Code*, an automatically controlled appliance is an appliance that is cycled through its operation by controls such as thermostats, pressure switches and timers, without manual intervention. A residential water heater is automatically controlled; a four-burner cooktop is not.

APPLIANCE TYPE.
Low-heat appliance (residential appliance). Any appliance in which the products of combustion at the point of entrance to the flue under normal operating conditions have a temperature of 1,000°F (538°C) or less.

❖ Residential appliances, including solid-fuel appliances, are in this category.

Medium-heat appliance. Any appliance in which the products of combustion at the point of entrance to the flue under normal operating conditions have a temperature of more than 1,000°F (538°C), but not greater than 2,000°F (1093°C).

❖ This classification includes industrial-type equipment such as furnaces, kilns, ovens, dryers and incinerators.

APPLIANCE, UNVENTED. An appliance designed or installed in such a manner that the products of combustion are not conveyed by a vent or chimney directly to the outside atmosphere.

❖ Direct-fired makeup air heaters, gas-fired infrared radiant unit heaters, gas-fired cooking equipment and small gas-fired room heaters are examples of appliances that are sometimes unvented. With the exception of direct-fired equipment, the appliances listed herein can also be of the vented type, depending on the design and type of appliance and the particular application. Unvented appliances must be designed, installed and operated to avoid both the depletion of the oxygen supply and the accumulation of toxic combustion byproducts (see Sections 501.8, 611, 612, 621 and 630).

APPLIANCE, VENTED. An appliance designed and installed in such a manner that all of the products of combustion are conveyed directly from the appliance to the outside atmosphere through an approved chimney or vent system.

❖ The majority of fuel-fired appliances are designed to vent the products of combustion to the outdoors through one or more specific types of vent or chimney (see Section 501.2).

APPROVED. Approved by the code official or other authority having jurisdiction.

❖ As related to the process of acceptance of mechanical installations, including materials, equipment and construction systems, this definition identifies where ultimate authority rests. Whenever this term is used, it means that only the enforcing authority can accept a specific installation or component as complying with the code. Research reports prepared and published by the International Code Council® may be used by code officials to aid in their review and approval of the material or method described in the report. Publishing a report does not indicate automatic "approval" for the material or method described in the report. When the code states that an item or method "shall be approved," it does not mean that the code official is obligated to allow it. Rather, it means that the code official must determine whether the item or method is acceptable; that is, the code official must make the decision to allow or disallow.

APPROVED AGENCY. An established and recognized agency that is approved by the code official and regularly engaged in conducting tests or furnishing inspection services.

❖ The word "approved" means "as approved by the code official." The basis for approval of an agency for a particular activity may include the capacity and capability of the agency to perform the work in accordance with Sections 301.4 through 301.4.2.3.

ATMOSPHERIC PRESSURE. The pressure of the weight of air and water vapor on the surface of the earth, approximately 14.7 pounds per square inch (psi) (101 kPa absolute) at sea level.

❖ This is the "standard" atmospheric pressure at sea level. The actual atmospheric pressure varies with climate conditions. Barometric pressure is a measure of the actual atmospheric pressure, usually given as inches of mercury. One standard atmosphere (14.7 psi) is equal to 29.9 inches of mercury (101.4 kPa) at 32°F (0°C). Absolute pressure includes atmospheric pressure. Gage pressure is the pressure above atmospheric pressure. If not otherwise defined, stated pressure is the pressure above atmospheric pressure (gage pressure). A vacuum (partial or complete) is a pressure below atmospheric pressure.

AUTOMATIC IGNITION. Ignition of gas at the burner(s) when the gas controlling device is turned on, including reignition if the flames on the burner(s) have been extinguished by means other than by the closing of the gas controlling device.

❖ As opposed to manual ignition, automatic ignition is accomplished by pilot burners, spark ignitors, hot surface ignitors and similar means.

BAFFLE. An object placed in an appliance to change the direction of or retard the flow of air, air-gas mixtures or flue gases.

❖ Baffles are used in gas-fired equipment to regulate the flow through the combustion chamber or flueways for purposes such as improving the mixing of fuel and air and affecting the speed and direction of flow of flue gases.

BAROMETRIC DRAFT REGULATOR. A balanced damper device attached to a chimney, vent connector, breeching or flue gas manifold to protect combustion equipment by controlling chimney draft. A double-acting barometric draft regulator is one whose balancing damper is free to move in either direction to protect combustion equipment from both excessive draft and backdraft.

❖ These units automatically open or close depending on the difference between the internal vent pressure and the atmospheric pressure. See the commentary for the definition of "Atmospheric Pressure." Excessive negative pressure in the vent will cause excessive draft; therefore, the draft regulator will open as a result of the higher atmospheric pressure on the exterior of the damper, allowing air to enter the vent and thereby increasing the internal vent pressure. With excessive positive internal vent pressure, a double-acting regulator will open to relieve the pressure, lowering the vent internal pressure [see commentary Figure 202(2)].

BOILER, LOW-PRESSURE. A self-contained appliance for supplying steam or hot water.

❖ Low-pressure boilers operate at pressures less than or equal to 15 pounds per square inch (psi) (103 kPa) for steam and 160 psi (1103 kPa) for water. High-pressure boilers operate at pressures exceeding those pressures.

Boilers are usually manufactured of steel, cast iron or copper and are used to transfer heat from the combustion of a fuel or from an electric-resistance element to water to make steam or pressurized hot water for heating or other process or power purposes.

Boilers are usually installed in closed systems where the heat transfer medium is recirculated and retained within the system.

Figure 202(2)
BAROMETRIC DRAFT REGULATOR

Boilers must be labeled and installed in accordance with the manufacturer's installation instructions and the applicable sections of the code. Boilers are rated in accordance with standards published by the American Society of Mechanical Engineers (ASME), the Hydronics Institute, the Steel Boiler Institute (SBI), the Canadian Standards Association (CSA) and the American Boiler Manufacturers Association (ABMA). Boilers can be classified by working temperature, working pressure, type of fuel used (or electric boilers), materials of construction and whether or not the heat transfer medium changes phase from a liquid to a vapor [see commentary Figure 202(3)].

Hot water heating boiler. A boiler in which no steam is generated, from which hot water is circulated for heating purposes and then returned to the boiler, and that operates at water pressures not exceeding 160 pounds per square inch gauge (psig) (1100 kPa gauge) and at water temperatures not exceeding 250°F (121°C) at or near the boiler outlet.

❖ Hot water heating boilers are normally part of a closed system in which the heated water is circulated through fan coils or radiators of various types, including convectors, finned tube units and baseboard units.

Hot water supply boiler. A boiler, completely filled with water, which furnishes hot water to be used externally to itself, and that operates at water pressures not exceeding 160 psig (1100 kPa gauge) and at water temperatures not exceeding 250°F (121°C) at or near the boiler outlet.

❖ Hot water supply boilers are normally part of open systems in which the heated water is supplied and used externally to the boiler. Large domestic (potable) water heating systems often use hot water supply boilers.

Steam heating boiler. A boiler in which steam is generated and that operates at a steam pressure not exceeding 15 psig (100 kPa gauge).

❖ Steam heating boilers are normally part of a closed system in which low-pressure steam can be circulated through a variety of terminal units including natural convection units, forced-convection units and radiant panel systems.

BRAZING. A metal-joining process wherein coalescence is produced by the use of a nonferrous filler metal having a melting point above 1,000°F (538°C), but lower than that of the base metal being joined. The filler material is distributed between the closely fitted surfaces of the joint by capillary action.

❖ Brazing is the act of producing a brazed joint and is often referred to as silver soldering. Silver soldering is more accurately described as silver brazing and employs high silver-bearing alloys primarily composed of silver, copper and zinc. Silver soldering (brazing) typi-

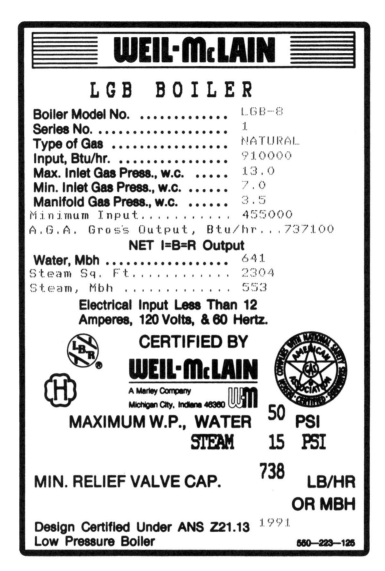

Figure courtesy of Weil-McLain

Figure 202(3)
LABEL FOR LOW PRESSURE BOILER

cally requires temperatures in excess of 1,000°F (538°C), and these solders are classified as "hard" solders.

Confusion has always been present with respect to the distinction between "silver solder" and "silver-bearing" solder. Silver solders are unique and can be further subdivided into soft and hard categories, which are determined by the percentages of silver and the other component elements of the particular alloy. The distinction is that silver-bearing solders (melting point less than 600°F [316°C]) are used in soft-soldered joints, and silver solders (melting point greater than 1,000°F [538°C]) are used in silver-brazed joints.

Brazed joints are considered to have superior strength and stress resistance and, because of the high melting point, are less likely to fail when exposed to fire.

Brazing with a filler metal conforming to AWS A5.8 produces a strong joint that will perform under extreme service conditions. The surfaces to be brazed must be cleaned to be free from oxides and impurities. Flux should be applied as soon as possible after the surfaces have been cleaned. Flux helps to remove residual traces of oxides, to promote wetting and to protect the surfaces from oxidation during heating. Care should be taken to prevent flux from entering the piping system during the brazing operation because flux that remains may corrode the pipe or contaminate the system.

Air and any other residual products should be removed from the pipe being brazed by purging the piping with a nonflammable gas such as carbon dioxide or nitrogen. Purging the system has several benefits, such as preventing oxidation from occurring on the inside of the pipe and preventing the creation of toxic gases that can result from the chemical breakdown of other products. Additionally, purging will eliminate the possibility of an explosion from any other flammable gas in the pipe that could ignite when mixed with air.

BROILER. A general term including salamanders, barbecues and other appliances cooking primarily by radiated heat, excepting toasters.

❖ The food being cooked can be above or below the gas flame and is usually placed on an open grill to allow draining away of grease from meat products.

BTU. Abbreviation for British thermal unit, which is the quantity of heat required to raise the temperature of 1 pound (454 g) of water 1°F (0.56°C) (1 Btu = 1055 J).

❖ Btu (J) is a unit of energy measurement. Fuel-fired appliances and equipment are rated based on their Btu/h (W) input or output.

BURNER. A device for the final conveyance of the gas, or a mixture of gas and air, to the combustion zone.

❖ There are several types of burners that vary in the method of providing combustion air and the pressures of fuel/air mixtures. Burner types include atmospheric ribbon, ported, porous ceramic, slotted and cone types and atmospheric or powered inshot and upshot gun burners.

Induced-draft. A burner that depends on draft induced by a fan that is an integral part of the appliance and is located downstream from the burner.

❖ See the commentary for the definition of "Draft/Mechanical or induced draft."

Power. A burner in which gas, air or both are supplied at pressures exceeding, for gas, the line pressure, and for air, atmospheric pressure, with this added pressure being applied at the burner.

❖ Power burners are typically used on larger appliances and employ a blower to introduce combustion air under pressure.

CHIMNEY. A primarily vertical structure containing one or more flues, for the purpose of carrying gaseous products of combustion and air from an appliance to the outside atmosphere.

❖ Chimneys differ from vents in their materials of construction and the type of appliance they are designed to serve. Chimneys are capable of venting much higher temperature flue gases than vents [see commentary Figure 202(4)].

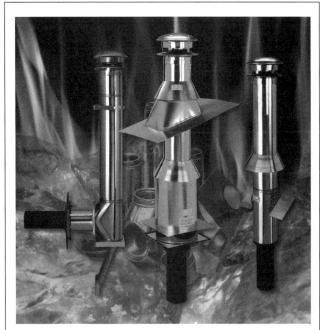

Photo courtesy of Selkirk L.L.C.

**Figure 202(4)
FACTORY-BUILT CHIMNEYS**

Factory-built chimney. A listed and labeled chimney composed of factory-made components, assembled in the field in accordance with manufacturer's instructions and the conditions of the listing.

❖ A factory-built chimney is a manufactured listed and labeled chimney that has been tested by an approved agency to determine its performance characteristics. Factory-built chimneys are manufactured in two basic designs: a double-wall mineral fiber insulated design and a triple-wall air-cooled design. Some chimneys employ both convection cooling and mineral fiber insulation. Both designs use stainless steel inner liners to resist the corrosive effects of combustion products [see commentary Figure 202(5)].

Masonry chimney. A field-constructed chimney composed of solid masonry units, bricks, stones or concrete.

❖ Masonry chimneys can have one or more flues within them and are field constructed of brick, stone, concrete or fire-clay materials. Masonry chimneys can stand alone or be part of a masonry fireplace.

Metal chimney. A field-constructed chimney of metal.

❖ Metal chimneys, being field constructed, are unlisted and often referred to as smokestacks. Metal chimneys are associated with industrial installations and equipment.

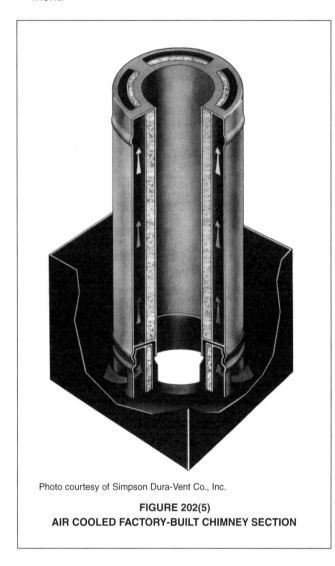

Photo courtesy of Simpson Dura-Vent Co., Inc.

**FIGURE 202(5)
AIR COOLED FACTORY-BUILT CHIMNEY SECTION**

CLEARANCE. The minimum distance through air measured between the heat-producing surface of the mechanical appliance, device or equipment and the surface of the combustible material or assembly.

❖ Clearances between sources of heat and combustibles are always airspace clearances.

CLOTHES DRYER. An appliance used to dry wet laundry by means of heated air. Dryer classifications are as follows:

❖ There are two types of clothes dryers.

Type 1. Factory-built package, multiple production. Primarily used in family living environment. Usually the smallest unit physically and in function output.

❖ Type 1 clothes dryers are used in dwelling units and in residential settings.

Type 2. Factory-built package, multiple production. Used in business with direct intercourse of the function with the public. Not designed for use in individual family living environment.

❖ Type 2 clothes dryers are intended for commercial, public or institutional use.

CODE. These regulations, subsequent amendments thereto or any emergency rule or regulation that the administrative authority having jurisdiction has lawfully adopted.

❖ The adopted regulations are generally referred to as "the code" and include not only the *International Fuel Gas Code* but also any adopted modifications to the code and all other related rules and regulations promulgated and enacted by the jurisdiction.

CODE OFFICIAL. The officer or other designated authority charged with the administration and enforcement of this code, or a duly authorized representative.

❖ The statutory power to enforce the code is normally vested in a building department (or the like) of a state, county or municipality whose designated enforcement officer is termed the "code official" [see commentary, Sections 103 and 104].

COMBUSTION. In the context of this code, refers to the rapid oxidation of fuel accompanied by the production of heat or heat and light.

❖ The primary components of combustion are fuel, oxygen and heat. The code regulates many aspects of combustion technology, including the process of combustion and providing sufficient air, the use of energy produced from combustion, the safe venting of the products of combustion, combustion efficiency, the containment and control of combustion and the fuel supplies for combustion equipment and appliances.

COMBUSTION AIR. Air necessary for complete combustion of a fuel, including theoretical air and excess air.

❖ The process of combustion requires a specific amount of oxygen to initiate and sustain the combustion reaction. Combustion air includes primary air, secondary air, draft hood dilution air and excess air. Combustion air includes the amount of atmospheric air required for complete combustion of a fuel and is related to the molecular composition of the fuel being burned, the design of the fuel-burning equipment and the percentage of oxygen in the combustion air. Too little combustion air will result in incomplete combustion of a fuel and the possible formation of carbon deposits (soot), carbon monoxide, toxic alcohols, ketones, aldehydes, nitrous oxides and other by-products. The required amount of combustion air is usually stated in terms of cubic feet per minute (m^3/s) or pounds per hour (kg/h).

DEFINITIONS

COMBUSTION CHAMBER. The portion of an appliance within which combustion occurs.

❖ Combustion chambers are either open to the atmosphere or are isolated (sealed) from the atmosphere in which the appliance is installed. Combustion chambers are often referred to as "fireboxes."

COMBUSTION PRODUCTS. Constituents resulting from the combustion of a fuel with the oxygen of the air, including inert gases, but excluding excess air.

❖ Such products include water, carbon dioxide, carbon monoxide, nitrous oxides and various trace compounds.

CONCEALED LOCATION. A location that cannot be accessed without damaging permanent parts of the building structure or finish surface. Spaces above, below or behind readily removable panels or doors shall not be considered as concealed.

❖ The space above a "drop-in" tile suspended ceiling system, for example, would not be considered as a concealed location.

CONCEALED PIPING. Piping that is located in a concealed location (see "Concealed location").

❖ This refers to gas piping in locations meeting the definition of "Concealed location."

CONDENSATE. The liquid that condenses from a gas (including flue gas) caused by a reduction in temperature or increase in pressure.

❖ Condensate forms when the temperature of a vapor is lowered to its dew point temperature. Air conditioning systems produce condensate when an airstream contacts cooling coils. The moisture in the air condenses on the cold surface of the coils and the air is "dehumidified." High efficiency (84 percent and up) fuel burning appliances produce condensate from the combustion gases. Condensate also forms within improperly designed chimneys and vents when the products of combustion (which contain water vapor) contact the colder inner walls of the vent or chimney. If the temperature of the products of combustion is lowered to the dew point temperature of the water vapor, condensate will form on the inside walls of the vent or chimney. Condensed steam in hydronic systems is also referred to as "condensate."

CONNECTOR. The pipe that connects an approved appliance to a chimney, flue or vent.

❖ In most cases, appliances are not located directly in line with the vertically rising chimney or vent; therefore, a vent connector is necessary to connect the appliance flue outlet to the vent. Vent connectors can be single- or double-wall pipes and are usually made from galvanized steel, stainless steel or aluminum sheet metal. Connectors are also made of corrugated aluminum. In many installations, the vent connectors must be constructed of the same material as the vent, as is typically done with Type B vent systems. There are many requirements in the code (Sections 503 and 504) regulating the location, length, size and type of connector.

CONSTRUCTION DOCUMENTS. All of the written, graphic and pictorial documents prepared or assembled for describing the design, location and physical characteristics of the elements of the project necessary for obtaining a mechanical permit.

❖ In order to determine that proposed construction is in compliance with code requirements, sufficient information must be submitted to the code official for review. This typically consists of the drawings (floor plans, elevations, sections, piping diagrams, etc.), specifications and product information describing the proposed work.

CONTROL. A manual or automatic device designed to regulate the gas, air, water or electrical supply to, or operation of, a mechanical system.

❖ A control is a device designed to respond to changes in temperature, pressure, liquid or gas flow rates, current, voltage, resistance, humidity and/or liquid levels.

CONVERSION BURNER. A unit consisting of a burner and its controls for installation in an appliance originally utilizing another fuel.

❖ Typical conversion burners are designed to convert an appliance such as a boiler from solid-fuel-fired to oil- or gas-fired.

COUNTER APPLIANCES. Appliances such as coffee brewers and coffee urns and any appurtenant water-heating equipment, food and dish warmers, hot plates, griddles, waffle bakers and other appliances designed for installation on or in a counter.

❖ Counter appliances are small low-energy input type appliances and need not be vented unless the aggregate input rating of all the appliances exceeds that allowed in Section 501.8.

CUBIC FOOT. The amount of gas that occupies 1 cubic foot (0.02832 m^3) when at a temperature of 60°F (16°C), saturated with water vapor and under a pressure equivalent to that of 30 inches of mercury (101 kPa).

❖ This definition applies specifically to a measurement of fuel gas.

DAMPER. A manually or automatically controlled device to regulate draft or the rate of flow of air or combustion gases.

❖ Dampers act as restrictors or valves in a gaseous flow stream (see Section 503.13).

DECORATIVE APPLIANCE, VENTED. A vented appliance wherein the primary function lies in the aesthetic effect of the flames.

❖ Such appliances include so-called "gas fireplaces" that are designed to simulate a wood-burning fireplace. The title of the standard for vented decorative gas-fired appliances (Z21.50) has been recently changed to "Vented Gas Fireplaces."

DECORATIVE APPLIANCES FOR INSTALLATION IN VENTED FIREPLACES. A vented appliance designed for installation within the fire chamber of a vented fireplace, wherein the primary function lies in the aesthetic effect of the flames.

❖ These appliances include gas log sets for installation in solid-fuel-burning fireplaces.

DEMAND. The maximum amount of gas input required per unit of time, usually expressed in cubic feet per hour, or Btu/h (1 Btu/h = 0.2931 W).

❖ Demand refers to the load that appliances place on the fuel-gas distribution system.

DESIGN FLOOD ELEVATION. The elevation of the "design flood," including wave height, relative to the datum specified on the community's legally designated flood hazard map.

❖ This term is used in Section 301.11 and refers to the worst case flood scenario that could be anticipated, such as might occur once every 100 years. The design flood elevation is the height to which flood waters will rise during passage or occurrence of the design flood. The datum specified on the flood hazard map is important because it may differ from that used locally for other purposes.

DILUTION AIR. Air that is introduced into a draft hood and is mixed with the flue gases.

❖ Dilution air is associated only with draft-hood-equipped Category I appliances. Fan-assisted Category I appliances do not involve dilution air because all of the air brought in by the fan passes directly through the combustion chamber of the appliance. Dilution air lowers the dew point of vent gases, reducing the possibility that condensation will occur inside the vent or chimney.

DIRECT-VENT APPLIANCES. Appliances that are constructed and installed so that all air for combustion is derived directly from the outside atmosphere and all flue gases are discharged directly to the outside atmosphere.

❖ These appliances have independent exhaust and intake pipes or have concentric pipes that vent combustion gases through the inner pipe and convey combustion air in the annular space between the inner and outer pipe walls [see commentary Figures 202(6) and 202(7)].

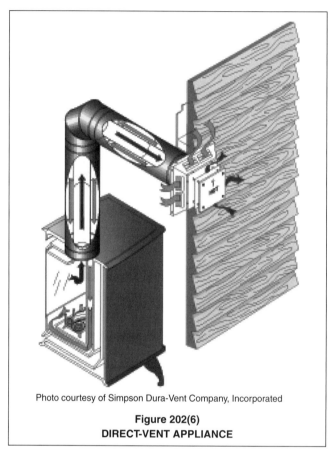

Photo courtesy of Simpson Dura-Vent Company, Incorporated

**Figure 202(6)
DIRECT-VENT APPLIANCE**

Figure courtesy of Lochinvar Corporation

**Figure 202(7)
DIRECT-VENT APPLIANCE**

DRAFT. The pressure difference existing between the equipment or any component part and the atmosphere, that causes a continuous flow of air and products of combustion through the gas passages of the appliance to the atmosphere.

❖ Draft is the negative static pressure measured relative to atmospheric pressure that is developed in chimneys and vents and in the flue-ways of fuel-burning appliances. Draft can be produced by hot flue-gas buoyancy ("stack effect"), mechanically by fans and exhausters or by a combination of both natural and mechanical means.

Mechanical or induced draft. The pressure difference created by the action of a fan, blower or ejector, that is located between the appliance and the chimney or vent termination.

❖ Induced draft systems use a fan or blower to boost or "induce" draft in a venting system that produces insufficient natural draft. Draft induction implies that the combustion gases are "pulled" through a passageway or conduit. Draft inducers produce negative pressures on the inlet (upstream) side of the fan or blower. They are separate field-installed units located between an appliance and its venting system. Draft inducers are used with natural draft venting systems to overcome the resistance of vent or chimney connectors and to compensate for the inability of the chimney or vent to produce sufficient and reliable draft.

Draft induction is also a design principle used in many fan-assisted appliances. Draft-inducer fans or blowers that are integral with fuel-fired appliances are necessary to overcome the internal resistance of the heat exchanger flue passageways. To attain higher thermal transfer efficiencies, some heat exchanger designs use flue passageways that retain the combustion gases longer over a greater surface area, thus extracting more heat. Such heat exchanger designs cannot rely upon natural (gravity) venting to overcome the higher resistance to flow and, therefore, must rely upon mechanical means to pull the combustion gases through the heat exchanger.

Induced draft systems are not to be confused with systems using power exhausters or other self-venting equipment.

Mechanical draft systems produce positive pressures on the discharge side of the fan or blower and can be said to "push" combustion gases through a passageway or conduit.

Natural draft. The pressure difference created by a vent or chimney because of its height, and the temperature difference between the flue gases and the atmosphere.

❖ Natural draft systems do not use mechanical devices such as fans or blowers but instead rely on the principle of buoyancy to carry the products of combustion to the atmosphere. Because of the difference in temperature and the resultant difference in density between the hot products of combustion and the ambient atmosphere, the gases within the chimney or flue will rise, creating a buoyant "draft." The phenomenon of natural draft is sometimes referred to as "stack effect" and is measured in inches of water column (kPa). The amount of draft is affected by the height of the chimney or vent and also by the ability of the chimney or vent to maintain the temperature differential between the combustion gases and the ambient air.

DRAFT HOOD. A nonadjustable device built into an appliance, or made as part of the vent connector from an appliance, that is designed to (1) provide for ready escape of the flue gases from the appliance in the event of no draft, backdraft or stoppage beyond the draft hood, (2) prevent a backdraft from entering the appliance, and (3) neutralize the effect of stack action of the chimney or gas vent upon operation of the appliance.

❖ Draft hoods are integral to or supplied with natural draft atmospheric-burner gas-fired appliances other than fan-assisted appliances. When classified, appliances equipped with draft hoods are classified by the manufacturer as Category I appliances. Because of minimum efficiency standards and the popularity of mechanical and other special proprietary venting systems, draft-hood-equipped appliances are becoming increasingly rare in the marketplace [see commentary Figure 202(8)].

Photo courtesy of A.O. Smith Water Products

**Figure 202(8)
DRAFT HOOD EQUIPPED APPLIANCE**

DRAFT REGULATOR. A device that functions to maintain a desired draft in the appliance by automatically reducing the draft to the desired value.

❖ Excessive draft reduces the combustion and thermal efficiency of an appliance. Draft regulators automatically adjust the draft by admitting air into the vent, thereby reducing the draft [see commentary Figure 202(2)].

DRIP. The container placed at a low point in a system of piping to collect condensate and from which the condensate is removable.

❖ These reservoirs, also referred to as "drip legs," are made up of pipe and fittings and are intended to collect liquid in piping systems where condensables are possible. A "drip leg" is distinct from a "sediment trap" even though they may be constructed identically.

DRY GAS. A gas having a moisture and hydrocarbon dew point below any normal temperature to which the gas piping is exposed.

❖ The dew point referred to herein is the temperature at which fuel gas hydrocarbons or water vapor begin to condense into liquid fuel or liquid water. Condensate will not form if the gas piping system temperature is maintained above the dew point of the fuel gas and the water vapor.
The gas supplier (utility company) should be contacted to determine the nature of the gas supplied.

DUCT FURNACE. A warm-air furnace normally installed in an air distribution duct to supply warm air for heating. This definition shall apply only to a warm-air heating appliance that depends for air circulation on a blower not furnished as part of the furnace.

❖ Duct furnaces are vented appliances and consist of burners, heat exchangers and related controls. Duct furnaces depend on an external air handler or blower.

DUCT SYSTEM. A continuous passageway for the transmission of air that, in addition to ducts, includes duct fittings, dampers, plenums, fans and accessory air-handling equipment.

❖ Duct systems are part of an air distribution system and include supply, return and relief/exhaust air systems.

EQUIPMENT. See "Appliance."

❖ See the commentaries for "Appliance (Equipment)"; "Appliance, Fan-Assisted Combustion"; "Appliance, Automatically Controlled"; "Appliance Type"; "Appliance, Unvented" and "Appliance, Vented."

FIREPLACE. A fire chamber and hearth constructed of noncombustible material for use with solid fuels and provided with a chimney.

❖ Fireplaces burn solid fuels (wood, coal, etc.) and are not referred to as appliances in the code. The IBC and the *International Residential Code* (IRC®) cover construction of masonry fireplaces.

Masonry fireplace. A hearth and fire chamber of solid masonry units such as bricks, stones, listed masonry units or reinforced concrete, provided with a suitable chimney.

❖ Masonry fireplaces must be constructed in accordance with the requirements found in the IBC or the IRC. These specific requirements are based on tradition and field experience and describe the conventional fireplace that has proven to be reliable where properly constructed, used and maintained.

Factory-built fireplace. A fireplace composed of listed factory-built components assembled in accordance with the terms of listing to form the completed fireplace.

❖ Factory-built fireplaces are solid-fuel-burning units having a fire chamber that is intended to be either open to the room or, if equipped with doors, operated with the doors either open or closed. Fireplaces are not referred to as appliances. The term "fireplace" describes a complete assembly, which includes the hearth, the fire chamber and a chimney. A factory-built fireplace is composed of factory-built components representative of the prototypes tested and is installed in accordance with the manufacturer's installation instructions to form the completed fireplace.

FIRING VALVE. A valve of the plug and barrel type designed for use with gas, and equipped with a lever handle for manual operation and a dial to indicate the percentage of opening.

❖ This term is not used in the code.

FLAME SAFEGUARD. A device that will automatically shut off the fuel supply to a main burner or group of burners when the means of ignition of such burners becomes inoperative, and when flame failure occurs on the burner or group of burners.

❖ These devices are primary safety controls and are provided on all automatically operated fuel-fired appliances.

FLOOD HAZARD AREA. The greater of the following two areas:
1. The area within a floodplain subject to a 1 percent or greater chance of flooding in any given year.
2. This area designated as a flood hazard area on a community's flood hazard map, or otherwise legally designated.

❖ The Federal Emergency Management Agency (FEMA) prepares flood insurance rate maps that delineate the land area that is subject to inundation by the 1-percent annual chance flood. Some states and local jurisdictions develop and adopt maps of flood hazard areas that are more extensive than the areas shown on

FEMA's map. For the purpose of the code, the flood hazard area within which the requirements are to be applied is the greater of the two delineated areas (see Section 301.11).

FLOOR FURNACE. A completely self-contained furnace suspended from the floor of the space being heated, taking air for combustion from outside such space and with means for observing flames and lighting the appliance from such space.

❖ These units supply heat through a floor grille placed directly over the unit's heat exchanger.

Gravity type. A floor furnace depending primarily upon circulation of air by gravity. This classification shall also include floor furnaces equipped with booster-type fans which do not materially restrict free circulation of air by gravity flow when such fans are not in operation.

❖ Typically floor furnaces are classified as gravity-type furnaces. Air circulates from a gravity-type floor furnace by convection.

Fan type. A floor furnace equipped with a fan which provides the primary means for circulating air.

❖ Floor furnaces may be equipped with factory-installed fans to circulate the air.

FLUE, APPLIANCE. The passage(s) within an appliance through which combustion products pass from the combustion chamber of the appliance to the draft hood inlet opening on an appliance equipped with a draft hood or to the outlet of the appliance on an appliance not equipped with a draft hood.

❖ Appliance flues are the passages through an appliance from the combustion chamber, through the heat exchanger and through the vent outlet.

FLUE COLLAR. That portion of an appliance designed for the attachment of a draft hood, vent connector or venting system.

❖ The flue collar size will be a determining factor in vent connector sizing.

FLUE GASES. Products of combustion plus excess air in appliance flues or heat exchangers.

❖ The exact composition of flue gases will depend on the fuel being burned. The primary components of flue gases are nitrogen, carbon dioxide, water vapor, particulates and a myriad of compounds and trace elements that vary with the nature of the fuel and purity of the combustion air. Carbon monoxide is also a component of flue gas.

FLUE LINER (LINING). A system or material used to form the inside surface of a flue in a chimney or vent, for the purpose of protecting the surrounding structure from the effects of combustion products and for conveying combustion products without leakage to the atmosphere.

❖ Flue liners must be resistant to heat and the corrosive action of the products of combustion. They provide insulation value to retard the transfer of heat to the chimney structure and limit exposure of the chimney structure to the harmful effects of combustion products. Flue liners are generally made of fire-clay tile, refractory brick, poured-in-place refractory materials, stainless steel alloys and aluminum. Flue liners used with gas-fired appliances are typically made of stainless steel or aluminum [see commentary Figure 202(9)].

FUEL GAS. Fuel gases include: a natural gas, manufactured gas, liquefied petroleum gas, hydrogen gas and mixtures of these gases.

❖ The nature of fuel gases makes proper design, installation and selection of materials and devices necessary to minimize the possibility of fire or explosion. Bringing fuel gases into a building is in itself a risk. The provisions of the code are intended to reduce that risk to a level comparable to that associated with other energy sources such as electricity.

  The two most commonly used fuel gases are natural gas and liquefied petroleum gas (LP-gas or LPG). These fuel gases have the following characteristics or properties.

  Natural gas: The principal constituent of natural gas is methane (CH_4). It can also contain small quantities of nitrogen, carbon dioxide, hydrogen sulfide, water vapor, other hydrocarbons (such as ethane and propane) and various trace elements. Natural gas is colorless, tasteless and odorless; however, an odorant is added to the gas so that it can be readily detected. Natural gas is lighter than air (specific gravity of 0.60 typical) and has the tendency to rise when escaping to the atmosphere. Natural gas has a rather narrow flammability range (approximately 3 to 15 percent volume in air) above and below which the gas-to-air mixture ratio will be too rich or too lean to support combustion. The heating value of natural gas is approximately 1,050 Btu per cubic foot (10.8 kW/m^3).

  LP-gas: Liquefied petroleum gases include commercial propane and commercial butane. LP-gas vapors are heavier than air (specific gravity of 1.52 typical) and tend to accumulate in low areas and near the floor. The ranges of flammability for LP-gases are narrower than those of natural gas (approximately 2 to 10 percent volume in air). Like natural gas, LP-gases are odorized to make them detectable. The heating value of propane is approximately 2,500 Btu per cubic foot (25.9 kW/m^3) of gas. The heating value of butane is approximately 3,300 Btu per cubic foot (34.2 kW/m^3) of gas.

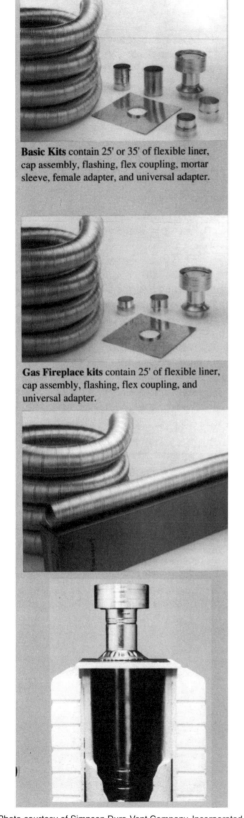

Figure 202(9)
CHIMNEY FLUE LINER SYSTEM FOR GAS APPLIANCE

Photo courtesy of Simpson Dura-Vent Company, Incorporated

FUEL GAS UTILIZATION EQUIPMENT. See "Appliance."

❖ See the commentary for "Appliance (Equipment)."

FURNACE. A completely self-contained heating unit that is designed to supply heated air to spaces remote from or adjacent to the appliance location.

❖ The single most distinguishing characteristic of furnaces is that they use air as the heat transfer medium. Furnaces can be fueled by gas, oil, solid fuel or electricity and use fans, blowers or gravity (convection) to circulate air to and from the unit. In the context of the code, the primary usage of the term "furnace" refers to heating appliances that combine a combustion chamber with related burner and control components, one or more heat exchangers, a flue gas conveying system and air-handling fans/blowers.

FURNACE, CENTRAL. A self-contained appliance for heating air by transfer of heat of combustion through metal to the air, and designed to supply heated air through ducts to spaces remote from or adjacent to the appliance location.

❖ The term "central furnace" has been adopted to identify a furnace that supplies conditioned air to remote locations from a central location. There are many types of furnaces. Most of these variations are a result of architectural design related to available space, desired location of the unit and size of the area to be conditioned.

Downflow furnace. A furnace designed with airflow discharge vertically downward at or near the bottom of the furnace.

❖ Downflow furnaces are typically installed with airflow directed into a plenum and ductwork installed in a crawl space.

Forced air furnace with cooling unit. A single-package unit, consisting of a gas-fired forced-air furnace of one of the types listed below combined with an electrically or fuel gas-powered summer air-conditioning system, contained in a common casing.

❖ A forced-air furnace with cooling unit is typically a package unit designed for outdoor installation on a rooftop or on a pad on grade.

Forced-air type. A central furnace equipped with a fan or blower which provides the primary means for circulation of air.

❖ All furnaces equipped with a fan/blower for the purpose of circulating air through the unit would be considered forced-air type furnaces.

Gravity furnace with booster fan. A furnace equipped with a booster fan that does not materially restrict free circulation of air by gravity flow when the fan is not in operation.

❖ These units can be equipped with a booster fan that may or may not operate continually, depending on the application, to assist in the circulation of air.

Gravity type. A central furnace depending primarily on circulation of air by gravity.

❖ Gravity type units are used for relatively small spaces where the natural flow of air by convection will condition the space.

Horizontal forced-air type. A furnace with airflow through the appliance essentially in a horizontal path.

❖ Horizontal forced-air type units are commonly installed in spaces above ceilings, suspended from ceiling and roof structures and installed in attics.

Multiple-position furnace. A furnace designed so that it can be installed with the airflow discharge in the upflow, horizontal or downflow direction.

❖ Many furnaces today are designed for multiple applications, allowing a single unit to be used in all of the common locations.

Upflow furnace. A furnace designed with airflow discharge vertically upward at or near the top of the furnace. This classification includes "highboy" furnaces with the blower mounted below the heating element and "lowboy" furnaces with the blower mounted beside the heating element.

❖ Upflow furnaces (typically made in high-boy and low-boy styles) are typically installed in basements and attics.

FURNACE, ENCLOSED. A specific heating, or heating and ventilating, furnace incorporating an integral total enclosure and using only outside air for combustion.

❖ "Enclosed furnaces" are believed to be extinct. Such furnaces had an integral housing that isolated the combustion chamber from the conditioned space thus allowing all combustion air to be obtained directly from the housing that opened to the outdoors.

FURNACE PLENUM. An air compartment or chamber to which one or more ducts are connected and which forms part of an air distribution system.

❖ The term "plenum" is no longer defined or used in this code. A furnace plenum is a box made of sheet metal or other duct material. Such a box is attached directly to a furnace to facilitate connection of supply and return ducts to the furnace air inlet and outlet. This term does not refer to cavities within a building, such as stud and joist spaces and spaces above dropped ceilings.

GAS CONVENIENCE OUTLET. A permanently mounted, manually operated device that provides the means for connecting an appliance to, and disconnecting an appliance from, the supply piping. The device includes an integral, manually operated valve with a nondisplaceable valve member and is designed so that disconnection of an appliance only occurs when the manually operated valve is in the closed position.

❖ These devices are listed and labeled and designed to mate with specialized appliance connectors. A gas outlet is the point at which the gas distribution system connects to a gas appliance. The outlets are intended for use with portable appliances such as space heaters and outdoor cooking appliances [(see commentary Figures 202(10) and 202(11)].

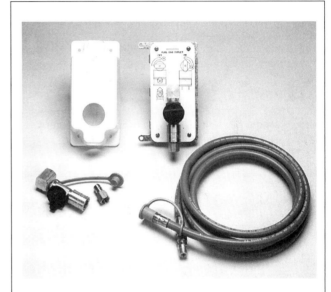

Photo courtesy of Gastite Division/Titeflex Corporation

**Figure 202(10)
GAS CONVENIENCE OUTLET**

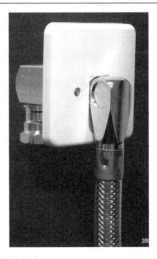

Photo courtesy of Maxitrol company

**Figure 202(11)
GAS CONVENIENCE OUTLET**

GASEOUS HYDROGEN SYSTEM. See Section 702.1.

GAS PIPING. An installation of pipe, valves or fittings installed on a premises or in a building and utilized to convey fuel gas.

❖ Gas piping includes all the components, fittings and piping needed to deliver the fuel gas from the point of delivery to the appliance or equipment connection. The point of delivery may be a regulator or meter that is typically installed by the gas utility. The point of delivery may be located at the user's property line, immediately outside the structure or in some instances in the structure.

GAS UTILIZATION EQUIPMENT. An appliance that utilizes gas as a fuel or raw material or both.

❖ See the commentary for "Appliance (Equipment)."

HAZARDOUS LOCATION. Any location considered to be a fire hazard for flammable vapors, dust, combustible fibers or other highly combustible substances. The location is not necessarily categorized in the building code as a high-hazard group classification.

❖ The environment in which mechanical equipment and appliances operate plays a significant role in the safe performance of the equipment installation. Locations that may contain ignitable or explosive atmospheres are classified as hazardous locations for the installation of mechanical equipment and appliances. For example, repair garages can be classified as hazardous locations because they can contain gasoline vapors from vehicles stored within them as well as other volatile chemicals. Public and private garages are not considered hazardous locations even though the presence of motor vehicles and the storage of fuel, paint, varnish, thinner, lawn- and home-maintenance products and other chemicals are a concern, as evidenced by Sections 305.2, 305.3, and 305.4.

HOUSE PIPING. See "Piping system."

❖ House piping is the distribution piping downstream of the point of delivery. House piping is an antiquated term.

HYDROGEN CUT-OFF ROOM. See Section 702.1.

HYDROGEN GENERATING APPLIANCE. See Section 702.1.

IGNITION PILOT. A pilot that operates during the lighting cycle and discontinues during main burner operation.

❖ Commonly referred to as "interrupted pilots," such pilots are ignited by a spark or hot surface device. Generally an electronic circuit controls the ignition system. When main burner ignition has been proven, the ignition pilot is shut off, and the main burner flame is supervised for the entire burn cycle.

IGNITION SOURCE. A flame, spark or hot surface capable of igniting flammable vapors or fumes. Such sources include appliance burners, burner ignitors, and electrical switching devices.

❖ This definition is important in the application of Section 305.3 regarding the elevation of ignition sources. By means of this definition, Section 305.3 applies to unintentional ignition sources such as electrical switching devices as well as intentional ignition sources such as pilot lights, spark ignitors, and hot surface ignitors.

In the context of Section 305.3, an "ignition source" is something capable of igniting flammable vapors that are present in the atmosphere in the locations listed in Section 305.3.

INCINERATOR. An appliance used to reduce combustible refuse material to ashes and which is manufactured, sold and installed as a complete unit.

❖ These appliances are believed to be extinct.

INDUSTRIAL AIR HEATERS, DIRECT-FIRED NONRECIRCULATING. A heater in which all the products of combustion generated by the burners are released into the air stream being heated. The purpose of the heater is to offset building heat loss by heating only outdoor air.

❖ These heaters have traditionally been called makeup air heaters because they heat 100 percent of outdoor air without recirculation of any air across the burner. These heaters are unvented (direct-fired) appliances regulated by Section 611 (see Figure 611.1).

INDUSTRIAL AIR HEATERS, DIRECT-FIRED RECIRCULATING. A heater in which all the products of combustion generated by the burners are released into the air stream being heated. The purpose of the heater is to offset building heat loss by heating outdoor air, and, if applicable, indoor air.

❖ These heaters are used in large spaces such as warehouses and are often referred to as "air turnover units" because they are typically vertical air handlers designed to prevent air stratification. These heaters are unvented (direct-fired) appliances regulated by Section 612.

INFRARED RADIANT HEATER. A heater that directs a substantial amount of its energy output in the form of infrared radiant energy into the area to be heated. Such heaters are of either the vented or unvented type.

❖ Those heaters include low-intensity tubular types and high-intensity ceramic burner element types. They heat objects (such as people) by direct radiation and do not heat the ambient air.

JOINT, FLANGED. A joint made by bolting together a pair of flanged ends.

❖ Flanges are commonly used with large piping at locations where the piping must be capable of being disassembled periodically and at connections to valves, regulators, devices and equipment. Full-face gaskets must be used with bronze and cast-iron flange fittings. Typically, materials for flange gaskets include metal or metal-jacketed asbestos, asbestos and aluminum "O" rings, or spiral-wound metal gaskets. Gaskets must be replaced whenever a joint is opened. The gasket material must be compatible with the piping contents to prevent chemical reaction.

JOINT, FLARED. A metal-to-metal compression joint in which a conical spread is made on the end of a tube that is compressed by a flare nut against a mating flare.

❖ Because the pipe end is expanded in a flared joint, only annealed and bending tempered soft-drawn copper tubing may be flared. Commonly used flaring tools employ a screw yoke and block assembly or an expander tool that is driven into the tube with a hammer. The flared tubing end is compressed between a fitting seat and a threaded nut to form a metal-to-metal seal.

JOINT, MECHANICAL. A general form of gas-tight joints obtained by the joining of metal parts through a positive-holding mechanical construction, such as flanged joint, threaded joint, flared joint or compression joint.

❖ Mechanical joints can take many forms, but most share a common characteristic, which is applying a radial pressure to the pipes or fittings they join. Mechanical joining means can be proprietary, which means that the manufacturer of the fittings and devices must provide adequate instructions for assembling the joint.

JOINT, PLASTIC ADHESIVE. A joint made in thermoset plastic piping by the use of an adhesive substance which forms a continuous bond between the mating surfaces without dissolving either one of them.

❖ Unlike solvent welded joints, adhesive (glue) joints are surface bonded.

JOINT, PLASTIC HEAT FUSION. A joint made in thermoplastic piping by heating the parts sufficiently to permit fusion of the materials when the parts are pressed together.

❖ Heat-fusion joints for plastic pipe are analogous to welded joints for steel pipe. Only polyethylene and polybutylene can be joined by heat fusion and only in accordance with the pipe manufacturer's instructions. The process involves heating the pipe and fittings with a special iron. Some fittings have integral heating elements that are connected to a special power supply. When the parts to be joined reach their melting points, they are assembled and allowed to fuse together.

JOINT, WELDED. A gas-tight joint obtained by the joining of metal parts in molten state.

❖ A welded joint is similar to a brazed joint. The primary differences are the temperature at which the joint is made, the type of filler metal used, and the fact that welding reaches the melting point of the base metal, whereas brazing temperatures are well below the melting point of the pipe and fittings.

LABELED. Devices, equipment, appliances or materials to which have been affixed a label, seal, symbol or other identifying mark of a nationally recognized testing laboratory, inspection agency or other organization concerned with product evaluation that maintains periodic inspection of the production of the above-labeled items and by whose label the manufacturer attests to compliance with applicable nationally recognized standards.

❖ When a product is labeled, the label indicates that the material has been tested for conformance to an applicable standard and that the component is subject to third-party inspection to verify that the minimum level of quality required by the standard is maintained. Labeling provides a readily available source of information that is useful for field inspection of installed products. The label identifies the product or material and provides other information that can be further investigated if there is question concerning the suitability of the product or material for the specific installation. The labeling agency performing the third-party inspection must be approved by the code official, and the basis for this approval may include, but is not necessarily limited to, the capacity and capability of the agency to perform the specific testing and inspection.

The applicable referenced standard often states the minimum identifying information that must be on a label, which typically includes, but is not necessarily limited to, the name of the manufacturer, the product name or serial number, installation specifications, applicable tests and standards and the approved testing and labeling agency [see commentary Figure 202(12)].

LIMIT CONTROL. A device responsive to changes in pressure, temperature or level for turning on, shutting off or throttling the gas supply to an appliance.

❖ Limit controls are safety devices used to protect equipment, appliances, property and persons. The controls act at their setpoint to limit a condition such as temperature or pressure and include high and low limits. This definition applies to controls for all types of fuel including gas, liquid and solid.

LIQUEFIED PETROLEUM GAS or LPG (LP-GAS). Liquefied petroleum gas composed predominately of propane, propylene, butanes or butylenes, or mixtures thereof that is gaseous under normal atmospheric conditions, but is capable of being liquefied under moderate pressure at normal temperatures.

Figure courtesy of The Trane Company, an American Standard Company

**Figure 202(12)
TYPICAL APPLIANCE LABEL**

❖ Liquefied petroleum gases are usually obtained as a byproduct of oil refinery operations or by stripping natural gas and are commercially available as butane, propane, or a mixture of the two. Propane has a boiling point of -40°F (-40°C) at atmospheric pressure and a heating value of approximately 2,500 Btu per cubic foot (25.9 kW/m^3). Butane has a boiling point of 32°F (0°C) at atmospheric pressure and a heating value of approximately 3,200 Btu per cubic foot (33.1 kW/m^3). LP-gas is odorized so that gas leakage can be detected.

LISTED. Equipment, appliances or materials included in a list published by a nationally recognized testing laboratory, inspection agency or other organization concerned with product evaluation that maintains periodic inspection of production of listed equipment, appliances or materials, and whose listing states either that the equipment, appliance or material meets nationally recognized standards or has been tested and found suitable for use in a specified manner. The means for identifying listed equipment, appliances or materials may vary for each testing laboratory, inspection agency or other organization concerned with product evaluation, some of which do not recognize equipment, appliances or materials as listed unless they are also labeled. The authority having jurisdiction shall utilize the system employed by the listing organization to identify a listed product.

❖ Section 301.3 requires all equipment and appliances regulated by the code to be listed and labeled unless otherwise approved in accordance with Section 105.2, which allows use of alternative materials, methods and equipment. The listing states that the equipment or material meets nationally recognized standards, or it has been found suitable for use in a specified manner.

LIVING SPACE. Space within a dwelling unit utilized for living, sleeping, eating, cooking, bathing, washing and sanitation purposes.

❖ The code uses this term to define locations that are not to be considered as part of a private garage for the purpose of requiring the elevation of ignition sources (see Section 305.3).

LOG LIGHTER. A manually operated solid fuel ignition appliance for installation in a vented solid fuel-burning fireplace.

❖ Log lighters are simple, manually operated burners used to start wood fires. Log lighters are not considered as decorative but rather as functional appliances. The heat produced by the log lighter flame raises the temperature of the wood fuel to its ignition temperature. A log lighter is designed to be turned off manually after the wood fire is capable of sustaining combustion.

LUBRICATED PLUG-TYPE VALVE. A valve of the plug and barrel type provided with means for maintaining a lubricant between the bearing surfaces.

❖ Valves of this type are rarely used today. They became rare with the advent of ball valves. This term is currently not used in the body of the code.

MAIN BURNER. A device or group of devices essentially forming an integral unit for the final conveyance of gas or a mixture of gas and air to the combustion zone, and on which combustion takes place to accomplish the function for which the appliance is designed.

❖ With the exception of small amounts of fuel for the pilot or ignition burners, the main burner consumes the entire fuel input into an appliance. Common types of atmospheric burners are the drilled port, slotted port, ribbon and single port.

MECHANICAL EXHAUST SYSTEM. Equipment installed in and made a part of the vent, which will provide a positive induced draft.

❖ Confusion may occur with the definition of this term. The definition in the IMC for the same term is different and is as follows: "A system for removing air from a room or space by mechanical means." The only place this term can be found in this code is in Section 501.8, and the IMC definition correctly defines its use in that section.

This term, as defined in this code, is currently not used in the body of the code (see Sections 503.2.1, 503.3.3, 503.3.4 and 505.1.1).

METER. The instrument installed to measure the volume of gas delivered through it.

❖ The meter is the actual point of commerce and is usually associated with the point of delivery. This code does not require or regulate meters. The gas-supplying utility generally installs meters for natural gas. Occasionally in locations where there are several tenants (large office buildings, mobile home parks and apartment complexes, for example), the gas utility installs one meter for the property owner and the owner installs individual tenant meters. Liquefied petroleum may be metered to individual tenants from a storage tank in the same manner [see commentary Figure 202(13)].

Photo courtesy of Washington Gas

Figure 202(13)
TYPICAL RESIDENTIAL GAS METER SETTING

MODULATING. Modulating or throttling is the action of a control from its maximum to minimum position in either predetermined steps or increments of movement as caused by its actuating medium.

❖ In the context of this code, modulating is associated with burner control valves. One type of automatic control is a modulating control that can infinitely vary the controlled variable over the range being controlled. Another common type of control is a two-position action control, which simply turns the variable being controlled, such as fuel gas, on and off. And another type of control is the floating action control, which regulates the controlled device by adjusting the controlled variable between the open or closed position using a sensing element, which must react faster than the actuating drive time or the system would be essentially the same as a two-position control.

OCCUPANCY. The purpose for which a building, or portion thereof, is utilized or occupied.

❖ The occupancy classification of a building is an indication of the level of hazard to which the occupants are exposed as a function of the actual building use. Occupancy in terms of a group classification is one of the primary considerations in the development and application of many code requirements that are designed to offset the hazards specific to each group designation.

OFFSET (VENT). A combination of approved bends that makes two changes in direction bringing one section of the vent out of line but into a line parallel with the other section.

❖ An offset results in the lateral displacement of a vertical vent. Vent offsets are measured in degrees of angle from vertical and are limited in venting systems because they restrict the flow of flue gas. An offset angle is measured between a line that is an extension of the original vertical vent and the new line of direction of the vent [see commentary Figure 202(14)].

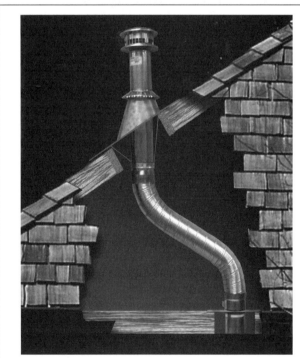

Photo courtesy of Selkirk L.L.C.

Figure 202(14)
VENT OFFSET

ORIFICE. The opening in a cap, spud or other device whereby the flow of gas is limited and through which the gas is discharged to the burner.

❖ An orifice is a precisely sized hole through which gas flows. The orifice is located in the spud or cap that is generally attached to a manifold or supply line. Installation of the orifice should be in accordance with the manufacturer's instructions to ensure proper sizing to prevent underfiring or overfiring of the gas appliance. Orifice size determines the Btu input to a burner.

OUTLET. A threaded connection or bolted flange in a pipe system to which a gas-burning appliance is attached.

❖ A gas outlet is analogous to an electrical receptacle outlet. See Section 404.13 for requirements covering the location of gas piping outlets.

OXYGEN DEPLETION SAFETY SHUTOFF SYSTEM (ODS). A system designed to act to shut off the gas supply to the main and pilot burners if the oxygen in the surrounding atmosphere is reduced below a predetermined level.

❖ An Oxygen Depletion Safety Shutoff System (ODS) is intended to protect the occupants from carbon monoxide build-up. The ODS measures the oxygen level in the air and shuts off the appliance if the level falls below a preset level, typically 18 percent. Monitoring oxygen levels will inversely monitor carbon monoxide levels because of the natural relationship between a decreasing oxygen level and an increasing carbon monoxide level [see commentary Figure 202(15)].

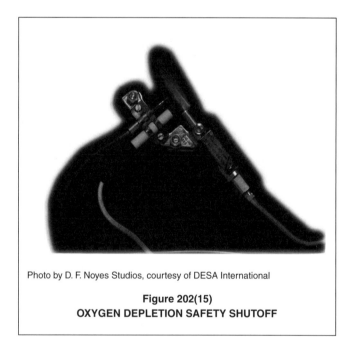

Photo by D. F. Noyes Studios, courtesy of DESA International

Figure 202(15)
OXYGEN DEPLETION SAFETY SHUTOFF

PILOT. A small flame that is utilized to ignite the gas at the main burner or burners.

❖ Pilot burners (pilot lights) can be continuously burning or can be intermittent or interrupted depending upon the type of appliance.

PIPING. Where used in this code, "piping" refers to either pipe or tubing, or both.

DEFINITIONS

Pipe. A rigid conduit of iron, steel, copper, brass or plastic.

Tubing. Semirigid conduit of copper, aluminum, plastic or steel.

❖ Piping includes tubing and pipe used to convey fuel gases. See the definition for "Piping System."

PIPING SYSTEM. All fuel piping, valves and fittings from the outlet of the point of delivery to the outlets of the equipment shutoff valves.

❖ A piping system includes tubing, pipe, fittings, valves and line pressure regulators used to convey fuel gas from the point of delivery to the appliance. This definition was revised to state that the piping system ends at the outlets of the appliance shutoff valves. This revision allows the last 6 feet (1829 mm) or less of gas conduit between the shutoff valve and the appliance inlet connection to be sized as a connector (see Section 409.5 and the definition for "Point of delivery").

PLASTIC, THERMOPLASTIC. A plastic that is capable of being repeatedly softened by increase of temperature and hardened by decrease of temperature.

❖ Polyvinyl chloride (PVC) is a type of thermoplastic.

POINT OF DELIVERY. For natural gas systems, the point of delivery is the outlet of the service meter assembly, or the outlet of the service regulator or service shutoff valve where a meter is not provided. Where a valve is provided at the outlet of the service meter assembly, such valve shall be considered to be downstream of the point of delivery. For undiluted liquefied petroleum gas systems, the point of delivery shall be considered the outlet of the first-stage pressure regulator that provides utilization pressure, exclusive of line gas regulators, in the system.

❖ It is necessary to understand where the point of delivery begins with respect to fuel gas systems because that is the location where the enforcement of the *International Fuel Gas Code* begins. For LP applications, the point of delivery is the outlet of the second-stage regulator in low pressure systems or the first regulator that reduces pressure to 2 psi (13.8 kPa) or less in elevated pressure systems (see Figure 401.1). NFPA 58 regulates piping and components upstream of the second stage regulator. Any valve located on the outlet of the service meter assembly is considered part of the piping system under the scope of this code.

PORTABLE FUEL CELL APPLIANCE. A fuel cell generator of electricity, which is not fixed in place. A portable fuel cell appliance utilizes a cord and plug connection to a grid-isolated load and has an integral fuel supply.

❖ As stated in Section 101.2.4, Item 19, the code does not regulate these appliances because they do not connect to a fuel piping system and are completely self-contained.

PRESSURE DROP. The loss in pressure due to friction or obstruction in pipes, valves, fittings, regulators and burners.

❖ The pressure in a fuel gas system is reduced as the gas flows in the system. This is caused by a loss of energy resulting from friction and turbulence. Fuel gas systems are designed so that the pressure drop does not result in the system pressure falling below the minimum pressure required for proper equipment operation.

PRESSURE TEST. An operation performed to verify the gas-tight integrity of gas piping following its installation or modification.

❖ Installed piping systems are tested on the job site to verify that the system is free of leaks.

PURGE. To free a gas conduit of air or gas, or a mixture of gas and air.

❖ Piping systems are purged (flushed) to remove gaseous, liquid or solid contaminants that could be harmful to the piping contents, the piping system or the system components or that could create a fire or explosion hazard.

QUICK-DISCONNECT DEVICE. A hand-operated device that provides a means for connecting and disconnecting an appliance or an appliance connector to a gas supply and that is equipped with an automatic means to shut off the gas supply when the device is disconnected.

❖ These devices are commonly installed with gas appliance connectors and used with commercial cooking appliances and other gas appliances that are routinely moved for cleaning or used in another location [see commentary Figure 202(16)].

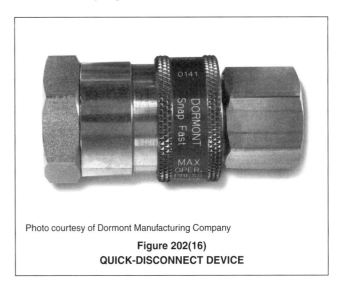

Photo courtesy of Dormont Manufacturing Company
Figure 202(16)
QUICK-DISCONNECT DEVICE

READY ACCESS (TO). That which enables a device, appliance or equipment to be directly reached, without requiring the removal or movement of any panel, door or similar obstruction (see "Access").

❖ Ready access can be described as the capability of being quickly reached or approached for the purpose of operation, inspection, observation or emergency action. Ready access does not require the removal or movement of any door, panel or similar obstruction, nor does it require surmounting physical obstructions or obstacles, including differences in elevation.

REGISTERED DESIGN PROFESSIONAL. An individual who is registered or licensed to practice their respective design profession as defined by the statutory requirements of the professional registration laws of the state or jurisdiction in which the project is to be constructed.

❖ Each state establishes legal qualifications for engineers and architects. Licensing and registration of engineers and architects is based on written or oral examinations offered by states or by reciprocity (licensing in other states).

REGULATOR. A device for controlling and maintaining a uniform supply pressure, either pounds-to-inches water column (MP regulator) or inches-to-inches water column (appliance regulator).

❖ In the context of the code, these devices are fuel-gas-pressure regulating devices.

REGULATOR, GAS APPLIANCE. A pressure regulator for controlling pressure to the manifold of equipment. Types of appliance regulators are as follows:

Adjustable.
1. Spring type, limited adjustment. A regulator in which the regulating force acting upon the diaphragm is derived principally from a spring, the loading of which is adjustable over a range of not more than 15 percent of the outlet pressure at the midpoint of the adjustment range.
2. Spring type, standard adjustment. A regulator in which the regulating force acting upon the diaphragm is derived principally from a spring, the loading of which is adjustable. The adjustment means shall be concealed.

Multistage. A regulator for use with a single gas whose adjustment means is capable of being positioned manually or automatically to two or more predetermined outlet pressure settings. Each of these settings shall be adjustable or nonadjustable. The regulator may modulate outlet pressures automatically between its maximum and minimum predetermined outlet pressure settings.

Nonadjustable.
1. Spring type, nonadjustable. A regulator in which the regulating force acting upon the diaphragm is derived principally from a spring, the loading of which is not field adjustable.
2. Weight type. A regulator in which the regulating force acting upon the diaphragm is derived from a weight or combination of weights.

❖ These regulators are integral to the appliance and serve to stabilize the appliance input pressure and supply fuel gas to the appliance at the desired pressure. Most appliances come equipped with appliance regulators because the appliances are designed to operate at a pressure lower than the delivery pressure, even when low pressure gas [0.5 psig or lower (3.4 kPa)] is supplied.

REGULATOR, LINE GAS PRESSURE. A device placed in a gas line between the service pressure regulator and the equipment for controlling, maintaining or reducing the pressure in that portion of the piping system downstream of the device.

❖ Line gas pressure regulators are used to reduce the gas pressure supplied by the service (point of delivery) regulator. They are typically located at the connection to each gas appliance and are commonly used as "pounds to inches" regulators in systems having elevated [14 inches wc to 5 psi typical (34.5 kPa)] supply pressures.

REGULATOR, MEDIUM-PRESSURE. A medium-pressure (MP) regulator reduces the gas piping pressure to the appliance regulator or to the appliance utilization pressure.

❖ With the increased use of high-pressure gas distribution systems, systems are designed with one or more regulators in addition to the service regulators. Appliances and equipment are also equipped with factory-installed regulators that control burner input (manifold) pressure. Medium pressure regulators are used to reduce the service pressure to a pressure suitable for delivery to the gas appliances and equipment, where the appliance regulator may further reduce the pressure (see Section 410.2).

REGULATOR, PRESSURE. A device placed in a gas line for reducing, controlling and maintaining the pressure in that portion of the piping system downstream of the device.

❖ See the definition and commentary for "Regulator."

REGULATOR, SERVICE PRESSURE. A device installed by the serving gas supplier to reduce and limit the service line pressure to delivery pressure.

❖ The utility company (gas supplier) installs this regulator to reduce the gas pressure from the utility distribution mains to that required for delivery to the meter and customer. The devices are usually located with the meter and are outdoors [see commentary Figure 202(17)].

RELIEF OPENING. The opening provided in a draft hood to permit the ready escape to the atmosphere of the flue products from the draft hood in the event of no draft, back draft, or stoppage beyond the draft hood, and to permit air into the draft hood in the event of a strong chimney updraft.

❖ Dilution air, a component of combustion air, enters the venting system through the draft hood opening. Combustion gases spilling from (exiting) a draft hood relief opening create an abnormal and potentially hazardous condition except for brief periods of spillage that might occur at the start of an appliance firing cycle. Typical periods of spillage at start-up last only a few seconds.

Photo courtesy of Washington Gas

**Figure 202(17)
SERVICE REGULATOR**

RELIEF VALVE (DEVICE). A safety valve designed to forestall the development of a dangerous condition by relieving either pressure, temperature or vacuum in the hot water supply system.

❖ A relief valve is a safety device designed to prevent the development of a potentially damaging or dangerous condition in a closed system. This is usually associated with a closed system that has a heat source as part of the system. Requirements for the installation of relief valves in water distribution systems and water heaters are given in Section 504 of the IPC.

RELIEF VALVE, PRESSURE. An automatic valve that opens and closes a relief vent, depending on whether the pressure is above or below a predetermined value.

❖ Pressure relief valves are intended to prevent harm to life and property. These valves are safety devices used to relieve abnormal pressures and to prevent the overpressurization of the vessel or system to which the valves are connected. Pressure relief valves are designed to operate (discharge) only when an abnormal pressure exists in a system. They are not to be used as regulating valves, nor are they intended to control or regulate flow or pressure. Pressure relief valves are set at the factory to begin opening at a predetermined pressure and are rated according to the maximum energy discharge per unit of time, usually in units of British thermal units per hour (Btu/h)(W).

Pressure relief valves must be properly sized and rated for the boiler, pressure vessel or system served; otherwise, a system malfunction could cause the pressure within the system to continue to rise, thereby creating a hazardous condition. Pressure relief valves do not open fully when they reach the factory-preset pressure; rather, the valves open an amount proportional to the forces produced by the pressure in the system. The valves open fully at a certain percentage above the preset pressure, then close when the pressure drops below the preset pressure. Pressure relief valves are sometimes combined in the same body with temperature relief valves, creating combination temperature and pressure relief valves.

RELIEF VALVE, TEMPERATURE.

Reseating or self-closing type. An automatic valve that opens and closes a relief vent, depending on whether the temperature is above or below a predetermined value.

Manual reset type. A valve that automatically opens a relief vent at a predetermined temperature and that must be manually returned to the closed position.

❖ Temperature relief valves are designed to open in response to excessive temperatures and to discharge heated water to limit the temperature of the water in the vessel, tank or system. Water heater control failure or improper adjustments could allow excessive temperature rises, which can heat water above the temperature at which it would vaporize at atmospheric pressure. Water heated to above 212°F (100°C) is referred to as superheated water and will flash to steam when the pressure is reduced below its vapor pressure. Very hot water and the presence of steam are hazardous; however, more importantly, higher temperatures produce higher pressures and superheated water increases the potential for and the magnitude of an explosion.

RELIEF VALVE, VACUUM. A valve that automatically opens and closes a vent for relieving a vacuum within the hot water supply system, depending on whether the vacuum is above or below a predetermined value.

❖ The vacuum relief valve is intended to prevent the possible reduction in pressure within the system (water heater tanks being of main concern) to below atmospheric pressure (that is, a partial vacuum). Many tanks are not designed to resist external pressures exceeding internal pressures; therefore, a vacuum relief valve is necessary to prevent atmospheric pressure from possibly collapsing or otherwise damaging the tank. A partial vacuum could also cause water to be siphoned from the tank thereby creating an overheating hazard in a water heater.

The vacuum relief valve operates by automatically opening the system to the atmosphere when a partial vacuum is created, thereby permitting air to enter and maintaining the internal pressure at atmospheric pressure.

RISER, GAS. A vertical pipe supplying fuel gas.

❖ A vertical gas pipe that distributes gas from a building main to upper floor levels is commonly referred to as a "riser."

ROOM HEATER, UNVENTED. See "Unvented room heater."

❖ See the commentary for "Unvented room heater" and commentary Figure 202(18).

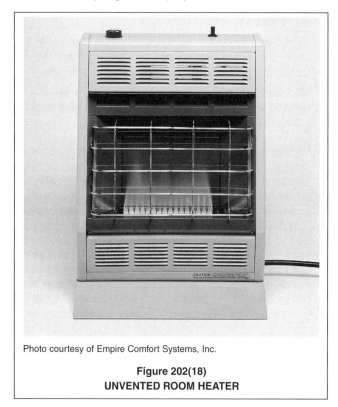

Photo courtesy of Empire Comfort Systems, Inc.

**Figure 202(18)
UNVENTED ROOM HEATER**

ROOM HEATER, VENTED. A free-standing heating unit used for direct heating of the space in and adjacent to that in which the unit is located (see also "Vented room heater").

❖ These heaters are typically small space-heating appliances designed to heat a single room.

ROOM LARGE IN COMPARISON WITH SIZE OF EQUIPMENT. Rooms having a volume equal to at least 12 times the total volume of a furnace or air-conditioning appliance and at least 16 times the total volume of a boiler. Total volume of the appliance is determined from exterior dimensions and is to include fan compartments and burner vestibules, when used. When the actual ceiling height of a room is greater than 8 feet (2438 mm), the volume of the room is figured on the basis of a ceiling height of 8 feet (2438 mm).

❖ This definition deals with providing adequate space for installing heat producing appliances, primarily for the prevention of excessive heat build-up and high temperatures. In rooms that satisfy this definition, appliances may be installed with clearances required by the listing and in accordance with the manufacturers' installation instructions. Equipment may be installed in smaller rooms, alcoves, and closets if the listing specifically states approval for such an installation and clearances are not reduced as would otherwise be allowed by Section 308. In effect, this definition serves as a definition for alcoves and closets (see Section 308.3).

SAFETY SHUTOFF DEVICE. See "Flame safeguard."

❖ See the commentary for "Flame safeguard."

SHAFT. An enclosed space extending through one or more stories of a building, connecting vertical openings in successive floors, or floors and the roof.

❖ A shaft is an enclosed, vertical passage that passes through stories and floor levels within a building. Shafts are sometimes referred to as "chases" and are usually constructed of fire-resistance-rated assemblies.

SPECIFIC GRAVITY. As applied to gas, specific gravity is the ratio of the weight of a given volume to that of the same volume of air, both measured under the same condition.

❖ Analogous to density, specific gravity values are used to describe how a gaseous substance will behave in, for example, air or piping systems.

STATIONARY FUEL CELL POWER PLANT. A self-contained package or factory-matched packages which constitute an automatically operated assembly of integrated systems for generating electrical energy and recoverable thermal energy that is permanently connected and fixed in place.

❖ See Chapter 7.

THERMOSTAT.

Electric switch type. A device that senses changes in temperature and controls electrically, by means of separate components, the flow of gas to the burner(s) to maintain selected temperatures.

Integral gas valve type. An automatic device, actuated by temperature changes, designed to control the gas supply to the burner(s) in order to maintain temperatures between predetermined limits, and in which the thermal actuating element is an integral part of the device.

   1. Graduating thermostat. A thermostat in which the motion of the valve is approximately in direct proportion to the effective motion of the thermal element induced by temperature change.

2. Snap-acting thermostat. A thermostat in which the thermostatic valve travels instantly from the closed to the open position, and vice versa.

❖ Thermostats most often consist of one or more thermally actuated on/off switches, but also include types that transmit varying voltage and signals. A thermostat is also referred to as a controller. The set point or desired unit of temperature is adjusted or set at the thermostat. The thermostat transmits a signal to the controlled device or appliance.

TRANSITION FITTINGS, PLASTIC TO STEEL. An adapter for joining plastic pipe to steel pipe. The purpose of this fitting is to provide a permanent, pressure-tight connection between two materials which cannot be joined directly one to another.

❖ Joining of piping of dissimilar materials such as plastic and steel requires specialized transition fittings. For example, such fittings are used with underground plastic fuel-gas piping and steel meter setting risers [see commentary Figure 202(19)].

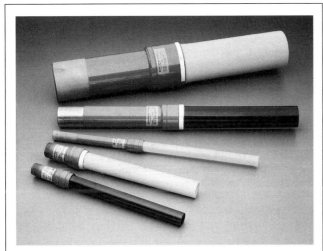

Photo courtesy of Perfection Corporation

**Figure 202(19)
TRANSITION FITTINGS**

UNIT HEATER.

High-static pressure type. A self-contained, automatically controlled, vented appliance having integral means for circulation of air against 0.2 inch (15 mm H_2O) or greater static pressure. Such appliance is equipped with provisions for attaching an outlet air duct and, where the appliance is for indoor installation remote from the space to be heated, is also equipped with provisions for attaching an inlet air duct.

Low-static pressure type. A self-contained, automatically controlled, vented appliance, intended for installation in the space to be heated without the use of ducts, having integral means for circulation of air. Such units are allowed to be equipped with louvers or face extensions made in accordance with the manufacturer's specifications.

❖ Unit heaters are similar to warm-air furnaces, except that ducts are not usually associated with unit heaters. These heaters are typically suspended from ceilings or roof structures.

UNLISTED BOILER. A boiler not listed by a nationally recognized testing agency.

❖ Boilers in new installations must be listed and gas equipment installed in existing unlisted boilers must be installed according to the manufacturers' instructions and the IMC. See the commentary for the definition of "Boiler, Low-pressure."

UNVENTED ROOM HEATER. An unvented heating appliance designed for stationary installation and utilized to provide comfort heating. Such appliances provide radiant heat or convection heat by gravity or fan circulation directly from the heater and do not utilize ducts.

❖ As opposed to vented heaters, unvented heaters discharge all products of combustion into the space being heated [see commentary Figure 202(18)].

VALVE. A device used in piping to control the gas supply to any section of a system of piping or to an appliance.

❖ The code requires valves to perform various functions in a fuel gas system as described in the definitions below. These valves are designed for the specific use intended and many are factory installed on equipment and are part of the listed equipment. However, there are valves that are not listed or manufactured in accordance with the standards listed in the code and if allowed must be evaluated in accordance with Section 105 of the code.

Automatic. An automatic or semiautomatic device consisting essentially of a valve and operator that control the gas supply to the burner(s) during operation of an appliance. The operator shall be actuated by application of gas pressure on a flexible diaphragm, by electrical means, by mechanical means, or by other approved means.

Automatic gas shutoff. A valve used in conjunction with an automatic gas shutoff device to shut off the gas supply to a water-heating system. It shall be constructed integrally with the gas shutoff device or shall be a separate assembly.

Equipment shutoff. A valve located in the piping system, used to isolate individual equipment for purposes such as service or replacement.

❖ This definition has been revised to make clear that the equipment/appliance shutoff valves are not considered as "emergency" valves.

Individual main burner. A valve that controls the gas supply to an individual main burner.

Main burner control. A valve that controls the gas supply to the main burner manifold.

Manual main gas-control. A manually operated valve in the gas line for the purpose of completely turning on or shutting off the gas supply to the appliance, except to pilot or pilots that are provided with independent shutoff.

Manual reset. An automatic shutoff valve installed in the gas supply piping and set to shut off when unsafe conditions occur. The device remains closed until manually reopened.

Service shutoff. A valve, installed by the serving gas supplier between the service meter or source of supply and the customer piping system, to shut off the entire piping system.

VENT. A pipe or other conduit composed of factory-made components, containing a passageway for conveying combustion products and air to the atmosphere, listed and labeled for use with a specific type or class of appliance.

❖ In code terminology, vents are distinguished from chimneys and usually are constructed of factory-made listed and labeled components intended to function as a system. Type B and BW vents are constructed of galvanized steel and aluminum sheet metal and are double wall and air insulated. Such vents are designed to vent gas-fired appliances and equipment that are equipped with draft hoods or are specifically listed (labeled) for use with Type B or BW vents. Type L vents are typically constructed of sheet steel and stainless steel. They are double wall and air insulated and are designed to vent gas- and oil-fired appliances and equipment. Some appliances are designed for use with corrosion-resistant special vents, such as those made of plastic pipe and special alloys of stainless steel [see commentary Figure 202(20)].

Photo courtesy of Selkirk L. L. C.

Figure 202(20)
TYPE B VENT

Special gas vent. A vent listed and labeled for use with listed Category II, III and IV appliances.

❖ These vents include high-temperature plastic pipe, stainless steel pipe and low-temperature (PVC and CPVC) plastic pipes.

Type B vent. A vent listed and labeled for use with appliances with draft hoods and other Category I appliances that are listed for use with Type B vents.

❖ See the commentary for "Vent."

Type BW vent. A vent listed and labeled for use with wall furnaces.

❖ See the commentary for "Vent."

Type L vent. A vent listed and labeled for use with appliances that are listed for use with Type L or Type B vents.

❖ See the commentary for "Vent."

VENT CONNECTOR. (See "Connector").

❖ See the commentary for "Connector."

VENT GASES. Products of combustion from appliances plus excess air plus dilution air in the vent connector, gas vent or chimney above the draft hood or draft regulator.

❖ It is critical that the materials used in venting systems be compatible with the vent gases being generated by the connecting appliances, equipment and venting system. The most important area of concern is the generation of condensate, which will be acidic and will corrode most metal venting systems. As newer equipment is produced and designed for higher efficiencies, which will result in lower vent-gas temperatures, the potential for generating condensate in the vent gases increases. Combustion gases that have not exited the appliance are called "flue gases."

VENTED APPLIANCE CATEGORIES. Appliances that are categorized for the purpose of vent selection are classified into the following four categories:

❖ Gas appliances are categorized for the purpose of matching the appliance to the required type of venting.

Category I. An appliance that operates with a nonpositive vent static pressure and with a vent gas temperature that avoids excessive condensate production in the vent.

❖ These appliances include draft-hood-equipped and fan-assisted types.

Category II. An appliance that operates with a nonpositive vent static pressure and with a vent gas temperature that is capable of causing excessive condensate production in the vent.

❖ The condition of nonpositive pressure (draft dependent) combined with low flue-gas temperatures makes it difficult to design such an appliance. At the time of this publication, no such appliances are known to exist in the marketplace.

Category III. An appliance that operates with a positive vent static pressure and with a vent gas temperature that avoids excessive condensate production in the vent.

❖ These appliances include mid-efficiency (approximately 78 to 83 percent) appliances that are typically sidewall-vented with stainless steel special vents.

Category IV. An appliance that operates with a positive vent static pressure and with a vent gas temperature that is capable of causing excessive condensate production in the vent.

❖ These appliances include high-efficiency (84 percent and higher) condensing-type furnaces, boilers and water heaters.

VENTED ROOM HEATER. A vented self-contained, free-standing, nonrecessed appliance for furnishing warm air to the space in which it is installed, directly from the heater without duct connections.

❖ These appliances are typically small space heaters that discharge all products of combustion to the outdoors.

VENTED WALL FURNACE. A self-contained vented appliance complete with grilles or equivalent, designed for incorporation in or permanent attachment to the structure of a building, mobile home or travel trailer, and furnishing heated air circulated by gravity or by a fan directly into the space to be heated through openings in the casing. This definition shall exclude floor furnaces, unit heaters and central furnaces as herein defined.

❖ Wall furnaces are designed to occupy very little room area and provide space heating for one or more rooms in small occupancies such as dwelling units, cottages and the like.

VENTING SYSTEM. A continuous open passageway from the flue collar or draft hood of an appliance to the outside atmosphere for the purpose of removing flue or vent gases. A venting system is usually composed of a vent or a chimney and vent connector, if used, assembled to form the open passageway.

❖ Venting systems fall into one of two categories: natural draft or mechanical draft systems. Natural draft chimneys and vents do not rely on any mechanical means to convey combustion products to the outdoors. Draft is produced by the temperature difference between the vent gases (combustion gases) and the ambient atmosphere. Hot gases are less dense and more buoyant; therefore, they are displaced by cooler (more dense) ambient gases, causing the hotter gases to rise and produce a draft.

Mechanical draft systems use fans or other mechanical means to cause the removal of flue or vent gases and also fall into one of two categories, based on positive or nonpositive static vent pressure. The positive vent pressure systems are referred to as forced-draft or power venting systems, and the nonpositive vent pressure systems are referred to as induced-draft venting systems.

Mechanical draft venting system. A venting system designed to remove flue or vent gases by mechanical means, that consists of an induced draft portion under nonpositive static pressure or a forced draft portion under positive static pressure.

❖ These systems do not depend on draft, but rather, employ fans or blowers.

Forced-draft venting system. A portion of a venting system using a fan or other mechanical means to cause the removal of flue or vent gases under positive static vent pressure.

❖ Power exhausters, including power draft venting systems, and some power burner systems are examples of forced-draft systems. Vents and chimneys must be listed for positive-pressure applications where used with forced-draft systems.

Induced draft venting system. A portion of a venting system using a fan or other mechanical means to cause the removal of flue or vent gases under nonpositive static vent pressure.

❖ Induced-draft venting is commonly accomplished with field-installed inducer fans designed to supplement natural draft chimneys or vents.

Natural draft venting system. A venting system designed to remove flue or vent gases under nonpositive static vent pressure entirely by natural draft.

❖ Natural draft chimneys and vents do not rely on any mechanical means to convey combustion products to the outdoors. Draft is produced by the temperature difference between the combustion gases (flue gases) and the ambient atmosphere. Hot gases are less dense and more buoyant; therefore, they rise as they are displaced by more dense ambient air.

WALL HEATER, UNVENTED-TYPE. A room heater of the type designed for insertion in or attachment to a wall or partition. Such heater does not incorporate concealed venting arrangements in its construction and discharges all products of combustion through the front into the room being heated.

❖ See the definition and commentary for "Unvented Room Heater."

WATER HEATER. Any heating appliance or equipment that heats potable water and supplies such water to the potable hot water distribution system.

DEFINITIONS

❖ A water heater is a closed pressure vessel or heat exchanger that has a heat source and that supplies potable (drinkable) water to the building's hot water distribution system. Large commercial water heaters are sometimes referred to as "hot water supply boilers." Water heaters can be of the storage type with an integral storage vessel, circulating type for use with an external storage vessel, tankless instantaneous type without storage capacity and point-of-use type with or without storage capacity. This code and the IPC regulate water heaters because these appliances have elements and installation requirements related to both mechanical and plumbing systems.

Bibliography

The following resource materials are referenced in this chapter or are relevant to the subject matter addressed in this chapter.

ASHRAE–93, *Handbook of Fundamentals*. Atlanta: American Society of Heating, Refrigerating and Air-Conditioning Engineers, Inc., 1993.

ASHRAE 15–01, *Safety Standard for Refrigeration Systems*. Atlanta: American Society of Heating, Refrigerating and Air-Conditioning Engineers, Inc., 2001.

ASTM E 84–01, *Test Method for Surface Burning Characteristics of Building Materials*. West Conshohocken, PA: American Society for Testing and Materials, 2001.

ASTM E 136–99e 01, *Test Method for Behavior of Materials in a Vertical Tube Furnace at 750EC*. West Conshohocken, PA: American Society for Testing and Materials, 2001.

AWS A5.8–92, *Specifications for Filler Metals for Brazing*. Miami: American Welding Society, 1992.

NFPA 211-00, *Standard for Chimneys, Fireplaces, Vents and Solid Fuel Burning Appliances*. Quincy, MA: National Fire Protection Association, 2000.

CHAPTER 3
GENERAL REGULATIONS

General Comments

A fundamental principle of the code is its dependence on the listing and labeling method of approval for appliances and equipment. Section 301.3 prohibits the installation of unlisted appliances except where approved in accordance with Section 105.

Purpose

Chapter 3 contains requirements for the safe and proper installation of gas-fired equipment and appliances to help assure protection of life and property

SECTION 301 (IFGC)
GENERAL

301.1 Scope. This chapter shall govern the approval and installation of all equipment and appliances that comprise parts of the installations regulated by this code in accordance with Section 101.2.

❖ This section states that this chapter governs the approval and installation of all gas-fired equipment and appliances that are regulated by the code. Section 101.2 establishes the scope of application of the code (see commentary, Section 101.2).

301.1.1 Other fuels. The requirements for combustion and dilution air for gas-fired appliances shall be governed by Section 304. The requirements for combustion and dilution air for appliances operating with fuels other than fuel gas shall be regulated by the *International Mechanical Code*.

❖ This code and the *International Mechanical Code®* (IMC®) each have a combustion air chapter that is specific to the fuels addressed in the respective code.

301.2 Energy utilization. Heating, ventilating and air-conditioning systems of all structures shall be designed and installed for efficient utilization of energy in accordance with the *International Energy Conservation Code*.

❖ This section states that all appliances and equipment must be designed and installed to use depletable energy sources efficiently. The *International Energy Conservation Code®* (IECC®) is the applicable document for regulating the efficiency and performance of appliances and HVAC systems. Special applications such as process heating or cooling should be designed for the maximum energy efficiency attainable.

301.3 Listed and labeled. Appliances regulated by this code shall be listed and labeled unless otherwise approved in accordance with Section 105. The approval of unlisted appliances in accordance with Section 105 shall be based upon approved engineering evaluation.

❖ Gas-fired appliances must be listed and labeled by an approved agency to show that they comply with the applicable national standards. The code requires listing and labeling for appliances such as boilers, furnaces, space heaters, direct-fired heaters, cooking appliances, clothes dryers, rooftop HVAC units, etc. The code also requires listing for system components as specifically stated in the text addressing those components. The label is the primary, if not the only, assurance to the installer, the inspector and the end user that a representative sample of an appliance model has been tested and evaluated by an approved agency and has been determined to perform safely and efficiently when installed and operated in accordance with its listing.

The presence of a label is part of the information that the code official is to consider in the approval of appliances. The only exception to the labeling requirement occurs when the code official approves a specific appliance in accordance with the authority granted in Section 105.

Approval of unlabeled appliances must be based on documentation that demonstrates compliance with applicable standards or, where no product standards exist, that the appliance is appropriate for the intended use and will provide the same level of performance as would be provided by listed and labeled appliances. A fundamental principle of the code is the reliance on the listing and labeling process to assure appliance performance; approvals granted in accordance with Section 105 must be well justified with supporting documentation. To the code official, the installer and the end-user, very little is known about the performance of an appliance that is not tested and built to an appliance standard.

301.4 Labeling. Labeling shall be in accordance with the procedures set forth in Sections 301.4.1 through 301.4.2.3.

❖ This section establishes the requirements for testing and labeling appliances by an approved agency. Included within this section are the requirements for testing the product and for approval of the testing agency, the testing equipment and the personnel who conduct the test. Also included is the information that must appear on a label.

301.4.1 Testing. An approved agency shall test a representative sample of the appliances being labeled to the relevant standard or standards. The approved agency shall maintain a record of all of the tests performed. The record shall provide sufficient detail to verify compliance with the test standard.

❖ When an approved agency labels an appliance, the agency is assuring that a representative sample of the appliance has been tested in accordance with an appropriate standard and has been determined to perform acceptably when installed and operated in accordance with the appliance's listing.

The basis for a label is the requirement for testing a representative, perhaps identical, sample of the appliance to indicate conformance to a required standard. This is an important premise in the code, because a code official will consider the presence of a label in the approval of an appliance. For this reason, the appliance must meet the requirements of the standard. Because the appliance tested is installed and operated in accordance with the manufacturer's instructions, these instructions must provide for proper installation and operation. This is important because the code requires that the labeled appliance be installed in accordance with the manufacturer's instructions, and operating instructions must be either attached to or shipped with each appliance.

There are numerous standards, not all of which are specifically referenced in the code, applicable to various appliances and equipment. For this reason, the approved agency determines the pertinent standards to be used for testing and then, in turn, as the basis for labeling. Each standard contains safety requirements for a given appliance or piece of equipment and specifies tests that must be performed. The labeling agency is required to maintain sufficient documentation in detail to demonstrate compliance with the test standard. The code official may require that copies of the test reports be submitted to determine the validity of the label.

Examples of standards that are used as a basis for testing and labeling include:

- ANSI Z21.47 Gas-Fired Central Furnaces
- ANSI Z83.8 Gas Unit Heaters and
- UL 795 Commercial – Industrial Gas Heating Equipment

301.4.2 Inspection and identification. The approved agency shall periodically perform an inspection, which shall be in- plant if necessary, of the appliances to be labeled. The inspection shall verify that the labeled appliances are representative of the appliances tested.

❖ The approved agency whose identification insignia appears on the label must perform periodic in-plant inspections. The primary objective of these inspections is to determine that the manufactured product is equivalent to the sample that was tested. Because the label is good only for the products that were tested, the in-plant inspections are intended to discover any design changes or production quality control problems. If any discrepancies are found, the labeling agency would discontinue labeling of the particular product, and the manufacturer would be required to resolve the problem and, if necessary, have the redesigned product retested before the labeling process is resumed.

301.4.2.1 Independent. The agency to be approved shall be objective and competent. To confirm its objectivity, the agency shall disclose all possible conflicts of interest.

❖ As a part of the basis for a code official's approval of a particular labeling agency, the agency must demonstrate both its independence from the manufacturer of the product and its competence to perform the required tests. The judgement of objectivity is linked to the financial and fiduciary independence of the agency. The competence of the agency is judged by its experience, organization and the experience of its personnel. As a hypothetical example, the Acme Inspection Agency is performing testing for gas-fired furnaces for the Real Hot Furnace Company. After some investigation, both Acme and Real Hot are found to be the subsidiaries of the same parent company. The inspection agency and the manufacturer clearly have a relationship that presents the potential for conflict of interest, and the objectivity of the inspection agency is sufficiently questionable for the code official to justify not approving Acme as a testing and labeling agency for equipment produced by the Real Hot Furnace Company.

301.4.2.2 Equipment. An approved agency shall have adequate equipment to perform all required tests. The equipment shall be periodically calibrated.

❖ Referring to the example in the commentary for Section 301.4.2.1, if the Acme Inspection Agency had only the facilities to test and label fire doors, the agency would not be qualified to test and label a gas-fired furnace. Although this example is oversimplified, the point is that the inspection agency must have all of the necessary equipment to perform the testing required by the applicable standard.

In addition to having the proper equipment, the agency must maintain records of the maintenance and calibration of their equipment to demonstrate that the equipment can be relied on to produce accurate, consistent and reproducible results. Testing apparatus, instruments and equipment must often be capable of measurements using very small units of measure within a specified tolerance. To produce accurate, dependable readings and reliable test results, testing apparatus, equipment and instruments must be routinely calibrated to a fixed reference. Having the proper testing equipment can be just as important as the competence of the testing personnel.

301.4.2.3 Personnel. An approved agency shall employ experienced personnel educated in conducting, supervising and evaluating tests.

❖ The competence of an inspection agency is based on the agency having the proper equipment to perform the test, as stated in Section 301.4.2.2, and also on the experience and abilities of its personnel. The best calibrated equipment can produce accurate results only when operated by experienced personnel who are trained to conduct, supervise and evaluate tests. For example, consider a newly formed agency that has employed individuals who do not have experience related to the testing to be conducted and have not been adequately trained. The capabilities and experience of supervisory personnel overseeing their work is also important.

301.5 Label information. A permanent factory-applied nameplate(s) shall be affixed to appliances on which shall appear in legible lettering, the manufacturer's name or trademark, the model number, serial number and, for listed appliances, the seal or mark of the testing agency. A label shall also include the hourly rating in British thermal units per hour (Btu/h) (W); the type of fuel approved for use with the appliance; and the minimum clearance requirements.

❖ This section requires that the label be a metal plate, tag or other permanent label. In general, label materials other than metal tags or plates usually consist of material that is similar in appearance to a decal, and the label, its adhesive and the printed information must be durable and water resistant. Because of the important information given by a label, the label is intended to be permanent, not susceptible to damage and legible for the life of the appliance to which it is attached. The standards that appliances are tested to usually specify the required label performance criteria, the method of attachment and the required label information. The code requires that the label be affixed permanently and prominently on the appliance or equipment and specifies the information that must appear on the label. The manufacturer may be required by the relevant standard or may voluntarily provide additional information on the label. Commentary Figures 301.5(1) and 301.5(2) show typical appliance labels.

301.6 Plumbing connections. Potable water supply and building drainage system connections to appliances regulated by this code shall be in accordance with the *International Plumbing Code*.

❖ Plumbing connections to appliances and equipment regulated by the code must be in accordance with the *International Plumbing Code®* (IPC®).

Section 624.2 of the IFGC requires that combination domestic water heating and hydronic supply water heating units be listed and installed according to their listing and manufacturers' installation instructions.

Hydronic systems normally require a means of supplying fill and makeup water to replace any water lost to evaporation, leakage or intentional draining. Where direct connections are made to the potable water supply, the connections must be isolated from the potable water source. This requirement is intended to protect the potable water system from contamination by backflow when a direct connection is made to a hydronic system.

Hydronic systems are normally pressurized, contain nonpotable water and fluids and can contain conditioning chemicals or antifreeze solutions. Low-temperature hydronic fluids and cooling towers have also been associated with disease-causing organisms such as the Legionnaires' disease bacterium. The potable water system must be protected from potential contamination resulting from connection to hydronic systems, water-wash filter systems, cooling towers, solar systems, water-cooled heat exchangers, cooking appliances, ice makers, humidifiers, evaporative coolers, etc.

In addition, water heaters are part of the potable water distribution system and, therefore, must comply with both this code and the IPC. A water heater installation is complex in that it has a fuel or power supply; a chimney or vent connection, if fuel-fired; a combustion air supply, if fuel-fired; connections to the plumbing potable water distribution system and controls and devices to prevent a multitude of potential hazards from conditions such as excessively high temperatures, pressures and ignition failure.

It is not uncommon for jurisdictions to issue both plumbing and mechanical permits for water heater installations or to require that the installer be licensed in both the plumbing and mechanical trades when performing such installations. See the commentary for Section 624. Water heaters are clearly under the purview of both plumbing and fuel gas codes. This section also refers to the IPC for the drainage associated with mechanical appliances and equipment, such as those addressed in Section 307.

301.7 Fuel types. Appliances shall be designed for use with the type of fuel gas to which they will be connected and the altitude at which they are installed. Appliances that comprise parts of the installation shall not be converted for the usage of a different fuel, except where approved and converted in accordance with the manufacturer's instructions. The fuel gas input rate shall not be increased or decreased beyond the limit rating for the altitude at which the appliance is installed.

❖ Appliances are usually designed by the manufacturer to operate on one specifically designated type of fuel. An element of information used for the approval of appliances is the label, which assures that the appliance has been tested in accordance with a valid standard and determined to perform acceptably when installed and operated in accordance with the appliance listing. See the commentary for Section 301.5. The fuel used in the appliance test must be the type of fuel specified by the manufacturer. When an appliance is converted to a different type of fuel, the original label that appears on the appliance is no longer valid. Because the original approval of the appliance is based in part on the label, the appliance is no longer approved for use.

Many gas appliances are listed and labeled for more than one fuel, most commonly natural gas and LP (propane). Such appliances can be converted from one fuel to the other using a manufacturer's conversion kit and following instructions provided by the manufacturer. These conversions must be done only with the approval of the code official to ensure that the instructions are complied with and the safety of the original installation is maintained. Fuel conversions that are not performed correctly can adversely affect the performance of burners, the venting of combustion gases and the proper clearance to combustibles.

Once a conversion has been completed, a supplemental label must be installed to update the information contained on the original label, thereby alerting any service personnel to the modifications that have been made.

Fuel-fired appliances are designed to operate with a maximum and minimum heat energy input capacity. This capacity is field adjusted to suit the elevation because of the change in air density at different elevations. Alteration of heat energy input beyond the allowable limits can result in hazardous overfiring or underfiring. Either condition can cause operation problems that include overheating, vent failure, corrosion, poor draft and poor combustion.

Figure courtesy of the Trane Company, an American Standard Company

Figure 301.5(1)
TYPICAL LABEL FOR A CATEGORY I GAS-FIRED FURNACE

GENERAL REGULATIONS FIGURE 301.5(2) - 301.10

Figure 301.5(2)
ROOF TOP HVAC UNIT LABEL

301.8 Vibration isolation. Where means for isolation of vibration of an appliance is installed, an approved means for support and restraint of that appliance shall be provided.

❖ Where vibration isolation connections are used in ducts and piping and where equipment is mounted with vibration dampers, support is required for the ducts, piping and equipment to maintain positioning and alignment and to prevent stress and strain on the vibration connectors and dampers.

301.9 Repair. Defective material or parts shall be replaced or repaired in such a manner so as to preserve the original approval or listing.

❖ Repair work must not alter the nature of appliances and equipment in a way that would invalidate the listing or conditions of approval. For example, replacement of safety control devices with different devices could alter the design and operation of an appliance from that intended by the manufacturer and the listing agency.

301.10 Wind resistance. Appliances and supports that are exposed to wind shall be designed and installed to resist the wind pressures determined in accordance with the *International Building Code*.

❖ Installations of equipment and appliances that are subject to wind forces must be designed to resist those forces. The wind pressures must be based on the wind provisions in the *International Building Code®* (IBC®) for the site. The requirements in the IBC are based on ASCE 7. The wind pressure requirements are based on the exposure of the building and wind speeds for that region.

301.11 Flood hazard. For structures located in flood hazard areas, the appliance, equipment and system installations regulated by this code shall be located at or above the design flood elevation and shall comply with the flood-resistant construction requirements of the *International Building Code*.

> **Exception:** The appliance, equipment and system installations regulated by this code are permitted to be located below the design flood elevation provided that they are designed and installed to prevent water from entering or accumulating within the components and to resist hydrostatic and hydrodynamic loads and stresses, including the effects of buoyancy, during the occurrence of flooding to the design flood elevation and shall comply with the flood-resistant construction requirements of the *International Building Code*.

❖ In areas designated as a special flood-hazard zone in accordance with the requirements of Section 1612 of the IBC, equipment and appliances must be protected from water or must be elevated above the anticipated flood level.

Exposure to water can cause deterioration of system components and serious appliance and equipment malfunctions. For example, appliance manufacturers require the replacement of appliances that have been submerged in floodwaters. Components such as pressure regulators, gas controls, motors and electronic circuitry can all be ruined by exposure to water. The exception allows locating appliances and system components below the design flood elevation if they are designed for submersion in floodwaters. Appliances and system components designed for submersion may be nonexistent and it would be extremely difficult to install effective barriers to keep out floodwaters.

301.12 Seismic resistance. When earthquake loads are applicable in accordance with the *International Building Code*, the supports shall be designed and installed for the seismic forces in accordance with that code.

❖ The IBC requires building systems to be designed for specified seismic forces. This section references the detailed equipment and component seismic support requirements contained in the IBC to bring these requirements to the attention of the design professional and the permit applicant.

Equipment and large-diameter piping must be braced for earthquake loads as stated in the IBC. The failure of the supports for these components has been shown to be a threat to health and safety in geographical areas where moderate- to high-magnitude earthquakes occur. The IBC specifies the geographical locations where earthquake design is required for certain piping and equipment and the size of the components that must be braced.

301.13 Ducts. All ducts required for the installation of systems regulated by this code shall be designed and installed in accordance with the *International Mechanical Code*.

❖ Chapter 6 of the IMC governs duct systems used for the movement of environmental air, governing the construction, installation, alteration, maintenance and repair of these systems (see the definition of "Duct system" in Chapter 2).

301.14 Rodentproofing. Buildings or structures and the walls enclosing habitable or occupiable rooms and spaces in which persons live, sleep or work, or in which feed, food or foodstuffs are stored, prepared, processed, served or sold, shall be constructed to protect against rodents in accordance with the *International Building Code*.

❖ This section references the IBC for the requirements to prevent rodent infestation of a building. Efforts must be made to protect the annular spaces around openings in exterior walls. The annular spaces can be sealed by any effective method, but the primary methods used are to fill the annular space with a sealant material or to place a metal collar around the penetrating pipe, duct, etc. The collar material must be durable for the weather exposure and strong enough to prevent the rodents from chewing through and entering the building.

Effort must also be made to cover ventilation and combustion air openings with wire cloth to prevent the entry of rodents through the opening.

301.15 Prohibited location. The appliances, equipment and systems regulated by this code shall not be located in an elevator shaft.

❖ The code views an elevator shaft as an unnecessarily risky location for fuel-gas system components. There is a potential for leaks to occur in fuel-gas system equipment and piping, which would release fuel gas into the shaft, creating a hazardous condition. Because the elevator is often relied on for firefighting access as well as egress for those with disabilities, the shaft must be maintained free from contaminants and potential hazards that could render the elevator inoperative or unsafe for the occupants. The location of fuel-gas-fired equipment in an elevator shaft could also create access problems for service of the gas-fired equipment and the elevator equipment.

SECTION 302 (IFGC)
STRUCTURAL SAFETY

[B] 302.1 Structural safety. The building shall not be weakened by the installation of any gas piping. In the process of installing or repairing any gas piping, the finished floors, walls, ceilings, tile work or any other part of the building or premises which is required to be changed or replaced shall be left in a safe structural condition in accordance with the requirements of the *International Building Code*.

❖ The installation of fuel-gas systems must not adversely affect the structural integrity of the building components. The IBC dictates the structural safety requirements that must be applied to any structural portion of the building that is penetrated, altered or removed during the installation, replacement or repair of fuel-gas systems.

[B] 302.2 Penetrations of floor/ceiling assemblies and fire-resistance-rated assemblies. Penetrations of floor/ceiling assemblies and assemblies required to have a fire-resistance rating shall be protected in accordance with the *International Building Code*.

❖ Penetrations of fire-resistance-rated assemblies can diminish the integrity of the assembly, allowing smoke and fire to pass through and cause it to fail prematurely.

[B] 302.3 Cutting, notching and boring in wood members. The cutting, notching and boring of wood members shall comply with Sections 302.3.1 through 302.3.4.

❖ The sections referenced provide prescriptive sizes and locations of acceptable cuts, notches and bored holes in wood framing members. Ideally, framing members should not be altered in any way, but because this is not always practical, the code permits alterations that do not significantly weaken the members.

[B] 302.3.1 Engineered wood products. Cuts, notches and holes bored in trusses, laminated veneer lumber, glued-laminated members and I-joists are prohibited except where the effects of such alterations are specifically considered in the design of the member.

❖ This section applies to engineered lumber products which are manufactured structural members. These products often have installation instructions that specifically state if, where and how the member can be altered, drilled, notched, etc. What appears to be a harmless cut, notch or boring might actually be structurally damaging to the member (see Section 302.4).

[B] 302.3.2 Joist notching. Notching at the ends of joists shall not exceed one-fourth the joist depth. Holes bored in joists shall not be within 2 inches (51 mm) of the top and bottom of the joist and their diameter shall not exceed one-third the depth of the member. Notches in the top or bottom of the joist shall not exceed one-sixth the depth and shall not be located in the middle one-third of the span.

❖ The code recognizes that at times floor and ceiling joists must be cut, notched or bored. The provisions are based on location and size limitations to assure that the structural member can support the load.

A notch or cut will decrease the allowable stress of the member. A notch or cut might place a lumber defect that was previously in a low-stress location in the interior of the board into a high-stress area at the edge of the notch or cut. Cutting and notching also cause shear stress concentrations in the member at the corners of the notch or cut. Because of these shear stress concentrations, the full extent of the damage to the member cannot be calculated. In consideration of the detrimental effects of cutting and notching, it is a good practice to avoid cutting, notching or boring adjacent to areas of the member that contain knots and, where possible, to use rounded or sloped edged notches to reduce the stress concentrations at the corners [see commentary Figure 302.3.2(1)].

Structural elements of engineered wood systems, such as metal-plate-connected parallel-chord wood trusses, pre-manufactured I-joists and metal-plate-connected wood roof trusses cannot be cut or notched unless structural analysis is provided by a qualified design professional.

Notches are not to exceed those permitted in this section unless a structural analysis is completed by a qualified design professional.

Holes cut and bored in joists and studs have the same effect as notches. Holes reduce the strength of the member. This section permits holes to be drilled in wood structural members within certain limitations. The limitations given in this section are not to be exceeded unless a structural analysis is submitted by a qualified design professional [see commentary Figure 302.3.2(2)].

[B] 302.3.3 Stud cutting and notching. In exterior walls and bearing partitions, any wood stud is permitted to be cut or notched to a depth not exceeding 25 percent of its width. Cutting or notching of studs to a depth not greater than 40 percent of the width of the stud is permitted in nonload-bearing partitions supporting no loads other than the weight of the partition.

❖ The load-sharing characteristics of a wall with studs, sheathing and plates allow for the stated limitation for notches in load-bearing studs. The maximum notch in a stud is not to exceed 25 percent of its depth unless the stud is reinforced to resist the anticipated load. See the commentary to Section 302.3.2 for further discussion of the effects of notching and cutting (see commentary Figure 302.3.4).

[B] 302.3.4 Bored holes. A hole not greater in diameter than 40 percent of the stud depth is permitted to be bored in any wood stud. Bored holes not greater than 60 percent of the depth of the stud are permitted in nonload-bearing partitions or in any wall where each bored stud is doubled, provided not more than two such successive doubled studs are so bored. In no case shall the edge of the bored hole be nearer than $^5/_8$ inch (15.9 mm) to the edge of the stud. Bored holes shall not be located at the same section of stud as a cut or notch.

❖ This section specifies the limits of bored holes in wood studs. The 40-percent limit is intended for studs in exterior walls and for load-bearing walls. The 60-percent limit for doubled studs is also intended to apply to studs in exterior walls and other load-bearing walls. Bored holes not greater than 60-percent are allowed in any nonload-bearing stud.

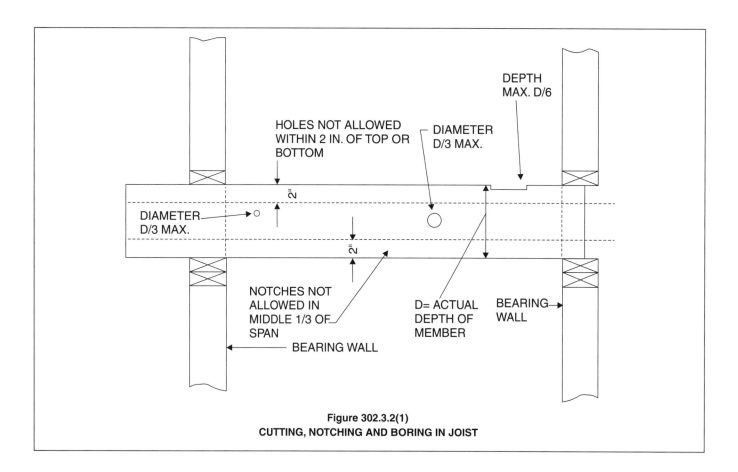

**Figure 302.3.2(1)
CUTTING, NOTCHING AND BORING IN JOIST**

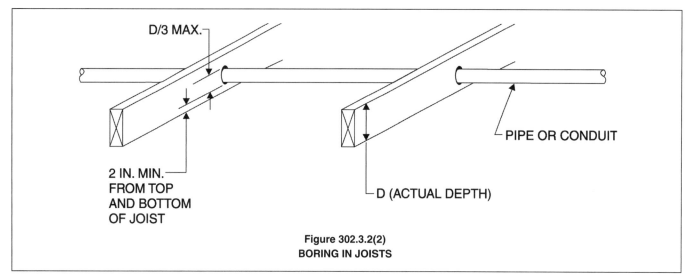

**Figure 302.3.2(2)
BORING IN JOISTS**

The bored-hole edge location limitations maintain the load-carrying capacity of the stud. These limits are intended to apply to load-bearing and nonload-bearing partitions and studs in exterior walls. A bored hole in the same cross section as a cut or notch could cause the stud to fail (see commentary Figure 302.3.4).

[B] 302.4 Alterations to trusses. Truss members and components shall not be cut, drilled, notched, spliced or otherwise altered in any way without the written concurrence and approval of a registered design professional. Alterations resulting in the addition of loads to any member (e.g., HVAC equipment, water heaters) shall not be permitted without verification that the truss is capable of supporting such additional loading.

❖ A truss is an engineered system of components that function as a structural unit. Trusses are susceptible to weakening by almost any type of alteration, such as boring, notching and cutting. Unlike dimensional solid lumber members, such as studs, rafters and joists, trusses cannot be altered in a code-prescribed way. The only exception is obtaining written approval of a

GENERAL REGULATIONS

specific alteration from a registered design professional.

The placement of a water heater in an attic, for example, can result in abnormal loads on roof trusses that must be accounted for structurally. Subjecting trusses to loads that they were not designed to carry is considered an alteration and is prohibited without a structural analysis to verify that the truss can bear the additional weight.

[B] 302.5 Cutting, notching and boring holes in structural steel framing. The cutting, notching and boring of holes in structural steel framing members shall be as prescribed by the registered design professional.

❖ The sections referenced provide prescriptive sizes and locations of acceptable cuts, notches and bored holes in steel framing members.

[B] 302.6 Cutting, notching and boring holes in cold-formed steel framing. Flanges and lips of load-bearing, cold-formed steel framing members shall not be cut or notched. Holes in webs of load-bearing, cold-formed steel framing members shall be permitted along the centerline of the web of the framing member and shall not exceed the dimensional limitations, penetration spacing or minimum hole edge distance as prescribed by the registered design professional. Cutting, notching and boring holes of steel floor/roof decking shall be as prescribed by the registered design professional.

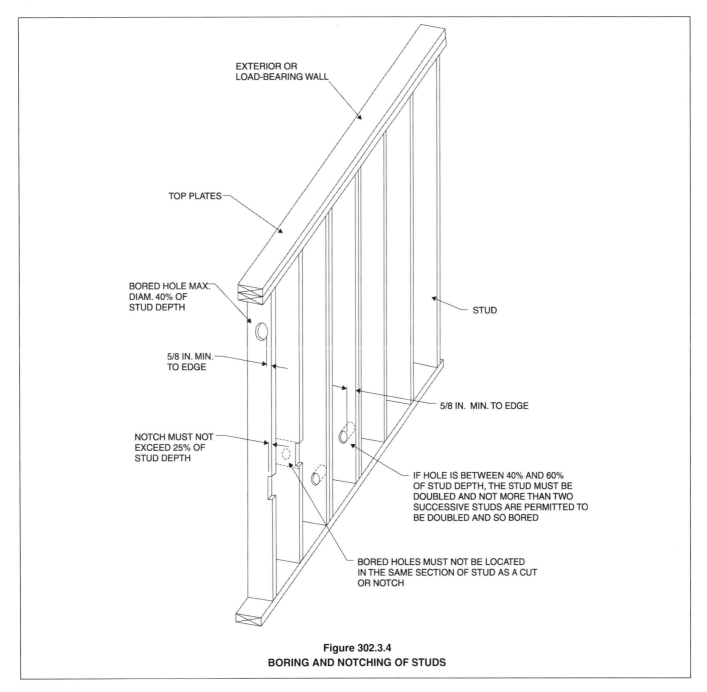

Figure 302.3.4
BORING AND NOTCHING OF STUDS

❖ This section does not allow any structural steel framing member to be cut, notched or bored without approval by a registered design professional. Unlike Section 302.3, which allows field cutting, notching and boring of wood studs and joists within the prescribed limits, this section requires that alterations to steel structural framing members be allowed only as dictated by a design professional prior to the alteration. The code official should not approve any alterations to structural members unless a signed drawing or revision is on the job site showing the allowed size and location of the cut, notch or bored hole.

[B] 302.7 Cutting, notching and boring holes in non-structural cold-formed steel wall framing. Flanges and lips of nonstructural cold-formed steel wall studs shall be permitted along the centerline of the web of the framing member, shall not exceed $1^1/_2$ inches (38 mm) in width or 4 inches (102 mm) in length, and the holes shall not be spaced less than 24 inches (610 mm) center to center from another hole or less than 10 inches (254 mm) from the bearing end.

❖ This section does not allow any cutting or notching of the flanges and lips of any cold-formed steel framing member. However, holes may be made in the web of nonstructural cold-formed steel wall studs without the involvement of a registered design professional if the prescribed size, location and spacing requirements are met.

SECTION 303 (IFGC)
APPLIANCE LOCATION

303.1 General. Appliances shall be located as required by this section, specific requirements elsewhere in this code and the conditions of the equipment and appliance listing.

❖ Section 303 is a consolidation of the code's generally applicable location requirements and limitations. The listing for an appliance or equipment will often contain location requirements parallel with or in addition to these sections.

303.2 Hazardous locations. Appliances shall not be located in a hazardous location unless listed and approved for the specific installation.

❖ Gas-fired appliances that are to be installed in a hazardous location must be approved and listed for use in that location. A hazardous location is defined as any location considered to be a potential fire hazard from flammable vapors, dust, combustible fibers or other highly combustible substances. This location is not necessarily classified in the IBC or the ICC *Electrical Code®* (ICC EC™) as a high-hazard occupancy classification. Appliances located in a hazardous location require special design, construction and installation because of the potential for an explosion or fire caused by the presence of dust, flammable vapors and highly combustible materials.

This section does not specify the type of appliance to be used in any particular hazardous location; therefore, the code official must evaluate information submitted by the designer on the intended occupancy of a space, the specification for the appliance and the location of the appliance in the space to determine the appropriate requirements. For example, these requirements may include spark-resistant materials for moving parts, such as fan blades; explosion-proof construction for electrical switches, starters and motors; or sealed combustion chambers for fuel-burning appliances.

See the commentary for Sections 305.3 and 305.6. Examples of appliances installed in a hazardous location are fuel-fired appliances in an industrial baking facility where fine flour dust is present and fuel-fired appliances in a paint shop where paint fumes and vapors are present.

303.3 Prohibited locations. Appliances shall not be located in, or obtain combustion air from, any of the following rooms or spaces:

1. Sleeping rooms.
2. Bathrooms.
3. Toilet rooms.
4. Storage closets.
5. Surgical rooms.

Exceptions:

1. Direct-vent appliances that obtain all combustion air directly from the outdoors.
2. Vented room heaters, wall furnaces, vented decorative appliances and decorative appliances for installation in vented solid fuel-burning fireplaces, provided that the room meets the required volume criteria of Section 304.5.
3. A single wall-mounted unvented room heater equipped with an oxygen depletion safety shutoff system and installed in a bathroom, provided that the input rating does not exceed 6,000 Btu/h (1.76kW) and the bathroom meets the required volume criteria of Section 304.5.
4. A single wall-mounted unvented room heater equipped with an oxygen depletion safety shutoff system and installed in a bedroom, provided that the input rating does not exceed 10,000 Btu/h (2.93 kW) and the bedroom meets the required volume criteria of Section 304.5.
5. Appliances installed in an enclosure in which all combustion air is taken from the outdoors, in accordance with Section 304.6. Access to such enclosure shall be through a solid weather-stripped door, equipped with an approved self-closing device.

❖ The intent of this section is to prevent fuel-fired appliances from being installed in rooms and spaces where the combustion process could pose a threat to the occupants. Potential threats include depleted oxygen levels; elevated levels of carbon dioxide, nitrous oxides, carbon monoxide, and other combustion gases; ignition of combustibles and elevated levels of flammable gases.

In small rooms such as bedrooms and bathrooms, the doors are typically closed when the room is occupied, which could allow combustion gases to build up to life-threatening levels. In bedrooms, sleeping occupants would not be alert to or aware of impending danger.

If an appliance obtains combustion air from a room or space, it communicates with the atmosphere in that room or space whether or not it is installed in that room or space. An appliance might be in a room, closet or alcove and obtain combustion air from an adjacent room, so Section 303.3 is worded to address both the actual location of an appliance and its source of combustion air. In other words, an appliance in a closet accessed from a bedroom is no different from an appliance located within the bedroom. It is not the intent of this section to prevent combustion air from being taken from a bedroom, bathroom, etc. as evidenced in Exceptions 2, 3, and 4. For example, the volume of a bedroom could be added to the volume of other rooms for the purpose of providing indoor combustion air for an appliance not installed in a location prohibited by this section if openings are installed to conjoin the space volumes in accordance with Section 304.5.3. The point being made is that if an appliance obtains combustion air from a room, the appliance must be considered to be in that room. In other words, the appliance combustion chamber would be open to the room if the appliance is able to draw combustion air from that room.

Exception 1 recognizes that direct-vent appliances have sealed combustion chambers and obtain all combustion air directly from the outdoors. The appliance combustion chambers do not communicate with the room atmosphere.

Exception 2 requires that the room be able to supply the necessary combustion air by infiltration as specified in Section 304.5.

Exceptions 3 and 4 allow the installation of a single wall-mounted unvented room heater in bathrooms and bedrooms if the heaters are equipped with oxygen depletion safety shutoff systems, are limited in Btu input rating and the space is capable of supplying indoor combustion air in accordance with Section 304.5. These exceptions specify "wall-mounted" heaters, which are less likely to come in contact with bed linens, towels, clothing and other combustibles. Exceptions 3 and 4 would not apply to room heaters that stand on the floor or fasten to a fireplace hearth or ventless firebox hearth (see commentary Figure 303.3).

Exception 5 would allow installation of fuel-fired appliances within a separate dedicated space that is accessed from the rooms and spaces listed in this section. A separated space containing the appliance must be open to the outdoors in accordance with Section 304.6, and the access door to the space must be weatherstripped to prevent communication between atmospheres in the separated spaces. The door must also be self-closing and not rely on occupants to keep it closed. The enclosure should not be used for storage or any other purpose. This exception can be used to avoid relocating an appliance when an existing appliance installed in a prohibited location needs to be replaced.

Photo by D. F. Noyes Studios, courtesy of DESA International

**Figure 303.3
UNVENTED WALL-MOUNTED ROOM HEATER**

303.4 Protection from physical damage. Appliances shall not be installed in a location where subject to physical damage unless protected by approved barriers meeting the requirements of the *International Fire Code*.

❖ Appliances must be installed with consideration for potential damage. Damage can result in property loss and can cause hazardous malfunctions, fire or explosions. Potential damage includes impact from vehicles such as fork-lifts, farm machinery and passenger cars and trucks. For example, an appliance installed in a garage might be subject to vehicle impact. Barriers must be approved and strong enough to resist the type of impact anticipated.

303.5 Indoor locations. Furnaces and boilers installed in closets and alcoves shall be listed for such installation.

❖ Fuel-fired appliances installed in spaces such as alcoves and closets can pose a potential fire hazard because of the heat emanating from the appliance and the inadequate ventilation afforded by the boundaries of small enclosures.

For appliances installed in closets or alcoves, the clearances stated in the listing (label) are necessary for ventilation cooling of the appliance and must not be reduced as provided for in Section 308. See the definition of "room large in comparison with size of equipment," which serves to define an alcove or closet.

303.6 Outdoor locations. Equipment installed in outdoor locations shall be either listed for outdoor installation or provided with protection from outdoor environmental factors that influence the operability, durability and safety of the equipment.

❖ Appliances installed outdoors must be specifically listed for outdoor installation or be protected from outdoor conditions that will affect the "operability, durability and safety of the equipment."

The concern is for weather and ambient temperatures. The manufacturer's instructions and listing must be consulted to determine whether a particular appliance is designed for or can be made suitable for outdoor installation. For example, furnaces cannot be installed outdoors in cold climates regardless of any weatherproof enclosure, unless the heat exchanger, burner assemblies and venting system are designed for exposure to temperatures below normal indoor temperatures. Cold ambient temperatures can cause harmful condensation to occur on heat exchanger surfaces of fuel-fired appliances.

Additionally, there may be local ordinances that govern the outdoor installation of appliances. Before installing an appliance outdoors, consult local zoning regulations, ordinances and subdivision covenants. Many of these regulations strictly limit the location of outdoor mechanical equipment and appliances. Also, for roof installations, the roof structure must be able to support all imposed static and dynamic structural loads. Substantiating data on the structural adequacy of the entire installation must be submitted and approved.

303.7 Pit locations. Appliances installed in pits or excavations shall not come in direct contact with the surrounding soil. The sides of the pit or excavation shall be held back a minimum of 12 inches (305 mm) from the appliance. Where the depth exceeds 12 inches (305 mm) below adjoining grade, the walls of the pit or excavation shall be lined with concrete or masonry, such concrete or masonry shall extend a minimum of 4 inches (102 mm) above adjoining grade and shall have sufficient lateral load-bearing capacity to resist collapse. The appliance shall be protected from flooding in an approved manner.

❖ Where installed in a depression, appliances could be damaged by corrosion or a malfunction could be caused by blockage of an opening by entry of soil. Because a depression could hold water, protection from flooding is also necessary.

SECTION 304 (IFGS)
COMBUSTION, VENTILATION AND DILUTION AIR

304.1 General. Air for combustion, ventilation and dilution of flue gases for gas utilization equipment installed in buildings shall be provided by application of one of the methods prescribed in Sections 304.5 through 304.9. Where the requirements of Section 304.5 are not met, outdoor air shall be introduced in accordance with one of the methods prescribed in Sections 304.6 through 304.9. Direct-vent appliances, gas appliances of other than natural draft design and vented gas appliances other than Category I shall be provided with combustion, ventilation and dilution air in accordance with the equipment manufacturer's instructions.

Exception: Type 1 clothes dryers that are provided with makeup air in accordance with Section 614.5.

❖ The provisions of Section 304 describe requirements for the combustion air necessary for the complete combustion of fuel gas, dilution of flue gases, and ventilation of gas-fired appliances and the space in which they are installed. An inadequate combustion air supply to gas-fired appliances can compromise safety by causing incomplete combustion, resulting in appliance malfunction and production of excess carbon monoxide.

Complete combustion of fuel gas is essential for the proper operation of gas-fired appliances. If insufficient quantities of oxygen are supplied, the combustion process will be incomplete, creating hazardous by-products [see commentary Figures 304.1(1) and 304.1(2)].

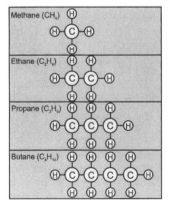 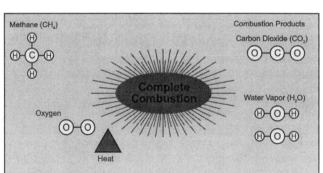

Figure courtesy of Reznor/Thomas & Betts Corporation

Figure 304.1(1)
THE CHEMISTRY OF IDEAL COMBUSTION

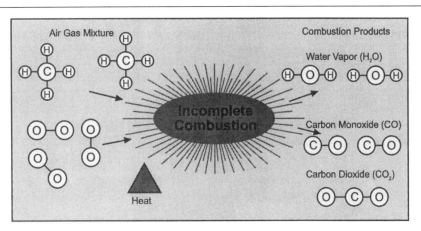

Figure courtesy of Reznor/Thomas & Betts Corporation

Figure 304.1(2)
THE CHEMISTRY OF INCOMPLETE COMBUSTION

Although not implied in the term, combustion air also serves other purposes in addition to supplying oxygen. Combustion air ventilates and cools appliances and the rooms or spaces that enclose them. Combustion air also plays an important role in producing and controlling draft in vents and chimneys.

Despite the fact that an adequate combustion air supply is extremely important, it is one aspect of gas-fired equipment installations that is often overlooked, ignored or compromised. Depending on the appliance type, location and building construction, supplying combustion air can either be easy or can involve complex designs and extraordinary methods. In any case, the importance of a proper combustion air supply cannot be overemphasized.

The methods of supplying combustion air range from simple (inherently more dependable) methods to more complex methods. This section offers five methods for providing combustion air [see commentary Figure 304.1(3)]:

1. All indoor air.
2. All outdoor air.
3. Combination indoor and outdoor air.
4. Mechanical combustion air supply.
5. Engineered design.

The last sentence of this code section clarifies that the section is intended to apply only to natural-draft atmospheric-burner-design appliances, Category I appliances and nondirect vent appliances (see commentary, Section 304.8). For example, the application of Section 304.6 with large power-burner-equipped boilers could result in excessively large openings in an outside wall, causing environmental problems within the boiler room. Direct-vent appliances supply their own combustion air through an outdoor air intake pipe or duct [see Figures 304.1(4) and 304.1(5)].

The exception recognizes that clothes dryers receive the required combustion air from the makeup air that compensates for the exhaust air flow from the appliance.

304.2 Appliance/equipment location. Equipment shall be located so as not to interfere with proper circulation of combustion, ventilation and dilution air.

❖ The provisions of Section 304 rely on the natural (gravity) movement of air; therefore, installations must allow unimpeded flow of combustion air from the source to the appliances being supplied. Other mechanical systems, appliances or equipment in the same room or building can adversely affect the combustion air supply. For example, when placed too close to the combustion air openings, gas-fired appliances or equipment will restrict the free circulation of air.

304.3 Draft hood/regulator location. Where used, a draft hood or a barometric draft regulator shall be installed in the same room or enclosure as the equipment served so as to prevent any difference in pressure between the hood or regulator and the combustion air supply.

❖ It is important that draft hoods and barometric draft regulators be located in the same pressure zone as the appliance combustion chamber.

Draft hoods and barometric regulators both serve to stabilize draft through an appliance and in some cases act to relieve backpressure/downdraft in the venting system. It is the draft in a Category I appliance that moves combustion air into the appliance and conveys the combustion gases to the venting system. Pressure differences between the atmospheres in which the draft hood/regulator and the combustion chamber are located could interfere with the function of the draft hood/regulator, resulting in poor combustion, excess draft, insufficient draft, combustion gas spillage and/or hazardous appliance operation. A draft hood/regulator is installed under the assumption that the combustion chamber it serves sees the same atmospheric pressure.

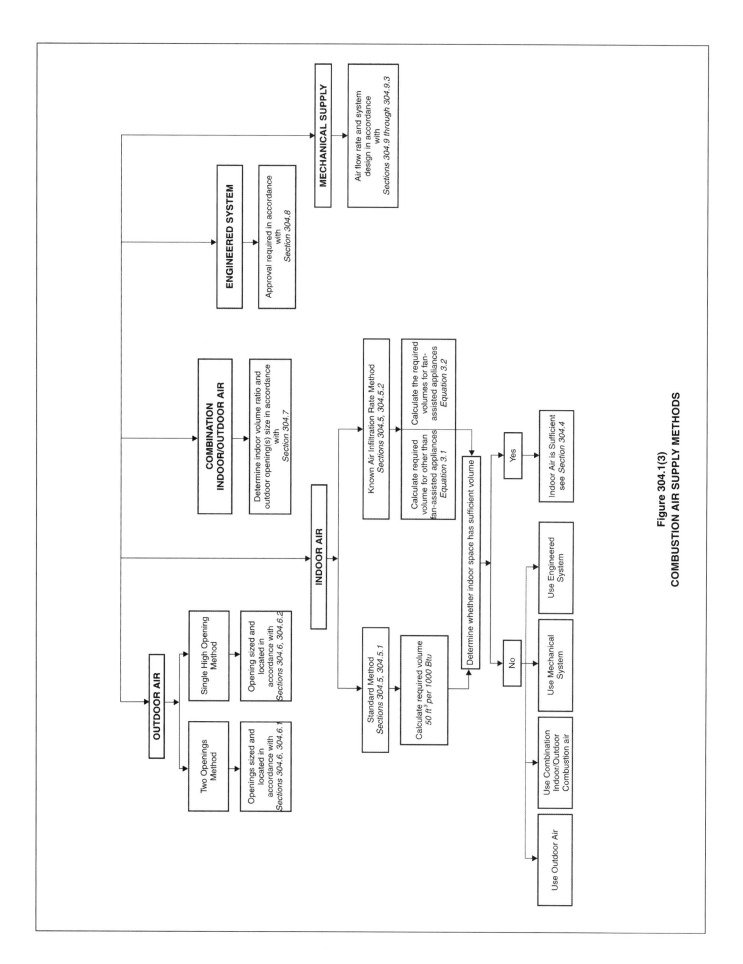

**Figure 304.1(3)
COMBUSTION AIR SUPPLY METHODS**

GENERAL REGULATIONS

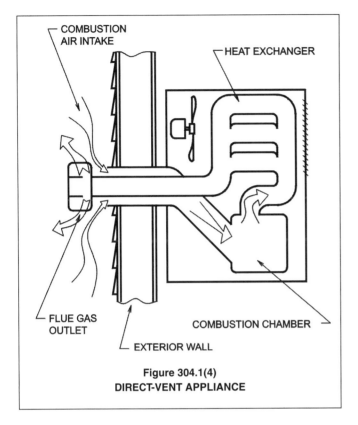

Figure 304.1(4)
DIRECT-VENT APPLIANCE

304.4 Makeup air provisions. Makeup air requirements for the operation of exhaust fans, kitchen ventilation systems, clothes dryers and fireplaces shall be considered in determining the adequacy of a space to provide combustion air requirements.

❖ The introduction of makeup air is critical to the proper operation of all fuel-burning appliances located in areas subject to the effects of exhaust systems. Too little makeup air will cause excessive negative pressures to develop, not only reducing the exhaust airflow but also interfering with appliance venting. Too little makeup air can cause loss of draft in appliance vents and chimneys or cause combustion by-products to discharge into the building. It is the draft produced in an appliance that causes combustion air to enter the combustion chamber of natural-draft atmospheric-burner appliances. It is the draft in the venting system that causes the combustion gases and dilution air to move through the venting system toward the vent terminal. A lack of combustion air can result in incomplete fuel combustion, appliance malfunction and flue gas spillage.

A significant reduction in building pressure could be created by fireplaces, exhaust fans, ventilation systems, clothes dryers and similar appliances and equipment. Gas-fired appliances are often in competition with other equipment or systems for the available combustion air and makeup air that flows through (infiltrates) the building envelope or is otherwise introduced into a building, room or space. The competition between powered exhaust equipment and natural-draft gas-fired appliance venting systems is an unfair contest at best. Eventually, the powered equipment will starve the natural-draft appliances unless provisions are made to compensate for the effect of the powered exhaust equipment/appliances. Supplying makeup air will prevent negative pressures within a space, thus negating the effect of exhaust fans and clothes dryers. In many cases, makeup air would be required because this is the only way to prevent exhaust systems from negatively affecting the appliance venting systems that draw combustion air from the same space from which the exhaust systems remove air.

Natural draft appliances also compete among themselves for combustion air. The appliance venting sys-

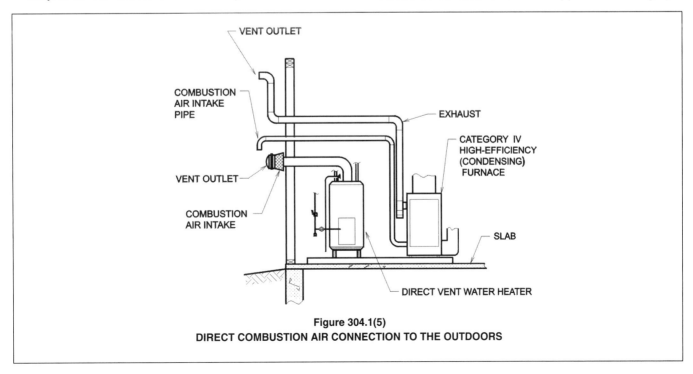

Figure 304.1(5)
DIRECT COMBUSTION AIR CONNECTION TO THE OUTDOORS

2003 INTERNATIONAL FUEL GAS CODE® COMMENTARY

tem that produces the strongest draft, such as a solid-fuel appliance or fireplace, can cause combustion air shortages for the appliance venting systems that produce a weaker draft.

Exhaust fans and similar equipment and appliances can produce significant negative building pressures that can interfere with the operation of vents and chimneys. This interference can cause reverse flow as outdoor air enters the building through the vents and chimneys because of the pressure difference. Any such interference with vents or chimneys could cause products of combustion to be discharged into the building and, therefore, must be avoided. The provisions of Section 304 could be woefully inadequate or nullified where compensating features are not installed to mitigate the effect of appliances or equipment removing air from any room or space. When the system depends on infiltration as the source of combustion air, it will probably be necessary to supply makeup air to offset the deficiency in many buildings. It is obvious that natural draft appliance installations cannot compete with exhaust fans and clothes dryers for the available infiltration air. Without makeup air, the gas-fired appliance venting systems and the mechanical exhaust systems will indeed compete for the same infiltration air, which could be insufficient to satisfy either need (see commentary Figure 304.4). Note that exhaust fans, clothes dryers and similar equipment that removes air from a space will have no negative impact on unvented appliances because they have no venting system. Direct-vent appliances are also immune.

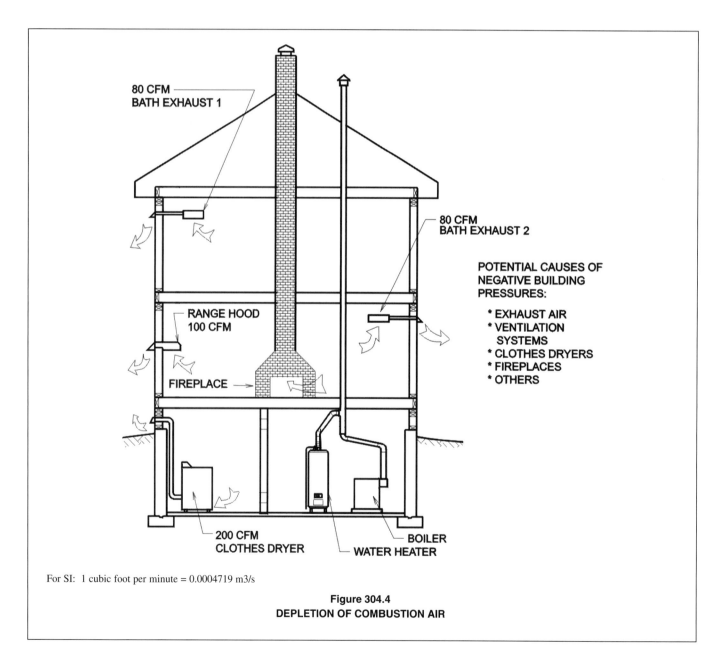

For SI: 1 cubic foot per minute = 0.0004719 m3/s

**Figure 304.4
DEPLETION OF COMBUSTION AIR**

304.5 Indoor combustion air. The required volume of indoor air shall be determined in accordance with Section 304.5.1 or 304.5.2, except that where the air infiltration rate is known to be less than 0.40 air changes per hour (ACH), Section 304.5.2 shall be used. The total required volume shall be the sum of the required volume calculated for all appliances located within the space. Rooms communicating directly with the space in which the appliances are installed through openings not furnished with doors, and through combustion air openings sized and located in accordance with Section 304.5.3, are considered to be part of the required volume.

❖ The terms "confined space," "unconfined space" and "unusually tight construction" are no longer used in the code. The required room volume is determined by two methods: one method based on the actual air infiltration rate of the building and the other based on the familiar fixed ratio of 50 ft³ per 1000 Btu/h (4.8 m³/kW). Air taken directly from inside the building is an acceptable source of combustion air if a sufficient volume of air is available for the appliances served and the building construction allows sufficient infiltration. The provisions of this section rely on building envelope infiltration as the only source of combustion air.

The method given in Section 304.5.1 is based on a conservative assumed air infiltration rate of at least 0.40 air change per hour (ACH); therefore, if the ACH is known to be less than 0.40, this method cannot be used. The air infiltration rate for most buildings is not known, and it is possible for the ACH rate of a building to be actually less than 0.40; however, research indicates that buildings having less than 0.40 ACH are rare. It is believed that 80 to 95 percent of U.S. homes have infiltration rates of 0.35 ACH or greater. A rate of 0.40 ACH or less is the quantitative expression of "tight construction." Testing also indicates that the 50 ft³/1000 Btu/h (4.8 m³/kW) convention is somewhat liberal; thus, a built-in safety factor exists. If the designer suspects that the building in question is extraordinarily tight (less than 0.40 ACH), calculations and/or testing should be done to verify the ACH rate. A conservative approach could be to use the method in Section 304.5.2 with a conservative ACH rate such as 0.25, 0.30 or 0.35. Commentary table 304.5.2 shows calculated required volumes for appliances up to 300,000 Btu/h (87.9 kW) input rating based on the standard method and known ACH method for multiple ACH rates.

The standard method can be calculated for each appliance and then all the volumes are added or can be calculated by adding all of the appliance inputs first and then calculating the total volume. The known-ACH-rate method must be performed in at least two distinct calculations where both fan-assisted and draft-hood-equipped appliances are present.

To increase the available volume, the code allows rooms and spaces to be coupled together by doorways without doors and by openings installed in accordance with Section 304.5.3.

304.5.1 Standard method. The minimum required volume shall be 50 cubic feet per 1,000 Btu/h (4.8 m³/kW) of the appliance input rating.

❖ This is the optional default method intended for use when the ACH rate is unknown. If the actual ACH rate is known to be greater than 0.40, the method of Section 304.5.2 will yield smaller required volumes.

If the ACH rate is known to be less than 0.40, this method must not be used because it is based on an assumed ACH rate of at least 0.40. The rate of 50 ft³/1,000 Btu/h (4.8 m³/kW) is a carryover from the previous editions of the code and was used for buildings that were not of unusually tight construction.

304.5.2 Known air-infiltration-rate method. Where the air infiltration rate of a structure is known, the minimum required volume shall be determined as follows:

For appliances other than fan-assisted, calculate volume using Equation 3-1.

$$\text{Required Volume}_{other} \geq \frac{21 ft^3}{ACH} \left(\frac{I_{other}}{1,000\ Btu/hr} \right)$$

(Equation 3-1)

For fan-assisted appliances, calculate volume using Equation 3-2.

$$\text{Required Volume}_{fan} \geq \frac{15 ft^3}{ACH} \left(\frac{I_{fan}}{1,000\ Btu/hr} \right)$$

(Equation 3-2)

where:

I_{other} = All appliances other than fan assisted (input in Btu/h).

I_{fan} = Fan-assisted appliance (input in Btu/h).

ACH = Air change per hour (percent of volume of space exchanged per hour, expressed as a decimal).

For purposes of this calculation, an infiltration rate greater than 0.60 ACH shall not be used in Equations 3-1 and 3-2.

❖ This method considers the actual or calculated ACH rate and requires space volumes to be commensurate with that rate. This method can also be used when the ACH rate is unknown by simply picking a conservative ACH rate (0.40 ACH or less) representing the lowest anticipated ACH rate for the given building. Equation 3-1 is for draft-hood-equipped appliances and reflects recent research and test results that show that such appliances need less air flow than previously assumed. The standard ratio of 50 ft³/1,000 Btu/h (4.8 m³/kW) that has been in codes for many years was based on an assumed combustion air flow need of 25 ft³/1,000 Btu/h with an assumed ACH rate of 0.50. The 25 ft³/1,000 Btu/h air flow consisted of the air required for stoichiometric combustion [10 ft³/ft³ of natural gas] plus excess air to assure complete combustion, draft hood dilution air and a safety factor allowance. The volume of 25 cubic feet (0.7 m³) divided by an ACH rate of 0.50 yields the familiar 50 ft³/1,000 Btu/h ratio. Equation 3-1 is based on a total air flow need of 21 cubic feet (0.6 m³) instead of 25 (0.7 m³) because research supports a less conservative revised volume.

Equation 3-2 is for fan-assisted appliances only and accounts for the fact that fan-assisted appliances have no draft hood and, therefore, do not need dilution air. Fan-assisted appliances need only air for stoichiometric combustion [10 ft³/ft³ (10 cm³/cm³)of natural gas] and excess air [5 ft³/ft³ (5 cm³/cm³) of natural gas]. Dividing by the ACH rate adjusts the required volume to correspond to the available infiltration rate.

For example, if the ACH rate is 0.35, 42.8 cubic feet (1.2 m³) would be required for each 1,000 Btu/h (ft³ natural gas), and if the ACH rate is 0.50, 30 cubic feet (0.8 m³) would be required for each 1,000 Btu/h. The greater the ACH rate, the lower the room volume needed.

Research and testing results indicate that the absolute minimum volume for draft-hood-equipped appliances must be 35 ft³/1,000 Btu/h and for fan-assisted appliances it must be 25 ft³/1,000 Btu/h. This is why this section limits the infiltration rate to a maximum of 0.60 ACH. Limiting the ACH rate to 0.60 results in a lower limit safety factor for the required volume, thereby preventing appliances from being installed in exceedingly small spaces. Commentary Table 304.5.2 is based on the methodology of Sections 304.5.1 and 304.5.2 and is provided for convenience.

304.5.3 Indoor opening size and location. Openings used to connect indoor spaces shall be sized and located in accordance with Sections 304.5.3.1 and 304.5.3.2 (see Figure 304.5.3).

❖ Sections 304.5.3.1 and 304.5.3.2 prescribe the size and location of openings used to conjoin (couple) spaces for the purpose of increasing the available volume. The last sentence of Section 304.5 speaks of this. These openings must be permanently open, except where interlocked motorized dampers are used in accordance with Section 304.10.

304.5.3.1 Combining spaces on the same story. Each opening shall have a minimum free area of 1 square inch per 1,000 Btu/h (2,200 mm²/kW) of the total input rating of all gas utilization equipment in the space, but not less than 100 square inches (0.06 m²). One opening shall commence within 12 inches (305 mm) of the top and one opening shall commence within 12 inches (305 mm) of the bottom of the enclosure. The minimum dimension of air openings shall be not less than 3 inches (76 mm).

❖ This section is applicable to adjacent spaces on the same floor level and provides for the familiar high and low openings. The opening configuration creates a thermosiphon air flow with the bottom opening acting as an inlet and the top opening acting as an outlet. This provision originated as a means to conjoin a small appliance enclosure (such as a furnace or boiler room) with other spaces, thereby effectively increasing the volume of the enclosures.

304.5.3.2 Combining spaces in different stories. The volumes of spaces in different stories shall be considered as communicating spaces where such spaces are connected by one or more openings in doors or floors having a total minimum free area of 2 square inches per 1,000 Btu/h (4402 mm²/kW) of total input rating of all gas utilization equipment.

❖ This section is a new concept that allows spaces on different floor levels to be conjoined, thereby effectively increasing the available volume for supplying combustion air. Installing a high and low opening configuration in accordance with the previous section would serve no purpose when vertically connecting adjacent stories; thus, a single opening is permitted. The opening or openings can be in a floor or in a door opening to an unenclosed stairway that connects the two stories. This method of conjoining spaces is particularly useful for dwelling units where the volume of a basement in which the appliances are located can be conjoined with the open spaces on upper stories by means of a louvered door at the basement stairs. The opening size requirement is simply the addition of the areas of the two openings required by Section 304.5.3.1. Bear in mind that openings in floors and stairway doors in other than dwelling units are strictly regulated by the IMC and IBC and would be prohibited or would require fire dampers in most cases. A combustion air opening in a floor would be considered by the IBC as a transfer opening in a horizontal assembly (see commentary Figure 304.5.3.2).

Example:

Determine whether indoor combustion air is adequate for the building shown in commentary Figure 304.5.3.2.

1. Starting with the fan-assisted boilers, use Equation 3-2 to calculate the required volume for each boiler.

$$\frac{15 \text{ ft}^3}{0.35} \times \frac{125,000}{1,000} = 42.8 \times 125 = 5,350 \text{ ft}^3$$

2. Use Equation 3-1 to calculate the required volume for each water heater.

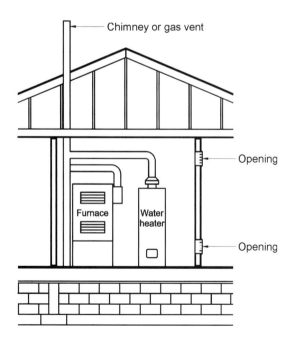

Figure 304.5.3
ALL AIR FROM INSIDE THE BUILDING (see Section 304.5.3)

GENERAL REGULATIONS

TABLE 304.5.2
CALCULATED VOLUMES

Standard Method Required Volume All Appliances		Known Air infiltration Rate Method, Minimum Space Volume for Appliance Other Than Fan-Assisted, For Specified Infiltration Rates (ACH*)				Known Air Infiltration Rate Method, Minimum Space Volume for Fan-Assisted Appliance, For specified Infiltration Rates (ACH*)			
Appliance Input Btu/hr	Space Volume (ft³)	Appliance Input Btu/hr	Space Volume (ft³) 0.25 ACH	Space Volume (ft³) 0.30 ACH	Space Volume (ft³) 0.35 ACH	Appliance Input Btu/hr	Space Volume (ft³) 0.25 ACH	Space Volume (ft³) 0.30 ACH	Space Volume (ft³) 0.35 ACH
5,000	250	5,000	420	350	300	5,000	300	250	214
10,000	500	10,000	840	700	600	10,000	600	500	429
15,000	750	15,000	1,260	1,050	900	15,000	900	750	643
20,000	1,000	20,000	1,680	1,400	1,200	20,000	1,200	1,000	857
25,000	1,250	25,000	2,100	1,750	1,500	25,000	1,500	1,250	1,071
30,000	1,500	30,000	2,520	2,100	1,800	30,000	1,800	1,500	1,286
35,000	1,750	35,000	2,940	2,450	2,100	35,000	2,100	1,750	1,500
40,000	2,000	40,000	3,360	2,800	2,400	40,000	2,400	2,000	1,714
45,000	2,250	45,000	3,780	3,150	2,700	45,000	2,700	2,250	1,929
50,000	2,500	50,000	4,200	3,500	3,000	50,000	3,000	2,500	2,143
55,000	2,750	55,000	4,620	3,850	3,300	55,000	3,300	2,700	2,357
60,000	3,000	60,000	5,040	4,200	3,600	60,000	3,600	3,000	2,571
65,000	3,250	65,000	5,460	4,550	3,900	65,000	3,900	3,250	2,786
70,000	3,500	70,000	5,800	4,900	4,200	70,000	4,200	3,500	3,000
75,000	3,750	75,000	6,300	5,250	4,500	75,000	4,500	3,750	3,214
80,000	4,000	80,000	6,720	5,600	4,800	80,000	4,800	4,000	3,429
85,000	4,250	85,000	7,140	5,950	5,100	85,000	5,100	4,250	3,643
90,000	4,500	90,000	7,500	6,300	5,400	90,000	5,400	4,500	3,857
95,000	4,750	95,000	7,900	6,650	5,700	95,000	5,700	4,750	4,071
100,000	5,000	100,000	8,400	7,000	6,000	100,000	6,000	5,000	4,286
105,000	5,250	105,000	8,820	7,350	6,300	105,000	6,300	5,250	4,500
110,000	5,500	110,000	9,240	7,700	6,600	110,000	6,600	5,500	4,714
115,000	5,700	115,000	9,660	8,050	6,900	115,000	6,900	5,750	4,929
120,000	6,000	120,000	10,080	8,400	7,200	120,000	7,200	6,000	5,143
125,000	6,250	125,000	10,500	8,750	7,500	125,000	7,500	6,250	5,357
130,000	6,500	130,000	10,920	9,100	7,800	130,000	7,800	6,500	5,571
135,000	6,750	135,000	11,340	9,450	8,100	135,000	8,100	6,700	5,786
140,000	7,000	140,000	11,760	9,800	8,400	140,000	8,400	7,000	6,000
145,000	7,250	145,000	12,180	10,150	8,700	145,000	8,700	7,250	6,214
150,000	7,500	150,000	12,600	10,500	9,000	150,000	9,000	7,500	6,429
160,000	8,000	160,000	13,440	11,200	9,600	160,000	9,600	8,000	6,857
170,000	8,500	170,000	14,280	11,900	10,200	170,000	10,200	8,500	7,286
180,000	9,000	180,000	15,120	12,600	10,800	180,000	10,800	9,000	7,714
190,000	9,500	190,000	15,960	13,300	11,400	190,000	11,400	9,500	8,143
200,000	10,000	200,000	16,800	14,000	12,000	200,000	12,000	10,000	8,571
210,000	10,500	210,000	17,640	14,700	12,600	210,000	12,600	10,500	9,000
220,000	11,000	220,000	18,480	15,400	13,200	220,000	13,200	11,000	9,429
230,000	11,500	230,000	19,320	16,100	13,800	230,000	13,800	11,500	9,857
240,000	12,000	240,000	20,160	16,800	14,400	240,000	14,400	12,000	10,286
250,000	12,500	250,000	21,000	17,500	15,000	250,000	15,000	12,500	10,714
260,000	13,000	260,000	21,840	18,200	15,600	260,000	15,600	13,000	11,143
270,000	13,500	270,000	22,680	18,900	16,200	270,000	16,200	13,500	11,571
280,000	14,000	280,000	23,520	19,600	16,800	280,000	16,800	14,000	12,000
290,000	14,500	290,000	24,360	20,300	17,400	290,000	17,400	14,500	12,429
300,000	15,000	300,000	25,200	21,000	18,000	300,000	18,000	15,000	12,857

*ACH=Air Change per Hour   *ACH=Air Change per Hour

Table courtesy of American Gas Association

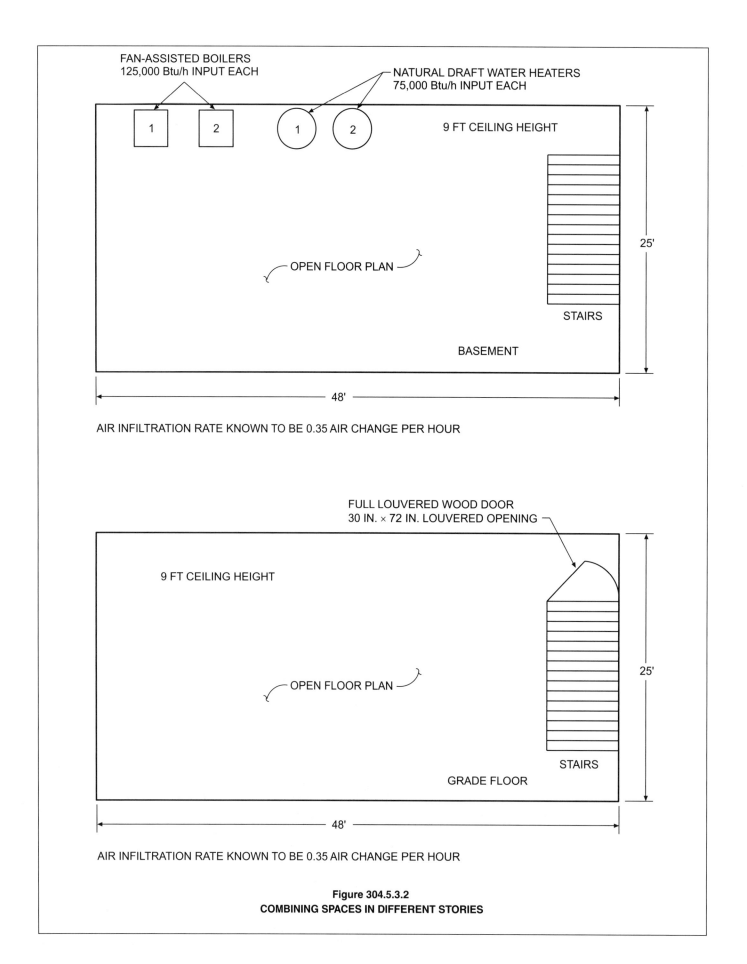

Figure 304.5.3.2
COMBINING SPACES IN DIFFERENT STORIES

$$\frac{21 \text{ ft}^3}{0.35} \times \frac{75,000}{1,000} = 60 \times 75 = 4,500 \text{ ft}^3$$

3. Total volumes required for all appliances.

Boiler	1	5,350
Boiler	2	5,350
Water Heater	1	4,500
Water Heater	2	4,500
		19,700 ft³

4. Determine whether louvered door opening area is adequate.

   - In accordance with Section 304.10, determine actual louvered opening area.

   30 in. × 72 in. = 2,160 in.² × 0.25 = 540 in.²

   - In accordance with Section 304.5.3.2, determine required area.

   $$\frac{400,000}{1,000} \text{ Btu/h} \times 2 = 800 \text{ in.}^2$$

5. Determine available (actual) volume in building.

   Basement volume = 48 ft × 25 ft × 9 ft = 10,800 ft³
   Grade floor volume = 48 ft × 25 ft × 9 ft = 10,800 ft³
   　　　　　　　　　　　　　　　　　　　　　　21,600 ft³

The combined volumes of both stories is greater than required; however, the louvered door opening area connecting the stories is inadequate. Possible solutions include adding outdoor openings in the basement in accordance with Section 304.7 or providing metal louvers in the stairway door Instead of wood.

For SI: 1 inch = 25.4 mm, 1 foot = 305.8 mm, 1 cubic foot = 0.0283 m³, 1 Btu per hour = 0.2931W

304.6 Outdoor combustion air. Outdoor combustion air shall be provided through opening(s) to the outdoors in accordance with Section 304.6.1 or 304.6.2. The minimum dimension of air openings shall be not less than 3 inches (76 mm).

❖ This section describes two methods for supplying combustion air from the outdoors: the traditional method of two direct openings or ducts to the outdoors and a newer method using one opening or duct to the outdoors.

Openings to spaces that are naturally ventilated with outdoor air, such as attic or crawl spaces, are considered as an acceptable alternative to a direct connection to the outdoors. Attic and crawl spaces can be acceptable sources of combustion air only where such spaces have adequate natural ventilation openings directly to the outdoors. Attic and crawl spaces ventilated by mechanical means are not an acceptable source of combustion air.

Combustion air ducts and openings that penetrate components of wall, floor, ceiling and roof assemblies must be installed with penetration protection as required by the IBC and the IMC.

When designing combustion air installations, the effect that openings to the outdoors can have on appliances, plumbing systems and building occupants must be considered. For example, depending on the location, openings to the outdoors can:

- Cause drafts that can blow out pilot lights or otherwise interfere with appliance ignition and operation.
- Cause freezing of plumbing piping or other water-containing components.
- Cause objectionable cold drafts that encourage the occupants to block or cover the openings.

In all cases, combustion air openings should be located to reduce the likelihood they will be accidentally or intentionally blocked or covered. A ceiling transfer opening that connects a furnace room with an attic is an example of a combustion air opening that is likely to be intentionally blocked by an occupant because of the drafts that can occur in cold climates.

Building occupants typically do not understand the need for or the importance of combustion air openings.

No side dimension of a square or rectangular opening and no diameter of a round opening can be less than 3 inches (76 mm). The smallest allowed square opening area would be 9 square inches (5806 mm²) and the smallest allowed round opening area would be 7 square inches (4516 mm²) (see commentary, Section 304.10).

304.6.1 Two-permanent-openings method. Two permanent openings, one commencing within 12 inches (305 mm) of the top and one commencing within 12 inches (305 mm) of the bottom of the enclosure, shall be provided. The openings shall communicate directly, or by ducts, with the outdoors or spaces that freely communicate with the outdoors.

Where directly communicating with the outdoors, or where communicating with the outdoors through vertical ducts, each opening shall have a minimum free area of 1 square inch per 4,000 Btu/h (550 mm²/kW) of total input rating of all equipment in the enclosure [see Figures 304.6.1(1) and 304.6.1(2)].

Where communicating with the outdoors through horizontal ducts, each opening shall have a minimum free area of not less than 1 square inch per 2,000 Btu/h (1,100 mm²/kW) of total input rating of all equipment in the enclosure [see Figure 304.6.1(3)].

❖ Two openings located as prescribed in this section are intended to induce a convective air current in the room or space by admitting cooler, denser air in the lower opening and allowing the escape of warmer, less dense air through the upper opening. The farther apart the openings, the greater the temperature differential and the greater the convective force behind the current. A component of combustion air is cooling (ventilation) air for the appliance enclosure. The two-opening method was created to ventilate the appliance enclosure in addition to supplying combustion air. This ventilation cools the appliances and would help remove any combustion gases that spilled from the appliances.

304.6.2 One-permanent-opening method. One permanent opening, commencing within 12 inches (305 mm) of the top of the enclosure, shall be provided. The equipment shall have clearances of at least 1 inch (25 mm) from the sides and back and 6 inches (152 mm) from the front of the appliance. The opening shall directly communicate with the outdoors or through a vertical or horizontal duct to the outdoors or spaces that freely communicate with the outdoors [see Figure 304.6.2] and shall have a minimum free area of 1 square inch per 3,000 Btu/h (734 mm^2/kW) of the total input rating of all equipment located in the enclosure, and not less than the sum of the areas of all vent connectors in the space.

❖ Research has shown that for modern appliances, a single opening to the outdoors will perform as well as the traditional two-opening method. The one-opening method described in this section depends on a reduced pressure being created in the enclosure by the draft created by the venting system. This reduced pressure causes combustion air to enter the enclosure through the single opening. The opening must be properly sized considering both sizing criteria: the square-inch-area-per-Btu/h ratio and the area minimum based on the sum of the areas of all vent connectors in the enclosure. This method allows for fewer openings, fewer ducts and fewer objections by the owners/occupants.

304.7 Combination indoor and outdoor combustion air. The use of a combination of indoor and outdoor combustion air shall be in accordance with Sections 304.7.1 through 304.7.3.

❖ This method of supplying combustion air is a combined application of Section 304.5 and Section 304.6. This method allows credit for the amount of infiltration that exists and makes up for the shortage with supplemental outdoor air. In other words, in addition to obtaining combustion air directly from the outdoors, this method relies on building infiltration for a portion of the total combustion air. In spaces where the volume is insufficient to satisfy the method of Section 304.5 or where smaller outdoor air openings than required by Section 304.6 are desired, this method allows infiltration and outdoor openings to supplement each other. If the appliances are enclosed in a small room such as a closet, openings to adjacent spaces as prescribed by Section 304.5.3 must be provided to couple the appliance room volume with any other space volume counted on to provide combustion air. Frequently, sufficient volume cannot be obtained in the appliance enclosure or by opening the appliance enclosure to adjacent spaces. This method of combining indoor and outdoor air is an alternative solution. Simply stated, this section uses ratios of what is required to what is actually supplied so that when combined, the indoor air component and the outdoor air component add up to the whole required.

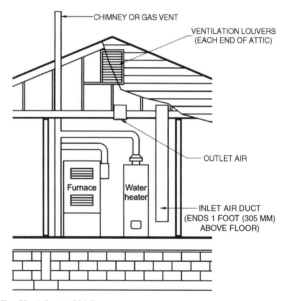

For SI: 1 foot = 304.8 mm.

Figure 304.6.1(2)
ALL AIR FROM OUTDOORS THROUGH VENTILATED ATTIC
(see Section 304.6.1)

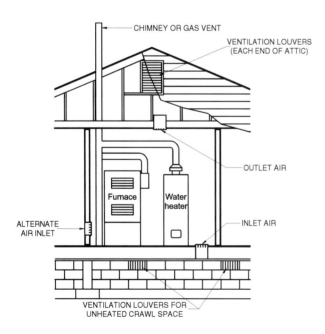

Figure 304.6.1(1)
ALL AIR FROM OUTDOORS–INLET AIR FROM VENTILATED CRAWL SPACE AND OUTLET AIR TO VENTILATED ATTIC

304.7.1 Indoor openings. Where used, openings connecting the interior spaces shall comply with Section 304.5.3.

❖ See the commentary for Section 304.5.3.

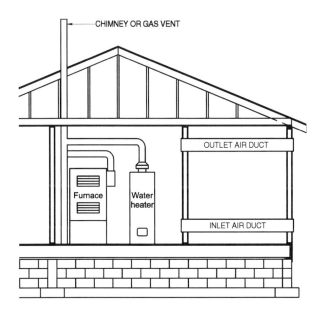

Figure 304.6.1(3)
APPLIANCES LOCATED IN CONFINED SPACES; ALL
AIR FROM OUTDOORS (See Section 304.11.1)

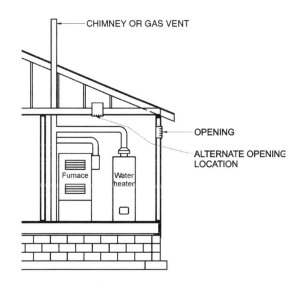

Figure 304.6.2
SINGLE COMBUSTION AIR OPENING,
ALL AIR FROM THE OUTDOORS
(see Section 304.6.2)

304.7.2 Outdoor opening location. Outdoor opening(s) shall be located in accordance with Section 304.6.

❖ See the commentary for Section 304.6.

304.7.3 Outdoor opening(s) size. The outdoor opening(s) size shall be calculated in accordance with the following:

1. The ratio of interior spaces shall be the available volume of all communicating spaces divided by the required volume.

2. The outdoor size reduction factor shall be one minus the ratio of interior spaces.

3. The minimum size of outdoor opening(s) shall be the full size of outdoor opening(s) calculated in accordance with Section 304.6, multiplied by the reduction factor. The minimum dimension of air openings shall be not less than 3 inches (76 mm).

❖ Although the principle has not changed, this method has been simplified compared to the same provision in the 2000 edition of the code. The intent is still the same; that is, the fraction of the required indoor volume plus the fraction of outdoor openings must be equal to or greater than 1. The indoor volume method and the outdoor air method can both be stand-alone methods; therefore, half of one and half of the other will work, as will three fourths of one and a quarter of the other, etc.

This section is expressed in the following equation:

$$\left[1 - \left(\frac{\text{available indoor volume}}{\text{volume required by Section 304.5}}\right)\right] \times \text{full size openings required by Section 304.6} = \text{reduced size outdoor air openings}$$

Example:
Given:

7,500 ft^3 of indoor volume is required if all indoor air is to be used.

4,950 ft^3 of indoor volume is available.

One opening to the outdoors is desired and must be 50 in.^{2} if all outdoor air is to be used.

$$\left[1 - \left(\frac{4950 \text{ ft}^3}{7500 \text{ ft}^3}\right)\right] \times 50 \text{ in.}^2 = \text{reduced outdoor opening size}$$

0.34 × 50 in.^{2} = 17 in.^{2} outdoor opening size

Either of the provisions of Section 304.6 when combined with the provision of this section can be used to satisfy the equation.

Example 1:

Room A is within a dwelling unit and contains a fuel-fired furnace and a fuel-fired water heater with input ratings of 150,000 and 50,000 Btu/h, respectively.

The size of Room A is 20 feet by 30 feet, and adjacent Room B is also 20 feet by 30 feet. Both rooms have 8-foot-high ceilings. Rooms A and B communicate through two openings located in accordance with Section 304.5.3, each having dimensions of 10 inches by 20 inches. Two 3-inch-diameter (round) direct openings to the outdoors are installed in Room A and are located in accordance with Section 304.6.1.

Question: Do the openings to the outdoors, when combined with the volumes of Rooms A and B, meet the combustion air demand of the appliances?

For **SI:** 1 inch = 25.4 mm, 1 foot = 305 mm, 1 Btu/h = 0.2931 W, 1 cubic foot = 0.0283 m^3, 1 square inch = 645 mm^2.

Given:

The air infiltration rate of the building is unknown.
Volume of Room A = 4,800 ft³
Volume of Room B = 4,800 ft³
Area of 3-inch-diameter direct openings = 7.07 in.²
Area of each opening between Rooms A and B = 200 in.²
Total input rating of appliances = 200,000 Btu/h

Step 1: Determine whether Rooms A and B meet the volume requirements of Section 304.5.

Available volume of Rooms A and B = 4,800 ft³ + 4,800 ft³ = 9,600 ft³

The required volume is: $\dfrac{200,000 \text{ Btu/h}}{1000 \text{ Btu/h}} \times 50 \text{ ft}^3 = 10,000 \text{ ft}^3$

Rooms A and B combined do not meet the volume requirements of Section 304.5.

Step 2: Determine whether the openings between Rooms A and B meet the requirements of Section 304.5.3.

The minimum required area is: $\dfrac{200,000 \text{ Btu/h}}{1000 \text{ Btu/h}} \times 1 \text{ in.}^2 = 200 \text{ in.}^2$

Actual area of each opening = 200 in.²

The area of each opening meets the size requirement of Section 304.5.3.

Step 3: Determine the required area of each direct outdoor opening in accordance with Section 304.6.1.

Required area of each opening: $\dfrac{200,000 \text{ Btu/h}}{4000 \text{ Btu/h}} \times 1 \text{ in.}^2 = 50 \text{ in.}^2$

Step 4: Determine whether the 3-inch-diameter direct openings combined with the volumes of Rooms A and B comply with Section 304.7.3.

$\left[1 - \left(\dfrac{9600 \text{ ft}^3}{10,000 \text{ ft}^3}\right)\right] \times 50 = (1 - 0.96) \times 50 = 0.04 \times 50 = 2 \text{ in.}^2$

The actual area of each direct opening (7.07 in.²) exceeds the required area of 2 in.².

Therefore, the combination of the 3-inch-diameter direct openings and the volumes of Rooms A and B satisfies the combustion air demand of the appliances.

Example 2:

A combination of indoor and outdoor combustion air is to be supplied for two natural draft boilers in a room as shown in Figure 304.7.3 (the air infiltration rate is not known).

1. In accordance with Section 304.5.1, the required volume of the room containing the boilers would be:

$\dfrac{90,000 \text{ Btu/h} + 90,000 \text{ Btu/h}}{1,000} \times 50 \text{ ft}^3$

$= \dfrac{180,000}{1,000} \times 50 \text{ ft}^3 = 180 \times 50 \text{ ft}^3 = 9,000 \text{ ft}^3$

2. The actual room volume is 20 ft × 50 ft × 8 ft = 8,000 ft³

3. In accordance with Section 304.7.3 Item 1, determine the ratio of available (actual) volume to the required volume.

$\dfrac{\text{Actual Volume}}{\text{Required Volume}} \dfrac{8,000 \text{ ft}^3}{9,000 \text{ ft}^3} = 0.89$

4. In accordance with Section 304.7.3 Item 2, determine outdoor opening size reduction factor.

1 − 0.89 = 0.11

5. In accordance with Section 304.7.3 Item 3, determine the size of the outdoor opening required as if all combustion air is to be supplied through the outdoor opening. Because a single high opening is built into an exterior wall, Section 304.6.2 applies and would require an area of 1 in.²/3000 But/h of total appliance input rating (Assuming 6-in. vent connectors, the total area of the boiler vent connectors is 56.6 in.²).

$\dfrac{90,000 \text{ Btu/h} + 90,000 \text{ Btu/h}}{3,000} = 60 \text{ in.}^2$

6. The required size of the outdoor air opening is 60 in.² × 0.11 = 6.6 in². Because the minimum dimensions must be not less than 3 inches, the minimum area of a square opening must be 9 in². This combination satisfies the combustion air demand of the appliances.

304.8 Engineered installations. Engineered combustion air installations shall provide an adequate supply of combustion, ventilation and dilution air and shall be approved.

❖ Potential sources of combustion air are the indoor atmosphere in which the appliance is located, outdoor openings and ducts that communicate with the outdoor atmosphere, a combination of these or any other approved arrangement for the dependable introduction of outdoor air into the building or directly to the fuel-burning appliance.

The provisions of Section 304 are intended for application only to appliances and equipment using natural draft atmospheric burners. Gas-fired appliances, especially those that are not of the atmospheric-burner type, are commonly supplied with combustion air by engineered systems. For example, special engineered systems are usually designed for appliances equipped with power burners. Power burners force combustion air into the burner under pressure and are, therefore, not dependent upon gravity airflow. In some applications, engineered combustion air supplies have advantages and can be preferable to large gravity (natural) openings to the outdoors. Engineered designs must be evaluated for code compliance, approved and installed in accordance with the appliance manufacturer's instructions [see commentary Figures 304.8(1) thru 304.8(3)].

304.9 Mechanical combustion air supply. Where all combustion air is provided by a mechanical air supply system, the combustion air shall be supplied from the outdoors at a rate not less than 0.35 cubic feet per minute per 1,000 Btu/h (0.034 m³/min per kW) of total input rating of all appliances located within the space.

❖ This method supplies combustion air by means of a fan/blower that runs when any of the served appliances are in a firing cycle. The fan/blower must be sized to supply the required air flow based on the simultaneous operation of all appliances that are served by the fans/blowers. The total Btu/h input rating of all fuel-burning appliances located in the room or space must be used because the potential exists for all of the appliances to be operating at the same time. This method is used where gravity openings to the outdoors are impractical or undesirable. A small fan-powered intake opening can substitute for comparatively large gravity openings, especially where freezing tempera-

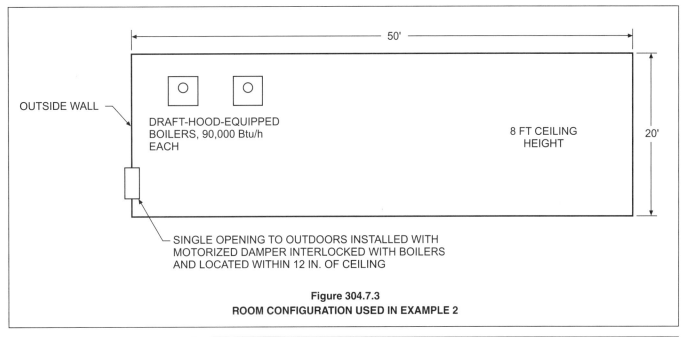

Figure 304.7.3
ROOM CONFIGURATION USED IN EXAMPLE 2

Photo courtesy of Tjernlund Products, Inc.

Figure 304.8(1)
MECHANICAL COMBUSTION AIR/MAKEUP AIR SUPPLY SYSTEM

Figure 304.8(2)
MECHANICAL COMBUSTION AIR SUPPLY SYSTEM

Figure 304.8(3)
MECHANICAL COMBUSTION AIR/MAKEUP AIR SUPPLY SYSTEM

tures are a problem or there is no room for the required size gravity openings. Commentary Figures 304.8(1), (2) and (3) illustrate factory-engineered mechanical combustion air supply units.

The prescribed air flow rate of 0.35 cfm/1000 Btu/h is equivalent to 1 cfm/2867 Btu/h and is based on a combustion air need of 21 cubic feet of volume for each cubic foot of natural gas (1000 Btu/h). The rate is derived as follows:

GENERAL REGULATIONS

$$\frac{2857 \text{ Btu/h}}{1000 \text{ Btu/ft}^3} = 2.857 \text{ ft}^3 \text{ of gas per hour}$$

2.857 ft³ of gas per hour × 21 ft³ of air per ft³ of gas
= 60 ft³ of air per/hour = 60 cfh

$$\frac{60 \text{ cfh}}{60 \text{ min/hr}} = 1 \text{ cfm}$$

Thus, 1 cfm is required for each 2857 Btu/h or 0.35 cfm/1000 Btu/h

304.9.1 Makeup air. Where exhaust fans are installed, makeup air shall be provided to replace the exhausted air.

❖ This section expresses the exact same concern as Section 304.4 (see commentary, Section 304.4). If the air that is supplied by the combustion air fan becomes makeup air for an exhaust fan, the combustion air will not be available to the appliance(s) for which the air was intended.

304.9.2 Appliance interlock. Each of the appliances served shall be interlocked with the mechanical air supply system to prevent main burner operation when the mechanical air supply system is not in operation.

❖ This section requires an interlock circuit as opposed to simple parallel operation wiring. In other words, the combustion air fan/blower must start when initiated by the means that controls the appliance operation cycles (e.g., thermostat), and only after the combustion air fan/blower is proven to be operating will the appliance be allowed to start a firing cycle. If the combustion air fan/blower fails during a firing cycle of an appliance, the appliance firing cycle must be terminated (see commentary Figure 304.9.2).

The following steps must occur in this order:
1. Call for heat.
2. Combustion air fan/blower starts.
3. Combustion air fan/blower is proven to be operating.
4. Proving controls allow appliance to fire.

304.9.3 Combined combustion air and ventilation air system. Where combustion air is provided by the building's mechanical ventilation system, the system shall provide the specified combustion air rate in addition to the required ventilation air.

❖ If combustion air is mechanically supplied through an HVAC system, the system must deliver the required combustion air rate above and beyond any outdoor air rate supplied for ventilation of the occupied spaces. The HVAC system must also be interlocked with the appliance(s) as required by Section 304.9.2. Neither continuous operation nor parallel operation of the HVAC system satisfies the requirement for a true interlock as required by Section 304.9.2.

304.10 Louvers and grilles. The required size of openings for combustion, ventilation and dilution air shall be based on the net free area of each opening. Where the free area through a design of louver, grille or screen is known, it shall be used in calculating the size opening required to provide the free area specified. Where the design and free area of louvers and grilles are not known, it shall be assumed that wood louvers will have 25-percent free area and metal louvers and grilles will have 75-percent free area. Screens shall have a mesh size not smaller than $^1/_4$ inch. Nonmotorized louvers and grilles shall be fixed in

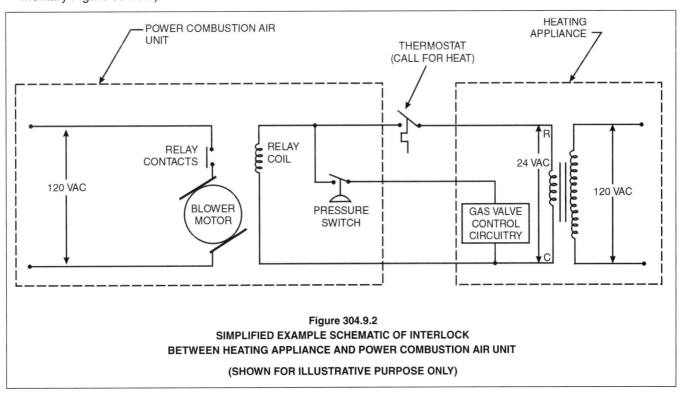

Figure 304.9.2
SIMPLIFIED EXAMPLE SCHEMATIC OF INTERLOCK
BETWEEN HEATING APPLIANCE AND POWER COMBUSTION AIR UNIT

(SHOWN FOR ILLUSTRATIVE PURPOSE ONLY)

the open position. Motorized louvers shall be interlocked with the equipment so that they are proven to be in the full open position prior to main burner ignition and during main burner operation. Means shall be provided to prevent the main burner from igniting if the louvers fail to open during burner start-up and to shut down the main burner if the louvers close during operation.

❖ This section recognizes that louvers and grilles are usually installed over air inlets and outlets to prevent rain, snow and animals from entering the building. When louvers or grilles are used, the solid portion of the louver or grille must be considered when determining the unobstructed (net clear) area of the opening.

Combustion air openings are sized based on there being a free, unobstructed area for the passage of air into the space where the fuel-burning appliances are located. Louvers or grilles placed over these openings reduce the area of the openings because of the area occupied by the solid portions of the grille or louver. The reduction in area must be considered because only the unobstructed area can be credited toward the required opening size.

The reduction in opening area caused by the presence of grilles or louvers will always require openings to be larger than determined from the sizing ratios of this chapter and larger than any duct of the minimum required size that might connect to these openings.

This section does not apply to grilles, louvers or screens that are an integral component of a labeled appliance that is installed in accordance with the manufacturer's instructions. Where the manufacturer states the free, unobstructed area of a grille or louver, it is acceptable to consider that area as actual without requiring further calculation.

The code also requires the free area of screens to be considered in the opening sizing calculations. If it is unknown, it will have to be calculated because there is no stated default area for screens as there is for louvers and grilles. The code prohibits screens having mesh sizes smaller than 1/4 inch (6 mm) (see errata for first printing). Screens with a mesh size of 1/4 inch (6 mm) and larger are not insect screens and would be suitable only for keeping out large debris, rodents and other small animals. Smaller size mesh is likely to become obstructed by lint, plant fibers, cottonwood seeds, etc. At best, placing screens in combustion air openings is risky, and grilles and louvers are preferred to keep out the unwanted.

Metal louvers are constructed so that the solid portion of the louver occupies approximately 25 percent of the opening area, leaving 75 percent of this area unobstructed. Wood louvers occupy approximately 75 percent of the opening area, leaving 25 percent of the area unobstructed. When the required amount of unobstructed area is determined, the effect that a louver has on that area must be taken into account to make certain the minimum amount of unobstructed opening is achieved. The material type, thickness and spacing all have a significant effect on the actual net free area of any louver or grille.

To prevent freezing of pipes, to save energy and maintain comfortable temperatures in appliance rooms, outdoor openings often have motorized dampers that close during the appliance(s) off-cycle. The damper motors and/or damper blades must be supervised so that an interlock can be established between the dampers and the appliances, thereby preventing appliance operation when the dampers are closed.

Combustion air openings must not be closable unless they open and close automatically using a mechanism that is interlocked with the appliances (see Figure 304.10). Damper supervision is easily accomplished by damper motor end switches or blade position travel switches that monitor the position of the drive motor shaft, the damper/motor linkage hardware or the damper blades. These switches provide feedback to the appliance control circuitry to allow firing only when the dampers are proven to be fully open.

304.11 Combustion air ducts. Combustion air ducts shall comply with all of the following:

1. Ducts shall be of galvanized steel complying with Chapter 6 of the *International Mechanical Code* or of equivalent corrosion-resistant material approved for this application.

    Exception: Within dwellings units, unobstructed stud and joist spaces shall not be prohibited from conveying combustion air, provided that not more than one required fireblock is removed.

2. Ducts shall terminate in an unobstructed space allowing free movement of combustion air to the appliances.

3. Ducts shall serve a single enclosure.

4. Ducts shall not serve both upper and lower combustion air openings where both such openings are used. The separation between ducts serving upper and lower combustion air openings shall be maintained to the source of combustion air.

5. Ducts shall not be screened where terminating in an attic space.

6. Horizontal upper combustion air ducts shall not slope downward toward the source of combustion air.

7. The remaining space surrounding a chimney liner, gas vent, special gas vent or plastic piping installed within a masonry, metal or factory-built chimney shall not be used to supply combustion air.

    Exception: Direct-vent gas-fired appliances designed for installation in a solid fuel-burning fireplace where installed in accordance with the listing and the manufacturer's instructions.

8. Combustion air intake openings located on the exterior of a building shall have the lowest side of such openings located not less than 12 inches (305 mm) vertically from the adjoining grade level.

❖ This section addresses the construction and installation of combustion air ducts.

Item 1 requires that combustion air ducts be constructed of galvanized sheet steel or an approved

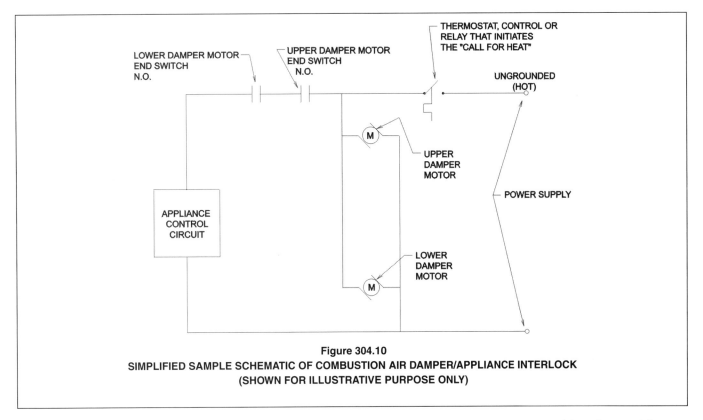

Figure 304.10
SIMPLIFIED SAMPLE SCHEMATIC OF COMBUSTION AIR DAMPER/APPLIANCE INTERLOCK
(SHOWN FOR ILLUSTRATIVE PURPOSE ONLY)

equivalent material. The intent is to cause ducts to be constructed of a material that is resistant to corrosion and physical damage, which would allow the ducts to remain in place undamaged for the life of the installation. It is doubtful that the authors of item 1 ever anticipated the use of flexible ducts and the higher friction losses associated with such rough-wall ducts. It is reasonable to assume that all of the sizing criteria in Section 304 are based on smooth-wall ducts. The exception expresses a concern for fire safety where stud cavities are used as a duct.

Item 2 is intended to prevent blockages and minimize resistance (friction loss) to airflow.

Item 3 prohibits a duct from serving more than one appliance enclosure because the duct and opening sizing criteria are based on a single duct serving a single opening in an appliance location. Multiple openings supplied by a single duct could result in unpredictable airflow and might not create the convective movements of air intended by the combustion air methods.

Item 4 recognizes that where both upper and lower combustion air openings are required, air circulation cannot occur if a single duct serves both combustion air openings. The lower opening functions as an air inlet and the upper opening as an air outlet, thus independent ducts are required (see commentary Figure 304.11).

Item 5 is intended to prevent obstruction or blockage by insulation materials.

Item 6 is intended to prevent trapping or sloping of the upper combustion air opening in a way that would impede the convective movement of air. The horizontal duct for the upper openings must slope upward as it travels away from the appliance(s) and toward the outdoors.

When a vent or chimney liner is installed within a chimney, an annular space will exist between the vent or liner and the interior chimney walls. According to Item 7, this space cannot be used as a conduit for conveying combustion air because of poor flow characteristics, the tendency for reverse flow to occur as the combustion air is warmed by the liner or vent and the adverse cooling effect on the vent or liner. The exception recognizes that there could be appliances that are designed and listed for the type of installation that this item intends to prohibit [see commentary Figure 503.5.6.1(2)].

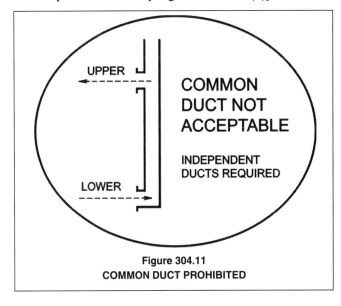

Figure 304.11
COMMON DUCT PROHIBITED

Item 8 intends to prevent combustion air openings from being obstructed by snow, leaves and vegetation. No part of a combustion air opening is allowed to be less than 12 inches (305 mm) above grade.

304.12 Protection from fumes and gases. Where corrosive or flammable process fumes or gases, other than products of combustion, are present, means for the disposal of such fumes or gases shall be provided. Such fumes or gases include carbon monoxide, hydrogen sulfide, ammonia, chlorine and halogenated hydrocarbons.

In barbershops, beauty shops and other facilities where chemicals that generate corrosive or flammable products, such as aerosol sprays, are routinely used, nondirect-vent-type appliances shall be located in an equipment room separated or partitioned off from other areas with provisions for combustion air and dilution air from the outdoors. Direct-vent appliances shall be installed in accordance with the appliance manufacturer's installation instructions.

❖ In many occupancies, the routine use of chemicals contaminates the indoor combustion air. These contaminants can combine with water vapor in the combustion gases to produce acids and other corrosive compounds that can destroy appliance components, vents, chimneys and connectors. For example, it is common to see vents and vent connectors for draft-hood-equipped water heaters in beauty shops that have deteriorated to the point of structural failure. Other examples are swimming pool heaters exposed to chlorine, boilers exposed to laundry and dry cleaning chemicals and boilers exposed to refrigerant leakage. Some contaminants are more toxic to life after they have passed through an appliance combustion chamber. The highly poisonous chemical "phosgene" (carbonyl chloride) can be produced when common refrigerants contaminate combustion air.

Obviously, choosing direct-vent appliances will eliminate this problem in most cases and will not require isolation enclosures for the appliances. It is possible that some atmospheres are unsuitable for any appliance, direct-vent or otherwise. The code provides an alternative to direct-vent appliances that requires the appliances to be isolated from the contaminated atmosphere by reasonably air-tight enclosures and requires that all combustion air be supplied for the appliance(s) from the outdoors.

SECTION 305 (IFGC)
INSTALLATION

305.1 General. Equipment and appliances shall be installed as required by the terms of their approval, in accordance with the conditions of listing, the manufacturer's instructions and this code. Manufacturers' installation instructions shall be available on the job site at the time of inspection. Where a code provision is less restrictive than the conditions of the listing of the equipment or appliance or the manufacturer's installation instructions, the conditions of the listing and the manufacturer's installation instructions shall apply.

Unlisted appliances approved in accordance with Section 301.3 shall be limited to uses recommended by the manufacturer and shall be installed in accordance with the manufacturer's instructions, the provisions of this code and the requirements determined by the code official.

❖ Manufacturers' installation instructions are evaluated by the listing agency verifying compliance with the applicable standard. The listing agency can require that the manufacturer alter, delete or add information to the instructions as necessary to achieve compliance with applicable standards and code requirements. Manufacturers' installation instructions are an enforceable extension of the code and must be in the hands of the code official when an inspection takes place.

When an appliance is tested to obtain a listing and label, the approved agency installs the appliance in accordance with the manufacturer's installation instructions. The appliance is tested under these conditions; thus, the installation instructions become an integral part of the labeling process.

The listing and labeling process assures that the appliance and its installation instructions are in compliance with applicable standards. Therefore, an installation in accordance with the manufacturer's instructions is required, except where the code requirements are more stringent. An inspector must carefully and completely read and comprehend the manufacturer's instructions to properly perform an installation inspection.

In some cases, the code will specifically address an installation requirement that is also addressed in the manufacturer's installation instructions. The code requirements may be the same or may exceed the requirements in the manufacturer's installation instructions, or the manufacturer's installation instructions could contain requirements that exceed those in the code. In all such cases, the more restrictive requirements would apply.

In addition to the installation requirements for the appliance or equipment itself, this section also regulates the connections to appliances and equipment by requiring compliance with all other applicable sections of this code and other ICC codes. These connections include but are not limited to fuel and electrical supplies, control wiring, hydronic piping, chimneys, vents and ductwork. Some overlap or coincidence of connection requirements may occur in the code and the manufacturer's installation instructions. In the unlikely event that the code was less restrictive (more lenient) than the manufacturer's instructions regarding an installation issue, the manufacturer's instructions must prevail. Where differences or conflicts occur, the requirements that provide the greatest level of safety always apply.

Even if an installation appears to be in compliance with the manufacturer's instructions, the installation cannot be complete or approved until all associated components, connections, and systems that serve the appliance or equipment are also in compliance with the applicable provisions of the code. For example, a gas-fired boiler installation must not be approved if the boiler is connected to a deteriorated, undersized, improper or otherwise unsafe chimney or vent. Likewise, the same installation must not be approved if the existing gas piping is in poor condition or if the electrical supply circuit is inadequate or unsafe.

In the case of replacement installations, the intent of this section is to require all new work associated with the installation to comply with the code without necessarily requiring full compliance for the existing, unchanged portions of the related ductwork, piping, electrical, venting and similar mechanical systems. In the case of an appliance replacement, such as a furnace, it is the expectation of both the manufacturer's instructions and the code that the new furnace be connected to a code-complying venting system, fuel supply system and electrical circuit, all of which are part of the furnace installation. If the ductwork system was such that the furnace temperature rise across the heat exchanger exceeded the maximum allowable temperature rise, the ductwork system might have to be altered as part of the furnace installation. Existing mechanical systems are accepted on the basis that they are free from hazard although not necessarily compliant with current codes. The code is not retroactive except where specifically stated that it applies to existing systems.

Manufacturer's installation instructions are often updated and changed for various reasons, such as changes in the appliance or equipment design, revisions to the product standards and as a result of field experience related to existing installations. The code official should stay abreast of any changes by reviewing the manufacturer's instructions for every installation.

Equipment and appliances must be installed in accordance with the manufacturer's installation instructions. The manufacturer's label and installation instructions must be consulted in determining whether or not an appliance or piece of equipment can be installed and operated in a particular hazardous location. Without the manufacturer's installation instructions, an inspector cannot fully perform his or her job, which is why the manufacturer's installation instructions must be available on the job site when the equipment is being inspected.

305.2 Hazardous area. Equipment and appliances having an ignition source shall not be installed in Group H occupancies or control areas where open use, handling or dispensing of combustible, flammable or explosive materials occurs.

❖ The installation of appliances and equipment having an ignition source within Group H occupancies or control areas where combustible, flammable or explosive materials are used, handled or dispensed is prohibited regardless of compliance with Section 305.3. The presence of an ignition source is too great a risk in such occupancies.

305.3 Elevation of ignition source. Equipment and appliances having an ignition source shall be elevated such that the source of ignition is not less than 18 inches (457 mm) above the floor in hazardous locations and public garages, private garages, repair garages, motor fuel-dispensing facilities and parking garages. For the purpose of this section, rooms or spaces that are not part of the living space of a dwelling unit and that communicate directly with a private garage through openings shall be considered to be part of the private garage.

Exception: Elevation of the ignition source is not required for appliances that are listed as flammable vapor resistant and for installation without elevation.

❖ To reduce the possibility of ignition of flammable vapors in hazardous areas, the potential sources of ignition must be elevated above the surface supporting the equipment or appliance. It is not the intent of this section for the installer to measure the elevation from the surface of a dedicated appliance stand or other structure built to support the appliance where that stand or structure does not afford room for the storage of flammable liquids. Some flammable and combustible liquids typically associated with hazardous locations give off vapors that are denser than air and tend to collect near the floor. The 18-inch (457 mm) height requirement is intended to reduce the possibility of flammable vapor ignition by keeping the ignition sources elevated above the anticipated level of accumulated vapors. The 18-inch (457 mm) value is a minimum requirement and must be increased when required by the manufacturer's installation instructions. This section will effectively prohibit the installation of most furnaces, boilers, space heaters, clothes dryers and water heaters directly on the floor of residential garages.

The accumulation of flammable vapors more than 18 inches (457 mm) deep is unlikely in most ventilated locations; therefore, maintaining all possible sources of ignition at least 18 inches (457 mm) above the floor will substantially reduce the risk of explosion and fire [see commentary Figures 305.3(1) and 305.3(2)].

In the context of this section, a source of ignition could be a pilot flame, burner, burner igniter or electrical component capable of producing a spark. The term "ignition source" is defined and can be interpreted as meaning an intentional source of ignition, such as for a burner, or an unintentional source of ignition for any flammable vapors that may be present. See the definition of "Ignition source."

An appliance installed in a closet or room that is accessible only from the garage must be considered as part of the garage for application of this section. Even though the room may be separated from the garage by walls and a door, there is no practical means of making the door vapor-tight, nor is there any assurance that the door will remain closed during normal use. An appliance room that is accessed only from the outdoors or only from the living space would not be considered as part of the garage. Rooms such as utility rooms or laundry rooms that communicate with both the garage and the living space and that are considered as part of the living space are not part of the garage. See the definition of "Living space." If a room opens to the garage and that room is not living space, it is part of the garage for the purpose of this section.

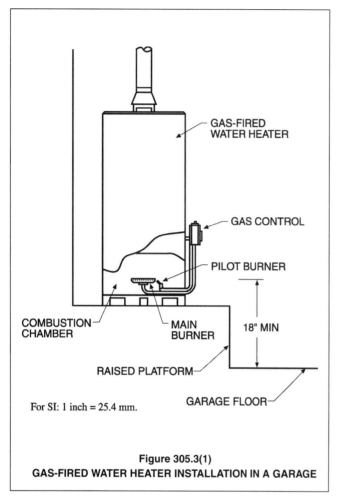

Figure 305.3(1)
GAS-FIRED WATER HEATER INSTALLATION IN A GARAGE

The exception recognizes that new technology exists that allows appliances, such as water heaters, to be tested and listed as being flammable-vapor-ignition resistant and suitable for installation without the 18-inch (457 mm) elevation requirement. This new technology was developed for conventional draft-hood-equipped water heaters, and now the standard for these appliances, ANSI Z21.10.1, mandates that by specified effective dates, all water heaters must comply with the flammable-vapor-ignition criteria in the standard. The technology involves variously configured flame arrestor screens having precisely engineered openings (slots). All combustion air for the appliance is passed through the flame arrestor screen. To pass through the small slots in the screen, the combustion air velocity is increased (Venturi principle) such that it exceeds the velocity at which flames travel through a flammable vapor. If the combustion air contains flammable vapors, the vapors will enter the appliance and be ignited within the combustion chamber. The flames will be contained (trapped) within the combustion chamber because the arrestor screen will not allow flames to exit the combustion chamber. The result is that vapor ignition is limited to the appliance interior and is not allowed to spread to the atmosphere surrounding the appliance. During such event, a thermal sensing device shuts down and locks out the appliance. Appliance manufacturers typically require the appliance to be replaced after a flammable-vapor-ignition event. The appliance standard includes testing requirements to determine whether the arrestor screen is subject to blockage by debris, including lint, plant fibers, dust particles, etc. If the arrestor screen openings become restricted by debris, the appliance will be deprived of combustion air and incomplete combustion and serious appliance malfunction can result. If an appliance shuts down as a result of the arrestor screen being blocked by debris, thus restricting the combustion air supply, a thermal device can be reset or replaced to restore operation of the appliance.

Figure 305.3(2)
SOURCES OF IGNITION ELEVATED IN PRIVATE GARAGE

If a water heater equipped with flammable-vapor-resistant technology states in the installation instructions or on the label that the appliance must be elevated so that the source of ignition is at least 18 inches (457 mm) above the floor, it must be installed as directed [see Figures 305.3(1) through 305.3(5)].

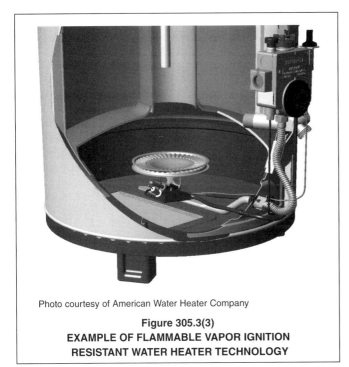

Photo courtesy of American Water Heater Company

Figure 305.3(3)
EXAMPLE OF FLAMMABLE VAPOR IGNITION RESISTANT WATER HEATER TECHNOLOGY

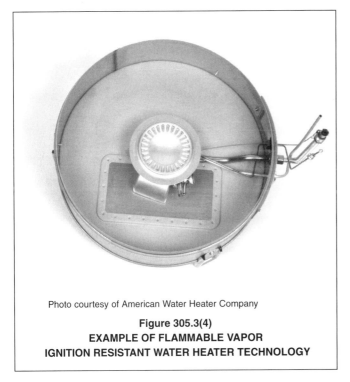

Photo courtesy of American Water Heater Company

Figure 305.3(4)
EXAMPLE OF FLAMMABLE VAPOR IGNITION RESISTANT WATER HEATER TECHNOLOGY

305.4 Public garages. Appliances located in public garages, motor fuel-dispensing facilities, repair garages or other areas frequented by motor vehicles shall be installed a minimum of 8 feet (2438 mm) above the floor. Where motor vehicles exceed 6 feet (1829 mm) in height and are capable of passing under an appliance, appliances shall be installed a minimum of 2 feet (610 mm) higher above the floor than the height of the tallest vehicle.

Exception: The requirements of this section shall not apply where the appliances are protected from motor vehicle impact and installed in accordance with Section 305.3 and NFPA 88B.

❖ Appliances located in motor vehicle areas must be installed to provide protection from vehicle impact. Protection is necessary because the resulting damage from vehicle impact could cause an explosion or fire.

The requirement that appliances be located a minimum of 8 feet (2438 mm) above the floor is intended to provide enough clearance for motor vehicles to pass under the appliance without impact. Although 8 feet (2438 mm) is the minimum height requirement, a clearance of at least 2 feet (610 mm) must always be maintained between the top of the highest vehicle and the bottom of any appliance that is installed where vehicles can pass underneath. If the vehicles that could pass under an appliance are 6 feet (1829 mm) or less in height, the height requirement of 8 feet (2438 mm) will result in a clearance of at least 2 feet (610 mm). If the vehicles that could pass under the appliance are greater than 6 feet (1829 mm) in height, the minimum installation height requirement must be increased to maintain a clearance of not less than 2 feet (610 mm). The height of the largest door or opening that the vehicles must pass through to gain access to the space could be used as a guide to determine the maximum vehicle height in some applications; however, the door height may not be the sole consideration in buildings such as warehouses, where forklifts or similar elevating vehicles are used. An exception to these requirements occurs when the appliance is installed in accordance with Section 305.3 and protected in accordance with NFPA 88B.

As a means of protecting appliances from motor vehicle impact, the code regulates elevation above the floor as an effective means of protection. Except for suspended unit heaters, the elevation of appliances and equipment above motor vehicles is often impractical, difficult to accomplish, and creates access problems. This section allows for an alternative means of protecting appliances located in public garages from vehicle impact.

Although the code does not specify alternative methods of protection, the most apparent method would be to locate the appliance where it could not reasonably be struck by a vehicle. A practical method of protection would be to place a formidable and permanent barrier between the motor vehicles and the appliance. This barrier could include an effectively located vehicle wheel stop that is anchored in place, an elevated platform higher than the vehicle's bumpers, or one or more concrete-filled steel pipes.

Where appliances are elevated in accordance with this section, compliance with Section 305.2 is automatically accomplished; however, alternative means of protection may or may not involve elevation of the unit. This section states, in effect, that regardless of the method of protection from vehicular impact, the appliances must be elevated in accordance with this section or elevated in accordance with Section 305.3. If the installation is in a public repair garage and elevated in accordance with

Section 305.3, the installation must also comply with NFPA 88B. NFPA 88B applies to motorized vehicle repair garages and introduces additional criteria and stipulations for nonresidential structures. NFPA 88B requires continuous mechanical ventilation of 0.75 cubic foot per minute (cfm) per square foot of floor area, and prohibits any operations involving the dispensing or transferring of Class I or II flammable and combustible liquids or liquefied petroleum gas.

In summary, appliances and equipment installed in private and public garages must at minimum comply with Section 305.3 and must be protected from vehicular impact. The area in which they are installed in public garages must also comply with NFPA 88B.

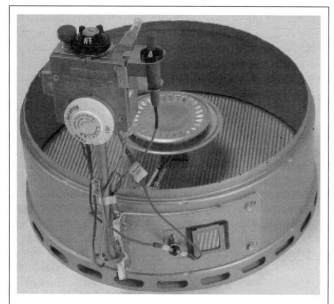

Photo courtesy of Bradford White Corporation
(Bradford White Defender Safety System™)

Figure 305.3(5)
EXAMPLE OF FLAMMABLE VAPOR IGNITION RESISTANT WATER HEATER TECHNOLOGY

305.5 Private garages. Appliances located in private garages shall be installed with a minimum clearance of 6 feet (1829 mm) above the floor.

Exception: The requirements of this section shall not apply where the appliances are protected from motor vehicle impact and installed in accordance with Section 305.3.

❖ Appliances located in a private garage or carport must be protected from vehicle impact. This section is applicable to appliances located in an area where motor vehicles can be operated and includes appliances under which a vehicle can pass and those located anywhere in a vehicle's path where impact is possible. The 6-foot (1829 mm) minimum height requirement is intended to provide adequate clearance above the typical automobile. With the popularity of conversion vans and recreational vehicles, which can be much higher than other automobiles, the 6-foot (1829 mm) minimum installation height above the floor may not provide adequate clearance; additional height might be necessary. The garage door height can be used as a guide in determining the maximum vehicle height [see commentary, Section 305.4 and Figures 305.5(1) through 305.5(5)].

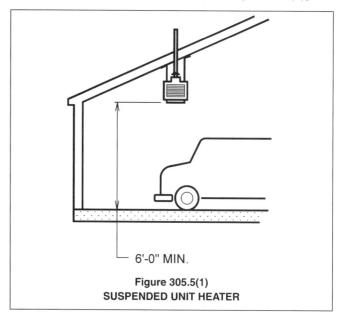

Figure 305.5(1)
SUSPENDED UNIT HEATER

305.6 Construction and protection. Boiler rooms and furnace rooms shall be protected as required by the *International Building Code*.

❖ The terms "boiler room" and "furnace room" are defined in the IMC. Section 302.1.1 of the IBC addresses boiler and furnace rooms and will require that they be treated as either incidental use areas in Table 302.1.1 or as an occupancy group in a mixed occupancy building (see Section 302.1.1 of the IBC). In many cases, the IBC will require furnace and boiler rooms to have 1-hour fire-resistance-rated enclosures or automatic fire-extinguishing systems. Note that the gross output of a boiler in Btu/h divided by 33,475 yields boiler horsepower rating. For example, a boiler having an input rating of 425,000 Btu/h and 80 percent thermal efficiency would have a gross output of 340,000 Btu/h, which is 10.2 hp.

305.7 Clearances from grade. Equipment and appliances installed at grade level shall be supported on a level concrete slab or other approved material extending above adjoining grade or shall be suspended a minimum of 6 inches (152 mm) above adjoining grade.

❖ This section addresses outdoor and crawl space installations where an appliance is resting on the earth and not supported by a building. The intent is to maintain level support and protect the appliance from corrosion and deterioration. The slab should be concrete or a material having comparable strength and longevity. Outdoor appliance installations must be designed to tolerate movement where frost heave will cause a slab to rise and fall.

GENERAL REGULATIONS FIGURE 305.5(2) - 306.2

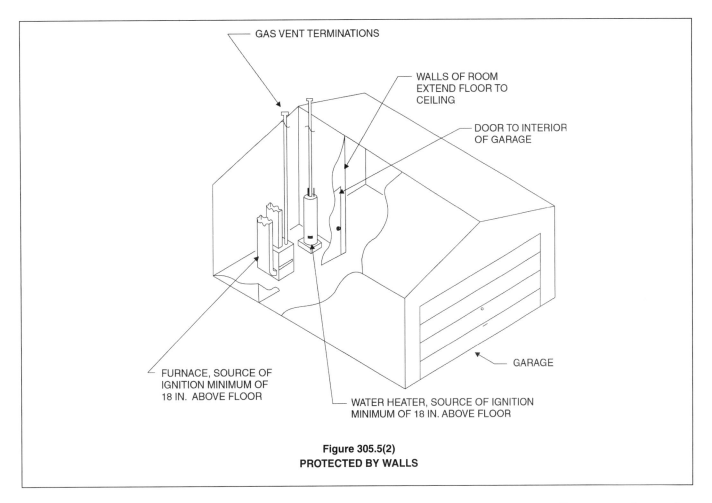

**Figure 305.5(2)
PROTECTED BY WALLS**

305.8 Clearances to combustible construction. Heat-producing equipment and appliances shall be installed to maintain the required clearances to combustible construction as specified in the listing and manufacturer's instructions. Such clearances shall be reduced only in accordance with Section 308. Clearances to combustibles shall include such considerations as door swing, drawer pull, overhead projections or shelving and window swing. Devices, such as door stops or limits and closers, shall not be used to provide the required clearances.

❖ Section 308 allows clearances to be reduced where protective assemblies are installed to decrease the transfer of heat from the source to the combustible material/assembly. Clearances must be measured from shelving, overhead projections and the possible positions of doors, windows and drawers. Doorstops and closers are easily defeated and subject to failure.

SECTION 306 (IFGC)
ACCESS AND SERVICE SPACE

[M] 306.1 Clearances for maintenance and replacement. Clearances around appliances to elements of permanent construction, including other installed appliances, shall be sufficient to allow inspection, service, repair or replacement without removing such elements of permanent construction or disabling the function of a required fire-resistance-rated assembly.

❖ Because equipment and appliances require routine maintenance, repairs and possible replacement, access is required. Additionally, access recommendations or requirements are usually stated in the manufacturer's installation instructions. As a result, the provisions stated here are intended to supplement the manufacturer's installation instructions.

The intent of this section is to assure access to components that require observation, inspection, adjustment, servicing, repair or replacement. Access is also necessary to conduct operating procedures such as startup or shutdown.

The code defines access as being able to be reached but first may require the removal of a panel, door or similar obstruction. An appliance or piece of equipment is not accessible if any portion of the structure's permanent finish materials such as drywall, plaster, paneling, built-in furniture, cabinets or any other similar permanently affixed building component must be removed.

The intent is to assure access to components such as controls, gauges, burners, filters, blowers and motors that require observation, inspection, adjustment, servicing, repair and replacement.

[M] 306.2 Appliances in rooms. Rooms containing appliances requiring access shall be provided with a door and an unobstructed passageway measuring not less than 36 inches (914 mm) wide and 80 inches (2032 mm) high.

Exception: Within a dwelling unit, appliances installed in a compartment, alcove, basement or similar space shall be provided with access by an opening or door and an unobstructed passageway measuring not less than 24 inches (610 mm) wide and large enough to allow removal of the largest appliance in the space, provided that a level service space of not less than 30 inches (762 mm) deep and the height of the appliance, but not less than 30 inches (762 mm), is present at the front or service side of the appliance with the door open.

❖ Access opening and passageway dimensions are specified to give service personnel reasonable access to appliances and to allow for the passage of system components. Quite often appliances such as furnaces, boilers and water heaters are installed in spaces with little or no forethought about future access for maintenance or replacement.

[M] 306.3 Appliances in attics. Attics containing appliances requiring access shall be provided with an opening and unobstructed passageway large enough to allow removal of the largest component of the appliance. The passageway shall not be less than 30 inches (762 mm) high and 22 inches (559 mm) wide and not more than 20 feet (6096 mm) in length when measured along the centerline of the passageway from the opening to the equipment. The passageway shall have continuous solid flooring not less than 24 inches (610 mm) wide. A level service space not less than 30 inches (762 mm) deep and 30 inches (762 mm) wide shall be present at the front or service side of the equipment. The clear access opening dimensions shall be a minimum of 20 inches by 30 inches (508 mm by 762 mm), where such dimensions are large enough to allow removal of the largest component of the appliance.

Exceptions:

1. The passageway and level service space are not required where the appliance is capable of being serviced and removed through the required opening.
2. Where the passageway is not less than 6 feet (1829 mm) high for its entire length, the passageway shall be not greater than 50 feet (15 250 mm) in length.

❖ There is not always sufficient room for equipment and appliances to be installed in spaces such as basements, alcoves, utility rooms and furnace rooms. In an effort to save floor space or simplify an installation, appliances and equipment are often installed in attics or in similar remote locations. Access to appliances and equipment could be difficult because of the lack of a walking surface, such as might occur in an attic or similar space with exposed ceiling joists. The intent of this section is to require a suitable access opening, passageway and work space that will allow reasonably

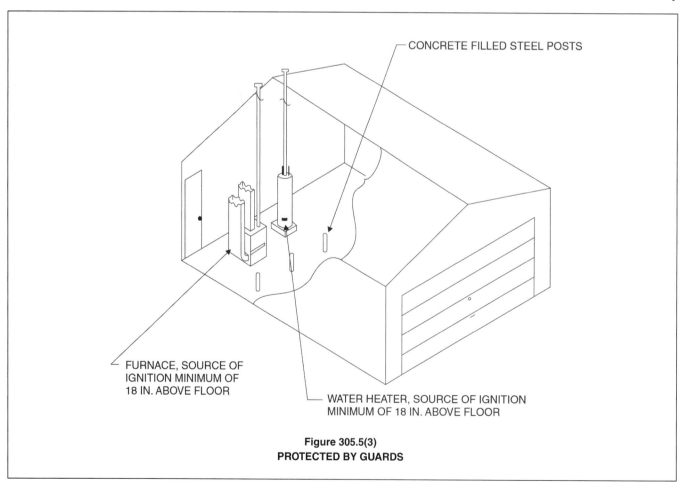

Figure 305.5(3)
PROTECTED BY GUARDS

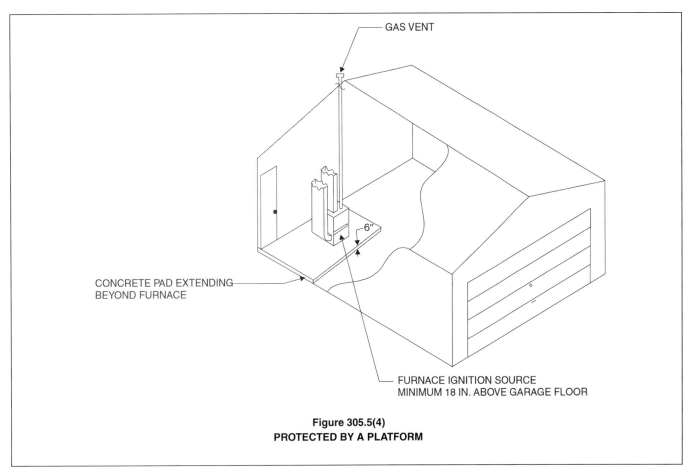

Figure 305.5(4)
PROTECTED BY A PLATFORM

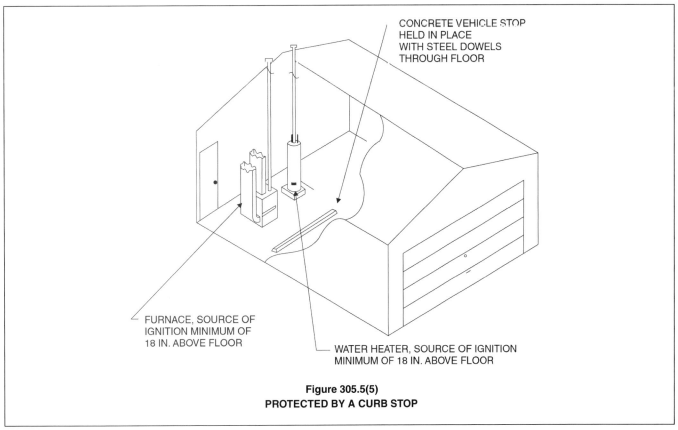

Figure 305.5(5)
PROTECTED BY A CURB STOP

easy access without endangering the service person. The longer the attic passageway, the more the service person will be exposed to extreme temperatures and risk of injury. Note that some appliances might not be listed for attic installation or might otherwise be unsuitable for such conditions (see commentary Figure 306.3).

Attic installations are out-of-sight and out-of-mind, therefore, providing safe and easy access is important to encourage service and occasional inspection of the appliances.

Exception 2 recognizes that a 6-foot minimum height passageway will allow quicker, less impeded travel, thus reducing risk to the personnel who must use the passageway.

[M] 306.3.1 Electrical requirements. A lighting fixture controlled by a switch located at the required passageway opening and a receptacle outlet shall be provided at or near the equipment location in accordance with the ICC *Electrical Code*.

❖ A lighting outlet and receptacle outlet will encourage and facilitate appliance maintenance. The receptacle will accommodate power tools, drop lights and diagnostic instruments. Also, these provisions will negate the need for extension cords, which can be hazardous to service personnel.

[M] 306.4 Appliances under floors. Under-floor spaces containing appliances requiring access shall be provided with an access opening and unobstructed passageway large enough to remove the largest component of the appliance. The passageway shall not be less than 30 inches (762 mm) high and 22 inches (559 mm) wide, nor more than 20 feet (6096 mm) in length when measured along the centerline of the passageway from the opening to the equipment. A level service space not less than 30 inches (762 mm) deep and 30 inches (762 mm) wide shall be present at the front or service side of the appliance. If the depth of the passageway or the service space exceeds 12 inches (305 mm) below the adjoining grade, the walls of the passageway shall be lined with concrete or masonry extending 4 inches (102 mm) above the adjoining grade and having sufficient lat-

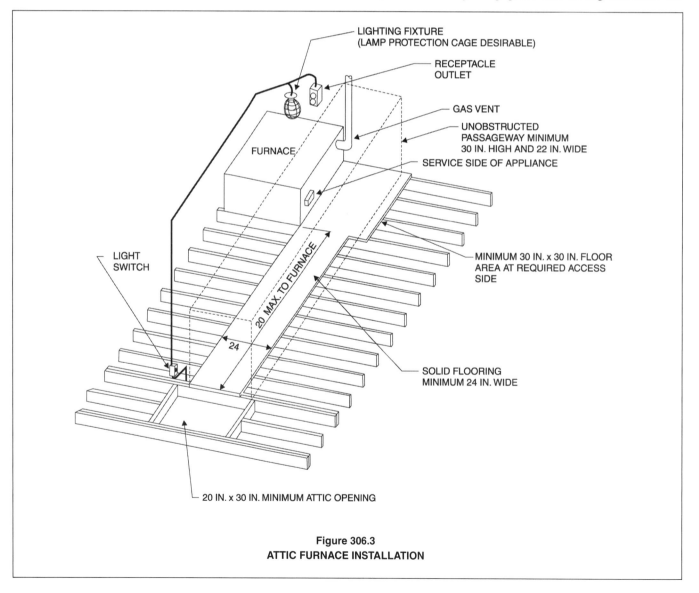

Figure 306.3
ATTIC FURNACE INSTALLATION

eral-bearing capacity to resist collapse. The clear access opening dimensions shall be a minimum of 22 inches by 30 inches (559 mm by 762 mm), where such dimensions are large enough to allow removal of the largest component of the appliance.

Exceptions:

1. The passageway is not required where the level service space is present when the access is open and the appliance is capable of being serviced and removed through the required opening.

2. Where the passageway is not less than 6 feet high (1829 mm) for its entire length, the passageway shall not be limited in length.

❖ The intent of this section, which applies to crawl spaces, is the same as for Section 306.3. The more difficult access to appliances and equipment is, the less likely it is that the appliance or equipment will be inspected and serviced. Attic and crawl space installations suffer from the "out-of-sight, out-of-mind" syndrome. See the commentary for Section 306.3. Passageways 6 feet or more in height would allow a person to walk erect or nearly so, thus allowing quicker egress in the event of an emergency (see commentary Figure 306.4).

[M] 306.4.1 Electrical requirements. A lighting fixture controlled by a switch located at the required passageway opening and a receptacle outlet shall be provided at or near the equipment location in accordance with the ICC *Electrical Code*.

❖ Appliances located in attics and under-floor spaces are generally not easy to access. The light switch, lighting fixture and receptacle outlet required by this section protect those who service the equipment and appliances. The receptacle will accommodate power tools, drop lights and diagnostic instruments. Also, these provisions will negate the need for extension cords, which can be hazardous to service personnel. The ICC EC requires ground fault circuit interrupter protection for receptacles located in a crawl space.

[M] 306.5 Appliances on roofs or elevated structures. Where appliances requiring access are installed on roofs or elevated structures at a height exceeding 16 feet (4877 mm), such access shall be provided by a permanent approved means of access, the extent of which shall be from grade or floor level to the appliance's level service space. Such access shall not require climbing over obstructions greater than 30 inches high (762 mm) or walking on roofs having a slope greater than four units vertical in 12 units horizontal (33-percent slope).

Permanent ladders installed to provide the required access shall comply with the following minimum design criteria.

1. The side railing shall extend above the parapet or roof edge not less than 30 inches (762 mm).

2. Ladders shall have a rung spacing not to exceed 14 inches (356 mm) on center.

3. Ladders shall have a toe spacing not less than 6 inches (152 mm) deep.

4. There shall be a minimum of 18 inches (457 mm) between rails.

5. Rungs shall have a minimum diameter of 0.75-inch (19 mm) and shall be capable of withstanding a 300-pound (136.1 kg) load.

6. Ladders over 30 feet (9144 mm) in height shall be provided with offset sections and landings capable of withstanding a load of 100 pounds per square foot (488.2 kg/m^2).

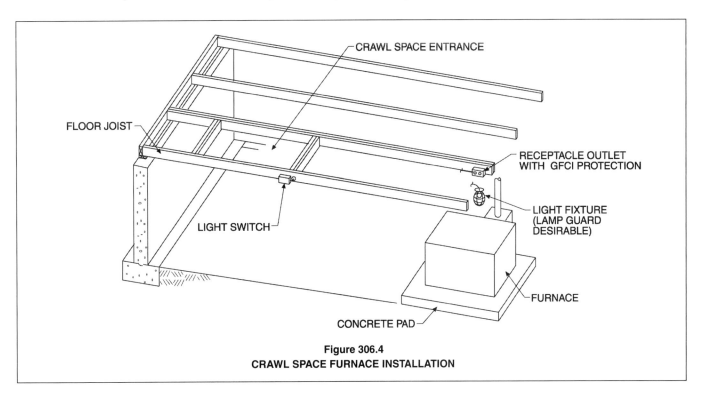

**Figure 306.4
CRAWL SPACE FURNACE INSTALLATION**

7. Ladders shall be protected against corrosion by approved means.

Catwalks installed to provide the required access shall be not less than 24 inches wide (610 mm) and shall have railings as required for service platforms.

Exception: This section shall not apply to Group R-3 occupancies.

❖ The requirements of this section are intended to prevent portable ladders from being the only means of access to elevated appliances and equipment located above the specified height. These provisions are for the protection of service and inspection personnel. The criteria for constructing and installing permanent ladders and catwalks are intended to make them safe to use.

[M] 306.5.1 Sloped roofs. Where appliances are installed on a roof having a slope of three units vertical in 12 units horizontal (25-percent slope) or greater and having an edge more than 30 inches (762 mm) above grade at such edge, a level platform shall be provided on each side of the appliance to which access is required by the manufacturer's installation instructions for service, repair or maintenance. The platform shall not be less than 30 inches (762 mm) in any dimension and shall be provided with guards in accordance with Section 306.6.

❖ A work space platform with guards will provide protection for service personnel and will facilitate inspection, servicing, and repair of appliances and equipment. (see commentary Figure 306.5.1). Working on a sloped surface is difficult and dangerous, and results in tools and materials sliding off roofs, possibly endangering persons below. Because of the additional expense and unattractive appearance of roof platforms, this section may have the effect of discouraging the installation of appliances on sloped roofs.

[M] 306.5.2 Electrical requirements. A receptacle outlet shall be provided at or near the equipment location in accordance with the ICC *Electrical Code*.

❖ A lighting outlet is not required for servicing roof mounted equipment. A receptacle outlet will encourage and facilitate appliance maintenance. The receptacle will accommodate power tools, drop lights, and diagnostic instruments. Also, these provisions will negate the need for extension cords, which can be hazardous to service personnel. The ICC EC requires ground fault circuit interrupter protection for rooftop receptacles.

[M] 306.6 Guards. Guards shall be provided where appliances, fans or other components that require service are located within 10 feet (3048 mm) of a roof edge or open side of a walking surface and such edge or open side is located more than 30 inches (762 mm) above the floor, roof or grade below. The guard shall extend not less than 30 inches (762 mm) beyond each end of

Photo courtesy of Reznor/Thomas & Betts Corporation

**Figure 306.5.1
ROOFTOP HVAC UNIT WITH WORK PLATFORM**

GENERAL REGULATIONS

such appliances, fans or other components and the top of the guard shall be located not less than 42 inches (1067 mm) above the elevated surface adjacent to the guard. The guard shall be constructed so as to prevent the passage of a 21-inch-diameter (533 mm) sphere and shall comply with the loading requirements for guards specified in the *International Building Code*.

❖ Equipment and appliances require routine inspection, maintenance and repair. Where the units requiring service are located within 10 feet (3048 mm) of a roof edge or open side with a drop greater than 30 inches (762 mm), a possibility exists for the service person to be injured by a fall. The requirements of this section are intended to protect service personnel from the hazard of falls. This section pertains to equipment and appliances regulated by the code that require routine inspection, maintenance and repair.

The guard serves as a warning and as a protective barrier and is required to be 42 inches (1067 mm) high and constructed to prevent the passage of a 21-inch (533 mm) sphere (see commentary Figure 306.6). The guard must extend 30 inches beyond each end of the equipment or appliance, to provide added protection for the installer or service technician by enlarging the protected area beyond the immediate area of the equipment or appliance. The requirement for the extension beyond the ends of the appliance has always been a logical interpretation of the intent of this section, though not specifically stated. The guardrail must be constructed to resist the imposed loading conditions.

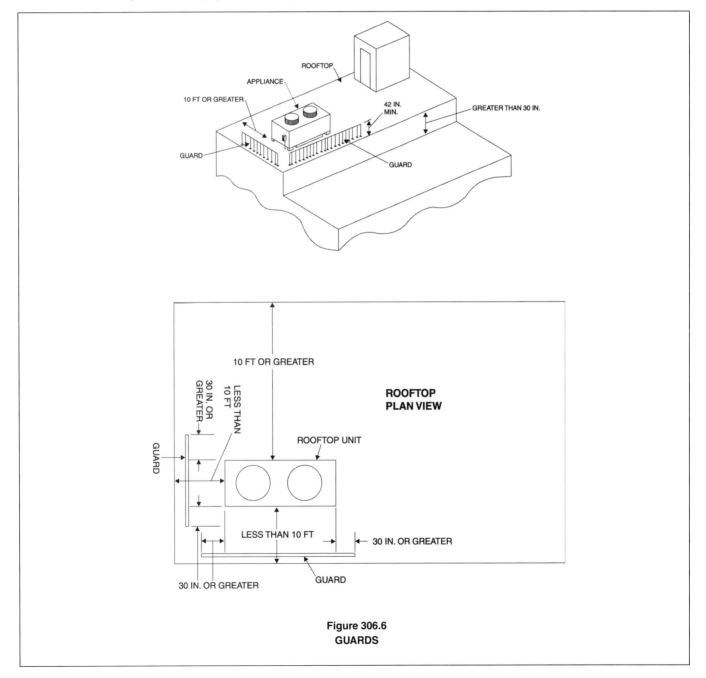

Figure 306.6
GUARDS

SECTION 307 (IFGC)
CONDENSATE DISPOSAL

307.1 Fuel-burning appliances. Liquid combustion by-products of condensing appliances shall be collected and discharged to an approved plumbing fixture or disposal area in accordance with the manufacturer's installation instructions. Condensate piping shall be of approved corrosion-resistant material and shall not be smaller than the drain connection on the appliance. Such piping shall maintain a minimum slope in the direction of discharge of not less than one-eighth unit vertical in 12 units horizontal (1-percent slope).

❖ This section contains detailed requirements for the disposal of condensate from appliances and equipment, such as evaporators, cooling coils and condensing fuel-fired appliances. Condensation must be collected from appliances and equipment in accordance with the manufacturers' installation instructions and this section. High-efficiency condensing-type appliances and some mid-efficiency appliances produce water as a combustion byproduct. This condensate is collected at various points in the appliance heat exchangers and venting system and must be disposed of. Because of impurities in the fuel and combustion air, the condensate can be acidic and thus corrosive to many materials. For example, such condensate can contain hydrochloric and sulfuric acids, although the acid solution would be weak. The slope requirement promotes drainage and prohibits low points (dips) in the piping that would trap water and air that would impede flow.

[M] 307.2 Drain pipe materials and sizes. Components of the condensate disposal system shall be cast iron, galvanized steel, copper, polybutylene, polyethylene, ABS, CPVC or PVC pipe or tubing. All components shall be selected for the pressure and temperature rating of the installation. Condensate waste and drain line size shall be not less than $^3/_4$-inch internal diameter (19 mm) and shall not decrease in size from the drain connection to the place of condensate disposal. Where the drain pipes from more than one unit are manifolded together for condensate drainage, the pipe or tubing shall be sized in accordance with an approved method. All horizontal sections of drain piping shall be installed in uniform alignment at a uniform slope.

❖ Condensate drains must be constructed of a material listed in this section. The materials listed are corrosion resistant. These drains must be straight runs of pipe without sags or dips that would trap liquid. If coiled stock tubing is used, it would be difficult to maintain a uniform slope, and the low points (dips) in the tubing would trap water, causing the drain to be double trapped (air bound). Air-bound drains will not allow flow. When drains are merged, the piping must be properly sized for the aggregate flow.

307.3 Traps. Condensate drains shall be trapped as required by the equipment or appliance manufacturer.

❖ The appliance or equipment manufacturer determines the need for a trap and often specifies the depth and configuration of the trap. The traps addressed in this section are unrelated to plumbing traps and serve a different purpose. Condensate drain traps do not directly connect in any way to the plumbing drain or the waste and vent system of a building. Condensate drain traps are installed to prevent air and combustion gases from being pushed or pulled through the drain piping. Such airflow can impede condensate flow, causing overflow or abnormal water depth in drain pans and components.

SECTION 308 (IFGS)
CLEARANCE REDUCTION

308.1 Scope. This section shall govern the reduction in required clearances to combustible materials and combustible assemblies for chimneys, vents, appliances, devices and equipment. Clearance requirements for air-conditioning equipment and central heating boilers and furnaces shall comply with Sections 308.3 and 308.4.

❖ Heat-producing appliances and mechanical equipment must be installed with the required minimum clearances to combustible materials indicated by their listing label. It is not uncommon to encounter practical or structural difficulties in maintaining clearances. Therefore, clearance reduction methods have been developed to allow, in some cases, reduction of the minimum prescribed clearance distance while achieving equivalent protection. An important understanding is that all prescribed clearances to combustibles are airspace clearances measured from the heat source to the face of the nearest combustible surface (see commentary Figure 308.1). When using the assemblies described in Table 308.2, the clearance is measured as described in Note "b." In the case of listed equipment, the required clearances are intended to be clear airspace, and therefore the space is not to be filled with insulation or any other material other than an assembly intended to allow clearance reduction. This is especially important where clearances are required from appliances and equipment that rely on the airspace for convection cooling to maintain their proper operating temperature.

The provisions contained in this section are based on the principles of heat transfer. Mechanical equipment or appliances producing heat can become hot, and many appliances have hot exterior surfaces by design. The heat energy is then radiated to objects surrounding the appliances or equipment. When mechanical equipment and appliances are tested, the minimum clearances are established so that radiant and, to a lesser extent, convective heat transfer do not represent an ignition hazard to adjacent surfaces and objects. This distance is called the "required clearance" to combustible materials. Appliance and equipment labels must specify minimum clearances in all directions.

This section permits the use of materials and systems to act as radiation shields, decreasing the amount of heat energy transferred to surrounding objects and reducing the required clearances between gas-fired appliances and equipment and combustible material to combustibles and does not change the fact that the adjacent surface is combustible.

Plaster and gypsum by themselves are classified as noncombustible materials. Under continued exposure to heat, however, these materials will gradually decompose as water molecules are driven out of the material. Plaster on wood lath, plasterboard, sheetrock and drywall are considered combustible material when the code provisions are applied.

Additionally, gypsum wallboard has a paper face that has a flame spread index that is measurable in the ASTM E 84 test. This alone identifies the need to classify gypsum wallboard as a combustible material for the purpose of requiring a separation from heat-producing equipment and appliances.

The code specifically prohibits the reduction of clearances in certain applications. For examples, see Sections 308.2, 308.3 and 308.4. The code does not contain a definition of "noncombustible materials"; however, the IMC does, and it is applicable to this code (see Section 201.3 in the IFGC). This definition follows.

NONCOMBUSTIBLE MATERIALS. Materials that, when tested in accordance with ASTM E 136, have at least three of four specimens tested meeting all of the following criteria:

1. The recorded temperature of the surface and interior thermocouples shall not at any time during the test rise more than 54°F (30°C) above the furnace temperature at the beginning of the test.

2. There shall not be flaming from the specimen after the first 30 seconds.

3. If the weight loss of the specimen during testing exceeds 50 percent, the recorded temperature of the surface and interior thermocouples shall not at any time during the test rise above the furnace air temperature at the beginning of the test, and there shall not be flaming of the specimen.

The definition of "noncombustible" in the IMC differs from the definition of "noncombustible" in the IBC. The difference is necessary because the IMC is concerned with the exposure to a continuous heat source, while the building code emphasizes the behavior of the material under actual fire conditions. An example of this difference is how gypsum wallboard is categorized. In the IMC, gypsum wallboard is considered to be a combustible material; the building code classifies it as a noncombustible material. Applying gypsum wallboard to a wood frame wall does not reduce the required clearance, nor does the type of gypsum board make any difference (that is, Class X, fire code or standard).

308.2 Reduction table. The allowable clearance reduction shall be based on one of the methods specified in Table 308.2 or shall utilize an assembly listed for such application. Where required clearances are not listed in Table 308.2, the reduced clearances shall be determined by linear interpolation between the distances listed in the table. Reduced clearances shall not be derived by extrapolation below the range of the table. The reduction of the required clearances to combustibles for listed and labeled appliances and equipment shall be in accordance with the requirements of this section except that such clearances shall not be reduced where reduction is specifically prohibited by the terms of the appliance or equipment listing [see Figures 308.2(1) through 308.2(3)].

❖ Another option for reducing the required clearances to combustible materials is to use one of the on-site field-constructed methods specified in Table 308.2. Note that the table intends to refer to 24 gage (0.024-inch thick) sheet metal.

308.3 Clearances for indoor air-conditioning equipment. Clearance requirements for indoor air-conditioning equipment shall comply with Sections 308.3.1 through 308.3.5.

❖ Requirements for clearances to combustibles are emphasized because of the potential fire hazard posed when those clearances are not observed. Maintaining an appropriate distance from the outer surfaces of an appliance or piece of equipment to combustible materials reduces the possibility of ignition of combustible materials.

The minimum clearances to combustibles are specified in the manufacturer's installation instructions for a labeled appliance. Because an approved agency tests appliances in accordance with these instructions, the clearances required are necessary for correct installation and operation of the appliance.

Reduction of the required clearances to combustibles is allowed only when the combustibles are protected by one of the methods outlined in Section 308 and clearance reduction is not prohibited by Section 308 or the appliance and equipment listing. See the commentary for Section 308. The manufacturer's specified minimum clearances and the clearances specified in Section 308 are all airspace clearances, and such spaces cannot be filled with insulation or any other material, even if the material is noncombustible. In some cases, the manufacturer's installation instructions will specify absolute minimum clearances that must not be reduced by any clearance reduction method.

The most common wall covering material–gypsum wallboard–is a combustible finish material for the purpose of the code. As a result, gypsum wallboard as well as all other combustible wall finishes must be separated from an appliance or equipment in accordance with the prescribed clearance to combustibles. Clearances to combustibles also apply to furnishings, window treatments and moveable items that can be placed within the required clearance range of appliances and equipment.

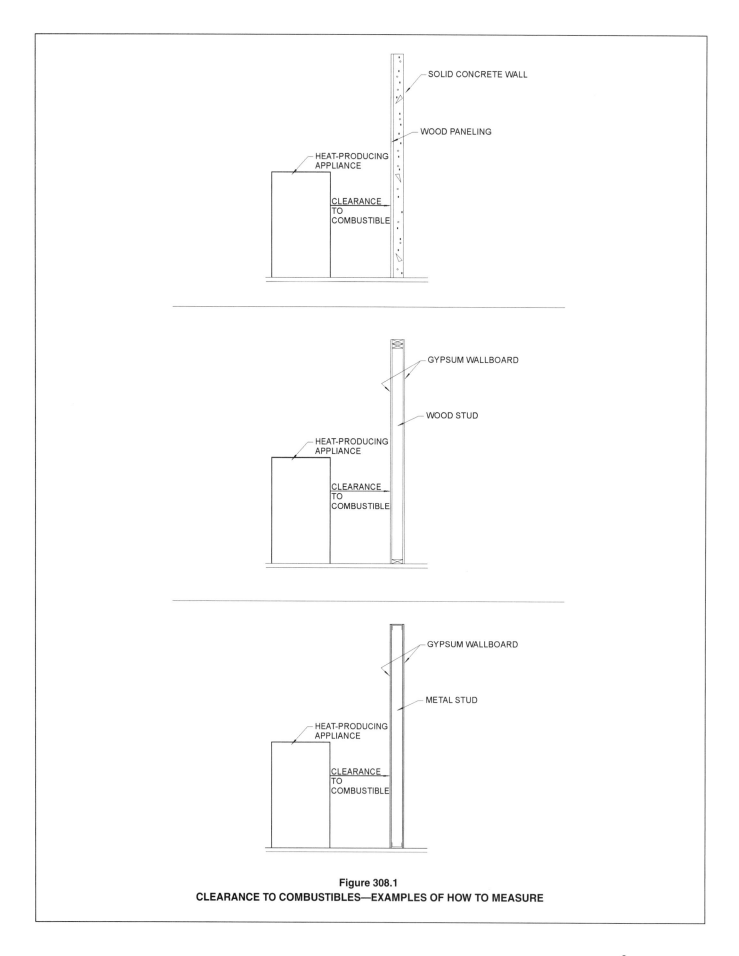

Figure 308.1
CLEARANCE TO COMBUSTIBLES—EXAMPLES OF HOW TO MEASURE

TABLE 308.2[a-k]
REDUCTION OF CLEARANCES WITH SPECIFIED FORMS OF PROTECTION

TYPE OF PROTECTION APPLIED TO AND COVERING ALL SURFACES OF COMBUSTIBLE MATERIAL WITHIN THE DISTANCE SPECIFIED AS THE REQUIRED CLEARANCE WITH NO PROTECTION [see Figures 308.2(1), 308.2(2), and 308.2(3)]	WHERE THE REQUIRED CLEARANCE WITH NO PROTECTION FROM APPLIANCE, VENT CONNECTOR, OR SINGLE-WALL METAL PIPE IS: (inches)									
	36		18		12		9		6	
	Allowable clearances with specified protection (inches)									
	Use Column 1 for clearances above appliance or horizontal connector. Use Column 2 for clearances from appliance, vertical connector, and single-wall metal pipe.									
	Above Col. 1	Sides and rear Col. 2	Above Col. 1	Sides and rear Col. 2	Above Col. 1	Sides and rear Col. 2	Above Col. 1	Sides and rear Col. 2	Above Col. 1	Sides and rear Col. 2
1. $3^1/_2$-inch-thick masonry wall without ventilated airspace	—	24	—	12	—	9	—	6	—	5
2. $^1/_2$-inch insulation board over 1-inch glass fiber or mineral wool batts	24	18	12	9	9	6	6	5	4	3
3. 0.024 inch (24 gage) sheet metal over 1-inch glass fiber or mineral wool batts reinforced with wire on rear face with ventilated airspace	18	12	9	6	6	4	5	3	3	3
4. $3^1/_2$-inch-thick masonry wall with ventilated airspace	—	12	—	6	—	6	—	6	—	6
5. 0.024 inch (24 gage) sheet metal with ventilated airspace	18	12	9	6	6	4	5	3	3	2
6. $^1/_2$-inch-thick insulation board with ventilated airspace	18	12	9	6	6	4	5	3	3	3
7. 0.024 inch (24 gage) sheet metal with ventilated air-space over 0.024 inch (24 gage) sheet metal with ventilated airspace	18	12	9	6	6	4	5	3	3	3
8. 1-inch glass fiber or mineral wool batts sandwiched between two sheets 0.024 inch (24 gage) sheet metal with ventilated airspace	18	12	9	6	6	4	5	3	3	3

For SI: 1 inch = 25.4 mm, °C = [(°F - 32)/1.8], 1 pound per cubic foot = 16.02 kg/m³, 1 Btu per inch per square foot per hour per °F = 0.144 W/m² · K.

a. Reduction of clearances from combustible materials shall not interfere with combustion air, draft hood clearance and relief, and accessibility of servicing.
b. All clearances shall be measured from the outer surface of the combustible material to the nearest point on the surface of the appliance, disregarding any intervening protection applied to the combustible material.
c. Spacers and ties shall be of noncombustible material. No spacer or tie shall be used directly opposite an appliance or connector.
d. For all clearance reduction systems using a ventilated airspace, adequate provision for air circulation shall be provided as described [see Figures 308.2(2) and 308.2(3)].
e. There shall be at least 1 inch between clearance reduction systems and combustible walls and ceilings for reduction systems using ventilated airspace.
f. Where a wall protector is mounted on a single flat wall away from corners, it shall have a minimum 1-inch air gap. To provide air circulation, the bottom and top edges, or only the side and top edges, or all edges shall be left open.
g. Mineral wool batts (blanket or board) shall have a minimum density of 8 pounds per cubic foot and a minimum melting point of 1500°F.
h. Insulation material used as part of a clearance reduction system shall have a thermal conductivity of 1.0 Btu per inch per square foot per hour per °F or less.
i. There shall be at least 1 inch between the appliance and the protector. In no case shall the clearance between the appliance and the combustible surface be reduced below that allowed in this table.
j. All clearances and thicknesses are minimum; larger clearances and thicknesses are acceptable.
k. Listed single-wall connectors shall be installed in accordance with the terms of their listing and the manufacturer's instructions.

❖ The column headings of Table 308.2 list required clearances without protection. The numbers to the right of each method indicate the permissible reduced clearance measured from the heat-producing appliances/equipment to the face of the combustible material.

The rationale behind the methods of protection listed in Table 308.2 is based on the ability of the protection to reduce radiant heat transmission from the appliance and equipment to the combustible material so that the temperature rise of the combustible material will remain below the maximum allowed.

Although the materials referred to in Table 308.2 are common construction materials, confusion often arises over what satisfies the requirement for "insulation board" (Item 6 in the table), sometimes referred to an inorganic insulating board, noncombustible mineral board or noncombustible insulating board. These products are not made of carbon-based compounds.

Carbon-based compounds are those found in cellulose (wood), plastics and other materials manufactured from raw materials that once existed as living organisms. Cement board materials must have a specified maximum "C" (conductance) value in addition to being noncombustible.

Note "h" specifies a maximum Thermal Conductivity of 1.0 Btu /inch × hour × square feet × °F. Conductivity is the amount of heat in Btu's that will flow each hour through a 1-foot-square slab of material, 1 inch thick with a 1°F temperature difference between both sides and is usually identified by the symbol k. Tables of k values usually do not include the area term in the dimensions for conductivity, and it must be understood that the value must be multiplied by the area to obtain the total Btu value.

Thermal Conductance (overall) is the time rate of heat flow through a body not taking thickness into account and is usually identified by the symbol C (Btu/h × square feet × °F).

Thermal Resistance (overall) is the reciprocal of overall thermal conductance and is usually identified by the symbol R (hour × square feet × °F/Btu).

This translates into a minimum required insulation R-value of 1.0 (square feet × hour × °F)/Btu per inch of insulating material. The methods in Table 308.2 control heat transmission by reflecting heat radiation, retarding thermal conductance and providing convective cooling. Where sheet metal materials or metal plates are specified, the effectiveness of the protection can be enhanced by the reflective surface of the metal. Painting or otherwise covering the surface would reduce the metal's ability to reflect radiant heat and, depending on the color, could increase heat absorption. The airspace between the protected surface and the clearance-reduction assembly allows convection air currents to cool the protection assembly by carrying away heat that has been conducted through the assembly. Where a clearance-reduction assembly must be spaced 1 inch off the wall, the top, bottom and sides of the assembly must remain open as required by Notes "d" and "f" to permit unrestricted airflow (convection currents). If the openings were not provided, the air-cooling effect would not take place, and the protection assembly would not be as effective in limiting the temperature rise on the protected surfaces. Ideally, the protection assembly should be open on all sides to provide maximum ventilation. Figure 308.2(2) and commentary Figure 308.2 show assemblies incorporating airspace.

Spacers must be noncombustible. Spacers should not be placed directly behind the heat source because the location would increase the amount of heat conduction through the spacer, thus creating a "hot spot." Figure 308.2(2) specifically shows a noncombustible spacer arrangement.

The performance of a protective assembly when applied to a horizontal surface, such as a ceiling, will differ substantially from the same assembly placed in a vertical plane. Obviously, temperatures at a ceiling surface will be higher because of natural convection and because the air circulation between the method of protection and the protected ceiling surface will be substantially reduced or nonexistent. It is for these reasons that Table 308.2 is divided into two application groups.

The manufacturer's instructions or label for many appliances will state an absolute minimum clearance, regardless of any clearance reduction method used. Those clearance requirements take precedence over Table 308.2.

The methods in Table 308.2 are intended to be permanent installations properly supported to prevent displacement or deformation. Movement could adversely affect the performance of the protection method, thus posing a potential fire hazard.

The assemblies in Table 308.2 are the product of experience and testing. To achieve predictable and dependable performance, the components of the various assemblies cannot be mixed, matched, combined or otherwise rearranged to comprise new assemblies, and materials cannot be substituted for those prescribed in the table. Any alterations or substitutions could have an effect on the assembly, and its performance must be tested and approved.

Table 308.2 does not specify the method of measuring the reduced clearance. However, the intent is to measure the reduced clearance from the heat source to the combustible material, disregarding any intervening protection assembly.

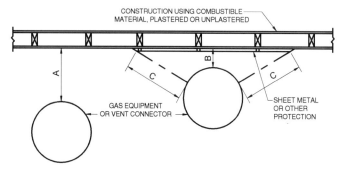

"A" equals the reduced clearance with no protection.

"B" equals the reduced clearance permitted in accordance with Table 308.2. The protection applied to the construction using combustible material shall extend far enough in each direction to made "C" equal to "A."

**Figure 308.2(1)
EXTENT OF PROTECTION NECESSARY TO
REDUCE CLEARANCES FROM GAS EQUIPMENT OR
VENT CONNECTIONS**

308.3.1 Equipment installed in rooms that are large in comparison with the size of the equipment. Air-conditioning equipment installed in rooms that are large in comparison with the size of the equipment shall be installed with clearances in accordance with the terms of their listing and the manufacturer's instructions.

❖ See the definition of and commentary for "Room Large in Comparison with Size of Equipment."

308.3.2 Equipment installed in rooms that are not large in comparison with the size of the equipment. Air-conditioning equipment installed in rooms that are not large in comparison with the size of the equipment, such as alcoves and closets, shall be listed for such installations and installed in accordance with the manufacturer's instructions. Listed clearances shall not be reduced by the protection methods described in Table 308.2, regardless of whether the enclosure is of combustible or noncombustible material.

❖ See the definition of and commentary for "Room Large in Comparison with Size of Equipment."

❖ Indirectly, the definition, along with this section, defines an alcove and a closet with respect to size.

An appliance must be listed for closet/alcove installation if it is installed in a space that does not meet the requirements in the above-named definition. Simply stated, if the room or space volume is less than the stated minimum volume in the definition, that room or space is considered to be a closet or alcove.

Fuel-fired appliances installed in small rooms or spaces can pose a potential fire hazard because of the heat emanating from the appliance and the inadequate ventilation afforded by the boundaries of small enclosures. For appliances installed in closets or alcoves, the clearances stated in the listing (label) are necessary for ventilation cooling of the appliance and must not be reduced as provided for in this section.

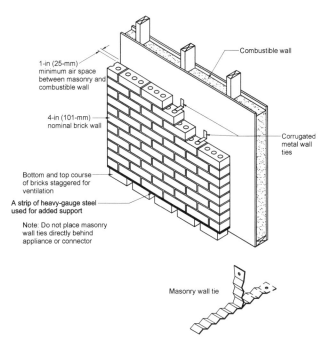

For SI: 1 inch = 25.4 mm.

**FIGURE 303.2(3)
MASONRY CLEARANCE REDUCTION SYSTEM**

308.3.3 Clearance reduction. Air-conditioning equipment installed in rooms that are large in comparison with the size of the equipment shall be permitted to be installed with reduced clearances to combustible material provided the combustible material or equipment is protected as described in Table 308.2.

❖ See the definition of and commentary for "Room Large in Comparison with Size of Equipment."

308.3.4 Plenum clearances. Where the furnace plenum is adjacent to plaster on metal lath or noncombustible material attached to combustible material, the clearance shall be measured to the surface of the plaster or other noncombustible finish where the clearance specified is 2 inches (51 mm) or less.

❖ Where the air-conditioning equipment plenum is adjacent to surfaces covered by noncombustible material as described in this section and the required clearance from combustibles is 2 inches or less, the required clearance must be measured to the surface of the adjacent material. Measuring to the combustible material would reduce the clearance between the equipment plenum and adjacent surface. Reducing this clearance would affect the cooling of the equipment and could cause overheating.

308.3.5 Clearance from supply ducts. Air-conditioning equipment shall have the clearance from supply ducts within 3 feet (914 mm) of the furnace plenum be not less than that specified from the furnace plenum. No clearance is necessary beyond this distance.

❖ Ducts within 3 feet of a supply plenum to which they connect will reach nearly the same temperature as the plenum.

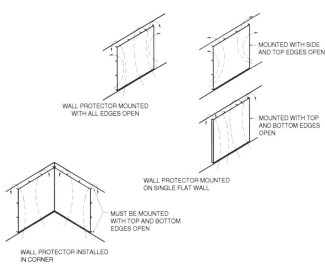

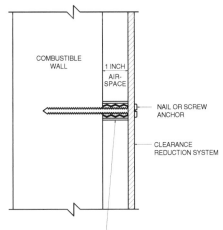

For SI: 1 inch = 25.4 mm.

**FIGURE 308.2(2)
WALL PROTECTOR CLEARANCE REDUCTION SYSTEM**

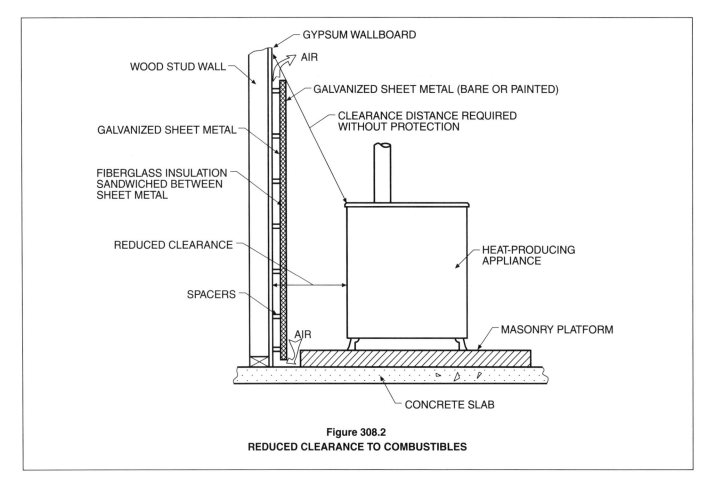

**Figure 308.2
REDUCED CLEARANCE TO COMBUSTIBLES**

308.4 Central-heating boilers and furnaces. Clearance requirements for central-heating boilers and furnaces shall comply with Sections 308.4.1 through 308.4.6. The clearance to this equipment shall not interfere with combustion air, draft hood clearance and relief, and accessibility for servicing.

❖ See the commentary for Section 308.3.

308.4.1 Equipment installed in rooms that are large in comparison with the size of the equipment. Central-heating furnaces and low-pressure boilers installed in rooms large in comparison with the size of the equipment shall be installed with clearances in accordance with the terms of their listing and the manufacturer's instructions.

❖ See the definition of and commentary for "Room Large in Comparison with Size of Equipment."

308.4.2 Equipment installed in rooms that are not large in comparison with the size of the equipment. Central-heating furnaces and low-pressure boilers installed in rooms that are not large in comparison with the size of the equipment, such as alcoves and closets, shall be listed for such installations. Listed clearances shall not be reduced by the protection methods described in Table 308.2 and illustrated in Figures 308.2(1) through 308.2(3), regardless of whether the enclosure is of combustible or noncombustible material.

❖ Indirectly, the definition, along with this section, defines an alcove and a closet with respect to size.

An appliance must be listed for closet/alcove installation if it is installed in a space that does not meet the requirements in the above-named definition. Simply stated, if the room or space volume is less than the stated minimum volume in the definition, that room or space is considered to be a closet or alcove.

Fuel-fired appliances installed in small rooms or spaces can pose a potential fire hazard because of the heat emanating from the appliance and the inadequate ventilation afforded by the boundaries of small enclosures. For appliances installed in closets or alcoves, the clearances stated in the listing (label) are necessary for ventilation cooling of the appliance and must not be reduced as provided for in this section.

308.4.3 Clearance reduction. Central-heating furnaces and low-pressure boilers installed in rooms that are large in comparison with the size of the equipment shall be permitted to be installed with reduced clearances to combustible material provided the combustible material or equipment is protected as described in Table 308.2.

❖ See the definition of commentary for "Room Large in Comparison with Size of Equipment."

308.4.4 Clearance for servicing equipment. Front clearance shall be sufficient for servicing the burner and the furnace or boiler.

❖ Because mechanical equipment and appliances require routine maintenance, repairs and possible replacement, access is required. Additionally, access recommendations or requirements are usually stated in the manufacturer's installation instructions.

The intent of this section is to assure access to components that require observation, inspection, adjustment, servicing, repair or replacement such as controls, gauges, burners, filters, blowers, and motors. Access is also necessary to conduct operating procedures such as startup or shutdown.

The code defines access as being able to be reached, but first may require the removal of a panel, door or similar obstruction. An appliance or piece of equipment is not accessible if any portion of the structure's permanent finish materials, such as drywall, plaster, paneling, built-in furniture, cabinets or any other similar permanently affixed building component must be removed.

308.4.5 Plenum clearances. Where the furnace plenum is adjacent to plaster on metal lath or noncombustible material attached to combustible material, the clearance shall be measured to the surface of the plaster or other noncombustible finish where the clearance specified is 2 inches (51 mm) or less.

❖ See the commentary for Section 308.3.4.

308.4.6 Clearance from supply ducts. Central-heating furnaces shall have the clearance from supply ducts within 3 feet (914 mm) of the furnace plenum be not less than that specified from the furnace plenum. No clearance is necessary beyond this distance.

❖ See the commentary for Section 308.2.

SECTION 309 (IFGC)
ELECTRICAL

309.1 Grounding. Gas piping shall not be used as a grounding electrode.

❖ The ICC EC prohibits metal underground gas piping from being used as a grounding electrode. Metal gas piping must be bonded to the electrical service grounding electrode system, but it cannot serve as a grounding electrode. See Section 310.1 for clarification of gas piping bonding means.

309.2 Connections. Electrical connections between equipment and the building wiring, including the grounding of the equipment, shall conform to the ICC *Electrical Code*.

❖ Field-installed power wiring and control wiring for appliances and equipment must be installed in accordance with the ICC EC Administrative Provisions, which reference NFPA 70.

The power wiring includes all the wiring, disconnects, overcurrent protection devices, starters and related hardware used to supply electrical power to the appliance or equipment. The control wiring includes all the wiring, devices and related hardware that connect the main unit to all external controls and accessories, such as temperature and pressure sensors, thermostats, exhausters, equipment contactors, interlock controls and remote damper motors. The internal factory wiring of appliances and equipment is not covered by this section unless it is specifically addressed in the ICC EC Administrative Provisions; however, the wiring is covered by the testing and review performed by an approved agency as part of the labeling process.

The mechanical or electrical code official responsible for the inspection of appliances and equipment must be familiar with the applicable sections of the ICC EC.

SECTION 310 (IFGS)
ELECTRICAL BONDING

310.1 Gas pipe bonding. Each above-ground portion of a gas piping system that is likely to become energized shall be electrically continuous and bonded to an effective ground-fault current path. Gas piping shall be considered to be bonded where it is connected to gas utilization equipment that is connected to the equipment grounding conductor of the circuit supplying that equipment.

❖ This section is consistent with Section 250.104(B) of the 2002 edition of NFPA 70. In the majority of cases, the gas-piping system will be connected to at least one, if not several, appliances that use electrical power. Those appliances will be connected to an equipment grounding conductor that will serve to bond the gas piping to the grounding electrode system for the building. Where the appliance branch circuits include grounding conductors properly connected to the appliances and one or more such appliances are connected to the gas-piping system, the gas-piping system is automatically bonded as required by this section.

Bibliography

The following resource material is referenced in this chapter or is relevant to the subject matter addressed in this chapter.

ANSI Z21.47–00, *Gas-Fired Central Furnaces with Addendum Z21.47a-00.* New York, NY: American National Standards Institute, 2000.

ANSI Z83.8–96, *Gas Fired Unit Heaters with Addendum 283.8a-1997.* New York, NY: American National Standards Institute, 1997.

ANSI Z223.1–02, National Fuel Gas Code. New York, NY: American National Standards Institute, 2002.

ASCE 7–02, *Minimum Design Loads for Buildings and Other Structures*. Reston, VA: American Society of Civil Engineers, 2002.

ASTM E 136–99, *Standard Test Method for Behavior of Materials in a Vertical Tube Furnace at 750EC*. West Conshohocken, PA: American Society for Testing and Materials, 1999.

NFPA 70, *National Electrical Code*. Quincy, MA: National Fire Protection Association, 2002.

NFPA 88B–97, *Repair Garages*. Quincy, MA: National Fire Protection Association, 1997.

UL 795–99, *Commercial-Industrial Gas Heating Equipment*. Northbrook, IL: Underwriters Laboratories, 1999.

Topical Report on Combustion-Air Issues Related to Residential Gas Appliances in Confined Spaces and Unusually Tight Construction. Gas Technology Institute, July 1997 - January 2001.

CHAPTER 4
GAS PIPING INSTALLATIONS

General Comments

The nature of fuel gases makes proper design, installation and selection of materials and devices necessary to minimize the risk to an acceptable level.

The two most commonly used fuel gases are natural gas and liquified petroleum gas (LP or LP-gas). These fuel gases have the following characteristics or properties.

Natural gas: The principal component of natural gas is methane (CH_4). It can also contain small quantities of nitrogen, carbon dioxide, hydrogen, hydrogen sulfide, water vapor, other hydrocarbons and various trace elements. Natural gas is colorless, tasteless and odorless; however, an odorant is added to the gas so that it can be readily detected. Natural gas is lighter than air (specific gravity < 1) and has the tendency to rise when escaping to the atmosphere. Natural gas has a rather narrow flammability range above and below which the gas-to-air mixture ratio is too rich or too lean to support combustion. See Figure 400. The heating value of natural gas varies depending on the source, but it averages 1,050 British thermal units (Btu) per cubic foot (10.9 kW/m^3).

Liquified petroleum gas: Liquified petroleum gases (LP or LP-gas) include commercial propane ($C_3 H_8$) and commercial butane ($C_4 H_{10}$). LP-gas vapors are heavier than air (specific gravity > 1) and tend to accumulate in low areas and near the floor. The ranges of flammability for LP-gases are narrower than those of natural gas. See Figure 400. Like natural gas, LP-gases are odorized to make them detectable.

The heating value of propane is approximately 2,500 Btu per cubic foot (25.9 kW/m^3) of gas. The heating value of butane is approximately 3,300 Btu per cubic foot (34.2 kW/m^3) of gas.

Purpose

This chapter is intended to regulate the design and installation of fuel-gas distribution piping and systems from the point of delivery of the fuel gas to the appliances and equipment that consume the fuel. The intent is to minimize the hazards associated with the use and distribution of highly flammable/explosive fuel gases.

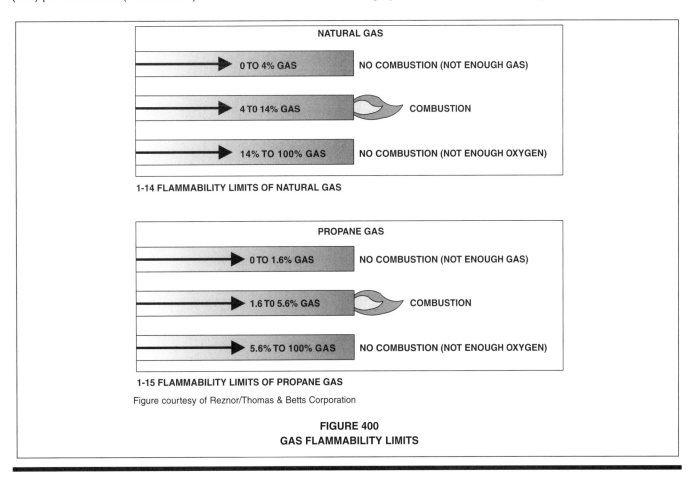

Figure courtesy of Reznor/Thomas & Betts Corporation

**FIGURE 400
GAS FLAMMABILITY LIMITS**

SECTION 401 (IFGC)
GENERAL

401.1 Scope. This chapter shall govern the design, installation, modification and maintenance of piping systems. The applicability of this code to piping systems extends from the point of delivery to the connections with the equipment and includes the design, materials, components, fabrication, assembly, installation, testing, inspection, operation and maintenance of such piping systems.

❖ This chapter regulates aspects of fuel-gas distribution systems, including design, installation, testing, repair and maintenance. The applicability of this chapter is limited to the consumer side of the public utility company's gas distribution system.

The code governs all piping and system components from the end point of the gas purveyor's (utility company) service line to the supplied appliances and equipment. Typically, the gas utility company service line terminates at the service pressure regulator and meter setting. In other words, the code does not apply to piping and components that are owned by the gas utility company. The construction of utility-owned gas piping is governed by the federal Department of Transportation (DOT) regulations wherever the piping is located. However, in cases where the utility-owned piping runs through, in or on a building, it must do so in a way that does not jeopardize the structural integrity or fire safety of the building. Therefore, the relationship of the utility-owned piping to the building (prohibited locations, penetrations, etc.) is governed by the *International Building Code®* (IBC®). The *International Fuel Gas Code®* (IFGC®) contains a reference (see Section 401.1.1) to the IBC to make sure this is recognized by the user. No other provision within the IFGC applies to utility-owned gas piping.

Although LP-gas storage systems generally remain under the ownership of the gas supplier, these systems are governed by the code through the reference to NFPA 58. See the Commentary for Section 401.2. An LP-gas piping system begins at the outlet side of the first-stage pressure regulator. The piping between the first- and second-stage (final) regulator is governed by NFPA 58, and the piping downstream of the second-stage regulator is governed by this code. The first-stage pressure regulator reduces the storage container pressure to 10 pounds per square inch gauge (psig) (138 kPa) or less.

The termination point of gas utility services varies depending on the particular company's policies, but typically the service terminates at the property line or outdoors, immediately adjacent to the building or structure served by the service. See Commentary Figures 401.1(1) and 401.1(2).

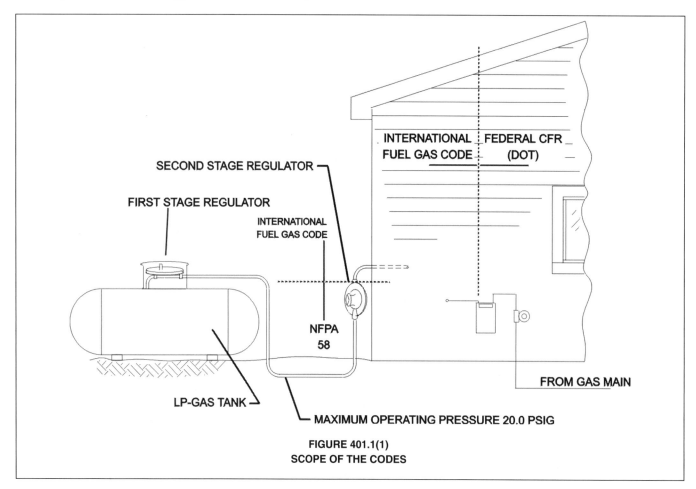

**FIGURE 401.1(1)
SCOPE OF THE CODES**

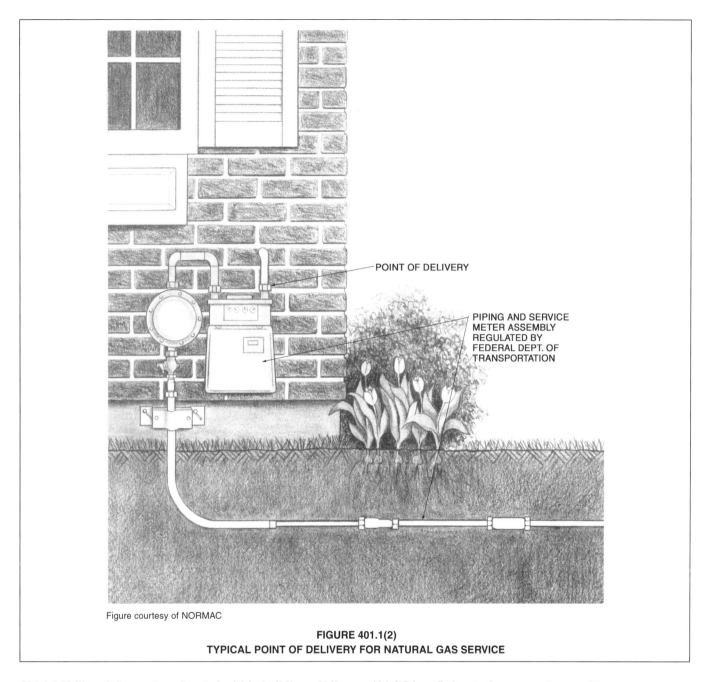

FIGURE 401.1(2)
TYPICAL POINT OF DELIVERY FOR NATURAL GAS SERVICE

401.1.1 Utility piping systems located within buildings. Utility service piping located within buildings shall be installed in accordance with the structural safety and fire protection provisions of the *International Building Code*.

❖ Piping and components upstream of the point of delivery are designed, installed, tested, owned and maintained by the gas utility company and are regulated by federal DOT regulations, not this code. This code does, however, regulate two aspects of the installation related to protection of the structure from a structural and fire safety standpoint. By reference, the IBC would regulate the piping penetrations through structural components and through fire-resistance-rated assemblies. The IBC would also regulate piping support to handle structural loads and seismic forces.

401.2 Liquefied petroleum gas storage. The storage system for liquefied petroleum gas shall be designed and installed in accordance with the *International Fire Code* and NFPA 58.

❖ LP-gas tanks, containers, cylinders, storage vessels and related components must comply with NFPA 58. The storage system includes storage vessels of any type and the vessel appurtenances such as shutoff valves, pressure gauges, liquid level indicators, pressure regulators, pressure relief valves and filling connections. The first-stage and second-stage pressure regulators and interconnecting piping are also addressed by the standard and could therefore be considered part of the LP-gas storage system. If the piping between the first-stage regulator and the point of delivery [second-stage or 2-psi (13.7 kPa) regulator] is not placed under the control of NFPA 58, it would be unregulated because Section 401.1 states that the IFGC coverage begins at the point of delivery (see definition of "point of delivery"). Second stage 2-psi regulators reduce the outlet pressure of the first stage regulator to 2-psi (13.7 kPa) and a line pressure regulator downstream of the 2-psi regulator reduces the pressure from 2-psi to a maximum of 14 inches w.c. (3.5 kPa). Second-stage regulators reduce the outlet pressure of the first-stage regulator to a maximum of 14 inches w.c. NFPA 58 addresses all aspects of LP-gas storage systems such as container design and construction, container appurtenances, safety devices, container location, and installation. A referenced standard is enforceable only in the context in which it is referenced. Any portions of NFPA 58 that extend beyond the scope of the storage system are not applicable because of the context of the reference in this section.

401.3 Modifications to existing systems. In modifying or adding to existing piping systems, sizes shall be maintained in accordance with this chapter.

❖ The gas volume demand on any portion of a distribution system cannot be increased beyond the capacity of the piping serving that portion. Modifications and additions to existing piping systems can have a detrimental effect on the existing piping system's ability to serve the connected loads. Commonly, new gas-fired equipment or appliances will be added to existing distribution piping without any regard for the effect of the extra load. To prevent dangerously low gas pressures, existing piping must be able to adequately supply additional loads, or piping must be increased in size.

401.4 Additional appliances. Where an additional appliance is to be served, the existing piping shall be checked to determine if it has adequate capacity for all appliances served. If inadequate, the existing system shall be enlarged as required or separate piping of adequate capacity shall be provided.

❖ This section parallels Section 401.8 and makes it clear that when any appliances are added to an existing gas piping system, the system must be carefully evaluated to verify that adding an appliance will not create a dangerous low gas pressure condition within the gas piping system. The evaluation must be conducted on the completed system as it will be configured after the appliance is added, and if the results indicate that the existing piping system cannot adequately serve the gas demand of the new appliance, the system must be revised accordingly. Undersized (overloaded) piping will cause pressure losses that can result in serious appliance malfunction.

401.5 Identification. For other than black steel pipe, exposed piping shall be identified by a yellow label marked "Gas" in black letters. The marking shall be spaced at intervals not exceeding 5 feet (1524 mm). The marking shall not be required on pipe located in the same room as the equipment served.

❖ The intent of this section is to prevent gas piping from being mistaken for other piping. A case of mistaken identity could lead to a dangerous condition if gas piping were to be cut or opened unintentionally. It is not uncommon for fuel-gas piping systems and various other gas or liquid piping systems in a building to be constructed of the same materials and thus have the same appearance. The method of identifying fuel gas piping must be permanent, legible and conspicuous. Black steel pipe is exempt from the identification requirements because it is the traditional gas pipe material and is generally recognized as such. CSST is manufactured in compliance with the marking requirements of this section, and some copper tubing is manufactured with a similar yellow plastic jacket. See Commentary Figures 403.5 and 403.5.4(1). Note that black steel pipe that has been painted another color is still black steel pipe.

401.6 Interconnections. Where two or more meters are installed on the same premises but supply separate consumers, the piping systems shall not be interconnected on the outlet side of the meters.

❖ If gas distribution systems with independent meters were connected downstream of the meters, it would be impossible to determine the actual consumption of each of the interconnected systems. As the result of building renovations, remodeling and additions, or as the result of intentionally valved cross-connections between separately metered systems, it is possible for a consumer to be paying for fuel gas consumed by another consumer. Accurate consumption metering can occur only if each metered distribution system is independent of all other metered distribution systems on the premises. Also, a system thought to be shut off could be backfed from an interconnected system, thereby maintaining pressure in a system that was intended to be shut off.

401.7 Piping meter identification. Piping from multiple meter installations shall be marked with an approved permanent identification by the installer so that the piping system supplied by each meter is readily identifiable.

❖ This provision allows service and emergency personnel to readily locate the shutoff valve that serves each piping system. The identification also helps to prevent interconnections between systems and between spaces when additions and/or alterations are made in the piping systems.

401.8 Minimum sizes. All pipe utilized for the installation, extension and alteration of any piping system shall be sized to supply the full number of outlets for the intended purpose and shall be sized in accordance with Section 402.

❖ Undersized gas piping systems are not capable of delivering the required volume of fuel at the required pressure. Inadequate gas pressure can cause hazardous operation of appliances. Appliances depend on proper gas pressure to maintain the design Btu (J) input rate, to maintain the required minimum manifold pressure and to produce the required gas velocity in the burners. If the gas pressure to an appliance is too low, the result can be incomplete combustion, burner malfunction and flashback, soot production and appliance malfunction and damage. Gas piping must be sized to maintain the required minimum gas pressures at the appliance inlet.

When designing a gas piping distribution system, all connected appliances and equipment are assumed to operate simultaneously, except as provided for in Section 402.2. In other words, the system is designed for maximum demand. The code contains provisions for sizing gas piping systems using an exact equation or design tables. The tabular method is most often used because it is simpler and less likely to produce errors.

Factors that affect the sizing of gas piping systems include the specific gravity of the gas, the length of the pipe and the number of fittings installed, the maximum gas demand, the allowable pressure loss through the system and any diversity factor that is applicable.

The design of gas piping installations should also take into account the possibility of future increases in gas demand (load). The further the gas meter or point of delivery is from the building, the more extensive will be the disruption of the property in the event that a larger yard line between the meter and the building becomes necessary in the future as a result of increased gas demand within the building, a change in fuel gas type or a change in service pressure.

SECTION 402 (IFGS)
PIPE SIZING

402.1 General considerations. Piping systems shall be of such size and so installed as to provide a supply of gas sufficient to meet the maximum demand without undue loss of pressure between the point of delivery and the gas utilization equipment.

❖ This section is a summation of the provisions of Sections 401.8 and 402.5 and is intended to give guidance to the designer (see commentary, Sections 401.8 and 402.5). "Undue loss of pressure" is not defined but must be considered as pressure loss (drop) in excess of the allowable loss described in Section 402.5.

402.2 Maximum gas demand. The volume of gas to be provided, in cubic feet per hour, shall be determined directly from the manufacturer's input ratings of the gas utilization equipment served. Where an input rating is not indicated, the gas supplier, equipment manufacturer or a qualified agency shall be contacted, or the rating from Table 402.2 shall be used for estimating the volume of gas to be supplied.

The total connected hourly load shall be used as the basis for pipe sizing, assuming that all equipment could be operating at full capacity simultaneously. Where a diversity of load can be established, pipe sizing shall be permitted to be based on such loads.

❖ To determine the demand volume required by an appliance in cubic feet (m^3) of gas per hour, the input rating of the appliance in British thermal units per hour (Btu/h) (W) as specified by the appliance manufacturer must be known. If the average heating value of the gas is known [about 1,000 Btu/ft^3 (10.4 kW/m^3) for natural gas], the volume of gas required per hour can be calculated. The heating value can be obtained for the gas supplier. The input rating of all appliances must be shown on the appliance label (see commentary, Section 301.5). When the input rating of the appliances is unknown at the time of piping system design, the gas demand will have to be estimated based on input from appliance manufacturers, gas utilities or other source (see Table 402.2). However, when the actual appliance demands are determined upon appliance selection, the approximate calculations done in accordance with this section must be checked to verify that the piping system is, in fact, adequate to supply the known demand for all appliances. The designer must be aware of the fact that a system designed in accordance with an estimated load could be undersized relative to the actual connected load.

TABLE 402.2
APPROXIMATE GAS INPUT FOR TYPICAL APPLIANCES

APPLIANCE	INPUT BTU/H (Approx.)
Space Heating Units	
Hydronic boiler	
Single family	100,000
Multifamily, per unit	60,000
Warm-air furnace	
Single family	100,000
Multifamily, per unit	60,000
Space and Water Heating Units	
Hydronic boiler	
Single family	120,000
Multifamily, per unit	75,000
Water Heating Appliances	
Water heater, automatic instantaneous	
Capacity at 2 gal./minute	142,800
Capacity at 4 gal./minute	285,000
Capacity at 6 gal./minute	428,400
Water heater, automatic storage, 30- to 40-gal. tank	35,000
Water heater, automatic storage, 50-gal. tank	50,000
Water heater, domestic, circulating or side-arm	35,000
Cooking Appliances	
Built-in oven or broiler unit, domestic	25,000
Built-in top unit, domestic	40,000
Range, free-standing, domestic	65,000
Other Appliances	
Barbecue	40,000
Clothes dryer, Type 1 (domestic)	35,000
Gas fireplace, direct-vent	40,000
Gas light	2,500
Gas log	80,000
Refrigerator	3,00

For SI: 1 British thermal unit per hour = 0.293 W, 1 gallon = 3.785 L, 1 gallon per minute = 3.785 L/m.

In all cases, the piping must have the capacity to supply the actual connected load of appliances. As also stated in Section 401.8, the load on the piping system must be based on the simultaneous operation of all appliances at full output. The only exception to this would be the demonstration of actual load diversity. For example, the designer could have access to data that demonstrates that the total load for cooking ranges or water heaters in a multiple-family dwelling (apartment building) is less than the sum of all of the appliance demands at full output. The code official would have to be convinced that any such diversity factor had been established. Any known diversity factors would likely come from the serving gas supplier.

The following is an example of how gas demand is calculated for a representative appliance: an appliance has a nameplate (label) input rating of 120,000 Btu/h (35.2 kW). Using natural gas with an average heating value of 1,000 Btu/ft³ (10.4 kW/m³), the volume of gas required for the appliance is: 120,000/1,000 = 120 cubic feet per hour (3.52 m³/h).

402.3 Sizing. Gas piping shall be sized in accordance with one of the following:

1. Pipe sizing tables or sizing equations in accordance with Section 402.4.
2. The sizing tables included in a listed piping system's manufacturer's installation instructions.
3. Other approved engineering methods.

❖ Using the sizing tables is generally simpler and less time consuming than using the equations found in Section 402.4. The gas system pressure, allowable pressure drop, maximum gas demand and the specific gravity of the fuel gas must be known in order to use tables. The allowable pressure drop is the designer's choice because the intent of the code is to provide the appliance or equipment with the required rate of gas flow at the required minimum pressure. The tables are based on a given pressure drop and cannot be used if the designer wishes to use any other pressure drop. The sizing tables are based on using smooth-bore pipe or tube and can be used with the piping materials listed in Section 403. Pipe materials with higher resistance to flow cannot be sized by these tables. CSST is sized by CSST-specific tables. The pipe sizes in the pipe sizing tables are based on the traditional steel pipe size designation as used in the material standards. More information may be obtained from a variety of pipe design publications, including the Piping Handbook by Sabin Crocker and the ASHRAE Handbook of HVAC Systems and Equipment. Each of the tables presents different piping material and design parameters and represents the most common system design applications. The tables assume that the system being sized is constructed of a single type pipe or tubing material throughout; thus, a single table cannot be used for systems constructed of multiple materials. For example, it is common to find hybrid systems composed of both steel pipe and CSST. Such systems would have to be sized from two different tables based on the longest run of the entire system.

The code does not limit sizing methodology to the tables or equations. Manufacturers of gas piping/tubing may provide sizing charts, tables or slide calculators that, in effect, become part of the installation instructions. This is the case for CSST, which is a listed piping system.

Section 402.4 describes a sizing method that uses an exact thermodynamic flow equation to determine the required pipe sizes. Although this is an exact method, the degree of accuracy it affords may not always prove valuable, because pipe sizes are standardized and the pipe diameter calculated may not correspond to an available pipe size. For example, a calculated pipe diameter of 0.80 inch (20.3 mm) will require the selection of the next larger standard pipe size [1-inch (25.4 mm) pipe].

There are two equations in Section 402.4 to be used, depending on the gas pressure within the piping system.

As stated in Appendix A, A.2.2 and A.2.3, an equivalent length of pipe should be added to the length of any piping run having four or more fittings (see Appendix A, Table A.2.2). Many designers simply add 50 percent of the actual length of piping as an all-inclusive fitting allowance (i.e., actual pipe length x 1.5). This is a conservative estimate and is used because the necessary number and type of fittings is usually not known when the system is being designed.

402.4 Sizing tables and equations. Where Tables 402.4(1) through 402.4(33) are used to size piping or tubing, the pipe length shall be determined in accordance with Section 402.4.1, 402.4.2 or 402.4.3.

Where Equations 4-1 and 4-2 are used to size piping or tubing, the pipe or tubing shall have smooth inside walls and the pipe length shall be determined in accordance with Section 402.4.1, 402.4.2 or 402.4.3.

1. Low-pressure gas equation [Less than 1.5 pounds per square inch (psi) (10.3 kPa)]:

$$D = \frac{Q^{0.381}}{19.17 \left(\frac{\Delta H}{C_r \times L} \right)^{0.206}}$$ (Equation 4-1)

2. High-pressure gas equation [1.5 psi (10.3 kPa) and above]:

$$D = \frac{Q^{0.381}}{18.93 \left[\frac{(P_1^2 - P_2^2) \times Y}{C_r \times L} \right]^{0.206}}$$ (Equation 4-2)

where:

D = Inside diameter of pipe, inches (mm).

Q = Input rate appliance(s), cubic feet per hour at 60°F (16°C) and 30-inch mercury column

P_1 = Upstream pressure, psia (P_1 + 14.7)

P_2 = Downstream pressure, psia (P_2 + 14.7)

L = Equivalent length of pipe, feet

ΔH = Pressure drop, inch water column (27.7 inch water column = 1 psi)

❖ The length of pipe or tube to be used with the tables and equations is determined by the sizing method being used. Section 402.4.1 involves only one length whereas Sections 402.4.2 and 402.4.3 involve multiple lengths.

TABLE 402.4
C_r AND Y VALUES FOR NATURAL GAS AND UNDILUTED PROPANE AT STANDARD CONDITIONS

GAS	EQUATION FACTORS	
	C_r	Y
Natural gas	0.6094	0.9992
Undiluted propane	1.2462	0.9910

For SI: 1 cubic foot = 0.028 m³, 1 foot = 305 mm, 1-inch water column = 0.249 kPa, 1 pound per square inch = 6.895 kPa, 1 British thermal unit per hour = 0.293 W.

Example 1:

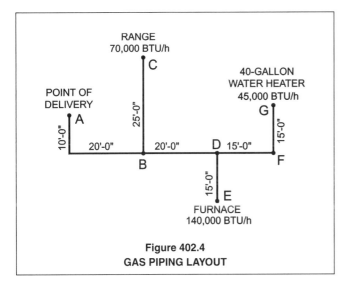

Figure 402.4
GAS PIPING LAYOUT

Determine the required pipe sizes for the gas distribution system in Commentary Figure 402.4 using the equation method outlined in this section. The system is designed to operate at less than 0.5 pound per square inch (psig), using steel pipe. Assume the following:

Fuel gas	Natural
Specific gravity	0.60
Heating value	1,000 Btu per cubic feet
Maximum allowable pressure drop	½ inch of water column
Negligible number of fittings	

For SI: 1 inch = 25.4 mm, 1 foot = 304.8 mm, 1 Btu/h = 0.2931 W, 1 Btu/ft³ = 10.35 W/m³, 1 inch water column = 0.2488 kPa, 1 cubic foot = 0.20832 m³/h, 1 pound per square inch gauge = 6.895 kPa.

To use equations in this section, the gas volume of each appliance must be known.

Gas volume of an appliance = Input rating of an appliance (Btu/h)/Heating value of the fuel gas (Btu/ft³)

Appliance	Input rating (Btu/h)	Gas volume (ft³/h)
Range	70,000	70
Furnace	140,000	140
Water heater	45,000	45

Using the equation for maximum demand quantity (Q), we can solve for the minimum pipe diameter (D) by rearranging it to read as follows:

$D = Q^{0.381}/19.17 \: [(\Delta H)/(C_r \times L)]^{0.206}$

Starting at the most remote appliance from the point of delivery (Point G - water heater), determine the minimum pipe size for pipe segment D.

$D = (45)^{0.381}/19.17 \: [(½)/(0.609 \times 80)]^{0.206}$
$= 0.574$ inch

Repeat the calculation for each pipe segment. The gas piping system is sized as follows:

Pipe segment	Flow rate (Q)	Calculated pipe diameter (D) (in.)	Nominal pipe size (in.)
A-B	255	1.11	1¼
B-C	70	0.679	¾
B-D	185	0.983	1
D-E	140	0.884	1
D-G	45	0.574	½

Most gas pipe sizing applications will be completed using the sizing tables. In the event that circumstances warrant it, however, pipe sizing may be done by calculation using the provisions of Appendix A or other approved methods. If in a particular design, the piping length, desired pressure loss, system supply pressure, specific gravity or pipe size does not fall within the parameters of a table, an approved alternate sizing procedure must be used.

In lieu of the methods specified in this section, piping sizes may be determined by the use of accurate gas flow, computer or pressure drop charts. Such sizing tools must be acceptable to the code official. Note that calculators, charts and computer programs can be designed to size gas piping for any pressure drop because the intent of the code is to make sure that the supply pressure at the appliance is greater than the minimum pressure required for proper operation of the appliance (see Section 402.5).

402.4.1 Longest length method. The pipe size of each section of gas piping shall be determined using the longest length of piping from the point of delivery to the most remote outlet and the load of the section.

❖ This section provides a step-by-step approach to proper application of the tables using the traditional longest-length method. When using the tables, the maximum pipe length from the point of delivery to the farthest outlet must be determined, including any allowance for the equivalent length of fittings. The designer will use the row in the table that equals the determined length or the next higher row if the determined length is between table values. The system will be designed using only values taken from this row. This is known as the longest-run method. Basing the sizing on the most demanding circuit (longest run) compensates for pressure losses throughout the entire system.

402.4.2 Branch length method. Pipe shall be sized as follows:

1. Pipe size of each section of the longest pipe run from the point of delivery to the most remote outlet shall be determined using the longest run of piping and the load of the section.

2. The pipe size of each section of branch piping not previously sized shall be determined using the length of piping from the point of delivery to the most remote outlet in each branch and the load of the section.

❖ This sizing method is a variation of the longest-length method, and the results can be less conservative. Since less headroom is built into this method, it is especially important to account for the equivalent length of fittings installed in the system (see Figure 402.4.2 and example). This method involves multiple piping lengths within a system for application of the tables or equations, whereas the longest-length method involves only one piping length per system.

402.4.3 Hybrid pressure. The pipe size for each section of higher pressure gas piping shall be determined using the longest length of piping from the point of delivery to the most remote line pressure regulator. The pipe size from the line pressure regulator to each outlet shall be determined using the length of piping from the regulator to the most remote outlet served by the regulator.

❖ Hybrid pressure piping systems convey gas at different pressures in different parts of the system. One or more pressure regulators are used to reduce the pressure in portions of the system. In some cases, a single gas piping system must deliver widely varying pressure to loads throughout a building. The most common hybrid pressure systems allow small economically sized piping to carry large loads over long distances by conveying gas at high pressures. Near the load being supplied, pressure regulators reduce the higher pressure to suit the load. For example, rooftop units on buildings are typically designed for a natural gas pressure of 14 inches w.c. (3.5 kPa) or less. Running hundreds of feet of pipe across expansive roofs would require large expensive pipe at such low pressure. At 5-psi (34.4 kPa) pressure, small, less expensive pipe can be run over long distances, and regulators at each rooftop unit reduce the 5-psi (34.4 kPa) pressure to less than ½ psi(3.4 kPa), which is typically 8 inches w.c. (2 kPa) (see Figure 402.4.3 and example).

Example:

1. In accordance with Section 405.4.2 Item 1, determine the size of Sections A, B, C, D and E (constructed of schedule 40 steel pipe) based on the load of each section and the longest run length of 85 ft. Because the system pressure is less than 0.5 psi (14 in. w.c.), Table 402.4(2) is chosen. Because the longest run length is between rows in the table, the 90-foot row must be chosen.

PIPE SECTION	LOAD MBH	SIZE
A	230	$1^{1}/_{4}''$
B	170	$1''$
C	135	$1''$
D	95	$^{3}/_{4}''$
E	20	$^{3}/_{8}''$

2. In accordance with Section 402.4.2 Item 2, determine the size of Branch Sections F, G, H, I and J (constructed of CSST) based on the load of each section and the length of piping and tubing from the point of delivery to the outlet on that section. Table 402.4(14) is chosen because the branch sections are constructed of CSST and the pressure is less than 0.5 psi (14 in. w.c). Where a length falls between entries in the table, use the next longer length row.

CSST BRANCH	LOAD MBH	LENGTH PIPING AND TUBING	SIZE EHD
F	20	85'	18
G	60	25'	19
H	35	40'	18
I	40	60'	23
J	75	75'	30

GAS PIPING INSTALLATIONS

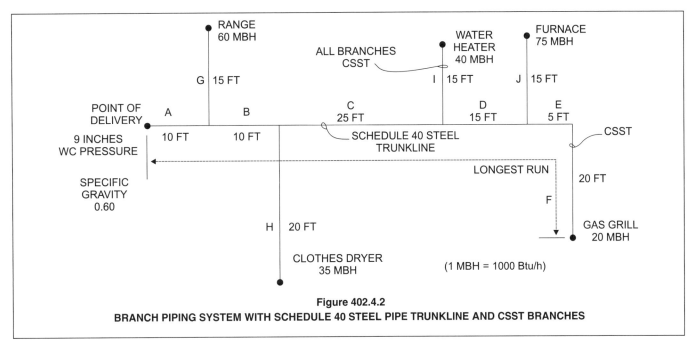

Figure 402.4.2
BRANCH PIPING SYSTEM WITH SCHEDULE 40 STEEL PIPE TRUNKLINE AND CSST BRANCHES

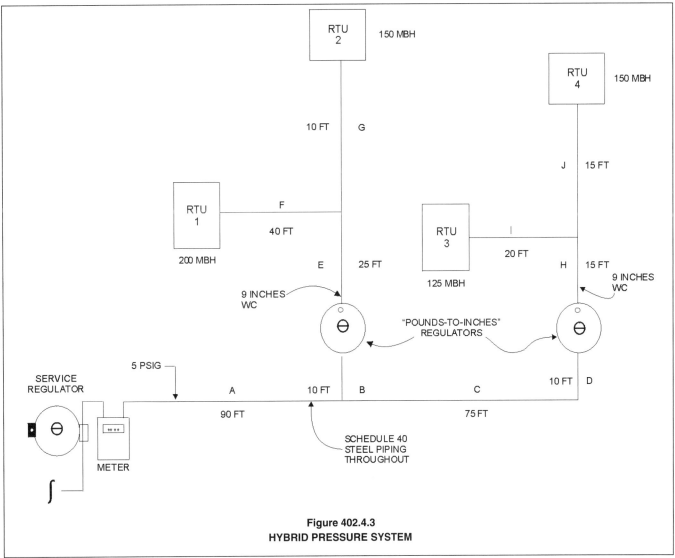

Figure 402.4.3
HYBRID PRESSURE SYSTEM

Example:

1. Using Table 402.4(4), determine the minimum required size of piping sections "A" through "D". The longest run of piping from the point of delivery to the most remote regulator 175 ft.

    - Section "A" serves a load of 625 MBH and in the 175 foot row of the table, ½ in. pipe is shown to have a capacity of 728 ft^3/hr or 728,000 Btu/h for gas with a heating value of 1,000 Btu/ft^3.
    - Because sections "B", "C" and "D" all serve lesser loads than section "A," they too can be ½ in. in size.
    - Because Table 402.4(4) is based upon a pressure drop of 3.5 psi, the available pressure at the inlets of the "pounds-to-inches" regulators under full load condition will be at least 1.5 psig.

2. Determine size of piping sections "E", "F" and "G," using Table 402.4(2) and the longest length of piping from the pounds-to-inches regulator to the most remote outlet. In this case, roof top unit 1 is the most remote at 65 ft from the regulator.

3. Determine size of piping sections "H", "I" and "J" in the same manner described in item 2 above.

NOTE: MBH = 1,000 Btu/h

402.5 Allowable pressure drop. The design pressure loss in any piping system under maximum probable flow conditions, from the point of delivery to the inlet connection of the equipment, shall be such that the supply pressure at the equipment is greater than the minimum pressure required for proper equipment operation.

❖ Between the point of delivery and the load, gas pressure losses will occur because of pipe friction. The design of the system must control these losses to assure that the pressure at the load (appliance) connection point will be in excess of the minimum required pressure as dictated by the appliance manufacturer. As stated in the commentary to Section 401.8, inadequate gas pressures can result in dangerous appliance operation. This section speaks of "maximum probable flow conditions," which must be taken to mean the total connected load of all appliances operating at full capacity except where a diversity-of-load factor is applied.

The designer can choose any pressure loss for a piping system if the goal of this section is met. For example, an appliance has a minimum required inlet pressure of 5 inches of water column (1.25 kPa), and the gas pressure at the point of delivery is 8 inches of water column (2 kPa). The maximum allowable pressure drop through the piping system would be just under 3 inches w.c. (0.75 kPa) because the pressure at the appliance connection must be greater than 5 inches w.c. A pressure of 5.1 inches w.c. is greater than 5.0 inches w.c. and would satisfy the intent of this section. Sizing tables such as Table 402.4 (1) and 402.4 (2) are based on very small pressure drops (0.3 and 0.5 inch w.c., respectively) and thus are quite conservative sizing methods that account for the friction losses through a limited number of fittings used in a piping system. The tables are optional sizing methods placed in the code for convenience. Piping systems can be designed for the exact pressure losses allowed by this section.

402.6 Maximum design operating pressure. The maximum design operating pressure for piping systems located inside buildings shall not exceed 5 pounds per square inch gauge (psig) (34 kPa gauge) except where one or more of the following conditions are met:

1. The piping system is welded.
2. The piping is located in a ventilated chase or otherwise enclosed for protection against accidental gas accumulation.
3. The piping is located inside buildings or separate areas of buildings used exclusively for:
    3.1. Industrial processing or heating;
    3.2. Research;
    3.3. Warehousing; or
    3.4. Boiler or mechanical equipment rooms.
4. The piping is a temporary installation for buildings under construction.

❖ Section 101.2.2 states that the IFGC applies to gas piping systems with an operating pressure of 125 psig (861.8 kPa) or less, but this section is specific to indoor piping systems. The pressure is limited to 5 psig (34.4 kPa) to minimize the consequences of a leak. Also, the higher the pressure, the more difficult it is to contain any gas, and thus leakage is more likely. Item 1 recognizes the superior strength and integrity of steel piping with welded joints. Item 2 allows the option of higher pressures where the piping is enclosed to prevent leakage from entering the building. Item 3 allows higher pressures in specific occupancies where the risks are acceptable because of low occupant density, the presence of trained maintenance and emergency response personnel and/or the well ventilated nature of the buildings. Gas-fired equipment in some of the occupancies listed might require that gas be supplied at higher than 5 psig pressure. Item 4 recognizes the unique nature of temporary piping on construction sites where small piping is desirable, thus making higher pressures necessary.

402.6.1 Liquefied petroleum gas systems. The operating pressure for undiluted LP-gas systems shall not exceed 20 psig (140 kPa gauge). Buildings having systems designed to operate below -5°F (-21°C) or with butane or a propane-butane mix shall be designed to either accommodate liquid LP-gas or prevent LP-gas vapor from condensing into a liquid.

Exception: Buildings or separate areas of buildings constructed in accordance with Chapter 7 of NFPA 58, and used exclusively to house industrial processes, research and experimental laboratories, or equipment or processing having similar hazards.

❖ The maximum operating pressure for LP-gas distribution systems applies to indoor and outdoor installations with the same concerns as expressed in the commentary to Section 402.6. In low-temperature environments, LP-gases might condense into the liquid state within the piping system. If the liquid fuel enters system components or appliances, dangerous malfunctions could result. At atmospheric pressure, 14.7 psia (sea level) propane will condense at 44°F (7°C) and butane will condense at 15°F (-9°C). Within the piping system the pressure could be higher than one atmosphere and LP-gas would therefore condense at a higher temperature. For example, at 20 psig (138 kPa), propane will condense at -5°F (-21°C), hence the -5°F threshold requirement of this section. The exception allows higher pressures in the specified occupancies if the buildings that house those occupancies are constructed in accordance with NFPA 58. Chapter 7 of NFPA 58 regulates buildings that house LP-gas distribution facilities and contains provisions for construction type, height, explosion venting, ventilation, heating systems and fire-rated separations, among others.

TABLE 402.4(1)
SCHEDULE 40 METALLIC PIPE

Gas	Natural
Inlet Pressure	0.50 psi or less
Pressure Drop	0.3 inch WC
Specific Gravity	0.60

	PIPE SIZE (in.)										
Nominal	$1/4$	$3/8$	$1/2$	$3/4$	1	$1 1/4$	$1 1/2$	2	$2 1/2$	3	4
Actual ID	0.364	0.493	0.622	0.824	1.049	1.380	1.610	2.067	2.469	3.068	4.026
Length (ft)	Maximum Capacity in Cubic Feet of Gas per Hour										
10	32	72	132	278	520	1,050	1,600	3,050	4,800	8,500	17,500
20	22	49	92	190	350	730	1,100	2,100	3,300	5,900	12,000
30	18	40	73	152	285	590	890	1,650	2,700	4,700	9,700
40	15	34	63	130	245	500	760	1,450	2,300	4,100	8,300
50	14	30	56	115	215	440	670	1,270	2,000	3,600	7,400
60	12	27	50	105	195	400	610	1,150	1,850	3,250	6,800
70	11	25	46	96	180	370	560	1,050	1,700	3,000	6,200
80	11	23	43	90	170	350	530	990	1,600	2,800	5,800
90	10	22	40	84	160	320	490	930	1,500	2,600	5,400
100	9	21	38	79	150	305	460	870	1,400	2,500	5,100
125	8	18	34	72	130	275	410	780	1,250	2,200	4,500
150	8	17	31	64	120	250	380	710	1,130	2,000	4,100
175	7	15	28	59	110	225	350	650	1,050	1,850	3,800
200	6	14	26	55	100	210	320	610	980	1,700	3,500

For SI: 1 inch = 25.4 mm, 1 foot = 304.8 mm, 1 cubic foot per hour = 0.0283 m^3/h, 1 pound per square inch = 6.895 kPa, 1-inch water column = 0.2488 kPa.

❖ See the commentary for Section 402.3.

TABLE 402.4(2)
GAS PIPING INSTALLATIONS

TABLE 402.4(2)
SCHEDULE 40 METALLIC PIPE

Gas	Natural
Inlet Pressure	0.50 psi or less
Pressure Drop	0.5 inch WC
Specific Gravity	0.60

	PIPE SIZE (in.)										
Nominal	1/4	3/8	1/2	3/4	1	1 1/4	1 1/2	2	2 1/2	3	4
Actual ID	0.364	0.493	0.622	0.824	1.049	1.380	1.610	2.067	2.469	3.068	4.026
Length (ft)	Maximum Capacity in Cubic Feet of Gas per Hour										
10	43	95	175	360	680	1,400	2,100	3,950	6,300	11,000	23,000
20	29	65	120	250	465	950	1,460	2,750	4,350	7,700	15,800
30	24	52	97	200	375	770	1,180	2,200	3,520	6,250	12,800
40	20	45	82	170	320	660	990	1,900	3,000	5,300	10,900
50	18	40	73	151	285	580	900	1,680	2,650	4,750	9,700
60	16	36	66	138	260	530	810	1,520	2,400	4,300	8,800
70	15	33	61	125	240	490	750	1,400	2,250	3,900	8,100
80	14	31	57	118	220	460	690	1,300	2,050	3,700	7,500
90	13	29	53	110	205	430	650	1,220	1,950	3,450	7,200
100	12	27	50	103	195	400	620	1,150	1,850	3,250	6,700
125	11	24	44	93	175	360	550	1,020	1,650	2,950	6,000
150	10	22	40	84	160	325	500	950	1,500	2,650	5,500
175	9	20	37	77	145	300	460	850	1,370	2,450	5,000
200	8	19	35	72	135	280	430	800	1,280	2,280	4,600

For SI: 1 inch = 25.4 mm, 1 foot = 304.8 mm, 1 cubic foot per hour = 0.0283 m^3/h, 1 pound per square inch = 6.895 kPa, 1-inch water column = 0.2488 kPa.

❖ See the commentary for Section 402.3.

TABLE 402.4(3)
SCHEDULE 40 METALLIC PIPE

Gas	Natural
Inlet Pressure	2.0 psi
Pressure Drop	1.0 psi
Specific Gravity	0.60

	PIPE SIZE (in.)								
Nominal	1/2	3/4	1	1 1/4	1 1/2	2	2 1/2	3	4
Actual ID	0.622	0.824	1.049	1.380	1.610	2.067	2.469	3.068	4.026
Length (ft)	Maximum Capacity in Cubic Feet of Gas per Hour								
10	1,506	3,041	5,561	11,415	17,106	32,944	52,505	92,819	189,326
20	1,065	2,150	3,932	8,072	12,096	23,295	37,127	65,633	133,873
30	869	1,756	3,211	6,591	9,876	19,020	30,314	53,589	109,307
40	753	1,521	2,781	5,708	8,553	16,472	26,253	46,410	94,663
50	673	1,360	2,487	5,105	7,650	14,733	23,481	41,510	84,669
60	615	1,241	2,270	4,660	6,983	13,449	21,435	37,893	77,292
70	569	1,150	2,102	4,315	6,465	12,452	19,845	35,082	71,558
80	532	1,075	1,966	4,036	6,048	11,647	18,563	32,817	66,937
90	502	1,014	1,854	3,805	5,702	10,981	17,502	30,940	63,109
100	462	934	1,708	3,508	5,257	10,125	16,138	28,530	58,194
125	414	836	1,528	3,138	4,702	9,056	14,434	25,518	52,050
150	372	751	1,373	2,817	4,222	8,130	12,960	22,911	46,732
175	344	695	1,271	2,608	3,909	7,527	11,999	21,211	43,265
200	318	642	1,174	2,413	3,613	6,959	11,093	19,608	39,997

For SI: 1 inch = 25.4 mm, 1 foot = 304.8 mm, 1 cubic foot per hour = 0.0283 m^3/h, 1 pound per square inch = 6.895 kPa.

❖ See the commentary for Section 402.3.

TABLE 402.4(4)
SCHEDULE 40 METALLIC PIPE

Gas	Natural
Inlet Pressure	5.0 psi
Pressure Drop	3.5 psi
Specific Gravity	0.60

	PIPE SIZE (in.)								
Nominal	1/2	3/4	1	1 1/4	1 1/2	2	2 1/2	3	4
Actual ID	0.622	0.824	1.049	1.380	1.610	2.067	2.469	3.068	4.026
Length (ft)	Maximum Capacity in Cubic Feet of Gas per Hour								
10	3,185	6,434	11,766	24,161	36,206	69,727	111,133	196,468	400,732
20	2,252	4,550	8,320	17,084	25,602	49,305	78,583	138,924	283,361
30	1,839	3,715	6,793	13,949	20,904	40,257	64,162	113,431	231,363
40	1,593	3,217	5,883	12,080	18,103	34,864	55,566	98,234	200,366
50	1,425	2,878	5,262	10,805	16,192	31,183	49,700	87,863	179,213
60	1,301	2,627	4,804	9,864	14,781	28,466	45,370	80,208	163,598
70	1,204	2,432	4,447	9,132	13,685	26,354	42,004	74,258	151,463
80	1,153	2,330	4,260	8,542	12,801	24,652	39,291	69,462	141,680
90	1,062	2,145	3,922	8,054	12,069	23,242	37,044	65,489	133,577
100	979	1,978	3,617	7,427	11,128	21,433	34,159	60,387	123,173
125	876	1,769	3,235	6,643	9,953	19,170	30,553	54,012	110,169
150	786	1,589	2,905	5,964	8,937	17,211	27,431	48,494	98,911
175	728	1,471	2,690	5,522	8,274	15,934	25,396	44,897	91,574
200	673	1,360	2,487	5,104	7,649	14,729	23,478	41,504	84,656

For SI: 1 inch = 25.4 mm, 1 foot = 304.8 mm, 1 cubic foot per hour = 0.0283 m^3/h, 1 pound per square inch = 6.895 kPa.

❖ See the commentary for Section 402.3.

GAS PIPING INSTALLATIONS TABLE 402.4(5)

**TABLE 402.4(5)
SCHEDULE 40 METALLIC PIPE**

	Gas	Natural
	Inlet Pressure	1.0 psi or less
	Pressure Drop	0.3 inch WC
	Specific Gravity	0.60

	PIPE SIZE (in.)												
Nominal	1	1¼	1½	2	2½	3	3½	4	5	6	8	10	12
Actual ID	1.049	1.380	1.610	2.067	2.469	3.068	3.548	4.026	5.047	6.065	7.981	10.020	11.938
Length (ft)	Maximum Capacity in Cubic Feet of Gas per Hour												
50	215	442	662	1,275	2,033	3,594	5,262	7,330	13,261	21,472	44,118	80,130	126,855
100	148	304	455	877	1,397	2,470	3,616	5,038	9,114	14,758	30,322	55,073	87,187
150	119	244	366	704	1,122	1,983	2,904	4,046	7,319	11,851	24,350	44,225	70,014
200	102	209	313	602	960	1,698	2,485	3,462	6,264	10,143	20,840	37,851	59,923
250	90	185	277	534	851	1,505	2,203	3,069	5,552	8,990	18,470	33,547	53,109
300	82	168	251	484	771	1,363	1,996	2,780	5,030	8,145	16,735	30,396	48,120
400	70	143	215	414	660	1,167	1,708	2,380	4,305	6,971	14,323	26,015	41,185
500	62	127	191	367	585	1,034	1,514	2,109	3,816	6,178	12,694	23,056	36,501
1,000	43	87	131	252	402	711	1,041	1,450	2,623	4,246	8,725	15,847	25,087
1,500	34	70	105	203	323	571	836	1,164	2,106	3,410	7,006	12,725	20,146
2,000	29	60	90	173	276	488	715	996	1,802	2,919	5,997	10,891	17,242

For SI: 1 inch = 25.4 mm, 1 foot = 304.8 mm, 1 cubic foot per hour = 0.0283 m^3/h, 1 pound per square inch = 6.895 kPa, 1-inch water column = 0.2488 kPa.

❖ See the commentary for Section 402.3.

TABLE 402.4(6)
SCHEDULE 40 METALLIC PIPE

Gas	Natural
Inlet Pressure	1.0 psi or less
Pressure Drop	0.5 inch WC
Specific Gravity	0.60

	PIPE SIZE (in.)												
Nominal	1	1¼	1½	2	2½	3	3½	4	5	6	8	10	12
Actual ID	1.049	1.380	1.610	2.067	2.469	3.068	3.548	4.026	5.047	6.065	7.981	10.020	11.938
Length (ft)	Maximum Capacity in Cubic Feet of Gas per Hour												
50	284	583	873	1,681	2,680	4,738	6,937	9,663	17,482	28,308	58,161	105,636	167,236
100	195	400	600	1,156	1,842	3,256	4,767	6,641	12,015	19,456	39,974	72,603	114,940
150	157	322	482	928	1,479	2,615	3,828	5,333	9,649	15,624	32,100	58,303	92,301
200	134	275	412	794	1,266	2,238	3,277	4,565	8,258	13,372	27,474	49,900	78,998
250	119	244	366	704	1,122	1,983	2,904	4,046	7,319	11,851	24,350	44,225	70,014
300	108	221	331	638	1,017	1,797	2,631	3,666	6,632	10,738	22,062	40,071	63,438
400	92	189	283	546	870	1,538	2,252	3,137	5,676	9,190	18,883	34,296	54,295
500	82	168	251	484	771	1,363	1,996	2,780	5,030	8,145	16,735	30,396	48,120
1,000	56	115	173	333	530	937	1,372	1,911	3,457	5,598	11,502	20,891	33,073
1,500	45	93	139	267	426	752	1,102	1,535	2,776	4,496	9,237	16,776	26,559
2,000	39	79	119	229	364	644	943	1,313	2,376	3,848	7,905	14,358	22,731

For SI: 1 inch = 25.4 mm, 1 foot = 304.8 mm, 1 cubic foot per hour = 0.0283 m^3/h, 1 pound per square inch = 6.895 kPa, 1-inch water column = 0.2488 kPa.

❖ See the commentary for Section 402.3.

GAS PIPING INSTALLATIONS

TABLE 402.4(7)

TABLE 402.4(7)
SEMI-RIGID COPPER TUBING

Gas	Natural
Inlet Pressure	0.5 psi or less
Pressure Drop	0.3 inch WC
Specific Gravity	0.60

Nominal		TUBE SIZE (in.)									
	K & L	1/4	3/8	1/2	5/8	3/4	1	1 1/4	1 1/2	2	2 1/2
	ACR	3/8	1/2	5/8	3/4	7/8	1 1/8	1 3/8	1 5/8	2 1/8	2 5/8
Outside		0.375	0.500	0.625	0.750	0.875	1.125	1.375	1.625	2.125	2.625
Inside		0.305	0.402	0.527	0.652	0.745	0.995	1.245	1.481	1.959	2.435
Length (ft)		Maximum Capacity in Cubic Feet of Gas per Hour									
10		20	42	85	148	210	448	806	1,271	2,646	4,682
20		14	29	58	102	144	308	554	873	1,819	3,218
30		11	23	47	82	116	247	445	701	1,461	2,584
40		10	20	40	70	99	211	381	600	1,250	2,212
50		8.4	17	35	62	88	187	337	532	1,108	1,960
60		7.6	16	32	56	79	170	306	482	1,004	1,776
70		7.0	14	29	52	73	156	281	443	924	1,634
80		6.5	13	27	48	68	145	262	413	859	1,520
90		6.1	13	26	45	64	136	245	387	806	1,426
100		5.8	12	24	43	60	129	232	366	761	1,347
125		5.1	11	22	38	53	114	206	324	675	1,194
150		4.7	10	20	34	48	103	186	294	612	1,082
175		4.3	8.8	18	31	45	95	171	270	563	995
200		4.0	8.2	17	29	41	89	159	251	523	926
250		3.5	7.3	15	26	37	78	141	223	464	821
300		3.2	6.6	13	23	33	71	128	202	420	744

Note: Table capacities are based on Type K copper tubing inside diameter (shown), which has the smallest inside diameter of the copper tubing products.
For SI: 1 inch = 25.4 mm, 1 foot = 304.8 mm, 1 cubic foot per hour = 0.0283 m^3/h, 1 pound per square inch = 6.895 kPa, 1-inch water column = 0.2488 kPa.

❖ See the commentary for Section 402.3.

TABLE 402.4(8)
SEMI-RIGID COPPER TUBING

Gas	Natural
Inlet Pressure	0.5 psi or less
Pressure Drop	0.5 inch WC
Specific Gravity	0.60

Nominal		TUBE SIZE (in.)									
	K & L	$1/4$	$3/8$	$1/2$	$5/8$	$3/4$	1	$1 1/4$	$1 1/2$	2	$2 1/2$
	ACR	$3/8$	$1/2$	$5/8$	$3/4$	$7/8$	$1 1/8$	$1 3/8$	$1 5/8$	$2 1/8$	$2 5/8$
Outside		0.375	0.500	0.625	0.750	0.875	1.125	1.375	1.625	2.125	2.625
Inside		0.305	0.402	0.527	0.652	0.745	0.995	1.245	1.481	1.959	2.435
Length (ft)		Maximum Capacity in Cubic Feet of Gas per Hour									
10		27	55	111	195	276	590	1,062	1,675	3,489	6,173
20		18	38	77	134	190	406	730	1,151	2,398	4,242
30		15	30	61	107	152	326	586	925	1,926	3,407
40		13	26	53	92	131	279	502	791	1,648	2,916
50		11	23	47	82	116	247	445	701	1,461	2,584
60		10	21	42	74	105	224	403	635	1,323	2,341
70		9.3	19	39	68	96	206	371	585	1,218	2,154
80		8.6	18	36	63	90	192	345	544	1,133	2,004
90		8.1	17	34	59	84	180	324	510	1,063	1,880
100		7.6	16	32	56	79	170	306	482	1,004	1,776
125		6.8	14	28	50	70	151	271	427	890	1,574
150		6.1	13	26	45	64	136	245	387	806	1,426
175		5.6	12	24	41	59	125	226	356	742	1,312
200		5.2	11	22	39	55	117	210	331	690	1,221
250		4.7	10	20	34	48	103	186	294	612	1,082
300		4.2	8.7	18	31	44	94	169	266	554	980

Note: Table capacities are based on Type K copper tubing inside diameter (shown), which has the smallest inside diameter of the copper tubing products.

For SI: 1 inch = 25.4 mm, 1 foot = 304.8 mm, 1 cubic foot per hour = 0.0283 m³/h, 1 pound per square inch = 6.895 kPa, 1-inch water column = 0.2488 kPa.

❖ See the commentary for Section 402.3.

GAS PIPING INSTALLATIONS

TABLE 402.4(9)
SEMI-RIGID COPPER TUBING
Use this Table to Size Tubing from House Line Regulator to the Appliance.

Gas	Natural
Inlet Pressure	0.5 psi or less
Pressure Drop	1.0 inch WC
Specific Gravity	0.60

Nominal		TUBE SIZE (in.)									
	K & L	$1/4$	$3/8$	$1/2$	$5/8$	$3/4$	1	$1^{1}/_{4}$	$1^{1}/_{2}$	2	$2^{1}/_{2}$
	ACR	$3/8$	$1/2$	$5/8$	$3/4$	$7/8$	$1^{1}/_{8}$	$1^{3}/_{8}$	$1^{5}/_{8}$	$2^{1}/_{8}$	$2^{5}/_{8}$
Outside		0.375	0.500	0.625	0.750	0.875	1.125	1.375	1.625	2.125	2.625
Inside		0.305	0.402	0.527	0.652	0.745	0.995	1.245	1.481	1.959	2.435
Length (ft)		Maximum Capacity in Cubic Feet of Gas per Hour									
10		39	80	162	283	402	859	1,546	2,437	5,076	8,981
20		27	55	111	195	276	590	1,062	1,675	3,489	6,173
30		21	44	89	156	222	474	853	1,345	2,802	4,957
40		18	38	77	134	190	406	730	1,151	2,398	4,242
50		16	33	68	119	168	359	647	1,020	2,125	3,760
60		15	30	61	107	152	326	586	925	1,926	3,407
70		13	28	57	99	140	300	539	851	1,772	3,134
80		13	26	53	92	131	279	502	791	1,648	2,916
90		12	24	49	86	122	262	471	742	1,546	2,736
100		11	23	47	82	116	247	445	701	1,461	2,584
125		9.8	20	41	72	103	219	394	622	1,295	2,290
150		8.9	18	37	65	93	198	357	563	1,173	2,075
175		8.2	17	34	60	85	183	329	518	1,079	1,909
200		7.6	16	32	56	79	170	306	482	1,004	1,776
250		6.8	14	28	50	70	151	271	427	890	1,574
300		6.1	13	26	45	64	136	245	387	806	1,426

Note: Table capacities are based on Type K copper tubing inside diameter (shown), which has the smallest inside diameter of the copper tubing products.
For SI: 1 inch = 25.4 mm, 1 foot = 304.8 mm, 1 cubic foot per hour = 0.0283 m^3/h, 1 pound per square inch = 6.895 kPa, 1-inch water column = 0.2488 kPa.

❖ See the commentary for Section 402.3.

**TABLE 402.4(10)
SEMI-RIGID COPPER TUBING**

Gas	Natural
Inlet Pressure	2.0 psi or less
Pressure Drop	17.0 inch WC
Specific Gravity	0.60

Nominal		TUBE SIZE (in.)									
	K & L	$1/4$	$3/8$	$1/2$	$5/8$	$3/4$	1	$1 1/4$	$1 1/2$	2	$2 1/2$
	ACR	$3/8$	$1/2$	$5/8$	$3/4$	$7/8$	$1 1/8$	$1 3/8$	$1 5/8$	$2 1/8$	$2 5/8$
Outside		0.375	0.500	0.625	0.750	0.875	1.125	1.375	1.625	2.125	2.625
Inside		0.305	0.402	0.527	0.652	0.745	0.995	1.245	1.481	1.959	2.435
Length (ft)		Maximum Capacity in Cubic Feet of Gas per Hour									
10		190	391	796	1,391	1,974	4,216	7,591	11,968	24,926	44,100
20		130	269	547	956	1,357	2,898	5,217	8,226	17,132	30,310
30		105	216	439	768	1,089	2,327	4,189	6,605	13,757	24,340
40		90	185	376	657	932	1,992	3,586	5,653	11,775	20,832
50		79	164	333	582	826	1,765	3,178	5,010	10,436	18,463
60		72	148	302	528	749	1,599	2,879	4,540	9,455	16,729
70		66	137	278	486	689	1,471	2,649	4,177	8,699	15,390
80		62	127	258	452	641	1,369	2,464	3,886	8,093	14,318
90		58	119	243	424	601	1,284	2,312	3,646	7,593	13,434
100		55	113	229	400	568	1,213	2,184	3,444	7,172	12,689
125		48	100	203	355	503	1,075	1,936	3,052	6,357	11,246
150		44	90	184	321	456	974	1,754	2,765	5,760	10,190
175		40	83	169	296	420	896	1,614	2,544	5,299	9,375
200		38	77	157	275	390	834	1,501	2,367	4,930	8,721
250		33	69	140	244	346	739	1,330	2,098	4,369	7,730
300		30	62	126	221	313	670	1,205	1,901	3,959	7,004

Note: Table capacities are based on Type K copper tubing inside diameter (shown), which has the smallest inside diameter of the copper tubing products.
For SI: 1 inch = 25.4 mm, 1 foot = 304.8 mm, 1 cubic foot per hour = 0.0283 m³/h, 1 pound per square inch = 6.895 kPa, 1-inch water column = 0.2488 kPa.

❖ See the commentary for Section 402.3.

GAS PIPING INSTALLATIONS

TABLE 402.4(11)

**TABLE 402.4(11)
SEMI-RIGID COPPER TUBING**

Gas	Natural
Inlet Pressure	2.0 psi or less
Pressure Drop	1.0 psi
Specific Gravity	0.60

Nominal		TUBE SIZE (in.)									
	K & L	¹/₄	³/₈	¹/₂	⁵/₈	³/₄	1	1¹/₄	1¹/₂	2	2¹/₂
	ACR	³/₈	¹/₂	⁵/₈	³/₄	⁷/₈	1¹/₈	1³/₈	1⁵/₈	2¹/₈	2⁵/₈
Outside		0.375	0.500	0.625	0.750	0.875	1.125	1.375	1.625	2.125	2.625
Inside		0.305	0.402	0.527	0.652	0.745	0.995	1.245	1.481	1.959	2.435
Length (ft)		Maximum Capacity in Cubic Feet of Gas per Hour									
10		245	506	1,030	1,800	2,554	5,455	9,820	15,483	32,247	57,051
20		169	348	708	1,237	1,755	3,749	6,749	10,641	22,163	39,211
30		135	279	568	993	1,409	3,011	5,420	8,545	17,798	31,488
40		116	239	486	850	1,206	2,577	4,639	7,314	15,232	26,949
50		103	212	431	754	1,069	2,284	4,111	6,482	13,500	23,885
60		93	192	391	683	969	2,069	3,725	5,873	12,232	21,641
70		86	177	359	628	891	1,904	3,427	5,403	11,253	19,910
80		80	164	334	584	829	1,771	3,188	5,027	10,469	18,522
90		75	154	314	548	778	1,662	2,991	4,716	9,823	17,379
100		71	146	296	518	735	1,570	2,826	4,455	9,279	16,416
125		63	129	263	459	651	1,391	2,504	3,948	8,223	14,549
150		57	117	238	416	590	1,260	2,269	3,577	7,451	13,183
175		52	108	219	383	543	1,160	2,087	3,291	6,855	12,128
200		49	100	204	356	505	1,079	1,942	3,062	6,377	11,283
250		43	89	181	315	448	956	1,721	2,714	5,652	10,000
300		39	80	164	286	406	866	1,559	2,459	5,121	9,060

Note: Table capacities are based on Type K copper tubing inside diameter (shown), which has the smallest inside diameter of the copper tubing products.
For SI: 1 inch = 25.4 mm, 1 foot = 304.8 mm, 1 cubic foot per hour = 0.0283 m^3/h, 1 pound per square inch = 6.895 kPa.

❖ See the commentary for Section 402.3.

TABLE 402.4(12)
SEMI-RIGID COPPER TUBING
Pipe Sizing Between Point of Delivery and the House Line Regulator. Total Load Supplied by a Single House Line Regulator Not Exceeding 150 Cubic Feet per Hour.[2]

Gas	Natural
Inlet Pressure	2.0 psi
Pressure Drop	1.5 psi
Specific Gravity	0.60

Nominal		TUBE SIZE (in.)									
	K & L	$1/4$	$3/8$	$1/2$	$5/8$	$3/4$	1	$1\,1/4$	$1\,1/2$	2	$2\,1/2$
	ACR	$3/8$	$1/2$	$5/8$	$3/4$	$7/8$	$1\,1/8$	$1\,3/8$	$1\,5/8$	$2\,1/8$	$2\,5/8$
Outside		0.375	0.500	0.625	0.750	0.875	1.125	1.375	1.625	2.125	2.625
Inside[1]		0.305	0.402	0.527	0.652	0.745	0.995	1.245	1.481	1.959	2.435
Length (ft)		Maximum Capacity in Cubic Feet of Gas per Hour									
10		303	625	1,272	2,224	3,155	6,739	12,131	19,127	39,837	70,481
20		208	430	874	1,528	2,168	4,631	8,338	13,146	27,380	48,441
30		167	345	702	1,227	1,741	3,719	6,696	10,557	21,987	38,900
40		143	295	601	1,050	1,490	3,183	5,731	9,035	18,818	33,293
50		127	262	533	931	1,321	2,821	5,079	8,008	16,678	29,507
60		115	237	483	843	1,197	2,556	4,602	7,256	15,112	26,736
70		106	218	444	776	1,101	2,352	4,234	6,675	13,903	24,597
80		98	203	413	722	1,024	2,188	3,939	6,210	12,934	22,882
90		92	191	388	677	961	2,053	3,695	5,826	12,135	21,470
100		87	180	366	640	908	1,939	3,491	5,504	11,463	20,280
125		77	159	324	567	804	1,718	3,094	4,878	10,159	17,974
150		70	145	294	514	729	1,557	2,803	4,420	9,205	16,286
175		64	133	270	473	671	1,432	2,579	4,066	8,469	14,983
200		60	124	252	440	624	1,333	2,399	3,783	7,878	13,938
250		53	110	223	390	553	1,181	2,126	3,352	6,982	12,353
300		48	99	202	353	501	1,070	1,927	3,038	6,327	11,193

Notes:

1. Table capacities are based on Type K copper tubing inside diameter (shown), which has the smallest inside diameter of the copper tubing products.
2. When this table is used to size the tubing upstream of a line pressure regulator, the pipe or tubing downstream of the line pressure regulator shall be sized using a pressure drop no greater than 1 inch w.c.

For SI: 1 inch = 25.4 mm, 1 foot = 304.8 mm, 1 cubic foot per hour = 0.0283 m^3/h, 1 pound per square inch = 6.895 kPa.

❖ See the commentary for Section 402.3.

GAS PIPING INSTALLATIONS

**TABLE 402.4(13)
SEMI-RIGID COPPER TUBING**

Gas	Natural
Inlet Pressure	5.0 psi or less
Pressure Drop	3.5 psi
Specific Gravity	0.60

Nominal		TUBE SIZE (in.)									
	K & L	1/4	3/8	1/2	5/8	3/4	1	1 1/4	1 1/2	2	2 1/2
	ACR	3/8	1/2	5/8	3/4	7/8	1 1/8	1 3/8	1 5/8	2 1/8	2 5/8
Outside		0.375	0.500	0.625	0.750	0.875	1.125	1.375	1.625	2.125	2.625
Inside		0.305	0.402	0.527	0.652	0.745	0.995	1.245	1.481	1.959	2.435
Length (ft)		Maximum Capacity in Cubic Feet of Gas per Hour									
10		511	1,054	2,144	3,747	5,315	11,354	20,441	32,229	67,125	118,758
20		351	724	1,473	2,575	3,653	7,804	14,049	22,151	46,135	81,622
30		282	582	1,183	2,068	2,934	6,267	11,282	17,788	37,048	65,545
40		241	498	1,013	1,770	2,511	5,364	9,656	15,224	31,708	56,098
50		214	441	898	1,569	2,225	4,754	8,558	13,493	28,102	49,719
60		194	400	813	1,421	2,016	4,307	7,754	12,225	25,463	45,049
70		178	368	748	1,308	1,855	3,962	7,134	11,247	23,425	41,444
80		166	342	696	1,216	1,726	3,686	6,636	10,463	21,793	38,556
90		156	321	653	1,141	1,619	3,459	6,227	9,817	20,447	36,176
100		147	303	617	1,078	1,529	3,267	5,882	9,273	19,315	34,172
125		130	269	547	955	1,356	2,896	5,213	8,219	17,118	30,286
150		118	243	495	866	1,228	2,624	4,723	7,447	15,510	27,441
175		109	224	456	796	1,130	2,414	4,345	6,851	14,269	25,245
200		101	208	424	741	1,051	2,245	4,042	6,374	13,275	23,486
250		90	185	376	657	932	1,990	3,583	5,649	11,765	20,815
300		81	167	340	595	844	1,803	3,246	5,118	10,660	18,860

Note: Table capacities are based on Type K copper tubing inside diameter (shown), which has the smallest inside diameter of the copper tubing products.
For SI: 1 inch = 25.4 mm, 1 foot = 304.8 mm, 1 cubic foot per hour = 0.0283 m^3/h, 1 pound per square inch = 6.895 kPa.

❖ See the commentary for Section 402.3.

TABLE 402.4(14)
CORRUGATED STAINLESS STEEL TUBING (CSST)

Gas	Natural
Inlet Pressure	0.5 psi or less
Pressure Drop	0.5 inch WC
Specific Gravity	0.60

Flow Designation	TUBE SIZE (EHD*)										
	13	15	18	19	23	25	30	31	37	46	62
Length (ft)	Maximum Capacity in Cubic Feet of Gas per Hour										
5	46	63	115	134	225	270	471	546	895	1,790	4,142
10	32	44	82	95	161	192	330	383	639	1,261	2,934
15	25	35	66	77	132	157	267	310	524	1,027	2,398
20	22	31	58	67	116	137	231	269	456	888	2,078
25	19	27	52	60	104	122	206	240	409	793	1,860
30	18	25	47	55	96	112	188	218	374	723	1,698
40	15	21	41	47	83	97	162	188	325	625	1,472
50	13	19	37	42	75	87	144	168	292	559	1,317
60	12	17	34	38	68	80	131	153	267	509	1,203
70	11	16	31	36	63	74	121	141	248	471	1,114
80	10	15	29	33	60	69	113	132	232	440	1,042
90	10	14	28	32	57	65	107	125	219	415	983
100	9	13	26	30	54	62	101	118	208	393	933
150	7	10	20	23	42	48	78	91	171	320	762
200	6	9	18	21	38	44	71	82	148	277	661
250	5	8	16	19	34	39	63	74	133	247	591
300	5	7	15	17	32	36	57	67	95	226	540

Note: Table includes losses for four 90-degree bends and two end fittings. Tubing runs with larger numbers of bends and/or fittings shall be increased by an equivalent length of tubing to the following equation: $L = 1.3n$ where L is additional length (ft) of tubing and n is the number of additional fittings and/or bends.

*EHD— Equivalent Hydraulic Diameter, which is a measure of the relative hydraulic efficiency between different tubing sizes. The greater the value of EHD, the greater the gas capacity of the tubing.

For SI: 1 foot = 304.8 mm, 1 cubic foot per hour = 0.0283 m^3/h, 1 pound per square inch = 6.895 kPa, 1-inch water column = 0.2488 kPa, 1 degree = 0.01745 rad.

❖ See the commentary for Section 402.3.

GAS PIPING INSTALLATIONS

TABLE 402.4(15)
CORRUGATED STAINLESS STEEL TUBING (CSST)

Gas	Natural
Inlet Pressure	0.5 psi or less
Pressure Drop	3.0 inch WC
Specific Gravity	0.60

Flow Designation	TUBE SIZE (EHD*)										
	13	15	18	19	23	25	30	31	37	46	62
Length (ft)	Maximum Capacity in Cubic Feet of Gas per Hour										
5	120	160	277	327	529	649	1,182	1,365	2,141	4,428	10,103
10	83	112	197	231	380	462	828	958	1,528	3,199	7,156
15	67	90	161	189	313	379	673	778	1,254	2,541	5,848
20	57	78	140	164	273	329	580	672	1,090	2,197	5,069
25	51	69	125	147	245	295	518	599	978	1,963	4,536
30	46	63	115	134	225	270	471	546	895	1,790	4,142
40	39	54	100	116	196	234	407	471	778	1,548	3,590
50	35	48	89	104	176	210	363	421	698	1,383	3,213
60	32	44	82	95	161	192	330	383	639	1,261	2,934
70	29	41	76	88	150	178	306	355	593	1,166	2,717
80	27	38	71	82	141	167	285	331	555	1,090	2,543
90	26	36	67	77	133	157	268	311	524	1,027	2,398
100	24	34	63	73	126	149	254	295	498	974	2,276
150	19	27	52	60	104	122	206	240	409	793	1,860
200	17	23	45	52	91	106	178	207	355	686	1,612
250	15	21	40	46	82	95	159	184	319	613	1,442
300	13	19	37	42	75	87	144	168	234	559	1,317

Note: Table includes losses for four 90-degree bends and two end fittings. Tubing runs with larger numbers of bends and/or fittings shall be increased by an equivalent length of tubing to the following equation: $L = 1.3n$ where L is additional length (ft) of tubing and n is the number of additional fittings and/or bends.

*EHD— Equivalent Hydraulic Diameter, which is a measure of the relative hydraulic efficiency between different tubing sizes. The greater the value of EHD, the greater the gas capacity of the tubing.

For SI: 1 foot = 304.8 mm, 1 cubic foot per hour = 0.0283 m^3/h, 1 pound per square inch = 6.895 kPa, 1-inch water column = 0.2488 kPa, 1 degree = 0.01745 rad.

❖ See the commentary for Section 402.3.

TABLE 402.4(16)
CORRUGATED STAINLESS STEEL TUBING (CSST)

Gas	Natural
Inlet Pressure	0.5 psi or less
Pressure Drop	6.0 inch WC
Specific Gravity	0.60

Flow Designation	TUBE SIZE (EHD*)										
	13	15	18	19	23	25	30	31	37	46	62
Length (ft)	Maximum Capacity in Cubic Feet of Gas per Hour										
5	173	229	389	461	737	911	1,687	1,946	3,000	6,282	14,263
10	120	160	277	327	529	649	1,182	1,365	2,141	4,428	10,103
15	96	130	227	267	436	532	960	1,110	1,758	3,607	8,257
20	83	112	197	231	380	462	828	958	1,528	3,119	7,156
25	74	99	176	207	342	414	739	855	1,371	2,786	6,404
30	67	90	161	189	313	379	673	778	1,254	2,541	5,848
40	57	78	140	164	273	329	580	672	1,090	2,197	5,069
50	51	69	125	147	245	295	518	599	978	1,963	4,536
60	46	63	115	134	225	270	471	546	895	1,790	4,142
70	42	58	106	124	209	250	435	505	830	1,656	3,837
80	39	54	100	116	196	234	407	471	778	1,548	3,590
90	37	51	94	109	185	221	383	444	735	1,458	3,386
100	35	48	89	104	176	210	363	421	698	1,383	3,213
150	28	39	73	85	145	172	294	342	573	1,126	2,626
200	24	34	63	73	126	149	254	295	498	974	2,276
250	21	30	57	66	114	134	226	263	447	870	2,036
300	19	27	52	60	104	122	206	240	409	793	1,860

Note: Table includes losses for four 90-degree bends and two end fittings. Tubing runs with larger numbers of bends and/or fittings shall be increased by an equivalent length of tubing to the following equation: $L = 1.3n$ where L is additional length (ft) of tubing and n is the number of additional fittings and/or bends.

*EHD— Equivalent Hydraulic Diameter, which is a measure of the relative hydraulic efficiency between different tubing sizes. The greater the value of EHD, the greater the gas capacity of the tubing.

For SI: 1 foot = 304.8 mm, 1 cubic foot per hour = 0.0283 m³/h, 1 pound per square inch = 6.895 kPa, 1-inch water column = 0.2488 kPa, 1 degree = 0.01745 rad.

❖ See the commentary for Section 402.3.

GAS PIPING INSTALLATIONS

TABLE 402.4(17)

TABLE 402.4(17)
CORRUGATED STAINLESS STEEL TUBING (CSST)

	Gas	Natural
	Inlet Pressure	2.0 psi
	Pressure Drop	1.0 psi
	Specific Gravity	0.60

Flow Designation	TUBE SIZE (EHD*)										
	13	15	18	19	23	25	30	31	37	46	62
Length (ft)	Maximum Capacity in Cubic Feet of Gas per Hour										
10	270	353	587	700	1,098	1,372	2,592	2,986	4,509	9,599	21,637
25	166	220	374	444	709	876	1,620	1,869	2,887	6,041	13,715
30	151	200	342	405	650	801	1,475	1,703	2,642	5,509	12,526
40	129	172	297	351	567	696	1,273	1,470	2,297	4,763	10,855
50	115	154	266	314	510	624	1,135	1,311	2,061	4,255	9,715
75	93	124	218	257	420	512	922	1,066	1,692	3,467	7,940
80	89	120	211	249	407	496	892	1,031	1,639	3,355	7,689
100	79	107	189	222	366	445	795	920	1,471	2,997	6,881
150	64	87	155	182	302	364	646	748	1,207	2,442	5,624
200	55	75	135	157	263	317	557	645	1,049	2,111	4,874
250	49	67	121	141	236	284	497	576	941	1,886	4,362
300	44	61	110	129	217	260	453	525	862	1,720	3,983
400	38	52	96	111	189	225	390	453	749	1,487	3,452
500	34	46	86	100	170	202	348	404	552	1,329	3,089

Notes:

1. Table does not include effect of pressure drop across the line regulator. Where regulator loss exceeds $^3/_4$ psi, DO NOT USE THIS TABLE. Consult with regulator manufacturer for pressure drops and capacity factors. Pressure drops across a regulator may vary with flow rate.
2. CAUTION: Capacities shown in table may exceed maximum capacity for a selected regulator. Consult with regulator or tubing manufacturer for guidance.
3. Table includes losses for four 90-degree bends and two end fittings. Tubing runs with larger numbers of bends and/or fittings shall be increased by an equivalent length of tubing to the following equation: $L = 1.3n$ where L is additional length (ft) of tubing and n is the number of additional fittings and/or bends.

*EHD— Equivalent Hydraulic Diameter, which is a measure of the relative hydraulic efficiency between different tubing sizes. The greater the value of EHD, the greater the gas capacity of the tubing.

For SI: 1 foot = 304.8 mm, 1 cubic foot per hour = 0.0283 m^3/h, 1 pound per square inch = 6.895 kPa, 1 degree = 0.01745 rad.

❖ See the commentary for Section 402.3.

TABLE 402.4(18)
CORRUGATED STAINLESS STEEL TUBING (CSST)

Gas	Natural
Inlet Pressure	5.0 psi
Pressure Drop	3.5 psi
Specific Gravity	0.60

Flow Designation	TUBE SIZE (EHD*)										
	13	15	18	19	23	25	30	31	37	46	62
Length (ft)	Maximum Capacity in Cubic Feet of Gas per Hour										
10	523	674	1,084	1,304	1,995	2,530	4,923	5,659	8,295	18,080	40,353
25	322	420	691	827	1,289	1,616	3,077	3,543	5,311	11,378	25,580
30	292	382	632	755	1,181	1,478	2,803	3,228	4,860	10,377	23,361
40	251	329	549	654	1,031	1,284	2,418	2,786	4,225	8,972	20,246
50	223	293	492	586	926	1,151	2,157	2,486	3,791	8,015	18,119
75	180	238	403	479	763	944	1,752	2,021	3,112	6,530	14,809
80	174	230	391	463	740	915	1,694	1,955	3,016	6,320	14,341
100	154	205	350	415	665	820	1,511	1,744	2,705	5,646	12,834
150	124	166	287	339	548	672	1,228	1,418	2,221	4,600	10,489
200	107	143	249	294	478	584	1,060	1,224	1,931	3,977	9,090
250	95	128	223	263	430	524	945	1,092	1,732	3,553	8,135
300	86	116	204	240	394	479	860	995	1,585	3,240	7,430
400	74	100	177	208	343	416	742	858	1,378	2,802	6,439
500	66	89	159	186	309	373	662	766	1,035	2,503	5,762

Notes:

1. Table does not include effect of pressure drop across the line regulator. Where regulator loss exceeds $^3/_4$ psi, DO NOT USE THIS TABLE. Consult with regulator manufacturer for pressure drops and capacity factors. Pressure drops across a regulator may vary with flow rate.
2. CAUTION: Capacities shown in table may exceed maximum capacity for a selected regulator. Consult with regulator or tubing manufacturer for guidance.
3. Table includes losses for four 90-degree bends and two end fittings. Tubing runs with larger numbers of bends and/or fittings shall be increased by an equivalent length of tubing to the following equation: $L = 1.3n$ where L is additional length (ft) of tubing and n is the number of additional fittings and/or bends.

*EHD— Equivalent Hydraulic Diameter, which is a measure of the relative hydraulic efficiency between different tubing sizes. The greater the value of EHD, the greater the gas capacity of the tubing.

For SI: 1 foot = 304.8 mm, 1 cubic foot per hour = 0.0283 m^3/h, 1 pound per square inch = 6.895 kPa, 1 degree = 0.01745 rad.

❖ See the commentary for Section 402.3.

GAS PIPING INSTALLATIONS

TABLE 402.4(19)

TABLE 402.4(19)
POLYETHYLENE PLASTIC PIPE

Gas	Natural
Inlet Pressure	1.0 psi or less
Pressure Drop	0.3 inch WC
Specific Gravity	0.60

	PIPE SIZE (in.)					
Nominal OD	1/2	3/4	1	1 1/4	1 1/2	2
Designation	SDR 9.33	SDR 11.0	SDR 11.00	SDR 10.00	SDR 11.00	SDR 11.00
Actual ID	0.660	0.860	1.077	1.328	1.554	1.943
Length (ft)	Maximum Capacity in Cubic Feet of Gas per Hour					
10	153	305	551	955	1,442	2,590
20	105	210	379	656	991	1,780
30	84	169	304	527	796	1,430
40	72	144	260	451	681	1,224
50	64	128	231	400	604	1,084
60	58	116	209	362	547	983
70	53	107	192	333	503	904
80	50	99	179	310	468	841
90	46	93	168	291	439	789
100	44	88	159	275	415	745
125	39	78	141	243	368	661
150	35	71	127	221	333	598
175	32	65	117	203	306	551
200	30	60	109	189	285	512

For SI: 1 inch = 25.4 mm, 1 foot = 304.8 mm, 1 cubic foot per hour = 0.0283 m^3/h, 1 pound per square inch = 6.895 kPa, 1-inch water column = 0.2488 kPa.

❖ See the commentary for Section 402.3.

TABLE 402.4(20)
POLYETHYLENE PLASTIC PIPE

Gas	Natural
Inlet Pressure	1.0 psi or less
Pressure Drop	0.5 inch WC
Specific Gravity	0.60

	PIPE SIZE (in.)					
Nominal OD	1/2	3/4	1	1 1/4	1 1/2	2
Designation	SDR 9.33	SDR 11.0	SDR 11.00	SDR 10.00	SDR 11.00	SDR 11.00
Actual ID	0.660	0.860	1.077	1.328	1.554	1.943
Length (ft)	Maximum Capacity in Cubic Feet of Gas per Hour					
10	201	403	726	1,258	1,900	3,415
20	138	277	499	865	1,306	2,347
30	111	222	401	695	1,049	1,885
40	95	190	343	594	898	1,613
50	84	169	304	527	796	1,430
60	76	153	276	477	721	1,295
70	70	140	254	439	663	1,192
80	65	131	236	409	617	1,109
90	61	123	221	383	579	1,040
100	58	116	209	362	547	983
125	51	103	185	321	485	871
150	46	93	168	291	439	789
175	43	86	154	268	404	726
200	40	80	144	249	376	675

For SI: 1 inch = 25.4 mm, 1 foot = 304.8 mm, 1 cubic foot per hour = 0.0283 m^3/h, 1 pound per square inch = 6.895 kPa, 1-inch water column = 0.2488 kPa.

❖ See the commentary for Section 402.3.

GAS PIPING INSTALLATIONS

TABLE 402.4(21)
POLYETHYLENE PLASTIC PIPE

Gas	Natural
Inlet Pressure	2.0 psi
Pressure Drop	1.0 psi
Specific Gravity	0.60

	PIPE SIZE (in.)					
Nominal OD	1/2	3/4	1	1 1/4	1 1/2	2
Designation	SDR 9.33	SDR 11.0	SDR 11.00	SDR 10.00	SDR 11.00	SDR 11.00
Actual ID	0.660	0.860	1.077	1.328	1.554	1.943
Length (ft)	Maximum Capacity in Cubic Feet of Gas per Hour					
10	1,858	3,721	6,714	11,631	17,565	31,560
20	1,277	2,557	4,614	7,994	12,072	21,691
30	1,026	2,054	3,706	6,420	9,695	17,419
40	878	1,758	3,172	5,494	8,297	14,908
50	778	1,558	2,811	4,869	7,354	13,213
60	705	1,412	2,547	4,412	6,663	11,972
70	649	1,299	2,343	4,059	6,130	11,014
80	603	1,208	2,180	3,776	5,703	10,246
90	566	1,134	2,045	3,543	5,351	9,614
100	535	1,071	1,932	3,347	5,054	9,081
125	474	949	1,712	2,966	4,479	8,048
150	429	860	1,551	2,688	4,059	7,292
175	395	791	1,427	2,473	3,734	6,709
200	368	736	1,328	2,300	3,474	6,241

For SI: 1 inch = 25.4 mm, 1 foot = 304.8 mm, 1 cubic foot per hour = 0.0283 m^3/h, 1 pound per square inch = 6.895 kPa.

❖ See the commentary for Section 402.3.

TABLE 402.4(22)
SCHEDULE 40 METALLIC PIPE
Pipe Sizing Between First Stage (High Pressure Regulator)
and Second Stage (Low Pressure Regulator)

Gas	Undiluted propane
Inlet Pressure	10.0 psi
Pressure Drop	1.0 psi
Specific Gravity	1.50

Nominal Inside	|	|	|	PIPE SIZE (in.)					
	1/2	3/4	1	1 1/4	1 1/2	2	3	3 1/2	4
Actual	0.622	0.824	1.049	1.38	1.61	2.067	3.068	3.548	4.026
Length (ft)	Maximum Capacity in Thousands of Btu/h								
30	1,834	3,835	7,225	14,834	22,225	42,804	120,604	176,583	245,995
40	1,570	3,283	6,184	12,696	19,022	36,634	103,222	151,132	210,539
50	1,391	2,909	5,480	11,252	16,859	32,468	91,484	133,946	186,597
60	1,261	2,636	4,966	10,195	15,275	29,419	82,891	121,364	169,071
70	1,160	2,425	4,568	9,379	14,053	27,065	76,258	111,654	155,543
80	1,079	2,256	4,250	8,726	13,074	25,179	70,944	103,872	144,703
90	1,012	2,117	3,988	8,187	12,267	23,624	66,564	97,460	135,770
100	956	2,000	3,767	7,733	11,587	22,315	62,876	92,060	128,247
150	768	1,606	3,025	6,210	9,305	17,920	50,492	73,927	102,987
200	657	1,374	2,589	5,315	7,964	15,337	43,214	63,272	88,144
250	582	1,218	2,294	4,711	7,058	13,593	38,300	56,077	78,120
300	528	1,104	2,079	4,268	6,395	12,316	34,703	50,810	70,782
350	486	1,015	1,913	3,927	5,883	11,331	31,926	46,744	65,119
400	452	945	1,779	3,653	5,473	10,541	29,701	43,487	60,581
450	424	886	1,669	3,428	5,135	9,890	27,867	40,802	56,841
500	400	837	1,577	3,238	4,851	9,342	26,323	38,541	53,691
600	363	759	1,429	2,934	4,395	8,465	23,851	34,921	48,648
700	334	698	1,314	2,699	4,044	7,788	21,943	32,127	44,756
800	310	649	1,223	2,511	3,762	7,245	20,413	29,888	41,637
900	291	609	1,147	2,356	3,530	6,798	19,153	28,043	39,066
1,000	275	575	1,084	2,225	3,334	6,421	18,092	26,489	36,902
1,500	221	462	870	1,787	2,677	5,156	14,528	21,272	29,633
2,000	189	395	745	1,529	2,291	4,413	12,435	18,206	25,362

For SI: 1 inch = 25.4 mm, 1 foot = 304.8 mm, 1 pound per square inch = 6.895 kPa, 1 British thermal unit per hour = 0.2931 W.

❖ See the commentary for Section 402.3.

GAS PIPING INSTALLATIONS

TABLE 402.4(23)
SCHEDULE 40 METALLIC PIPE

Gas	Undiluted propane
Inlet Pressure	2.0 psi
Pressure Drop	1.0 psi
Specific Gravity	1.50

	PIPE SIZE (in.)								
Nominal	1/2	3/4	1	1 1/4	1 1/2	2	2 1/2	3	4
Actual ID	0.622	0.824	1.049	1.380	1.610	2.067	2.469	3.068	4.026
Length (ft)	Maximum Capacity in Thousands of Btu/h								
10	2,676	5,595	10,539	21,638	32,420	62,438	99,516	175,927	358,835
20	1,839	3,845	7,243	14,872	22,282	42,913	68,397	120,914	246,625
30	1,477	3,088	5,817	11,942	17,893	34,461	54,925	97,098	198,049
40	1,264	2,643	4,978	10,221	15,314	29,494	47,009	83,103	169,504
50	1,120	2,342	4,412	9,059	13,573	26,140	41,663	73,653	150,229
60	1,015	2,122	3,998	8,208	12,298	23,685	37,750	66,735	136,118
70	934	1,952	3,678	7,551	11,314	21,790	34,729	61,395	125,227
80	869	1,816	3,422	7,025	10,526	20,271	32,309	57,116	116,499
90	815	1,704	3,210	6,591	9,876	19,020	30,314	53,590	109,307
100	770	1,610	3,033	6,226	9,329	17,966	28,635	50,621	103,251
125	682	1,427	2,688	5,518	8,268	15,923	25,378	44,865	91,510
150	618	1,293	2,435	5,000	7,491	14,427	22,995	40,651	82,914
175	569	1,189	2,240	4,600	6,892	13,273	21,155	37,398	76,280
200	529	1,106	2,084	4,279	6,411	12,348	19,681	34,792	70,964

For SI: 1 inch = 25.4 mm, 1 foot = 304.8 mm, 1 pound per square inch = 6.895 kPa, 1 British thermal unit per hour = 0.2931 W.

❖ See the commentary for Section 402.3.

TABLE 402.4(24)
SCHEDULE 40 METALLIC PIPE
Pipe Sizing Between Single or Second Stage
(Low Pressure Regulator) and Appliance

Gas	Undiluted propane
Inlet Pressure	11.0 inch WC
Pressure Drop	0.5 inch WC
Specific Gravity	1.50

	PIPE SIZE (in.)								
Nominal Inside	1/2	3/4	1	1 1/4	1 1/2	2	3	3 1/2	4
Actual	0.622	0.824	1.049	1.38	1.61	2.067	3.068	3.548	4.026
Length (ft)	Maximum Capacity in Thousands of Btu/h								
10	291	608	1,145	2,352	3,523	6,786	19,119	27,993	38,997
20	200	418	787	1,616	2,422	4,664	13,141	19,240	26,802
30	160	336	632	1,298	1,945	3,745	10,552	15,450	21,523
40	137	287	541	1,111	1,664	3,205	9,031	13,223	18,421
50	122	255	480	984	1,475	2,841	8,004	11,720	16,326
60	110	231	434	892	1,337	2,574	7,253	10,619	14,793
80	94	197	372	763	1,144	2,203	6,207	9,088	12,661
100	84	175	330	677	1,014	1,952	5,501	8,055	11,221
125	74	155	292	600	899	1,730	4,876	7,139	9,945
150	67	140	265	543	814	1,568	4,418	6,468	9,011
200	58	120	227	465	697	1,342	3,781	5,536	7,712
250	51	107	201	412	618	1,189	3,351	4,906	6,835
300	46	97	182	373	560	1,078	3,036	4,446	6,193
350	42	89	167	344	515	991	2,793	4,090	5,698
400	40	83	156	320	479	922	2,599	3,805	5,301

For SI: 1 inch = 25.4 mm, 1 foot = 304.8 mm, 1-inch water column = 0.2488 kPa, 1 British thermal unit per hour = 0.2931 W.

❖ See the commentary for Section 402.3.

GAS PIPING INSTALLATIONS

TABLE 402.4(25)
SEMI-RIGID COPPER TUBING
Sizing Between First Stage (High Pressure Regulator) and Second Stage (Low Pressure Regulator)

Gas	Undiluted propane
Inlet Pressure	10.0 psi
Pressure Drop	1.0 psi
Specific Gravity	1.50

Nominal		TUBE SIZE (in.)									
	K & L	$1/4$	$3/8$	$1/2$	$5/8$	$3/4$	1	$1 1/4$	$1 1/2$	2	$2 1/2$
	ACR	$3/8$	$1/2$	$5/8$	$3/4$	$7/8$	$1 1/8$	$1 3/8$	$1 5/8$	$2 1/8$	$2 5/8$
Outside		0.375	0.500	0.625	0.750	0.875	1.125	1.375	1.625	2.125	2.625
Inside		0.305	0.402	0.527	0.652	0.745	0.995	1.245	1.481	1.959	2.435
Length (ft)		Maximum Capacity in Thousands of Btu/h									
10		513	1,058	2,152	3,760	5,335	11,396	20,516	32,347	67,371	119,193
20		352	727	1,479	2,585	3,667	7,832	14,101	22,232	46,303	81,921
30		283	584	1,188	2,075	2,944	6,290	11,323	17,853	37,183	65,785
40		242	500	1,016	1,776	2,520	5,383	9,691	15,280	31,824	56,304
50		215	443	901	1,574	2,234	4,771	8,589	13,542	28,205	49,901
60		194	401	816	1,426	2,024	4,323	7,782	12,270	25,556	45,214
70		179	369	751	1,312	1,862	3,977	7,160	11,288	23,511	41,596
80		166	343	699	1,221	1,732	3,700	6,661	10,502	21,873	38,697
90		156	322	655	1,145	1,625	3,471	6,250	9,853	20,522	36,308
100		147	304	619	1,082	1,535	3,279	5,903	9,307	19,385	34,297
125		131	270	549	959	1,361	2,906	5,232	8,249	17,181	30,396
150		118	244	497	869	1,233	2,633	4,741	7,474	15,567	27,541
175		109	225	457	799	1,134	2,423	4,361	6,876	14,321	25,338
200		101	209	426	744	1,055	2,254	4,057	6,397	13,323	23,572
225		95	196	399	698	990	2,115	3,807	6,002	12,501	22,117
250		90	185	377	659	935	1,997	3,596	5,669	11,808	20,891
275		85	176	358	626	888	1,897	3,415	5,385	11,215	19,841
300		81	168	342	597	847	1,810	3,258	5,137	10,699	18,929

For SI: 1 inch = 25.4 mm, 1 foot = 304.8 mm, 1 pound per square inch = 6.895 kPa, 1 British thermal unit per hour = 0.2931 W.

Note: Table capacities are based on Type K copper tubing inside diameter (shown), which has the smallest inside diameter of the copper tubing products.

❖ See the commentary for Section 402.3.

TABLE 402.4(26)
SEMI-RIGID COPPER TUBING
Sizing Between Single or Second Stage
(Low Pressure Regulator) and Appliance

Gas	Undiluted propane
Inlet Pressure	11.0 inch WC
Pressure Drop	0.5 inch WC
Specific Gravity	1.50

Nominal		TUBE SIZE (in.)									
	K & L	1/4	3/8	1/2	5/8	3/4	1	1 1/4	1 1/2	2	2 1/2
	ACR	3/8	1/2	5/8	3/4	7/8	1 1/8	1 3/8	1 5/8	2 1/8	2 5/8
Outside		0.375	0.500	0.625	0.750	0.875	1.125	1.375	1.625	2.125	2.625
Inside		0.305	0.402	0.527	0.652	0.745	0.995	1.245	1.481	1.959	2.435
Length (ft)		Maximum Capacity in Thousands of Btu/h									
10		45	93	188	329	467	997	1,795	2,830	5,895	10,429
20		31	64	129	226	321	685	1,234	1,945	4,051	7,168
30		25	51	104	182	258	550	991	1,562	3,253	5,756
40		21	44	89	155	220	471	848	1,337	2,784	4,926
50		19	39	79	138	195	417	752	1,185	2,468	4,366
60		17	35	71	125	177	378	681	1,074	2,236	3,956
70		16	32	66	115	163	348	626	988	2,057	3,639
80		15	30	61	107	152	324	583	919	1,914	3,386
90		14	28	57	100	142	304	547	862	1,796	3,177
100		13	27	54	95	134	287	517	814	1,696	3,001
125		11	24	48	84	119	254	458	722	1,503	2,660
150		10	21	44	76	108	230	415	654	1,362	2,410
175		10	20	40	70	99	212	382	602	1,253	2,217
200		8.9	18	37	65	92	197	355	560	1,166	2,062
225		8.3	17	35	61	87	185	333	525	1,094	1,935
250		7.9	16	33	58	82	175	315	496	1,033	1,828
275		7.5	15	31	55	78	166	299	471	981	1,736
300		7.1	15	30	52	74	158	285	449	936	1,656

For SI: 1 inch = 25.4 mm, 1 foot = 304.8 mm, 1-inch water column = 0.2488 kPa, 1 British thermal unit per hour = 0.2931 W.

Note: Table capacities are based on Type K copper tubing inside diameter (shown), which has the smallest inside diameter of the copper tubing products.

❖ See the commentary for Section 402.3.

GAS PIPING INSTALLATIONS

TABLE 402.4(27)

TABLE 402.4(27)
SEMI-RIGID COPPER TUBING

Gas	Undiluted propane
Inlet Pressure	2.0. psi
Pressure Drop	1.0 psi
Specific Gravity	1.50

Nominal		TUBE SIZE (in.)									
	K & L	1/4	3/8	1/2	5/8	3/4	1	1 1/4	1 1/2	2	2 1/2
	ACR	3/8	1/2	5/8	3/4	7/8	1 1/8	1 3/8	1 5/8	2 1/8	2 5/8
Outside		0.375	0.500	0.625	0.750	0.875	1.125	1.375	1.625	2.125	2.625
Inside		0.305	0.402	0.527	0.652	0.745	0.995	1.245	1.481	1.959	2.435
Length (ft)		Maximum Capacity in Thousands of Btu/h									
10		413	852	1,732	3,027	4,295	9,175	16,517	26,042	54,240	95,962
20		284	585	1,191	2,081	2,952	6,306	11,352	17,899	37,279	65,954
30		228	470	956	1,671	2,371	5,064	9,116	14,373	29,936	52,963
40		195	402	818	1,430	2,029	4,334	7,802	12,302	25,621	45,330
50		173	356	725	1,267	1,798	3,841	6,915	10,903	22,708	40,175
60		157	323	657	1,148	1,629	3,480	6,266	9,879	20,575	36,401
70		144	297	605	1,057	1,499	3,202	5,764	9,088	18,929	33,489
80		134	276	562	983	1,394	2,979	5,363	8,455	17,609	31,155
90		126	259	528	922	1,308	2,795	5,031	7,933	16,522	29,232
100		119	245	498	871	1,236	2,640	4,753	7,493	15,607	27,612
125		105	217	442	772	1,095	2,340	4,212	6,641	13,832	24,472
150		95	197	400	700	992	2,120	3,817	6,017	12,533	22,173
175		88	181	368	644	913	1,950	3,511	5,536	11,530	20,399
200		82	168	343	599	849	1,814	3,267	5,150	10,727	18,978
225		77	158	321	562	797	1,702	3,065	4,832	10,064	17,806
250		72	149	304	531	753	1,608	2,895	4,564	9,507	16,819
275		69	142	288	504	715	1,527	2,750	4,335	9,029	15,974
300		66	135	275	481	682	1,457	2,623	4,136	8,614	15,240

For SI: 1 inch = 25.4 mm, 1 foot = 304.8 mm, 1 pound per square inch = 6.895 kPa, 1 British thermal unit per hour = 0.2931 W.

Note: Table capacities are based on Type K copper tubing inside diameter (shown), which has the smallest inside diameter of the copper tubing products.

❖ See the commentary for Section 402.3.

TABLE 402.4(28)
CORRUGATED STAINLESS STEEL TUBING (CSST)

Gas	Undiluted propane
Inlet Pressure	11.0 inch WC
Pressure Drop	0.5 inch WC
Specific Gravity	1.50

Flow Designation	TUBE SIZE (EHD*)										
	13	15	18	19	23	25	30	31	37	46	62
Length (ft)	Maximum Capacity in Thousands of Btu/h										
5	72	99	181	211	355	426	744	863	1,415	2,830	6,547
10	50	69	129	150	254	303	521	605	971	1,993	4,638
15	39	55	104	121	208	248	422	490	775	1,623	3,791
20	34	49	91	106	183	216	365	425	661	1,404	3,285
25	30	42	82	94	164	192	325	379	583	1,254	2,940
30	28	39	74	87	151	177	297	344	528	1,143	2,684
40	23	33	64	74	131	153	256	297	449	988	2,327
50	20	30	58	66	118	137	227	265	397	884	2,082
60	19	26	53	60	107	126	207	241	359	805	1,902
70	17	25	49	57	99	117	191	222	330	745	1,761
80	15	23	45	52	94	109	178	208	307	696	1,647
90	15	22	44	50	90	102	169	197	286	656	1,554
100	14	20	41	47	85	98	159	186	270	621	1,475
150	11	15	31	36	66	75	123	143	217	506	1,205
200	9	14	28	33	60	69	112	129	183	438	1,045
250	8	12	25	30	53	61	99	117	163	390	934
300	8	11	23	26	50	57	90	107	147	357	854

For SI: 1 foot = 304.8 mm, 1-inch water column = 0.2488 kPa, 1 British thermal unit per hour = 0.2931 W, 1 degree = 0.01745 rad.

Note: Table includes losses for four 90-degree bends and two end fittings. Tubing runs with larger numbers of bends and/or fittings shall be increased by an equivalent length of tubing to the following equation: $L = 1.3n$ where L is additional length (ft) of tubing and n is the number of additional fittings and/or bends.

*EHD—Equivalent Hydraulic Diameter, which is a measure of the relative hydraulic efficiency between different tubing sizes. The greater the value of EHD, the greater the gas capacity of the tubing.

❖ See the commentary for Section 402.3.

GAS PIPING INSTALLATIONS

TABLE 402.4(29)
CORRUGATED STAINLESS STEEL TUBING (CSST)

Gas	Undiluted propane
Inlet Pressure	2.0 psi
Pressure Drop	1.0 psi
Specific Gravity	1.50

Flow Designation	TUBE SIZE (EHD*)										
	13	15	18	19	23	25	30	31	37	46	62
Length (ft)	Maximum Capacity in Thousands of Btu/h										
10	426	558	927	1,106	1,735	2,168	4,097	4,720	7,128	15,174	34,203
25	262	347	591	701	1,120	1,384	2,560	2,954	4,564	9,549	21,680
30	238	316	540	640	1,027	1,266	2,331	2,692	4,176	8,708	19,801
40	203	271	469	554	896	1,100	2,012	2,323	3,631	7,529	17,159
50	181	243	420	496	806	986	1,794	2,072	3,258	6,726	15,357
75	147	196	344	406	663	809	1,457	1,685	2,675	5,480	12,551
80	140	189	333	393	643	768	1,410	1,629	2,591	5,303	12,154
100	124	169	298	350	578	703	1,256	1,454	2,325	4,738	10,877
150	101	137	245	287	477	575	1,021	1,182	1,908	3,860	8,890
200	86	118	213	248	415	501	880	1,019	1,658	3,337	7,705
250	77	105	191	222	373	448	785	910	1,487	2,981	6,895
300	69	96	173	203	343	411	716	829	1,363	2,719	6,296
400	60	82	151	175	298	355	616	716	1,163	2,351	5,457
500	53	72	135	158	268	319	550	638	1,027	2,101	4,883

For SI: 1 foot = 304.8 mm, 1 pound per square inch = 6.895 kPa, 1-inch water column = 0.2488 kPa, 1 British thermal unit per hour = 0.2931 W, 1 degree = 0.01745 rad.

Notes:

1. Table does not include effect of pressure drop across the line regulator. Where regulator loss exceeds ½ psi (based on 13 in. w.c. outlet pressure), DO NOT USE THIS TABLE. Consult with regulator manufacturer for pressure drops and capacity factors. Pressure drops across a regulator may vary with flow rate.
2. CAUTION: Capacities shown in table may exceed maximum capacity for a selected regulator. Consult with regulator or tubing manufacturer for guidance.
3. Table includes losses for four 90-degree bends and two end fittings. Tubing runs with larger numbers of bends and/or fittings shall be increased by an equivalent length of tubing to the following equation: $L = 1.3n$ where L is additional length (ft) of tubing and n is the number of additional fittings and/or bends.

*EHD—Equivalent Hydraulic Diameter, which is a measure of the relative hydraulic efficiency between different tubing sizes. The greater the value of EHD, the greater the gas capacity of the tubing.

❖ See the commentary for Section 402.3.

**TABLE 402.4(30)
CORRUGATED STAINLESS STEEL TUBING (CSST)**

Gas	Undiluted propane
Inlet Pressure	5.0 psi
Pressure Drop	3.5 psi
Specific Gravity	1.50

Flow Designation	TUBE SIZE (EHD*)										
	13	15	18	19	23	25	30	31	37	46	62
Length (ft)	Maximum Capacity in Thousands of Btu/h										
10	826	1,065	1,713	2,061	3,153	3,999	7,829	8,945	13,112	28,580	63,788
25	509	664	1,092	1,307	2,037	2,554	4,864	5,600	8,395	17,986	40,436
30	461	603	999	1,193	1,866	2,336	4,430	5,102	7,682	16,403	36,928
40	396	520	867	1,033	1,629	2,029	3,822	4,404	6,679	14,183	32,004
50	352	463	777	926	1,463	1,819	3,409	3,929	5,993	12,670	28,642
75	284	376	637	757	1,206	1,492	2,769	3,194	4,919	10,322	23,409
80	275	363	618	731	1,169	1,446	2,677	3,090	4,768	9,990	22,670
100	243	324	553	656	1,051	1,296	2,388	2,756	4,276	8,925	20,287
150	196	262	453	535	866	1,062	1,941	2,241	3,511	7,271	16,581
200	169	226	393	464	755	923	1,675	1,934	3,052	6,287	14,369
250	150	202	352	415	679	828	1,493	1,726	2,738	5,616	12,859
300	136	183	322	379	622	757	1,359	1,572	2,505	5,122	11,745
400	117	158	279	328	542	657	1,173	1,356	2,178	4,429	10,178
500	104	140	251	294	488	589	1,046	1,210	1,954	3,957	9,108

For SI: 1 foot = 305 mm, 1 pound per square inch = 6.895 kPa, 1 British thermal unit per hour = 0.2931 W, 1 degree = 0.01745 rad.

Notes:

1. Table does not include effect of pressure drop across line regulator. Where regulator loss exceeds 1 psi, DO NOT USE THIS TABLE. Consult with regulator manufacturer for pressure drops and capacity factors. Pressure drops across a regulator may vary with flow rate.
2. CAUTION: Capacities shown in table may exceed maximum capacity of selected regulator. Consult with tubing manufacturer for guidance.
3. Table includes losses for four 90-degree bends and two end fittings. Tubing runs with larger numbers of bends and/or fittings shall be increased by an equivalent length of tubing to the following equation: $L = 1.3n$ where L is additional length (ft) of tubing and n is the number of additional fittings and/or bends.

*EHD—Equivalent Hydraulic Diameter, which is a measure of the relative hydraulic efficiency between different tubing sizes. The greater the value of EHD, the greater the gas capacity of the tubing.

❖ See the commentary for Section 402.3.

GAS PIPING INSTALLATIONS

**TABLE 402.4(31)
POLYETHYLENE PLASTIC PIPE**

Gas	Undiluted propane
Inlet Pressure	11.0 inch WC
Pressure Drop	0.5 inch WC
Specific Gravity	1.50

	PIPE SIZE (in.)					
Nominal OD	$1/2$	$3/4$	1	$1 1/4$	$1 1/2$	2
Designation	SDR 9.33	SDR 11.0	SDR 11.00	SDR 10.00	SDR 11.00	SDR 11.00
Actual ID	0.660	0.860	1.077	1.328	1.554	1.943
Length (ft)	Maximum Capacity in Thousands of Btu/h					
10	340	680	1,227	2,126	3,211	5,769
20	233	467	844	1,461	2,207	3,965
30	187	375	677	1,173	1,772	3,184
40	160	321	580	1,004	1,517	2,725
50	142	285	514	890	1,344	2,415
60	129	258	466	807	1,218	2,188
70	119	237	428	742	1,121	2,013
80	110	221	398	690	1,042	1,873
90	103	207	374	648	978	1,757
100	98	196	353	612	924	1,660
125	87	173	313	542	819	1,471
150	78	157	284	491	742	1,333
175	72	145	261	452	683	1,226
200	67	135	243	420	635	1,141

For SI: 1 inch = 25.4 mm, 1 foot = 304.8 mm, 1 pound per square inch = 6.895 kPa, 1-inch water column = 0.2488 kPa, 1 British thermal unit per hour = 0.2931 W, 1 degree = 0.01745 rad.

Notes:

1. Table does not include effect of pressure drop across line regulator. If regulator loss exceeds 1 psi, **DO NOT USE THIS TABLE**. Consult with regulator manufacturer for pressure drops and capacity factors. Pressure drop across regulator may vary with the flow rate.
2. CAUTION: Capacities shown in table may exceed maximum capacity of selected regulator. Consult with tubing manufacturer for guidance.
3. Table includes losses for four 90-degree bends and two end fittings. Tubing runs with larger numbers of bends and/or fittings shall be increased by an equivalent length of tubing to the following equation: $L = 1.3n$ where L is additional length (feet) of tubing and n is the number of additional fittings and/or bends.

*EHD—Equivalent Hydraulic Diameter, which is a measure of the relative hydraulic efficiency between different tubing sizes. The greater the value of EHD, the greater the gas capacity of the tubing.

❖ See the commentary for Section 402.3.

TABLE 402.4(32)
POLYETHYLENE PLASTIC PIPE

Gas	Undiluted propane
Inlet Pressure	2.0 psi
Pressure Drop	1.0 psi
Specific Gravity	1.50

	PIPE SIZE (in.)					
Nominal OD	1/2	3/4	1	1 1/4	1 1/2	2
Designation	SDR 9.33	SDR 11.0	SDR 11.00	SDR 10.00	SDR 11.00	SDR 11.00
Actual ID	0.660	0.860	1.077	1.328	1.554	1.943
Length (ft)	Maximum Capacity in Thousands of Btu/h					
10	3,126	6,259	11,293	19,564	29,545	53,085
20	2,148	4,302	7,762	13,446	20,306	36,485
30	1,725	3,454	6,233	10,798	16,307	29,299
40	1,477	2,957	5,335	9,242	13,956	25,076
50	1,309	2,620	4,728	8,191	12,369	22,225
60	1,186	2,374	4,284	7,421	11,207	20,137
70	1,091	2,184	3,941	6,828	10,311	18,526
80	1,015	2,032	3,666	6,352	9,592	17,235
90	952	1,907	3,440	5,960	9,000	16,171
100	899	1,801	3,249	5,629	8,501	15,275
125	797	1,596	2,880	4,989	7,535	13,538
150	722	1,446	2,609	4,521	6,827	12,266
175	664	1,331	2,401	4,159	6,281	11,285
200	618	1,238	2,233	3,869	5,843	10,498

For SI: 1 inch = 25.4 mm, 1 foot = 304.8 mm, 1 pound per square inch = 6.895 kPa, 1 British thermal unit per hour = 0.2931 W.

❖ See the commentary for Section 402.3.

GAS PIPING INSTALLATIONS

**TABLE 402.4(33)
POLYETHYLENE PLASTIC TUBING**

Gas	Undiluted propane
Inlet pressure	11.0 inch WC
Pressure Drop	0.5 inch WC
Specific Gravity	1.50

	PLASTIC TUBING SIZE (CTS) (in.)	
Nominal OD	1/2	3/4
Designation	SDR 7.00	SDR 11.00
Actual ID	0.445	0.927
Length (ft)	Maximum Capacity in Thousands of Btu/h	
10	121	828
20	83	569
30	67	457
40	57	391
50	51	347
60	46	314
70	42	289
80	39	269
90	37	252
100	35	238
125	31	211
150	28	191
175	26	176
200	24	164
225	22	154
250	21	145
275	20	138
300	19	132
350	18	121
400	16	113

For SI: 1 inch = 25.4 mm, 1 foot = 304.8 mm, 1 British thermal unit per hour = 0.2931 W, 1-inch water column = 0.2488 kPa.

❖ See the commentary for Section 402.3.

SECTION 403 (IFGS)
PIPING MATERIALS

403.1 General. Materials used for piping systems shall comply with the requirements of this chapter or shall be approved.

❖ This section dictates what materials and components can be used to construct gas distribution systems and also specifies the allowable applications of those materials.

403.2 Used materials. Pipe, fittings, valves and other materials shall not be used again except where they are free of foreign materials and have been ascertained to be adequate for the service intended.

❖ Ideally, gas pipe installations should be constructed of new materials. However, this section recognizes that there are occasions when used piping materials may be perfectly acceptable for an installation if they meet a number of very strict criteria that intend to reduce the potential for poor performance. To judge equivalency of reused piping materials, compliance with approval criteria for new materials should be used.

403.3 Other materials. Material not covered by the standards specifications listed herein shall be investigated and tested to determine that it is safe and suitable for the proposed service, and, in addition, shall be recommended for that service by the manufacturer and shall be approved by the code official.

❖ This section echoes the intent of Section 105.2.

403.4 Metallic pipe. Metallic pipe shall comply with Sections 403.4.1 through 403.4.4.

❖ This section addresses the traditional metal pipes used in piping installations.

403.4.1 Cast iron. Cast-iron pipe shall not be used.

❖ Because cast iron pipe is brittle compared to the malleable metals, it is more likely to fail under stress and therefore is not suitable for conveying fuel gases.

403.4.2 Steel. Steel and wrought-iron pipe shall be at least of standard weight (Schedule 40) and shall comply with one of the following standards:

1. ASME B 36.10, 10M
2. ASTM A 53; or
3. ASTM A 106.

❖ Steel pipe must be Schedule 40 or heavier, must comply with one of the listed standards and can be black iron or galvanized. Contrary to popular belief, natural gas does not adversely react with the zinc coating on galvanized pipe.

403.4.3 Copper and brass. Copper and brass pipe shall not be used if the gas contains more than an average of 0.3 grains of hydrogen sulfide per 100 standard cubic feet of gas (0.7 milligrams per 100 liters). Threaded copper, brass and aluminum-alloy pipe shall not be used with gases corrosive to such materials.

❖ The intent of this section is to prohibit the use of a piping and tubing material where it would be subject to corrosion from the fuel gas. Gas piping materials must be chemically compatible with the fuel gas that will be conveyed in the piping. For example, natural gas contains trace amounts of hydrogen sulfide and, depending on the concentration of the sulfur compound, can be corrosive to copper and brass pipe and tubing. The chemical reaction can create a precipitant on the pipe interior walls. The precipitant can flake off and be carried into appliance gas controls, causing hazardous operation. This is one of the reasons why the code requires sediment traps at connections to appliances. The code considers natural gas to be corrosive if it contains more than an average of 0.3 grain (0.0194 g) of hydrogen sulfide per 100 cubic feet (2.83 m^3) of gas. Most pipeline-quality gas does not exceed this limit; however, the gas supplier must be consulted to obtain the chemical composition of the gas that will be conveyed by copper pipe or tubing.

403.4.4 Aluminum. Aluminum-alloy pipe shall comply with ASTM B 241 (except that the use of alloy 5456 is prohibited), and shall be marked at each end of each length indicating compliance. Aluminum-alloy pipe shall be coated to protect against external corrosion where it is in contact with masonry, plaster, or insulation, or is subject to repeated wettings by such liquids as water, detergents, or sewage. Aluminum-alloy pipe shall not be used in exterior locations or underground.

❖ Aluminum pipe is not commonly used for gas service. Because aluminum is a fairly reactive metal, it corrodes readily when exposed to a variety of materials and the environment.

403.5 Metallic tubing. Seamless copper, aluminum alloy and steel tubing shall not be used with gases corrosive to such materials.

❖ Aluminum and steel tubing are used in the manufacturing of appliances but are rarely used for gas distribution. Copper tubing is widely used in gas distribution systems and is used as an alternative to CSST in parallel distribution manifold systems. As with copper pipe, copper tubing cannot be used with gases that will cause it to corrode. Commentary Figure 403.5 illustrates a type of sheathed copper tubing that is designed for gas distribution.

403.5.1 Steel tubing. Steel tubing shall comply with ASTM A 254 or ASTM A 539.

❖ Steel tubing is used only in industrial applications and in the construction of appliances.

403.5.2 Copper and brass tubing. Copper tubing shall comply with Standard Type K or L of ASTM B 88 or ASTM B 280.

Copper and brass tubing shall not be used if the gas contains more than an average of 0.3 grains of hydrogen sulfide per 100 standard cubic feet of gas (0.7 milligrams per 100 liters).

❖ ASTM B 88 applies to copper water tube, and type K and L refers to wall thickness. Type M tubing is the thinnest wall and is not allowed for gas service.

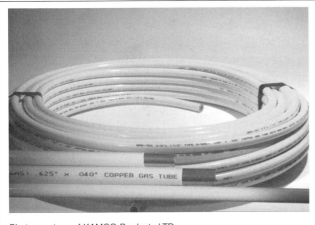

Photo courtesy of KAMCO Products LTD.

Figure 403.5
PLASTIC SHEATHED COPPER TUBING

403.5.3 Aluminum tubing. Aluminum-alloy tubing shall comply with ASTM B 210 or ASTM B 241. Aluminum-alloy tubing shall be coated to protect against external corrosion where it is in contact with masonry, plaster or insulation, or is subject to repeated wettings by such liquids as water, detergent or sewage.

Aluminum-alloy tubing shall not be used in exterior locations or underground.

❖ See the commentary for Section 403.4.4.

403.5.4 Corrugated stainless steel tubing. Corrugated stainless steel tubing shall be tested and listed in compliance with the construction, installation and performance requirements of ANSI LC 1/CSA 6.26.

❖ ANSI LC-1 contains material and performance criteria and installation requirements for corrugated stainless steel tubing (CSST) gas distribution systems. CSST systems are used as an alternative to more traditional gas piping systems for the distribution of natural gas within buildings.

Corrugated stainless steel tubing is a semirigid stainless steel tubing with a plastic jacket and is available in coil lengths varying from 100 to 250 feet (30 480 to 76 200 mm) and in the following standard sizes: $^3/_8$, $^1/_2$, $^3/_4$, and 1 inch (9.5, 12.7, 19.1, and 25.4 mm) (larger sizes may be available) [see Commentary Figure 403.5.4(1)]. Special proprietary fittings are required for connection of the tubing [see Commentary Figures 403.5.4(2) through 403.5.4(4)]. CSST can be used in low-pressure systems [up to 0.5 psig (14 inches of water column) (3.45 kPa)] and in higher pressure systems operating at pressures up to 5 psig (34.5 kPa). Commentary Figure 403.5.4(5) shows a typical manifold installation using CSST and shows a multiport manifold with a pressure regulator and shutoff valve installed. To prevent damage or puncture by screws and nails, specialized steel shield plates and protective flexible conduit are used to protect CSST systems that pass through or near structural members. Such protection must employ the parts specified by the CSST manufacturer [See commentary Figures 403.5.4(1) and 403.5.4(6)].

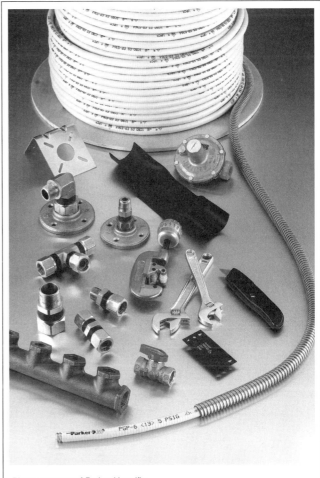

Photo courtesy of Parker Hannifin

Figure 403.5.4(1)
CSST AND SYSTEM COMPONENTS

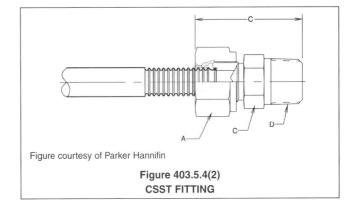

Figure courtesy of Parker Hannifin

Figure 403.5.4(2)
CSST FITTING

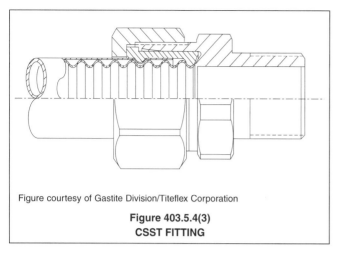

Figure courtesy of Gastite Division/Titeflex Corporation

**Figure 403.5.4(3)
CSST FITTING**

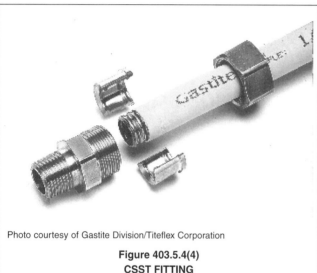

Photo courtesy of Gastite Division/Titeflex Corporation

**Figure 403.5.4(4)
CSST FITTING**

Photo courtesy of Titeflex Corporation

**Figure 403.5.4(5)
MULTIPORT MANIFOLD INSTALLATION USING CSST**

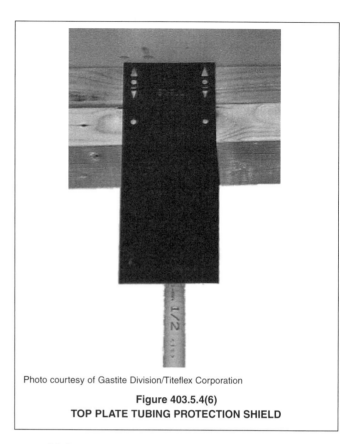

Photo courtesy of Gastite Division/Titeflex Corporation

**Figure 403.5.4(6)
TOP PLATE TUBING PROTECTION SHIELD**

CSST systems are usually designed as parallel distribution (manifold) systems, whereas traditional gas distribution systems are typically designed as branch (series) systems. See Commentary Figure 403.5.4(7).

Medium-pressure CSST systems, typically 2 psig (13.8 kPa) maximum, are designed to allow the use of small diameter tubing for long runs. Significant distribution pressure losses can be compensated for by providing sufficiently high supply pressures. Medium-pressure distribution systems require the installation of pressure regulators, commonly called pounds-to-inches regulators, to reduce the supply pressure to the pressure required for delivery to the gas-fired equipment [see commentary, Section 410 and commentary Figure 403.5.4(8)]. Note that systems with pressures over 2 psi (13.8 kPa) that supply appliances designed for a maximum inlet pressure of 14 inches w.c. (3.5 kPa) must have over-pressure protection devices (OPD) in addition to the pounds-to-inches line pressure regulators.

Section 404.3 prohibits tubing fittings in concealed locations but does not apply to CSST fittings. In accordance with the manufacturer's installation instructions and the material listing, CSST fittings can be concealed in the building construction. The tubing, fittings, shields and outlet terminals are all part of a system and must be installed together as a system. The manufacturer's installation instructions are to be considered as a complete system installation manual, which also includes the proprietary sizing criteria. Note that unless specific to CSST, the sizing tables in Section 402 do not apply to and cannot be used for CSST systems.

GAS PIPING INSTALLATIONS

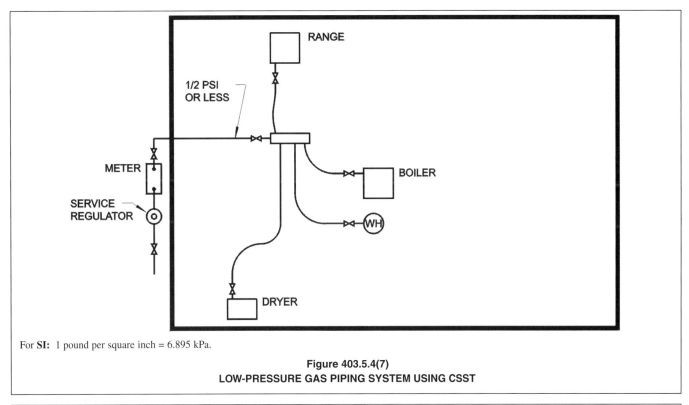

For **SI:** 1 pound per square inch = 6.895 kPa.

Figure 403.5.4(7)
LOW-PRESSURE GAS PIPING SYSTEM USING CSST

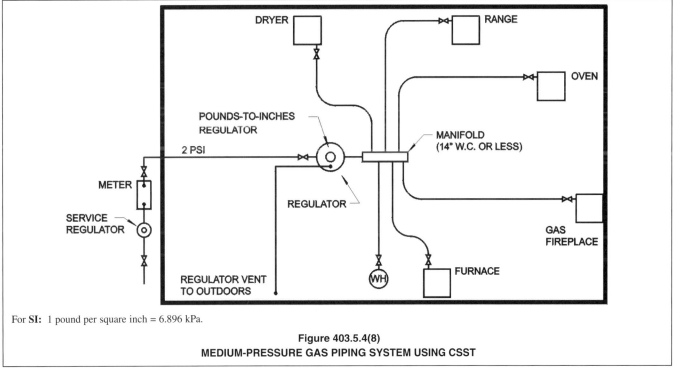

For **SI:** 1 pound per square inch = 6.896 kPa.

Figure 403.5.4(8)
MEDIUM-PRESSURE GAS PIPING SYSTEM USING CSST

Installation of CSST is not allowed underground (directly buried) unless it is of a type specially designed and listed for underground applications. Some manufacturers' installation instructions allow CSST to be "indirectly" buried by installing the tubing within a nonmetallic, water-tight conduit, such as PVC, PE or ABS plastic pipe. See Sections 404.6 and 404.11 regarding piping buried in slabs and beneath buildings. CSST is allowed by the manufacturers' installation instructions to connect directly to fixed-in-place, non-portable, nonmoveable appliances such as furnaces, boilers and water heaters. CSST is not intended for direct connection to moveable appliances such as clothes dryers or ranges nor is it intended to connect to appliances suspended on chains such as radiant heaters. It is also not allowed to connect directly to log lighters installed in solid-fuel-burning fireplaces. Direct connection is allowed for gas fireplace appliances in accordance with the CSST and fireplace manufacturers' instructions. At range and clothes dryer locations, the CSST must terminate at a CSST termination fitting assembly, and a listed gas appliance connector must be used to connect the appliance to the termination fitting assembly. CSST must not be exposed to impact, vibration or repeated movement. "Drops" to fixed- in-place, nonmoveable appliances must be well supported and routed to avoid being disturbed. In other words "flying runs" of CSST are prohibited. CSST, like copper tubing, is not flexible, but rather is semirigid, allowing it to be formed into shape as it is installed. CSST, like any metallic tubing, must not be repeatedly bent in the same location because this can result in kinking, work hardening and structural failure.

Generally, the sale of CSST materials is restricted and only individuals who have been specially trained and certified are allowed to purchase and install CSST systems.

Traditionally, the utility company (gas supplier) service regulator delivers natural gas to the building at a pressure of approximately 6 to 10 inches w.c. (1.49 to 2.49 kPa). Where medium-pressure [2 psig (13.8 kPa)] systems are used, the service regulator is set to deliver gas at a pressure of 2 psig (13.8 kPa). A medium-pressure gas distribution system must be properly designed to deliver gas to the connected appliances at pressures not exceeding the maximum inlet (supply) pressure for which the appliance is designed. A hazardous condition could develop if gas is supplied to appliances or equipment at pressures greater than the design inlet pressure of the appliances and equipment.

ANSI LC-1/CSA 6.26 addresses CSST systems and contains installation requirements. Installers and inspectors must have copies of the standard and the CSST manufacturer's installation instructions to properly install and inspect CSST systems. Manufacturers' instructions contain detailed requirements for protection hardware designed to prevent damage to installed CSST systems. As with most tubing systems, the most common installation errors involve abuse of the CSST tubing, lack of support and improperly installed puncture protection devices or the omission of protection devices. CSST manufacturers publish design and installation guides that are very helpful and should be on the book shelves of inspectors, installers and designers [see commentary Figures 403.5.4(9) and 403.5.4(10)]. Some CSST manufacturers offer preassembled manifolds in a rough-in cabinet to simplify installation [see commentary Figure 403.5.4(11)]. CSST transitions to rigid pipe appliance connections and listed appliance connectors must be made with fittings that prevent stress and rotational forces from being applied to the CSST. Commentary Figures 403.5.4(12) through (14) illustrate typical fitting assemblies used to make the transition from CSST to appliance connections or connectors.

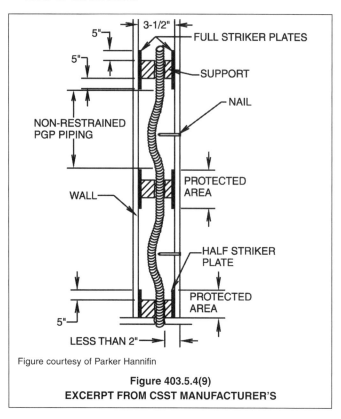

Figure courtesy of Parker Hannifin

Figure 403.5.4(9)
EXCERPT FROM CSST MANUFACTURER'S

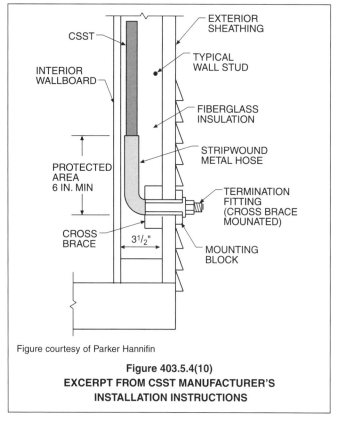

Figure courtesy of Parker Hannifin

Figure 403.5.4(10)
EXCERPT FROM CSST MANUFACTURER'S INSTALLATION INSTRUCTIONS

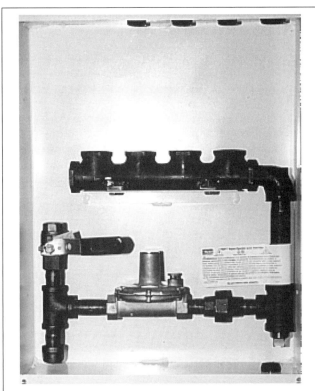

Photo courtesy of Parker Hannifin

Figure 403.5.4(11)
MANIFOLD FOR 2 PSI PARALLEL DISTRIBUTION SYSTEM

Photo courtesy of Gastite Division/Titeflex Corporation

Figure 403.5.4(12)
"STUB-OUT" TERMINATION TRANSITION FITTING

❖ The installation of plastic pipe and tubing is limited to areas that are both outside the building and underground because of the potential hazard associated with the use of a material that has lower resistance to physical damage and heat compared to metallic pipe. Plastic pipe and tubing are widely used for underground gas distribution systems because of their ease of installation and inherent resistance to corrosion.

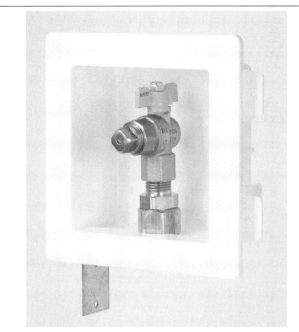

Photo courtesy of Gastite Division/Titeflex Corporation

Figure 403.5.4(13)
"VALVE BOX" TERMINATION TRANSITION FITTING

Photo courtesy of Gastite Division/Titeflex Corporation

Figure 403.5.4(14)
TERMINATION TRANSITION FITTING

403.6 Plastic pipe, tubing and fittings. Plastic pipe, tubing and fittings shall be used outside, underground, only, and shall conform to ASTM D 2513. Pipe shall be marked "gas" and "ASTM D 2513."

403.6.1 Anodeless risers. Plastic pipe, tubing and anodeless risers shall comply with the following:

1. Factory-assembled anodeless risers shall be recommended by the manufacturer for the gas used and shall be leak tested by the manufacturer in accordance with written procedures.
2. Service head adapters and field-assembled anodeless risers incorporating service head adapters shall be recommended by the manufacturer for the gas used, and shall be designed and certified to meet the requirements of Category I of ASTM D 2513, and U.S. Department of Transportation, Code of Federal Regulations, Title 49, Part 192.281(e). The manufacturer shall provide the user with qualified installation instructions as prescribed by the U.S. Department of Transportation, Code of Federal Regulations, Title 49, Part 192.283(b).

❖ This section is an exception to the absolute prohibition of plastic gas piping aboveground. Plastic pipe is allowed to rise from underground within a steel riser conduit constructed as an anodeless riser assembly. Such risers are coated to resist corrosion and do not use a sacrificial anode for cathodic corrosion protection, hence the name "anodeless riser." The steel riser protects the PE plastic pipe from physical damage and terminates the plastic pipe with a threaded steel pipe transition fitting. A sweeping 90-degree (1.57 rad) bend is placed on the underground end of rigid risers to prevent them from rotating in the ground when threaded connections are made at the terminal end. The plastic pipe could be damaged by rotational and bending stresses transmitted by the riser; therefore, a horizontal portion of steel piping is installed at the bottom of the riser to limit such movement. These risers can eliminate the need for an underground plastic-to-metal transition fitting and simplify installation. They are used extensively by utility companies for service laterals and can be used for other installations such as supply runs to out-buildings, pool heaters, gas grills and yard lights.

Service head adapters are also a type of terminal transition fitting that allow plastic pipe to be pulled through a retired steel gas service lateral and terminated at the point where the concentric piping passes through the building foundation wall. The adapter makes the transition from plastic pipe to steel pipe and seals the end of the retired steel service lateral [see commentary Figures 202(1), 403.6.1(1) 403.6.1(2) and 403.6.1(3)].

403.6.2 LP-gas systems. The use of plastic pipe, tubing and fittings in undiluted liquefied petroleum gas piping systems shall be in accordance with NFPA 58.

❖ NFPA 58 limits plastic piping for LP-gas service to only polyethylene pipe and tubing. The PE pipe and tubing must be marked as being in compliance with ASTM D 2513 and is allowed for LP-gas service only if so recommended by the material manufacturer.

403.7 Workmanship and defects. Pipe, tubing and fittings shall be clear and free from cutting burrs and defects in structure or threading, and shall be thoroughly brushed, and chip and scale blown.

Defects in pipe, tubing and fittings shall not be repaired. Defective pipe, tubing and fittings shall be replaced (see Section 406.1.2).

❖ This section stresses that proper installation procedures are necessary to secure a safe installation. In the context of this section, good workmanship is an enforceable requirement. Burrs result from the cutting of pipe or tube and cause flow restriction. Pipe scale, dirt, construction site debris, metal chips from cutting and threading operations and excess thread compound are potentially harmful contaminates that might cause damage to valves, regulators and appliances. Damaged and/or improperly cut threads can cause leakage or joint failure. Defective materials must be replaced because attempts to repair them are always makeshift at best, and any cost savings would not be worth the risk.

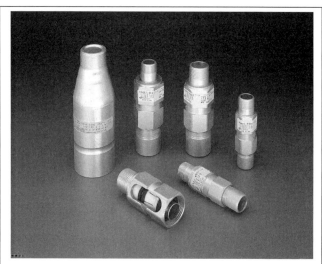

Photo courtesy of Perfection Corporation

**Figure 403.6.1(1)
SERVICE HEAD ADAPTERS**

403.8 Protective coating. Where in contact with material or atmosphere exerting a corrosive action, metallic piping and fittings coated with a corrosion-resistant material shall be used. External or internal coatings or linings used on piping or components shall not be considered as adding strength.

❖ Protective coatings in the form of wrappings, tapes, enamels, epoxies, sleeves, casings and factory-applied coverings are a common method of isolating metallic piping materials from the atmosphere or from the earth in which they are buried. Application of these materials under strictly controlled factory conditions improves the quality control process and reduces the likelihood of coating defects. Corrosion can be caused by weather exposure, burial in the soil or construction materials that are in contact with the pipe. Cinders (also known as dross) are the waste residue from steel production and carry the waste products skimmed from the molten metal during the smelting process. Cinders are also the waste residue of coal-fired appliances and

GAS PIPING INSTALLATIONS

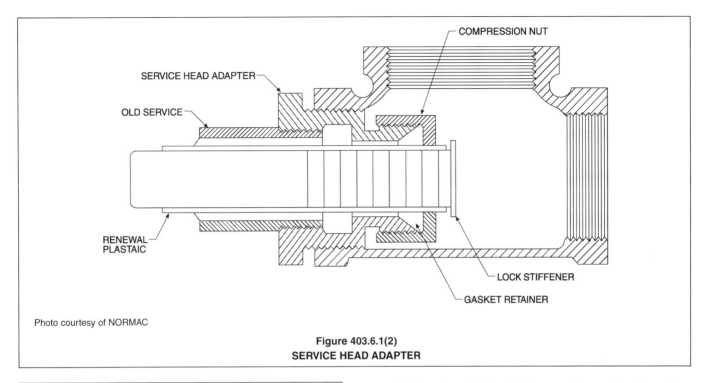

Figure 403.6.1(2)
SERVICE HEAD ADAPTER

Figure 403.6.1(3)
FLEXIBLE ANODELESS RISER

equipment. These materials are extremely corrosive to metallic piping and may not be used in any form (backfill or cinder blocks, for example) where in direct contact with the pipe. Protection is usually provided by a factory-applied coating or by field wrapping the pipe with a protective covering, such as a coal-tar-based or plastic wrapping. Galvanized piping in contact with soil would require additional coating or protection. The zinc coating does not usually provide long-term protection in underground applications. Where possible, a piping material is chosen that is not subject to the type of corrosion of the application.

Corrosion can also be caused by galvanic action that takes place where dissimilar metals are joined in a current-carrying medium, such as soil or water. For example, if steel and copper pipe are joined in a medium that conducts electrical current, the steel pipe will corrode at an accelerated rate because of the electrochemical process between the dissimilar metals. In this case, the soil acts as an electrolyte; the steel will be a sacrificial anode, and the copper will be a cathode. Because a cell is created, current will flow through the metal junction and the soil, resulting in the gradual deterioration of the steel. To protect against galvanic corrosion, dielectric fittings and couplings are used to join the piping, thus breaking the circuit of the cell.

This section is commonly interpreted as requiring priming and painting of black steel pipe that is exposed to the weather as the minimum form of protection.

403.9 Metallic pipe threads. Metallic pipe and fitting threads shall be taper pipe threads and shall comply with ASME B1.20.1.

❖ Threads must be taper cut in accordance with the stated standard. The standard regulates tapered and straight threads, but this section specifies tapered threads. Tapered pipe threads, when made up, form a metal-to-metal (interference fit) seal. Pipe-joint compound or PTFE tape is to be applied to male threads only to decrease the possibility of tape fragments or compound entering the piping system. Such debris in piping systems can block orifices, restrict flow or interfere with the operation of controls. The primary purpose of pipe-thread compounds is to act as a lubricant to allow proper tightening and to achieve a metal-to-metal seal. They also fill in small imperfections on the threaded surfaces. This section prohibits the use of straight tapped fittings such as couplings because the threaded joint will not provide the same seal as a joint made with taper-tapped fittings. Steel taper-threaded couplings are easily identified because they look exactly like the run of a tee fitting (see commentary Figure 403.9). The common designation "NPT" is misinterpreted as "nominal pipe threads" and actually stands for "national standard, pipe, tapered."

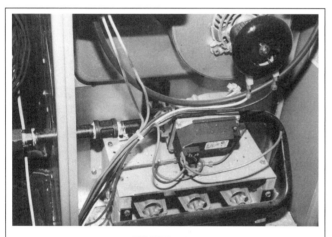

**FIGURE 403.9
TAPERED THREAD COUPLING
IN CONNECTION TO FURNACE**

403.9.1 Damaged threads. Pipe with threads that are stripped, chipped, corroded or otherwise damaged shall not be used. Where a weld opens during the operation of cutting or threading, that portion of the pipe shall not be used.

❖ Damaged or improperly cut threads and defective welded seam pipe must be eliminated from any gas piping system because of the potential for leaks.

403.9.2 Number of threads. Field threading of metallic pipe shall be in accordance with Table 403.9.2.

403.9.3 Thread compounds. Thread (joint) compounds (pipe dope) shall be resistant to the action of liquefied petroleum gas or to any other chemical constituents of the gases to be conducted through the piping.

❖ Joint compounds in both paste and tape forms are commonly misapplied and used for the wrong application. Some compounds chemically react with the gas being conveyed, which could result in leakage. The label on the compound container will specify the applications for which the compound is suitable. Thread compounds act as a lubricant for the threads during assembly and also act as a sealant for the life of the joint. Care must be taken to keep all compounds out of the piping system interior because the contamination can cause damage to components and appliances. Pipe-joint compound or tape is limited to application on the male threads only to decrease the possibility of tape fragments or compound entering the piping system. Such debris in piping systems can block orifices, restrict flow or interfere with the operation of safety controls. It is important to leave the first one or two threads on the end of the male threads bare to help prevent the compound from entering the piping system.

Once made from lead compounds and linseed oil, joint compound formulations now commonly contain PTFE (Teflon). The primary purpose of pipe-thread compounds is to act as a lubricant to allow proper tightening and to achieve a metal-to-metal seal. They also fill in small imperfections on the threaded surfaces. Pipe-thread compounds and tapes must be compatible with both the piping material and the contents of the piping.

LP-gases can be a solvent for some pipe joint compound formulations. Therefore, a type must be chosen that will not react with the gas.

**TABLE 403.9.2
SPECIFICATIONS FOR THREADING METALLIC PIPE**

IRON PIPE SIZE (inches)	APPROXIMATE LENGTH OF THREADED PORTION (inches)	APPROXIMATE NUMBER OF THREADS TO BE CUT
$^1/_2$	$^3/_4$	10
$^3/_4$	$^3/_4$	10
1	$^7/_8$	10
$1^1/_4$	1	11
$1^1/_2$	1	11
2	1	11
$2^1/_2$	$1^1/_2$	12
3	$1^1/_2$	12
4	$1^5/_8$	13

For SI: 1 inch = 25.4 mm.

❖ The table sets forth the required number of threads per unit length of pipe based on the pipe size. These specifications are intended to be both maximums and minimums because both undercutting and overcutting can result in faulty joints. Proper threading of pipe requires skill, practice, and the proper well-maintained tools.

403.10 Metallic piping joints and fittings. The type of piping joint used shall be suitable for the pressure-temperature conditions and shall be selected giving consideration to joint tightness and mechanical strength under the service conditions. The joint shall be able to sustain the maximum end force caused by the internal pressure and any additional forces caused by temperature expansion or contraction, vibration, fatigue or the weight of the pipe and its contents.

❖ This section is in performance language and requires the designer and/or installer to choose the type of joining means that is appropriate for the application. Besides being leak-tight, the joints must be able to withstand the conditions of service. Many joining methods are addressed in the code, but not all of them are suitable for all applications. For example, PE pipe for LP-gas use is limited to a maximum pressure of 30 psig (206.8 kPa), and most tubing materials would be unsuitable for uses involving equipment vibration that is transferred to the tubing. Also, some fittings might not be suitable for exposure to extreme changes in temperature.

403.10.1 Pipe joints. Pipe joints shall be threaded, flanged, brazed or welded. Where nonferrous pipe is brazed, the brazing materials shall have a melting point in excess of 1,000°F (538°C). Brazing alloys shall not contain more than 0.05-percent phosphorus.

❖ Joints in piping materials can be made by four methods: threaded (screwed), brazed, welded and flanged. Flanged joints must also involve threading or welding as the means of attaching the flanges to the pipe. The limit on phosphorus content is related to corrosion that can occur from the reaction of phosphorus with a trace contaminant (sulphur) in the gas (see commentary, Section 403.10.2).

403.10.2 Tubing joints. Tubing joints shall be either made with approved gas tubing fittings or brazed with a material having a melting point in excess of 1,000°F (538°C). Brazing alloys shall not contain more than 0.05-percent phosphorus.

❖ Tubing joints can be made by two methods: brazing and tubing fittings. Tubing fittings include flare, compression and CSST.

Brazing is the act of producing a brazed joint. Brazing is often referred to as silver soldering. Silver soldering is more accurately described as silver brazing and employs high-silver-bearing alloys primarily composed of silver, copper, and zinc. Silver soldering (brazing) typically requires temperatures in excess of 1,000°F (538°C), and such solders are classified as "hard" solders (see the commentary for "Brazed joint").

Confusion has always been present with respect to the distinction between "silver solder" and "silver-bearing" solder. Silver solders are unique and can be further subdivided into soft and hard categories, which are determined by the percentages of silver and the other component elements of the particular alloy. The distinction is that silver-bearing solders [melting point less than 600°F (316°C)] are used in soft-soldered joints, and silver solders [melting point greater than 1,000°F (538°C)] are used in silver-brazed joints.

Brazed joints are considered to have superior strength and stress resistance and, because of the high melting point, are less likely to fail when exposed to fire.

Brazing with a filler metal conforming to AWS A5.8 produces a strong joint that will perform under extreme service conditions. The surfaces to be brazed must be cleaned free of oxides and impurities. Flux should be applied as soon as possible after the surfaces have been cleaned. Flux helps to remove residual traces of oxides, to promote wetting and to protect the surfaces from oxidation during heating. Care should be taken to prevent flux from entering the piping system during the brazing operation because flux that remains may corrode the pipe or contaminate the system.

Air should be removed from the pipe being brazed by purging the piping with a nonflammable gas such as carbon dioxide or nitrogen. Purging the system has several benefits, such as preventing oxidation from occurring on the inside of the pipe. Mechanical joints are usually proprietary joints that are developed and marketed by individual manufacturers. Many types of mechanical joints employ a sleeve or ferrule that is compressed around the circumference of the pipe or tube. Mechanical joints must be specifically designed for and compatible with the type of pipe or tube to be joined.

403.10.3 Flared joints. Flared joints shall be used only in systems constructed from nonferrous pipe and tubing where experience or tests have demonstrated that the joint is suitable for the conditions and where provisions are made in the design to prevent separation of the joints.

❖ Flared joints are typically used with copper and aluminum tubing and are prohibited for steel tubing. Flared joints require the use of specialized tools. Because the pipe end is expanded in a flared joint, only annealed and bending tempered (drawn) copper tubing may be flared. Commonly used flaring tools employ a screw yoke and block assembly or an expander tool that is driven into the tube with a hammer. The flared tubing end is compressed between a fitting seat and a threaded nut to form a metal-to-metal seal.

403.10.4 Metallic fittings. Metallic fittings, including valves, strainers and filters, shall comply with the following:

1. Threaded fittings in sizes larger than 4 inches (102 mm) shall not be used except where approved.
2. Fittings used with steel or wrought-iron pipe shall be steel, brass, bronze, malleable iron or cast iron.
3. Fittings used with copper or brass pipe shall be copper, brass or bronze.
4. Fittings used with aluminum-alloy pipe shall be of aluminum alloy.
5. Cast-iron fittings:
   5.1. Flanges shall be permitted.
   5.2. Bushings shall not be used.
   5.3. Fittings shall not be used in systems containing flammable gas-air mixtures.
   5.4. Fittings in sizes 4 inches (102 mm) and larger shall not be used indoors except where approved.

5.5. Fittings in sizes 6 inches (152 mm) and larger shall not be used except where approved.
6. Aluminum-alloy fittings. Threads shall not form the joint seal.
7. Zinc aluminum-alloy fittings. Fittings shall not be used in systems containing flammable gas-air mixtures.
8. Special fittings. Fittings such as couplings, proprietary-type joints, saddle tees, gland-type compression fittings, and flared, flareless or compression-type tubing fittings shall be: used within the fitting manufacturer's pressure-temperature recommendations; used within the service conditions anticipated with respect to vibration, fatigue, thermal expansion or contraction; installed or braced to prevent separation of the joint by gas pressure or external physical damage; and shall be approved.

❖ Item 1 limits threaded fittings to a maximum size of 4 inches because larger pipe threads are impractical, difficult to cut and difficult to make leak-tight.

Item 2 allows use of cast iron fittings with steel, but cast iron pipe is prohibited. Malleable iron (steel) fittings are used almost exclusively. Items 3 and 4 require fittings for nonferrous pipes to be of a material consistent with the pipe material. Item 5 places limitations on cast iron fittings because of the brittle nature of cast iron. Bushings have been known to split from overtightening. Fittings sizes 4 inches and up are allowed only outdoors, and fittings 6 inches and up are not allowed anywhere. Item 8 speaks of special fittings of the mechanical type. None of the joining methods addressed in the code use a nonmetallic or elastomeric sealing element except those used outdoors only.

403.11 Plastic pipe, joints and fittings. Plastic pipe, tubing and fittings shall be joined in accordance with the manufacturer's instructions. Such joint shall comply with the following:

1. The joint shall be designed and installed so that the longitudinal pull-out resistance of the joint will be at least equal to the tensile strength of the plastic piping material.
2. Heat-fusion joints shall be made in accordance with qualified procedures that have been established and proven by test to produce gas-tight joints at least as strong as the pipe or tubing being joined. Joints shall be made with the joining method recommended by the pipe manufacturer. Heat fusion fittings shall be marked "ASTM D 2513."
3. Where compression-type mechanical joints are used, the gasket material in the fitting shall be compatible with the plastic piping and with the gas distributed by the system. An internal tubular rigid stiffener shall be used in conjunction with the fitting. The stiffener shall be flush with the end of the pipe or tubing and shall extend at least to the outside end of the compression fitting when installed. The stiffener shall be free of rough or sharp edges and shall not be a force fit in the plastic. Split tubular stiffeners shall not be used.
4. Plastic piping joints and fittings for use in liquefied petroleum gas piping systems shall be in accordance with NFPA 58.

❖ This section contains general requirements for the proper joining of plastic gas pipe materials. Although plastic pipe materials offer many advantages such as ease of installation, corrosion resistance, lighter weight and lower cost, these materials require different handling and installation methods than steel or other metal pipe or tubing. In all cases, the pipe material manufacturer's installation instructions must be strictly adhered to in order to reduce the likelihood of creating improper joints, which could lead to gas leakage or piping or joint failure. Dissimilar types of plastic materials may not be joined by heat fusion methods. Because of the resulting reduction in wall thickness, plastic pipe and tubing must not be field-threaded as a joining method. All approved joints must be designed to account for the forces acting upon the piping system, such as expansion, contraction and external loads imposed by burial in underground installations. The longitudinal pull-out resistance of the joint should be approximately the same as the tensile strength of the materials being joined.

Heat-fusion joints for plastic pipe are analogous to the welding of steel pipe. Heat fusion must be performed in accordance with the pipe manufacturer's instructions. The process involves heating the pipe and fittings with a special iron. When the parts to be joined reach their melting points, they are assembled and allowed to fuse. Electrical fusion fittings are available and employ an integral heating element that is connected to a power supply unit that allows the fitting to be heated at the required rate and duration.

Mechanical compression joints employ an elastomeric seal that must be compatible not only with the piping material but also with the gas that is conveyed in the system. The manufacturers' instructions normally require an internal stiffener insert used in conjunction with the fitting to provide additional support. The stiffener insert prevents the pipe from deforming under the compression force exerted by the seal and compression nut assembly.

Category 1 fittings in accordance with ASTM D 2513 are designed to provide a gas-tight seal and resist axial (pull-out) forces equal to the yield strength of the piping material [see commentary Figures 403.11(1) through 403.11(4)].

403.12 Flanges. All flanges shall comply with ASME B16.1, ASME B16.20, AWWA C111/A21.11 or MSS SP-6. The pressure-temperature ratings shall equal or exceed that required by the application.

❖ Flanges provide the same function as a union and are used for large piping at locations where it is necessary to disassemble and reassemble piping connections to valves, meters, regulators and other sections of piping. A flange consists of two mating fittings that bolt together to form a butt joint. A gasket is compressed between the flanges as the through-bolts are tightened (see commentary Figure 403.12).

Photo courtesy of Perfection Corporation

**Figure 403.11(1)
MECHANICAL JOINTS**

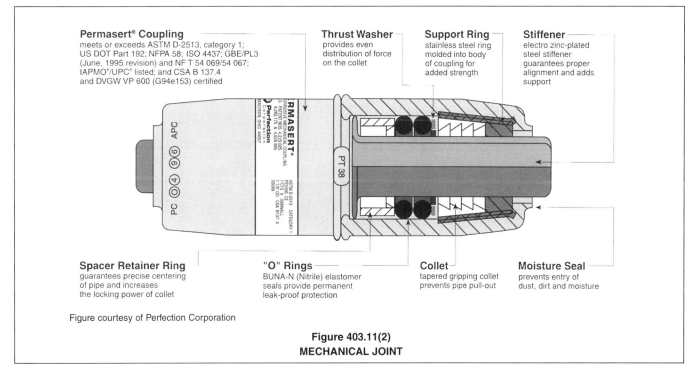

Figure courtesy of Perfection Corporation

**Figure 403.11(2)
MECHANICAL JOINT**

403.12.1 Flange facings. Standard facings shall be permitted for use under this code. Where 150-pound (1034 kPa) pressure-rated steel flanges are bolted to Class 125 cast-iron flanges, the raised face on the steel flange shall be removed.

❖ Because cast iron fittings are brittle, a joint failure could occur from bolt tightening if the two flange surfaces are not in full contact with each other and the intervening gasket. In this application, Section 403.13 requires gaskets to cover the entire surface area of the flanges. Any partial raised face on a flange would create a fulcrum over which the cast iron flange could be broken.

403.12.2 Lapped flanges. Lapped flanges shall be used only above ground or in exposed locations accessible for inspection.

❖ Lapped flanges are held on the pipe by flaring the end of the pipe. Such joints are rarely used compared to gasket-type flanges that are threaded or welded onto the pipe.

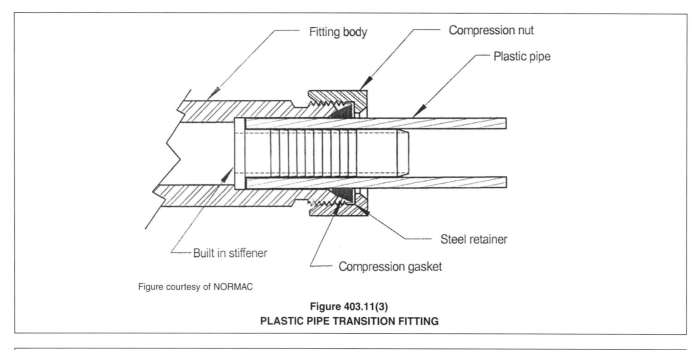

Figure 403.11(3)
PLASTIC PIPE TRANSITION FITTING

Figure 403.11(4)
PLASTIC PIPE TRANSITION FITTING

403.13 Flange gaskets. Material for gaskets shall be capable of withstanding the design temperature and pressure of the piping system, and the chemical constituents of the gas being conducted, without change to its chemical and physical properties. The effects of fire exposure to the joint shall be considered in choosing material. Acceptable materials include metal or metal-jacketed asbestos (plain or corrugated), asbestos, and aluminum "O" rings and spiral wound metal gaskets. When a flanged joint is opened, the gasket shall be replaced. Full-face gaskets shall be used with all bronze and cast-iron flanges.

❖ Similar to a ground joint union, a flanged joint creates a seal by compressing a softer metal between the two flange fitting surfaces. This author assumes that asbestos gaskets are no longer made. Once compressed, the gasket in a disassembled joint is spent and must be discarded (see commentary, Section 403.12.1).

Figure 403.12
FLANGED JOINTS

SECTION 404 (IFGC)
PIPING SYSTEM INSTALLATION

404.1 Prohibited locations. Piping shall not be installed in or through a circulating air duct, clothes chute, chimney or gas vent, ventilating duct, dumbwaiter or elevator shaft.

❖ An air supply duct or areas that create a shaft that provide a path for the gas to travel are considered to be locations where the problem of a gas leak could be compounded by spreading the gas or any resultant fire throughout the building. Also, in some locations, the gas piping could be subject to deterioration because of temperature or corrosive conditions such as in air ducts or chimneys.

Locating fuel gas piping in certain areas or atmospheres may cause corrosion of the pipe, which in turn may cause a leak in the piping system. Within a supply air duct, the conditioned air may cause moisture to condense on or within the pipe, thereby causing corrosion of the pipe. Fuel gas piping located within a chimney or a vent will be subjected to high temperatures and the corrosive effects of flue gases.

Additionally, piping located in a dumbwaiter or elevator shaft may be subject to an additional hazard resulting in mechanical damage to the pipe. Damage from a dumbwaiter or elevator impacting the fuel gas piping may not only cause a leak that would allow the gas to escape and travel up the shaft but may also put any occupants within an elevator in immediate danger. The elevator codes also intend to prohibit nonessential piping within elevator shafts.

Gas piping is not prohibited in concealed spaces used to convey environmental air; for example, above ceiling and below floor plenums, stud cavities and joist spaces. Gas piping is often installed in return air plenums above a suspended ceiling with no history of problems. The piping system is assumed not to leak in a ceiling air plenum or any other location, for that matter. In fact, when viewed logically, it would be more hazardous for a leak to occur in a location where the air is static and an explosive gas/air mixture could be created. A circulating air duct is not considered to be a plenum. These ducts are supply and return air conduits constructed as components of a ductwork system. On the other hand, a furnace plenum is part of a ductwork system and would be treated as a circulating air duct (see definition of "Furnace plenum").

404.2 Piping in solid partitions and walls. Concealed piping shall not be located in solid partitions and solid walls, unless installed in a chase or casing.

❖ As with the alternative installation requirements in Section 404.6, this section allows installation of gas piping within solid walls or partitions only if the piping is installed within a chase or casing to protect the pipe from stress and from corrosive effects of wall materials such as concrete.

404.3 Piping in concealed locations. Portions of a piping system installed in concealed locations shall not have unions, tubing fittings, right and left couplings, bushings, compression couplings and swing joints made by combinations of fittings.

Exceptions:
1. Tubing joined by brazing.
2. Fittings listed for use in concealed locations.

❖ A concealed location is a location that requires the removal of permanent construction in order to gain access (see the definition of "Concealed location"). The space above a dropped ceiling having readily removable lay-in panels or other locations that have removable access panels are not considered concealed locations for the purposes of this section. Concealed locations include wall, floor and ceiling cavities bounded by permanent finish materials such as gypsum board, masonry or paneling. Unions and mechanical joint tubing fittings are not permitted in concealed locations because they are more likely to loosen and leak than other joining means and fittings. Tubing fittings include flare, compression and similar proprietary-type fittings, all of which are mechanical joints. Joints for CSST systems are mechanical tubing joints; however, these joints use specialized proprietary fittings that are listed for concealment as part of the CSST system. Such fittings undergo stringent testing to determine their ability to remain leak-tight.

Right and left couplings are somewhat archaic and were used as a union is now used. The coupling is made with right-hand threads on one end and left-hand threads on the other, and the corresponding male pipe threads can be used to connect pipe and fittings within a section of existing piping. If the coupling was concealed, people would not be aware of such a fitting in a piping system and would not realize that they were loosening a joint as they applied the traditional tightening torque to an exposed section of piping. Bushings have been known to split after assembly and are not trusted in concealed locations.

Swings joints were used to control the forces that develop in piping systems as a result of expansion, contraction and movement in structural components. If the joint "swings" (rotates), a leak could develop. Brazing is a commonly used method of joining copper tubing and is as reliable as welded joints for steel pipe.

404.4 Piping through foundation wall. Underground piping, where installed below grade through the outer foundation or basement wall of a building, shall be encased in a protective pipe sleeve. The annular space between the gas piping and the sleeve shall be sealed.

❖ Because of the likelihood of the fuel gas piping settling after installation and because of the abrasive nature of foundation wall materials such as concrete, an approved sleeve is required where piping passes through foundation walls. If the elevation of the underground piping is above the frost-depth line, additional protection may be necessary because of the vertical movement of the pipe caused by frost heave action. The preferred method for bringing outdoor gas piping into the building is to penetrate the exterior wall at a point above grade.

The pipe and protective sleeve must be adequately sealed to prevent the possibility of water or insect entry into the building. A gas pipe that enters or exits the building above grade is preferable from the standpoint that any gas leakage in the underground piping will not be channeled into the building through the surrounding soil and the wall penetration. Gas leakage can travel in the annular space that exists between the outside wall of the pipe and the soil that surrounds it. With an above-ground wall penetration, any outdoor piping leakage will probably vent to the outdoors instead of finding its way into the building. Underground penetrations of building foundations must be sealed to eliminate the possibility of gas entering the building (see Commentary Figure 404.4).

404.5 Protection against physical damage. In concealed locations, where piping other than black or galvanized steel is installed through holes or notches in wood studs, joists, rafters or similar members less than 1 inch (25 mm) from the nearest edge of the member, the pipe shall be protected by shield plates. Shield plates shall be a minimum of $^1/_{16}$-inch-thick (1.6 mm) steel, shall cover the area of the pipe where the member is notched or bored, and shall extend a minimum of 4 inches (102 mm) above sole plates, below top plates and to each side of a stud, joist or rafter.

❖ This section is intended to minimize the possibility that nails or screws will be driven into the gas pipe or tube. Because nails and screws sometimes miss the stud, rafter, joist, or sole or top plates, the shield plates must extend parallel to the pipe or tube not less than 4 inches (105 mm) beyond the member on each side or not less than 4 inches (105 mm) above or below sole or top wall plates, respectively. Commentary Figures 404.5 and 403.5.4(6) show typical shield plates. Black and galvanized steel pipe (Schedule 40) each have wall thicknesses greater than the required thickness for the shield plates, which makes these piping materials inherently resistant to nail and screw penetrations. This section does not apply to CSST tubing, because CSST systems have their own protection requirements dictated by the manufacturer's instructions and ANSI LC-1/CSA 6.26. The protection requirements for CSST systems are, overall, more stringent than those of this section. Before studs, rafters, joists or other structural members are drilled or notched, Section 302 and the IBC should be consulted.

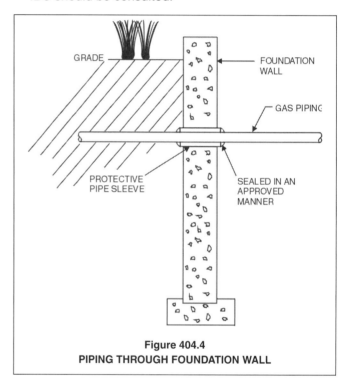

Figure 404.4
PIPING THROUGH FOUNDATION WALL

404.6 Piping in solid floors. Piping in solid floors shall be laid in channels in the floor and covered in a manner that will allow access to the piping with a minimum amount of damage to the building. Where such piping is subject to exposure to excessive moisture or corrosive substances, the piping shall be protected in an approved manner. As an alternative to installation in channels, the piping shall be installed in a casing of Schedule 40 steel, wrought iron, PVC or ABS pipe with tightly sealed ends and joints. Both ends of such casing shall extend not less than 2 inches (51 mm) beyond the point where the pipe emerges from the floor.

❖ Piping must not be installed in any solid concrete or masonry floor construction. The potential for pipe damage from slab settlement, cracking or the corrosive action of the floor material makes it imperative that one of the installation methods in this section be used. This section does not intend to allow any direct encasement of gas piping in solid concrete or masonry floor systems.

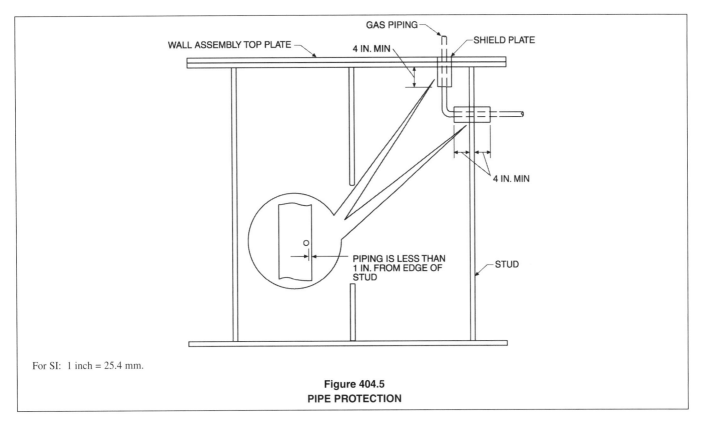

For SI: 1 inch = 25.4 mm.

**Figure 404.5
PIPE PROTECTION**

Gas piping installed within a solid floor system must be safeguarded by installation in a sealed casing or in a floor channel with a removable cover for pipe access [see commentary Figures 404.6(1) and (2)]. Either of these methods should provide reasonable protection of the pipe from the effects of settling, cracking and being in contact with corrosive materials.

Because casings constructed of metal could corrode where installed within a concrete slab on grade, consideration should be given to corrosion protection for the steel casing, or an alternate material should be chosen.

404.7 Above-ground outdoor piping. All piping installed outdoors shall be elevated not less than $3^1/_2$ inches (152 mm) above ground and where installed across roof surfaces, shall be elevated not less than $3^1/_2$ inches (152 mm) above the roof surface. Piping installed above ground, outdoors, and installed across the surface of roofs shall be securely supported and located where it will be protected from physical damage. Where passing through an outside wall, the piping shall also be protected against corrosion by coating or wrapping with an inert material. Where piping is encased in a protective pipe sleeve, the annular space between the piping and the sleeve shall be sealed.

❖ Gas piping in any location must be properly supported and protected from physical damage. Protection from damage is especially important where piping is run outdoors near grade or across roof surfaces. See Sections 407 and 415. Piping passing through an outside wall must be protected where the material of the wall could corrode or abrade the piping (concrete, masonry or stucco, for example). To help protect piping from corrosion resulting from exposure to moisture, the piping must always be located at least 3 ½ inches (89 mm) above the earth and roof surfaces. The distance of 3 ½ inches (89 mm) was chosen because it corresponds to pressure treated wood 4-inch by 4-inch (102 mm by 102 mm) lumber which is commonly used to support piping run across roof surfaces.

Where piping is supported on 4-inch by 4-inch (102 mm by 102 mm) wood blocks or similar supports, the piping must be attached to the blocking with suitable clamps or straps. It is common to see wood blocking or piping that has been displaced because the piping was not properly fastened to the blocking. Section 415.1 dictates the minimum required spacing for blocking.

404.8 Protection against corrosion. Metallic pipe or tubing exposed to corrosive action, such as soil condition or moisture, shall be protected in an approved manner. Zinc coatings (galvanizing) shall not be deemed adequate protection for gas piping underground. Ferrous metal exposed in exterior locations shall be protected from corrosion in a manner satisfactory to the code official. Where dissimilar metals are joined underground, an insulating coupling or fitting shall be used. Piping shall not be laid in contact with cinders.

❖ See the commentary for Section 403.8.

404.8.1 Prohibited use. Uncoated threaded or socket welded joints shall not be used in piping in contact with soil or where internal or external crevice corrosion is known to occur.

❖ Pits, crevices and other surface imperfections can create concentrated corrosion cells where the corrosion effect is amplified. Pipe or fitting failure could occur

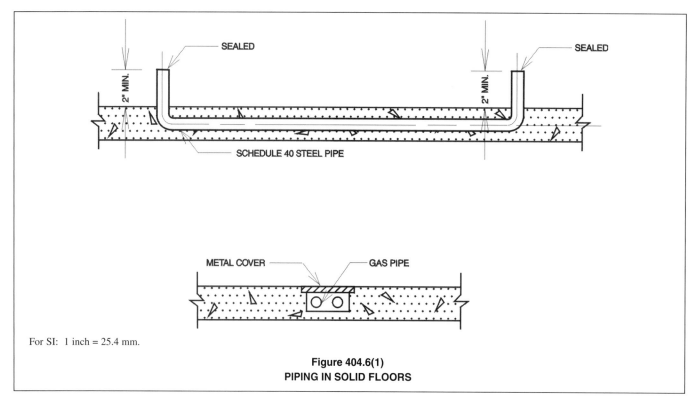

Figure 404.6(1)
PIPING IN SOLID FLOORS

more quickly than if the corrosion was uniformly distributed over the entire surface of the pipe or fitting (see commentary, Section 404.8.2).

404.8.2 Protective coatings and wrapping. Pipe protective coatings and wrappings shall be approved for the application and shall be factory applied.

Exception: Where installed in accordance with the manufacturer's installation instructions, field application of coatings and wrappings shall be permitted for pipe nipples, fittings and locations where the factory coating or wrapping has been damaged or necessarily removed at joints.

❖ Protective coatings in the form of wrappings, tapes, enamels, epoxies, sleeves, casings and factory-applied coverings are a common method of isolating metallic piping materials from the atmosphere or from the earth in which they are buried. Corrosion can be caused by soil or construction materials that are in contact with the pipe. Protection is usually provided by a factory-applied coating or by field wrapping the pipe with a protective covering such as a coal-tar-based or plastic wrapping. Where possible, a piping material is chosen that is not subject to the type of corrosion typical of the application. Application of these materials under strictly controlled factory conditions improves the quality control process and reduces the likelihood of coating defects. In the event that coatings or wrappings must be applied in the field, the exception states that they must be installed in accordance with the manufacturers' installation instructions. Coatings and wrappings should be applied only by persons trained and experienced in such work in order to improve reliability of the coating/wrapping process. Field application is permitted only in cases where the factory-applied material may have been damaged during transit or installation of the pipe, where the applied coating was removed for pipe welding or threading of the pipe or for short sections (nipples) of pipe used in the installation. Corrosion can be concentrated on the pipe or fitting where there are flaws in the coating or wrapping; therefore, improperly applied coatings and wrappings can cause pipe or fitting failures to occur faster than failures would occur in unprotected metals. It is evident why plastic piping is used extensively in underground installations.

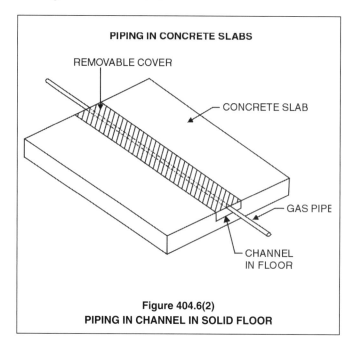

Figure 404.6(2)
PIPING IN CHANNEL IN SOLID FLOOR

404.9 Minimum burial depth. Underground piping systems shall be installed a minimum depth of 12 inches (305 mm) below grade, except as provided for in Section 404.9.1.

❖ The depth of 12 inches (305 mm) is considered sufficient to avoid possible harm to the pipe from the use of hand tools such as spades and shovels. However, if the piping is located in an area subject to surface loads such as vehicular traffic, the 12-inch (305 mm) depth may not be sufficient to protect the piping from those loads (see commentary Figure 404.9).

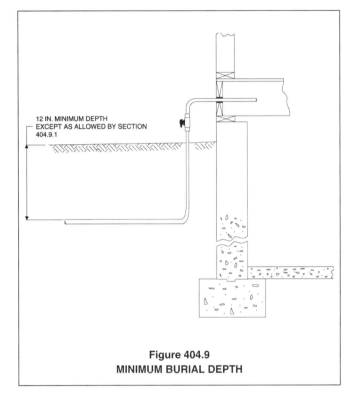

Figure 404.9
MINIMUM BURIAL DEPTH

404.9.1 Individual outside appliances. Individual lines to outside lights, grills or other appliances shall be installed a minimum of 8 inches (203 mm) below finished grade, provided that such installation is approved and is installed in locations not susceptible to physical damage.

❖ Gas piping may be installed within 8 inches (203 mm) of the ground surface where it serves individual outdoor appliances and is not likely to be subjected to damage such as might occur from vehicular traffic, gardening, future excavation, etc. Each individual installation must be reviewed and approved by the code official. The intent is to allow shallow installations for small distribution lines that serve outdoor gas lights, cooking appliances, pool heaters and similar loads only where the piping is unlikely to be disturbed.

404.10 Trenches. The trench shall be graded so that the pipe has a firm, substantially continuous bearing on the bottom of the trench.

❖ Where trenches have nonuniform depth or peaks and valleys in the bottom, the piping could lack continuous support and could be subjected to stresses from the backfill and surface loads.

404.11 Piping underground beneath buildings. Piping installed underground beneath buildings is prohibited except where the piping is encased in a conduit of wrought iron, plastic pipe, or steel pipe designed to withstand the superimposed loads. Such conduit shall extend into an occupiable portion of the building and, at the point where the conduit terminates in the building, the space between the conduit and the gas piping shall be sealed to prevent the possible entrance of any gas leakage. Where the end sealing is capable of withstanding the full pressure of the gas pipe, the conduit shall be designed for the same pressure as the pipe. Such conduit shall extend not less than 4 inches (102 mm) outside the building, shall be vented above grade to the outdoors, and shall be installed so as to prevent the entrance of water and insects. The conduit shall be protected from corrosion in accordance with Section 404.8.

❖ This section prohibits the installation of gas piping beneath buildings to reduce the potential for an inaccessible pipe failure caused by settling of the structure. The prohibition also reduces the potential for corrosion-caused failure of piping embedded in soil or fill. See Section 404.6 for gas pipe installations in floor slabs. Where underground installation under a building is unavoidable, the piping must be encased in another pipe to channel any leakage to the outdoors (see commentary Figure 404.11).

404.12 Outlet closures. Gas outlets that do not connect to appliances shall be capped gas tight.

Exception: Listed and labeled flush-mounted-type quick-disconnect devices and listed and labeled gas convenience outlets shall be installed in accordance with the manufacturer's installation instructions.

❖ Unused fuel gas outlets must be capped or plugged gas-tight, regardless of whether a shutoff valve is provided at the outlet. A closed valve alone is not dependable and poses an unnecessary risk of leakage. The exception recognizes that listed gas outlet devices are built with an inherent safety feature that will automatically shut off the gas flow if the mating connector is disengaged or that will not allow the appliance connector to be disengaged until the integral shutoff valve is manually closed.

404.13 Location of outlets. The unthreaded portion of piping outlets shall extend not less than 1 inch (25 mm) through finished ceilings and walls and where extending through floors or outdoor patios and slabs, shall not be less than 2 inches (51 mm) above them. The outlet fitting or piping shall be securely supported. Outlets shall not be placed behind doors. Outlets shall be located in the room or space where the appliance is installed.

Exception: Listed and labeled flush-mounted-type quick-disconnect devices and listed and labeled gas convenience outlets shall be installed in accordance with the manufacturer's installation instructions.

❖ This section regulates the location, installation and termination of gas piping system outlets to reduce the like-

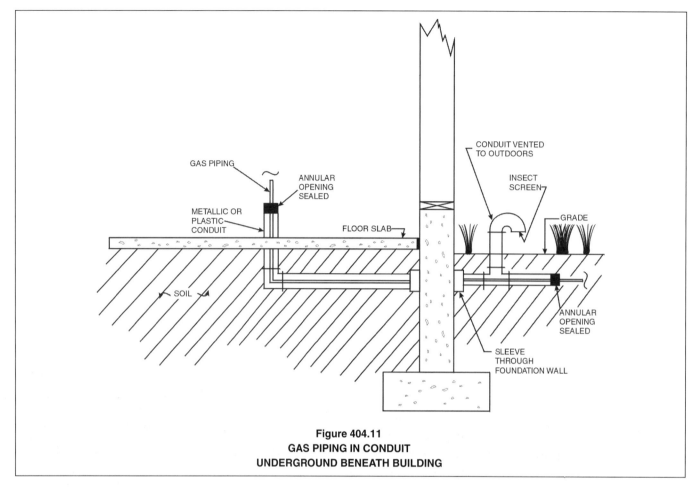

**Figure 404.11
GAS PIPING IN CONDUIT
UNDERGROUND BENEATH BUILDING**

lihood of physical damage and to provide clearances for the use of tools. When making connections to piping outlets, "back-up" wrenches are used to prevent piping from rotating, loosening or being damaged. Sufficient pipe length is necessary to allow the application of tools. Gas outlet devices that have been tested and labeled for installation methods other than those addressed in this section must be installed in accordance with the terms of their testing and listing as contained in the manufacturer's installation instructions. Gas outlets for CSST systems must be installed with the termination fitting designed specifically for that purpose and provided by the CSST manufacturer.

404.14 Plastic pipe. The installation of plastic pipe shall comply with Sections 404.14.1 through 404.14.3.

❖ This section places restrictions on the location and operating pressures for plastic pipe in addition to stating installation requirements specific to plastic pipe.

404.14.1 Limitations. Plastic pipe shall be installed outside underground only. Plastic pipe shall not be used within or under any building or slab or be operated at pressures greater than 100 psig (689 kPa) for natural gas or 30 psig (207 kPa) for LP-gas.

Exceptions:

1. Plastic pipe shall be permitted to terminate above ground outside of buildings where installed in pre-manufactured anodeless risers or service head adapter risers that are installed in accordance with the manufacturer's installation instructions.

2. Plastic pipe shall be permitted to terminate with a wall head adapter within buildings where the plastic pipe is inserted in a piping material for fuel gas use in buildings.

❖ Because of the potential hazard associated with the use of a material that has lower resistance to physical damage and heat as compared to metallic pipe, installation of plastic pipe and tubing is limited to areas that are both outside of the building and underground. Plastic pipe and tubing are widely used for underground gas distribution systems because of their ease of installation and inherent resistance to corrosion. Polyethylene (PE) pipe is the only allowable plastic pipe for use with LP-gas, and the code user is directed to the referenced standard, NFPA 58, which requires that PE piping materials comply with ASTM D 2513. The exception makes it clear that if an equivalent level of physical protection is installed, plastic pipe may terminate above ground outside of the building. It is common practice for gas utility companies to install pre-manufactured riser assemblies at their meter settings. Such risers typically consist of a steel pipe with a corrosion-resistant coating in which a length of PE pipe is pre-installed for coupling to the service lateral. A 90-degree (1.57 rad) sweeping bend is built into the underground end of the riser assembly. Plastic piping is not allowed in or under a concrete slab. The code official might determine that it is not the intent of this section to prevent an outdoor slab from being placed over an existing customer-owned (private) underground gas line, such as in the case of sidewalks and open patios. The prohibition on piping within a slab is applicable to all slabs. Exceptions 1 and 2 relate to Section 403.6.1, see the commentary for that section (see commentary Figure 404.14.1).

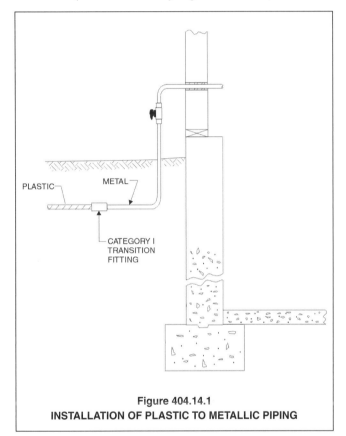

Figure 404.14.1
INSTALLATION OF PLASTIC TO METALLIC PIPING

404.14.2 Connections. Connections made outside and underground between metallic and plastic piping shall be made only with transition fittings categorized as Category I in accordance with ASTM D 2513.

❖ Because installation of plastic piping is allowed only where both outside and underground, the same requirement holds for joining plastic and metallic piping. Mechanical compression joints employ an elastomeric seal that must be compatible not only with the piping material but also with the gas that is conveyed in the system. The manufacturer's instructions normally require use of an internal stiffener insert in conjunction with the fitting for additional support. The stiffener insert prevents the pipe from deforming under the compression force exerted by the seal and compression nut assembly.

Category 1 fittings are designed to make a gas-tight seal and resist axial (pull-out) forces equal to the yield strength of the piping material [see commentary Figure 403.11(3)].

404.14.3 Tracer. A yellow insulated copper tracer wire or other approved conductor shall be installed adjacent to underground nonmetallic piping. Access shall be provided to the tracer wire or the tracer wire shall terminate above ground at each end of the nonmetallic piping. The tracer wire size shall not be less than 18 AWG and the insulation type shall be suitable for direct burial.

❖ To avoid piping damage and a hazardous condition, the location of underground piping must be known prior to excavation work in the vicinity of such piping.

In the past, gas utility companies, installers and contractors relied on metal detectors to locate gas piping underground. With the advent of nonmetallic (plastic) piping, this is no longer possible. A tracer wire provides a means for locating nonmetallic gas piping without having to excavate, thereby avoiding the possibility of damaging the pipe. An electrical current can be passed through the wire to allow a metal detector to locate the piping. Insulated wire is required because copper wire or other approved conductors may be susceptible to corrosion in some soils.

Yellow is the standardized color used to identify gas piping. Tapes that contain a conducting strip and also serve as a visual warning are also available.

404.15 Prohibited devices. A device shall not be placed inside the piping or fittings that will reduce the cross-sectional area or otherwise obstruct the free flow of gas.

Exception: Approved gas filters.

❖ Devices such as filters and flow-measuring instruments could create considerable pressure drop and are prohibited except where the piping system is designed to tolerate such losses. An unaccounted for restriction in the piping could make an otherwise properly designed system fail to supply the required pressure.

404.16 Testing of piping. Before any system of piping is put in service or concealed, it shall be tested to ensure that it is gas tight. Testing, inspection and purging of piping systems shall comply with Section 406.

❖ A pressure test is required after every installation, alteration, addition or repair to the fuel-gas piping system. The location of a leak may be difficult to determine, especially if it is concealed in the building construction. If a leak is found, the leaking component must be repaired or replaced before the system is concealed or put into operation. Section 406 specifies testing pressures based on the type of system, the design working pressure or other parameters. The testing duration is based on the total cubic feet of pipe volume, and the piping system must sustain the test pressure for the duration without exhibiting any sign of leakage.

SECTION 405 (IFGS)
PIPING BENDS AND CHANGES IN DIRECTION

405.1 General. Changes in direction of pipe shall be permitted to be made by the use of fittings, factory bends, or field bends.

❖ This section permits making changes in direction of gas piping through the use of pipe fittings compatible with the piping material or, if recommended by the pipe manufacturer, by bending the pipe. If bending is the chosen method, it must be accomplished in accordance with Sections 405.2 and 405.3. In practice, steel gas piping is rarely bent to accomplish changes in direction because pipe failure at a seam is a possibility for steel pipe that is bent.

405.2 Metallic pipe. Metallic pipe bends shall comply with the following:

1. Bends shall be made only with bending equipment and procedures intended for that purpose.
2. All bends shall be smooth and free from buckling, cracks, or other evidence of mechanical damage.
3. The longitudinal weld of the pipe shall be near the neutral axis of the bend.
4. Pipe shall not be bent through an arc of more than 90 degrees (1.6 rad).
5. The inside radius of a bend shall be not less than six times the outside diameter of the pipe.

❖ Pipe is bent to accomplish changes in direction without the use of fittings. Pipe must be bent with the appropriate bending tools and materials. Some materials are not tempered or intended to be bent; therefore, consideration must be given to choosing the proper materials. Bending tools are designed to make bends without damaging the pipe. Pipe intended to be bent has a minimum bend radius which must be observed to avoid damage to the pipe. For example, bending welded seam pipe with the seam located outside of the neutral axis of the bend may result in a split seam because of the stresses induced from the bend. In other words, the welded seam must not be oriented in a bend that would place it in tension or compression. If pipe is to be bent, it should be stated in the pipe specifications that the pipe is suitable for bending. Rigid gas piping is not commonly bent because of the perceived risk of pipe stress failures at the bend and because bending can be more labor intensive than using fittings.

405.3 Plastic pipe. Plastic pipe bends shall comply with the following:

1. The pipe shall not be damaged and the internal diameter of the pipe shall not be effectively reduced.
2. Joints shall not be located in pipe bends.
3. The radius of the inner curve of such bends shall not be less than 25 times the inside diameter of the pipe.
4. Where the piping manufacturer specifies the use of special bending equipment or procedures, such equipment or procedures shall be used.

❖ Because of the material characteristics of plastic pipe, the code defers to the manufacturer's installation and bending instructions to achieve the performance level contemplated by this section. For example, coiled plastic pipe and tubing have limitations regarding bending beyond or against the natural curvature of the coil. A joint in a bend would be subject to bending stresses and pipe/fitting misalignments that could cause joint structural failure and/or leakage. Thus, a 1-inch (2.54 mm) ID plastic pipe would have a minimum bending radius of 25 inches (635 mm) measured to the inner curve of the bend.

405.4 Mitered bends. Mitered bends are permitted subject to the following limitations:

1. Miters shall not be used in systems having a design pressure greater than 50 psig (340 kPa gauge). Deflections caused by misalignments up to 3 degrees (0.05 rad) shall not be considered as miters.
2. The total deflection angle at each miter shall not exceed 90 degrees (1.6 rad).

❖ Mitered joints are associated with welded steel piping. Such joints do not use fittings. Changes in direction are made by cutting the pipe ends at the desired angles and welding them together. Mitered bends are rarely used as they are difficult to make, have high flow resistance, and have a poor appearance.

405.5 Elbows. Factory-made welding elbows or transverse segments cut therefrom shall have an arc length measured along the crotch at least 1 inch (25 mm) in pipe sizes 2 inches (51 mm) and larger.

❖ Welding elbows of various angles up to 90 degrees (1.57 rad) are commonly used to make changes in direction in steel piping systems. The minimum length of the arc on the inside radius of the fitting must be one inch to provide enough metal and room to make a proper weld.

SECTION 406 (IFGS)
INSPECTION, TESTING AND PURGING

406.1 General. Prior to acceptance and initial operation, all piping installations shall be inspected and pressure tested to determine that the materials, design, fabrication, and installation practices comply with the requirements of this code.

❖ Before any gas piping system is put to use, it must be inspected, tested and approved. See Sections 406.1.1 through 406.6.4. Pressure testing is distinct from leakage testing as covered in Section 406.6.

406.1.1 Inspections. Inspection shall consist of visual examination, during or after manufacture, fabrication, assembly, or pressure tests as appropriate. Supplementary types of nondestructive inspection techniques, such as magnetic-particle, radiographic, ultrasonic, etc., shall not be required unless specifically listed herein or in the engineering design.

❖ The inspection of piping installations is intended to be a visual observation of the system and the testing procedure. The designer may require that welded joints in, for example, very high pressure applications be examined by a method that is capable of discovering internal defects that are not detectable by visual observation.

406.1.2 Repairs and additions. In the event repairs or additions are made after the pressure test, the affected piping shall be tested.

Minor repairs and additions are not required to be pressure tested provided that the work is inspected and connections are tested with a noncorrosive leak-detecting fluid or other approved leak-detecting methods.

❖ Some time after the initial test of the piping, a repair could be made or a branch could be added or extended. The repaired, added or extended portion of the piping system needs to be tested without requiring the unchanged portions to be tested again. Minor work on an existing system is allowed without pressure testing the minor work if the work is visually inspected and tested by a leak-detecting method. Leak-detecting methods include bubble test fluids and electronic sensors. Some bubble test fluids could be corrosive to some piping; therefore, a noncorrosive fluid designed for this purpose should always be used. Methods other than bubble test fluids must be acceptable to the code official.

406.1.3 New branches. Where new branches are installed from the point of delivery to new appliances, only the newly installed branches shall be required to be pressure tested. Connections between the new piping and the existing piping shall be tested with a noncorrosive leak-detecting fluid or other approved leak-detecting methods.

❖ Often, because it is convenient or because the existing piping is fully loaded, a new piping branch will be run from the point of delivery to serve a new appliance installation. The new branch is typically taken from a tee fitting installed immediately downstream of the meter. The entire new branch piping must be tested, including the tee fitting in the existing piping from which the new branch is supplied. The point of connection to the existing piping can be tested by a means other than pressure testing, although all other piping in the new branch must be isolated from the existing piping and pressure tested as required for new work. Section 406.1.4 requires that the new branch piping be disconnected from the existing piping during pressure testing, except where an assembly consisting of two closed valves in series with an intermediate open valved port is installed to isolate the new piping from the existing piping (see commentary Figure 406.1.4).

406.1.4 Section testing. A piping system shall be permitted to be tested as a complete unit or in sections. Under no circumstances shall a valve in a line be used as a bulkhead between gas in one section of the piping system and test medium in an adjacent section, unless two valves are installed in series with a valved "telltale" located between these valves. A valve shall not be subjected to the test pressure unless it can be determined that the valve, including the valve-closing mechanism, is designed to safely withstand the test pressure.

❖ Depending on the progression of a job, it may be desirable to test portions of a system as they are completed. It is also possible that portions of a system will be put in service before the entire system is completed. To prevent a test medium from leaking into piping containing fuel gas or vice versa, portions of piping under test must be isolated from portions that are in service. A single

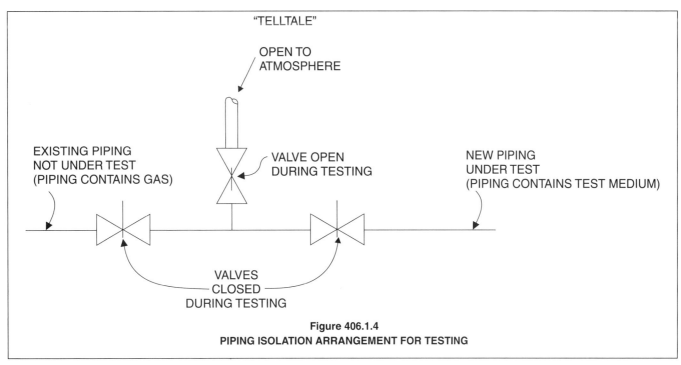

Figure 406.1.4
PIPING ISOLATION ARRANGEMENT FOR TESTING

valve cannot be depended on to establish this isolation because all valves have some "leak-through" allowance, and also the valve could be accidentally opened. To eliminate such risk, two valves in series must be used, and a valved open port (tell-tale) must be installed between the valves. The open port will release any gases that have passed through either isolation valve and also will serve to indicate any leakage or accidental opening of an isolation valve (see commentary Figure 406.1.4).

406.1.5 Regulators and valve assemblies. Regulator and valve assemblies fabricated independently of the piping system in which they are to be installed shall be permitted to be tested with inert gas or air at the time of fabrication.

❖ Assemblies of valves, regulators and/or over- pressure devices might be factory or shop assembled and may be tested independently of the piping system in which they will be installed.

406.2 Test medium. The test medium shall be air, nitrogen, carbon dioxide or an inert gas. Oxygen shall not be used.

❖ The gas that is forced into the piping system for pressure testing must be chemically unreactive to prevent undesirable reactions with the piping system components and the fuel to be conveyed in the piping. Oxygen is not inert, and any residual amount could form an explosive mixture with the fuel gas. The typical test gas is air, which is also not inert but is readily available and free. Nitrogen and carbon dioxide are considered inert for the purpose of testing piping systems and are inexpensive compared to most truly inert gases. If the piping system must be purged with an inert gas (see Section 406.7.2), testing with an inert gas would have an advantage over testing with air.

406.3 Test preparation. Pipe joints, including welds, shall be left exposed for examination during the test.

Exception: Covered or concealed pipe end joints that have been previously tested in accordance with this code.

❖ This section is consistent with Section 107.1 and requires piping joints to remain exposed during testing to allow locations of defects. This section speaks only of joints, but Section 107.1 requires that the piping and joints be left exposed until inspected and approved.

406.3.1 Expansion joints. Expansion joints shall be provided with temporary restraints, if required, for the additional thrust load under test.

❖ If the piping system uses expansion joints to accommodate piping movement, those joints might need to be restrained to resist the additional thrust loading caused by the movement of the pressurized test gas. Generally, testing does not produce noticeable dynamic thrust forces because of the slow rate at which the test gas is put in and taken out of the piping system. The thrust force produced by the static pressure of the test gas is of concern. Expansion joints are rarely used because expansion loops and offsets perform the same function.

406.3.2 Equipment isolation. Equipment that is not to be included in the test shall be either disconnected from the piping or isolated by blanks, blind flanges, or caps. Flanged joints at which blinds are inserted to blank off other equipment during the test shall not be required to be tested.

❖ Some equipment can be damaged by test pressure and is therefore disconnected from the piping system or is otherwise isolated by blank-offs placed between the flanges.

406.3.3 Equipment disconnection. Where the piping system is connected to equipment or components designed for operating pressures of less than the test pressure, such equipment or equipment components shall be isolated from the piping system by disconnecting them and capping the outlet(s).

❖ This requirement is often overlooked in the field, which could lead to appliance damage and serious consequences. Appliances are typically designed for a maximum inlet gas pressure of $^1/_2$ psig (3.5 kPa) or less, above which damage to controls and regulators could occur. The test pressure will be at least 3 psig (20.7 kPa) and oftentimes much higher in accordance with general practice or local traditions or requirements. Installers routinely pressurize piping systems to pressures between 50 and 100 psig (345 and 690 kPa), often at the insistence of the local authority. It is obvious that there is a good potential to expose appliances to extreme pressures during a test, which is why the code requires the precaution of disconnecting the appliances from the supply piping. Disconnection is normally accomplished by opening a union or disconnecting the appliance connector. The appliance shutoff valve must not be used as a means of appliance isolation because the valve can leak through and/or can be accidentally

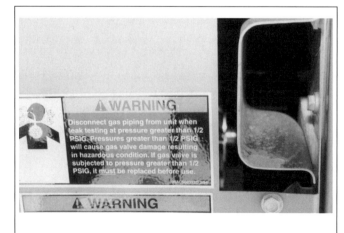

Figure 406.3.3
MANUFACTURER'S WARNING REGARDING
PRESSURE TESTING OF GAS PIPING

opened or left open (see Section 406.3.4 and commentary Figure 406.3.3).

406.3.4 Valve isolation. Where the piping system is connected to equipment or components designed for operating pressures equal to or greater than the test pressure, such equipment shall be isolated from the piping system by closing the individual equipment shutoff valve(s).

❖ If the appliances and other system components are not subject to damage by the test pressure, the equipment shutoff valve can serve to isolate the equipment during the test. Unlike Section 406.3.3, this section does not require disconnection of the appliance/equipment.

406.3.5 Testing precautions. All testing of piping systems shall be done with due regard for the safety of employees and the public during the test. Bulkheads, anchorage, and bracing suitably designed to resist test pressures shall be installed if necessary. Prior to testing, the interior of the pipe shall be cleared of all foreign material.

❖ Compressed gases store energy in direct proportion to the pressure. This stored energy can be destructive and dangerous to personnel if released quickly because of joint, component or pipe failure or an intentional action. Debris will be found in all piping systems after installation. This debris can be harmful to appliances and system components and can act as projectiles during testing and purging operations. Piping is typically purged before testing to flush out foreign materials.

406.4 Test pressure measurement. Test pressure shall be measured with a manometer or with a pressure-measuring device designed and calibrated to read, record, or indicate a pressure loss caused by leakage during the pressure test period. The source of pressure shall be isolated before the pressure tests are made. Mechanical gauges used to measure test pressures shall have a range such that the highest end of the scale is not greater than five times the test pressure.

❖ Test pressures must be measured with a manometer or other pressure-measuring device that is designed and calibrated to indicate a pressure loss caused by leakage during the pressure test period. Pressure- measuring devices must be graduated so that small variations in test pressure will be readily detectable. Small pressure variations might go undetected in instruments having broader ranges. Gauges should have scale increments that are small enough to detect small changes in pressure. Mechanical gauges generally achieve their best accuracy near mid-range of their scale. All test instruments should have recently been calibrated and their accuracy verified before starting the test. Although no piping system can be absolutely leak free, the intent of the code is to make the system free of leaks that are measurable with the testing procedures and instruments prescribed in the code and attainable in the field. Spring-type mechanical pressure gauges must have a high-end scale reading of no greater than 5 times the test pressure. If the test pressure was 3 psig (20.7 kPa), for example, the gauge would have to have a maximum range of 0 to 15 psig (0 to 104 kPa). Such gauges are a wise choice because the test reading will be higher in the range of the gauge where the gauge accuracy is better. Also, the gauge can have much smaller increments than the typical 100 psig (690 kPa) gauge, thus allowing greater discernability in detecting a leak [see commentary Figure 406.4(1)]. All spring-type mechanical gauges are susceptible to inaccuracies caused by conditions of use and mishandling and should be calibrated regularly. Manometers are very accurate and do not need to be calibrated. In all cases, the intent of this section is to require an instrument that is capable of indicating the very small pressure changes that occur as a result of a detectable leak. Test instrument sensitivity is especially important considering the short duration of the test.

The source of pressure in the piping system under test must be disconnected from the system to prevent the source from invalidating the test. For example, the compressed air tank or inert gas cylinder could be adding gas to the piping system at the same rate as a leak is letting gas escape; the test would then be meaningless.

Figures 406.4(2), (3) and (4) show examples of test instruments that can be considerably more accurate than mechanical spring gauges. Electronic instruments can provide high resolution readings, making very small leaks as well as regular leaks detectable within short time periods. Figures 406.4 (3) and (4) show instruments that are designed to give accurate pressure readings and give positive indication of a leak. The liquid manometer-type instrument, shown in Figure 406.4 (3), functions as a manometer that readily indicates test pressure and leakage. A separate feature of the instrument allows small leaks to be recognized quickly. While the system is under test pressure, an integral valve can be closed that will isolate the liquid reservoir from the piping system, thereby trapping the test pressure in the manometer's liquid reservoir. A small diameter vertical dip tube, having its open end submerged in the liquid, rises above the liquid level in the reservoir. The top end of the dip tube is connected to the piping system under test. If there is no leakage, the pressures at the upper end and lower (submerged) end of the dip tube are equal and liquid will not rise. If leakage occurs, the pressure in the piping system will fall below the pressure trapped in the reservoir at the start of the test procedure and liquid will be forced up into the small diameter dip tube, indicating the leak. The use of a small diameter dip tube allows a small change in liquid level in the reservoir to be seen as a comparatively large movement of liquid in the dip tube.

406.4.1 Test pressure. The test pressure to be used shall be no less than $1^1/_2$ times the proposed maximum working pressure, but not less than 3 psig (20 kPa gauge), irrespective of design pressure. Where the test pressure exceeds 125 psig (862 kPa gauge), the test pressure shall not exceed a value that produces a hoop stress in the piping greater than 50 percent of the specified minimum yield strength of the pipe.

❖ The minimum test pressure will never be less than 3 psig (20.7 kPa). The majority of residential and small commercial occupancies are served by piping systems with a pressure of less than ½ psig (14 inches water column) (3.5 kPa), and for such systems the test pressure must be 3 psig (20.7 kPa) or higher. The test pressure must also be not less than 1.5 times the pressure at which the piping system is designed to operate. If the piping system is designed to operate at 5 psig (34.5 kPa), the test pressure must be not less than 1.5 x 5 = 7.5 psig (51.7 kPa). In very high pressure applications, the test pressure might approach the structural limits of the piping material, and, therefore, the test pressure must be limited. Hoop stress is the result of internal pressure that tends to expand the pipe walls outward along the circumference of the pipe.

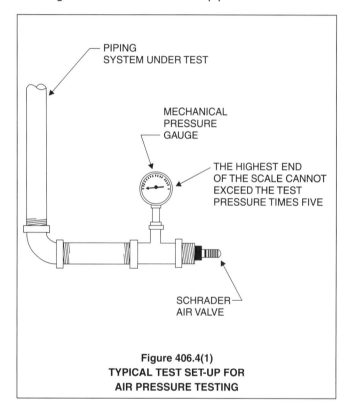

Figure 406.4(1)
TYPICAL TEST SET-UP FOR AIR PRESSURE TESTING

406.4.2 Test duration. Test duration shall be not less than $^1/_2$ hour for each 500 cubic feet (14 m^3) of pipe volume or fraction thereof. When testing a system having a volume less than 10 cubic feet (0.28 m^3) or a system in a single-family dwelling, the test duration shall be not less than 10 minutes. The duration of the test shall not be required to exceed 24 hours.

❖ As the piping system becomes larger, the test duration must be longer so that any leak can be detected. For example, a leakage rate of 1 cubic foot per hour (472 cm^3/min) would produce a more easily detectable pressure drop in a small 10 cubic foot (0.3 m^3) volume system than it would in a large 500 cubic foot (14.2 m^3) volume system. It would require a longer test period to detect the leak in the large volume system because the leakage of 1 cubic foot represents a much smaller fraction of the total gas volume in the system. If a system has a volume of 501 cubic feet, the minimum test duration would be 1 hour. To put this in perspective, 500 cubic feet of piping system internal volume would equate to 83,333 feet (25 400m) of 1-inch (25 mm) schedule 40 steel pipe or 9,747 feet (2971 m) of 3-inch (76 mm) schedule 40 steel pipe.

Because 10 cubic feet (0.3 m^3) equates to 1,667 feet (508.1 m) of 1-inch (25 mm) pipe, it is obvious that single-family dwellings fall in the 10 cubic foot or less volume category. A 10-minute test period is not much time to detect a pressure drop, which makes it even more important to use a pressure instrument that is capable of indicating very small drops in pressure (see commentary, Section 406.4).

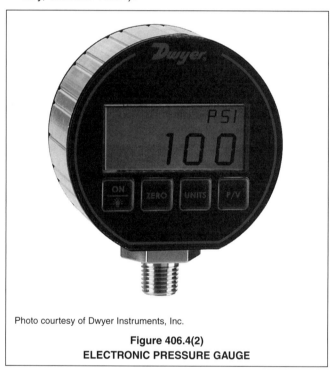

Photo courtesy of Dwyer Instruments, Inc.
Figure 406.4(2)
ELECTRONIC PRESSURE GAUGE

406.5 Detection of leaks and defects. The piping system shall withstand the test pressure specified without showing any evidence of leakage or other defects.

Any reduction of test pressures as indicated by pressure gauges shall be deemed to indicate the presence of a leak unless such reduction can be readily attributed to some other cause.

❖ Any pressure drop, no matter how small, is considered a failure of the test, except where the drop can be shown to result from a change in temperature or other cause. This would be practically impossible to demonstrate in the field, especially for short test durations. A true leak should show up as a continuous drop in test pressure. If the rate of pressure drop slows as the test progresses, it may be caused by cooling of the test medium, which causes it to contract. For example, warm air coming from a compressor will contract when introduced into a cold piping system, thereby causing a test instrument to indicate a drop in pressure. To help eliminate any guess work, the temperature of the test medium should be allowed to stabilize within the piping system before the test begins.

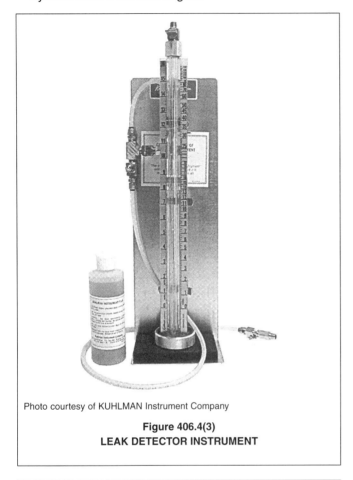

Photo courtesy of KUHLMAN Instrument Company

Figure 406.4(3)
LEAK DETECTOR INSTRUMENT

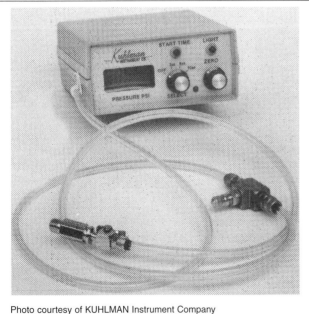

Photo courtesy of KUHLMAN Instrument Company

Figure 406.4(4)
ELECTRONIC LEAK DETECTOR INSTRUMENT

406.5.1 Detection methods. The leakage shall be located by means of an approved gas detector, a noncorrosive leak detection fluid, or other approved leak detection methods. Matches, candles, open flames, or other methods that could provide a source of ignition shall not be used.

❖ Once the pressure measuring instrument indicates leakage, the leak or leaks must be located for repair or replacement. Electronic sensors and bubble fluids are used to find leaks that are not readily found by human senses. Many people who have worked with and/or inspected gas piping installations can recall instances of personal injury, fires and close calls that were caused by searching for leaks with an open flame.

406.5.2 Corrections. Where leakage or other defects are located, the affected portion of the piping system shall be repaired or replaced and retested.

❖ Where a leak is detected by a method named in Section 406.5.1, the defect must be corrected. Corrections include tightening of fittings or threaded joints and replacing defective pipe, tubing or fittings.

406.6 System and equipment leakage test. Leakage testing of systems and equipment shall be in accordance with Sections 406.6.1 through 406.6.4.

❖ Leakage testing is different from pressure testing. Sections 406.1 through 406.5 addressed the pressure testing of newly installed or altered piping systems for the purpose of locating defects in the installation or materials. This section addresses leakage tests that are intended to discover open outlets, defective appliance connections and defects that have developed since the initial installation. See Section 406.6.2.

406.6.1 Test gases. Leak checks using fuel gas shall be permitted in piping systems that have been pressure tested in accordance with Section 406.

❖ Piping systems are pressure tested with air or an inert gas, and leakage testing is done with the fuel gas at whatever pressure the system is designed to operate. See Section 406.6.2.

406.6.2 Before turning gas on. Before gas is introduced into a system of new gas piping, the entire system shall be inspected to determine that there are no open fittings or ends and that all valves at unused outlets are closed and plugged or capped.

❖ This section requires that a precautionary measure be taken to prevent fuel gas from escaping from the piping system when the gas is turned on in the system for the first time. Because the piping system has never been used, it is not to be fully trusted as being gas tight. Because anything could have been done to the system since it was pressure tested, an additional visual check is justified before the gas is turned on. The condition of appliances may also be unknown; thus, all appliance shutoff valves must be closed until the appliances are ready to be put in service. Often, parties other than those who installed and tested the piping system are involved with connecting to the outlets in the system. This makes it likely that what was gas tight at the initial pressure testing is no longer gas tight as a result of incomplete or improper connections at the system outlets.

406.6.3 Test for leakage. Immediately after the gas is turned on into a new system or into a system that has been initially restored after an interruption of service, the piping system shall be tested for leakage. Where leakage is indicated, the gas supply shall be shut off until the necessary repairs have been made.

❖ This section is a logical progression from the previous section. After the visual check for the status of outlets and equipment shutoff valves, the gas is introduced into the system, and the system is tested for leakage. This testing is typically accomplished by examining all the outlets, observing a gas meter for indication of flow, monitoring the system pressure for any drop or any combination of these. Because all outlet valves are closed and unused outlets are capped, any gas flow or pressure drop indicates leakage, and the gas must be turned off. This section would also allow leakage testing with the appliance shutoff valves open and the appliances shut down to prevent their operation. Testing in this manner would verify that the appliance gas controls were not faulty. A leakage test is also required after gas service is restored.

406.6.4 Placing equipment in operation. Gas utilization equipment shall be permitted to be placed in operation after the piping system has been tested and determined to be free of leakage and purged in accordance with Section 406.7.2.

❖ Gas appliances cannot be put into service until the gas piping system that supplies them has been pressure tested, tested for leakage and purged of air. See Section 406.7.2.

406.7 Purging. Purging of piping shall comply with Sections 406.7.1 through 406.7.4.

❖ Purging is intended to prevent a flammable gas/air mixture from being created in the piping. A flammable mixture could be a fire/explosion hazard in the piping, room or space, or in an appliance combustion chamber.

406.7.1 Removal from service. Where gas piping is to be opened for servicing, addition, or modification, the section to be worked on shall be turned off from the gas supply at the nearest convenient point, and the line pressure vented to the outdoors, or to ventilated areas of sufficient size to prevent accumulation of flammable mixtures.

The remaining gas in this section of pipe shall be displaced with an inert gas as required by Table 406.7.1.

TABLE 406.7.1
LENGTH OF PIPING REQUIRING PURGING WITH INERT GAS FOR SERVICING OR MODIFICATION

NOMINAL PIPE SIZE (inches)	LENGTH OF PIPING REQUIRING PURGING
$2^1/_2$	> 50 feet
3	> 30 feet
4	> 15 feet
6	> 10 feet
8 or larger	Any length

For SI: 1 inch = 25.4 mm, 1 foot = 304.8 mm.

❖ If gas piping is opened for some reason, the residual gas in the affected piping must be disposed of in a manner that will not create a fire/explosion hazard. Large piping systems are commonly constructed of welded steel, and the act of welding would be a source of ignition. The larger and longer the piping, the more residual gas there will be because of the greater internal volume of the piping system. The table dictates what volume of piping must be purged with an inert gas based on the internal diameter and length of the piping. Purging would be performed by introducing an unreactive gas such as dry nitrogen into one end of the piping at a sufficiently high rate to flush the fuel gas out of the other end of the piping. Air cannot be used to purge piping because it contains oxygen.

406.7.2 Placing in operation. Where piping full of air is placed in operation, the air in the piping shall be displaced with fuel gas, except where such piping is required by Table 406.7.2 to be purged with an inert gas prior to introduction of fuel gas. The air can be safely displaced with fuel gas provided that a moderately rapid and continuous flow of fuel gas is introduced at one end of the line and air is vented out at the other end. The fuel gas flow shall be continued without interruption until the vented gas is free of air. The point of discharge shall not be left unattended during purging. After purging, the vent shall then be closed. Where required by Table 406.7.2, the air in the piping shall first be displaced with an inert gas, and the inert gas shall then be displaced with fuel gas.

❖ Piping that has been left open for some period of time because it is out of service or has never been in service will contain air. This air must be eliminated from the piping to avoid creating a flammable gas/air mixture that could ignite in the piping or an appliance combustion chamber. Gas/air mixtures can also be out of the flammability range and could cause delayed ignition in appliance combustion chambers, resulting in personal injury and/or appliance damage. In small volume systems, the air can be purged by introduction of fuel gas at a rate sufficient to flush the air from the system. Table 406.7.2 serves to distinguish small volume systems from large volume systems and requires large volume systems to be purged with an inert gas prior to introduction of fuel gas. Purging with an inert gas reduces the possibility of forming a flammable gas/air mixture.

TABLE 406.7.2
LENGTH OF PIPING REQUIRING PURGING WITH INERT GAS BEFORE PLACING IN OPERATION

NOMINAL PIPE SIZE (inches)	LENGTH OF PIPING REQUIRING PURGING
3	> 30 feet
4	> 15 feet
6	> 10 feet
8 or larger	Any length

For SI: 1 inch = 25.4 mm, 1 foot = 304.8 mm.

❖ See the Commentary for Section 406.7.2.

406.7.3 Discharge of purged gases. The open end of piping systems being purged shall not discharge into confined spaces or areas where there are sources of ignition unless precautions are taken to perform this operation in a safe manner by ventilation of the space, control of purging rate, and elimination of all hazardous conditions.

❖ Obviously, mixtures of fuel gas, air and inert gases resulting from the purging of piping are potentially hazardous and must be discharged to the outdoors or to adequately ventilated areas. Consideration must be given to sources of ignition as well as ventilation where purged gases are discharged indoors.

406.7.4 Placing equipment in operation. After the piping has been placed in operation, all equipment shall be purged and then placed in operation, as necessary.

❖ For the same reason that piping systems must be purged of foreign gases, the appliances served must also be purged before being placed in operation.

SECTION 407 (IFGC)
PIPING SUPPORT

407.1 General. Piping shall be provided with support in accordance with Section 407.2.

❖ Often, piping support is inadequate, poorly installed or overlooked entirely, despite the fact that support is critical in preventing stresses and strains in piping, fittings, joints, connectors and appliance gas trains (see Section 407.2).

407.2 Design and installation. Piping shall be supported with pipe hooks, metal pipe straps, bands, brackets, or hangers suitable for the size of piping, of adequate strength and quality, and located at intervals so as to prevent or damp out excessive vibration. Piping shall be anchored to prevent undue strains on connected equipment and shall not be supported by other piping. Pipe hangers and supports shall conform to the requirements of MSS SP-58 and shall be spaced in accordance with Section 415. Supports, hangers, and anchors shall be installed so as not to interfere with the free expansion and contraction of the piping between anchors. All parts of the supporting equipment shall be designed and installed so they will not be disengaged by movement of the supported piping.

❖ As with all piping systems, the support of the fuel gas system is as important as any other part of the overall design. Proper supports are necessary to maintain piping alignment and slope, to support the weight of the pipe and its contents, to control movement, and to resist dynamic loads, such as thrust. Inadequate support can cause piping to fail under its own weight, resulting in fire, explosion or property damage. Building design must take into consideration the structural loads created by the support of piping systems.

Hangers or supports must not react with or be detrimental to the pipe they support. Hangers or supports for metallic pipe must be of a material that is compatible with the pipe to prevent any corrosive action. For example, copper, copper-clad or specially coated hangers are required if the piping system is constructed of copper tubing. Hangers and supports are constructed of noncombustible materials to prevent premature failure in a fire. This section is commonly interpreted to prohibit combustible hooks, bands, brackets and hangers. Gas piping must not depend on other piping for support; thus, it must be supported from the structure independently. Gas controls for appliances can be stressed and damaged by unsupported piping drops that bear on or produce bending moments in the controls. A simple check for proper support at an appliance connected with rigid piping is to open the union and observe any movement of the piping. If the piping drops or moves in any direction, it is evident that stress is being applied to the appliance controls. Eliminating such stress is one of the advantages of using flexible gas connectors (see Section 415).

SECTION 408 (IFGC)
DRIPS AND SLOPED PIPING

408.1 Slopes. Piping for other than dry gas conditions shall be sloped not less than $^1/_4$ inch in 15 feet (6.3 mm in 4572 mm) to prevent traps.

❖ The required minimum slope (0.14%) is intended for gas known to contain enough water vapor to cause condensate to form inside the piping. Such conditions are rarely, if ever, encountered today.

408.2 Drips. Where wet gas exists, a drip shall be provided at any point in the line of pipe where condensate could collect. A drip shall also be provided at the outlet of the meter and shall be installed so as to constitute a trap wherein an accumulation of condensate will shut off the flow of gas before the condensate will run back into the meter.

❖ Drips, often referred to as "drip legs," are distinct from sediment traps. Modern gas supplies and distribution systems are typically dry; thus, drips would be required only when recommended by the gas supplier.

408.3 Location of drips. Drips shall be provided with ready access to permit cleaning or emptying. A drip shall not be located where the condensate is subject to freezing.

❖ A drip is not the same as a sediment trap; they are required for different reasons (see Section 408.4).

408.4 Sediment trap. Where a sediment trap is not incorporated as part of the gas utilization equipment, a sediment trap shall be installed downstream of the equipment shutoff valve as close to the inlet of the equipment as practical. The sediment trap shall be either a tee fitting with a capped nipple in the bottom opening of the run of the tee or other device approved as an effective sediment trap. Illuminating appliances, ranges, clothes dryers and outdoor grills need not be so equipped.

❖ In addition to the code requirement, most appliance manufacturers require the installation of a sediment trap (dirt leg) to protect the appliance from debris in the gas. Sediment traps are necessary to protect appliance gas controls from the dirt, soil, pipe chips, pipe joint tapes and compounds and construction site debris that enters the piping during installation and repairs. Hazardous appliance operation could result from debris entering gas controls and burners. Despite the fact that utilities supply clean gas, debris can enter the piping prior to and during installation on the utility side of the system and on the customer side.

Sediment traps are designed to cause the gas flow to change direction 90 degrees at the sediment collection point, thus causing the solid or liquid contaminants to drop out of the gas flow [see commentary Figure 408.4(1)]. The nipple and cap must not be placed in the branch opening of a tee fitting because this would not create a change in direction of flow and would allow debris to pass over the collection point. Commentary Figure 408.4(2) illustrates an improper sediment trap that is prohibited by this section. The code does not specify a minimum length for the capped nipple. Three inches (76 mm) minimum is customary. The sediment trap must be downstream of the appliance shutoff valve and as close to the appliance inlet as practical. The sediment trap must be downstream of the appliance shutoff valve is to make sure it is within 6 feet (1829 mm) of the appliance inlet. If there is 6 feet (1829 mm) or less of piping between a sediment trap and the appliance inlet served, the intent of the code has been met, regardless of the shutoff valve location. Manufactured sediment traps are available that have the configuration of a straight section of pipe and are equipped with cleanout openings. Although it would be wise to install sediment traps at all appliance connections, they are not mandated by code for gas lights, ranges, clothes dryers and outdoor grills. These appliances are susceptible to harm from debris in gas, especially ranges and clothes dryers, and the appliance manufacturer may require sediment traps where the code does not. The code's logic is that these appliances are manually operated rather than automatically operated; therefore, the user would be aware of a problem.

SECTION 409 (IFGC)
SHUTOFF VALVES

409.1 General. Piping systems shall be provided with shutoff valves in accordance with this section.

❖ This section addresses shutoff valves for meters, buildings, tenant spaces, appliances and fireplaces.

409.1.1 Valve approval. Shutoff valves shall be of an approved type. Shutoff valves shall be constructed of materials compatible with the piping. Shutoff valves installed in a portion of a piping system operating above 0.5 psig shall comply with ASME B 16.33. Shutoff valves installed in a portion of a piping system operating at 0.5 psig or less shall comply with ANSI Z 21.15 or ASME B 16.33.

❖ The code official must approve the type of shutoff valve used. For example, he or she might not approve a valve that requires a wrench to operate. Shutoff valves must be tested to the specified standard and so identified. ANSI standard Z21.15 is applicable to appliance shutoff valves and appliance connector shutoff valves. Ball valves are used almost exclusively today because of their low pressure losses, ease of operation and reliability (see commentary Figures 409.1.1(1) and 409.1.1(2)].

409.1.2 Prohibited locations. Shutoff valves shall be prohibited in concealed locations and furnace plenums.

❖ This section does not prohibit shutoff valves in spaces above ceilings and below floors used for conveying environmental air. See the definition of "Concealed location." A valve in a concealed location would be of no value and a potential leak source. The concern for valves in furnace plenums is temperature extremes and lack of access.

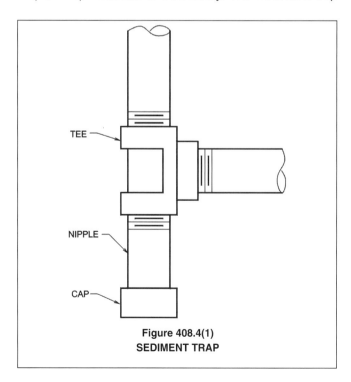

Figure 408.4(1)
SEDIMENT TRAP

Figure 408.4(2)
PROHIBITED SEDIMENT TRAP CONFIGURATION

409.1.3 Access to shutoff valves. Shutoff valves shall be located in places so as to provide access for operation and shall be installed so as to be protected from damage.

❖ As expressed in the previous section, a shutoff valve has no utility if it is installed where it cannot be accessed. Location is especially important for valves installed for emergency use, such as a ball valve installed indoors upstream of all appliances to allow the building owner to shutoff all gas in the event of an emergency. Outdoor valves are particularly subject to damage.

Photo courtesy of Dormont Manufacturing Company
Figure 409.1.1(1)
TYPICAL BALL-TYPE GAS SHUTOFF VALVE

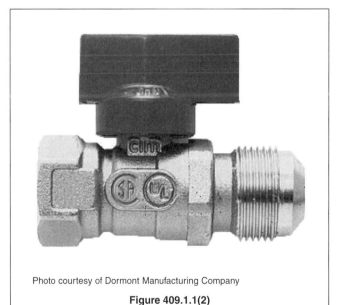

Photo courtesy of Dormont Manufacturing Company
Figure 409.1.1(2)
BALL-TYPE GAS SHUTOFF VALVE DESIGNED TO MATE WITH A CONNECTOR

409.2 Meter valve. Every meter shall be equipped with a shutoff valve located on the supply side of the meter.

❖ This valve is typically installed and owned by the gas supplier because it is upstream of the point of delivery for utility-owned meters. These valves require a wrench to operate and are capable of being locked off.

409.3 Shutoff valves for multiple-house line systems. Where a single meter is used to supply gas to more than one building or tenant, a separate shutoff valve shall be provided for each building or tenant.

❖ A separate shutoff valve for each building or tenant space will allow isolation of any building or space in the event of an emergency or when work must be performed on the piping system.

409.3.1 Multiple tenant buildings. In multiple tenant buildings, where a common piping system is installed to supply other than one- and two-family dwellings, shutoff valves shall be provided for each tenant. Each tenant shall have access to the shutoff valve serving that tenant's space.

❖ Each tenant space and dwelling unit in buildings having more than two dwellings must have its own shutoff valve to allow isolation from the common piping system. The tenant spaces and dwelling units may or may not have separate meters (see Section 409.3). The shutoff valves must be within the tenant space or dwelling unit, or the valves must be located in a common space that can be accessed by all of the occupants.

409.3.2 Individual buildings. In a common system serving more than one building, shutoff valves shall be installed outdoors at each building.

❖ This section is similar to Section 409.3 and requires the shutoff valve for each building to be located outside of the building, thus allowing quick access in an emergency.

409.3.3 Identification of shutoff valves. Each house line shutoff valve shall be plainly marked with an identification tag attached by the installer so that the piping systems supplied by such valves are readily identified.

❖ The term "house line" is used to describe the main supply pipe to a building or a dwelling unit. Utilities refer to the house line as the piping on the downstream side of the utility-owned meter. In the case of multiple house lines supplying different spaces in a building, the required shutoff valves must be permanently tagged to indicate which space is supplied by each valve. Without tags, it might not be apparent which space each valve controls. For both service work and emergency action, the load the valve serves must be identified. The tags should be designed and attached to be permanent and are typically brass medallions attached with small chains.

409.4 MP Regulator valves. A listed shutoff valve shall be installed immediately ahead of each MP regulator.

❖ See the definition of "Regulator, Medium-pressure." This regulator (also known as "pounds-to-inches" regulator) is placed between the service regulator and the appliance regulator and is typically associated with manifold parallel distribution systems using CSST or copper tubing. A shutoff valve ahead of the MP regulator will allow the regulator to be conveniently serviced, adjusted or replaced. The valve also serves to isolate the higher pressure system from the lower pressure system. For purposes of sizing, the piping downstream of an MP regulator is considered as a separate system, with the MP regulator being the source of supply for that system.

409.5 Equipment shutoff valve. Each appliance shall be provided with a shutoff valve separate from the appliance. The shutoff valve shall be located in the same room as the appliance, not further than 6 feet (1829 mm) from the appliance, and shall be installed upstream from the union, connector or quick disconnect device it serves. Such shutoff valves shall be provided with access.

Exception: Shutoff valves for vented decorative appliances and decorative appliances for installation in vented fireplaces shall not be prohibited from being installed in an area remote from the appliance where such valves are provided with ready access. Such valves shall be permanently identified and shall serve no other equipment. Piping from the shutoff valve to within 3 feet (914 mm) of the appliance connection shall be sized in accordance with Section 402.

❖ Each appliance, regardless of type, size or location, must have its own dedicated shutoff valve to isolate the appliance from its gas supply. It is not the intent of this section to allow a single shutoff valve to control more than one appliance. Equipment/appliance shutoff valves are not considered to be emergency valves, but rather are considered service valves that allow the appliance to be serviced, repaired, replaced or shut down. See the definition of "Valve, Equipment shutoff." The location requirements of this section are intended to make the valve conspicuous and easy to operate. The code does not require the equipment shutoff valve to be provided with "ready access." Because such a valve is not an emergency valve, there is no reason to require ready access, and simply providing access is sufficient. Common practice has been to install the appliance shutoff valve behind ranges and clothes dryers and in the control compartment of decorative gas fireplace appliances. This section allows the practice because the valves are accessible. To avoid the use of an exposed valve or concealed "key-handle" valve, it is accepted practice to locate the shutoff valve for gas fireplace appliances in the control compartment in the bottom of the unit, behind the control access panel (see commentary Figure 409.5).

This section calls for a "separate" shutoff valve, which means a valve other than the valve that is integral with the appliance gas control. The intent is that the separate shutoff be located with no more than 6 feet (1829 mm) of piping between the outlet of the shutoff valve and the inlet of the appliance. This section works with the definition of "piping system" and Section 402.3 to allow the piping downstream of the shutoff valve to be considered and sized as a connector.

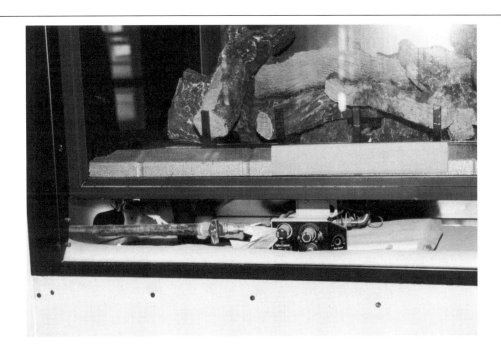

Figure 409.5
SHUTOFF VALVE IN CONTROL COMPARTMENT OF GAS FIREPLACE

The exception allows locating shutoff valves for the stated appliances remote from the appliance if the valves are permanently tagged as to their function, serve only the single appliance and have ready access. In the exception, "ready access" is intended because the shutoff valve will be more difficult to locate being remote from the appliance served. The exception is an alternative to using concealed "key-handle" type valves located in the wall or floor adjacent to the appliance. Unvented appliances are not addressed by the exception.

The shutoff valve for a log lighter is also the appliance operating valve; therefore, that valve cannot be located remotely. Because the remote shutoff valve allowed by the exception could be more than 6 feet (1829 mm) from the appliance, the piping from the shutoff valve outlet to within 3 feet (914 mm) of the appliance inlet must be treated and sized as gas piping, not as a connector.

409.5.1 Shutoff valve in fireplace. Equipment shutoff valves located in the firebox of a fireplace shall be installed in accordance with the appliance manufacturer's instructions.

❖ This section addresses appliances such as decorative gas log sets that are installed in solid-fuel-burning fireplaces. The manual shutoff may be in the firebox of the fireplace if it is allowed by the appliance manufacturer's instructions and the valve is located as directed by those instructions. There is typically an area adjacent to the appliance controls or inlet connection that is shielded from radiant heat, and this "cool zone" is where the manual shutoff valve must be located. A shutoff valve must never be located in the firebox if solid fuel is burned.

SECTION 410 (IFGC)
FLOW CONTROLS

410.1 Pressure regulators. A line pressure regulator shall be installed where the appliance is designed to operate at a lower pressure than the supply pressure. Access shall be provided to pressure regulators. Pressure regulators shall be protected from physical damage. Regulators installed on the exterior of the building shall be approved for outdoor installation.

❖ All gas supply systems employ regulators at some point in the distribution system, and the type and location of these devices depends on the design pressures of the system and utilization equipment. A line pressure regulator reduces the pressure from the service delivery pressure to the pressure of the distribution system. Additional pressure regulators could be required at other points in the system or at the appliance to reduce the pressure even further [see commentary Figure 410.1(1)]. With the increased use of medium-pressure gas distribution systems, systems are designed with one or more regulators in addition to the service regulators. Appliances and gas utilization equipment are also equipped with factory-installed regulators that control burner input (manifold) pressure.

Pressure regulators are extremely important devices that are relied on to prevent overpressurization of appliances and equipment. Typical gas appliance controls are not designed to tolerate a pressure of more than 14 inches of water column (3.5 kPa). The stan-

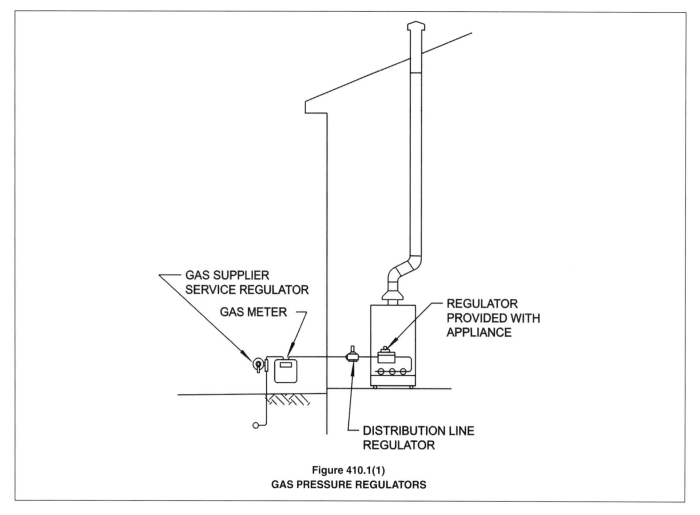

Figure 410.1(1)
GAS PRESSURE REGULATORS

dards that regulate the design and function of appliance gas controls require that the controls be tested for resistance to permanent damage resulting from exposure to excessive gas supply pressure. Each function of the gas controls must be tested for exposure to $2^1/_2$ psi (17 kPa) pressure for a period of 1 minute, after which, the control must function normally. Excess pressures could lead to severe appliance malfunction. Depending on the application, gas pressure regulators may contain relief mechanisms that are designed to vent gas from the downstream side in the event that the downstream (load) side pressure exceeds a predetermined maximum. This feature is a safeguard intended to prevent the overpressurization of the piping and utilization equipment supplied by the regulator. Regulators with relief devices must be installed outdoors, or the relief vent must be piped to the outdoors to prevent gas from being discharged into the building.

Access to pressure regulators is required for the maintenance, adjustment and servicing of the devices. The definition of "Access" appears in Chapter 2. Protecting regulators from physical damage essentially means that regulators are to be installed in locations that are not prone to impact from equipment or vehicles, or they are to be shielded from impact. Because of the range of environmental conditions encountered outside the building, regulators must be evaluated and approved for installation outdoors. Regulators exposed to the weather must be protected from moisture and precipitation. Figure 410.1(2) shows a service regulator with discharge piping connected to the integral relief valve port. This regulator establishes the pressure at the point of delivery. Figure 410.1(3) illustrates the basic operating principle of gas pressure regulators. Outlet pressure acts upon the flexible diaphragm and adjustment spring, causing the valve to close on a rise in outlet pressure and open on a fall in outlet pressure. The vent allows the diaphragm to move freely by admitting or emitting air above the diaphragm. Atmospheric pressure and the adjustment spring force act together to establish the outlet pressure setting of the regulator.

410.2 MP regulators. MP pressure regulators shall comply with the following:

1. The MP regulator shall be approved and shall be suitable for the inlet and outlet gas pressures for the application.

2. The MP regulator shall maintain a reduced outlet pressure under lockup (no-flow) conditions.

3. The capacity of the MP regulator, determined by published ratings of its manufacturer, shall be adequate to supply the appliances served.

4. The MP pressure regulator shall be provided with access. Where located indoors, the regulator shall be vented to the outdoors or shall be equipped with a leak-limiting device, in either case complying with Section 410.3.

5. A tee fitting with one opening capped or plugged shall be installed between the MP regulator and its upstream shutoff valve. Such tee fitting shall be positioned to allow connection of a pressure-measuring instrument and to serve as a sediment trap.

6. A tee fitting with one opening capped or plugged shall be installed not less than 10 pipe diameters downstream of the MP regulator outlet. Such tee fitting shall be positioned to allow connection of a pressure-measuring instrument.

❖ This section states the minimum requirements for medium-pressure (MP) gas regulators installed in 2-psi (13.8 kPa) gas piping systems or portions of systems. These requirements intend to keep the regulators accessible for service and repair, protect them from physical damage, require that they are suitable for the use intended, are properly sized and installed, and, if installed indoors, are properly vented to the outdoors (see commentary, Section 410.3). Item 2 requires a "lockup"-type regulator capable of holding the approximate outlet pressure setpoint indefinitely under no-flow conditions. Section 409.4 requires a gas shutoff valve immediately upstream of the MP regulator to allow isolation of the regulator during testing, repair or replacement of the regulator. Items 5 and 6 require a tee fitting upstream and downstream of the MP regulator to facilitate the connection of pressure-measuring instruments for calibrating the regulator. These fittings should have a $^1/_8$-inch (3.2 mm) threaded tapping for connection of common gauges and manometers. Item 5 also requires a sediment trap on the inlet side of the regulator. Because a lot of faith is placed in the integrity of this regulator, it deserves the protection from debris afforded by a simple tee, nipple and cap arrangement; see Section 408.4.

Where a line pressure regulator serves appliances rated for $^1/_2$ psi (3.5 kPa) or less inlet pressure and the regulator's inlet pressure is over 2 psi (13.8 kPa), a downstream over-pressure protection device must be installed in accordance with ANSI Z21.80, the standard for line pressure regulators (see commentary Figure 410.2). The OPD is designed to limit the downstream (outlet) pressure to 2 psi.

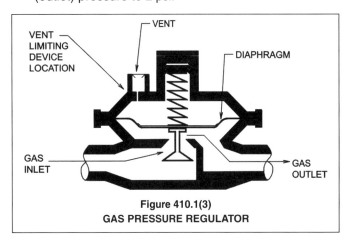

Figure 410.1(3)
GAS PRESSURE REGULATOR

Photo courtesy of Washington Gas

Figure 410.1(2)
SERVICE REGULATOR

Photo courtesy of MAXITROL Company

Figure 410.2
PRESSURE REGULATOR/OVERPRESSURE
PROTECTION DEVICE ASSEMBLIES

410.3 Venting of regulators. Pressure regulators that require a vent shall have an independent vent to the outside of the building. The vent shall be designed to prevent the entry of water or foreign objects.

Exception: A vent to the outside of the building is not required for regulators equipped with and labeled for utilization with approved vent-limiting devices installed in accordance with the manufacturer's instructions.

❖ All regulators use synthetic or natural rubber diaphragms, one side of which is vented to the atmosphere. The devices must vent one side of the diaphragm to the atmosphere to allow unimpeded movement of the diaphragm within its enclosure. Thus, these devices, in effect, see atmospheric pressure as a reference point for controlling the fuel-gas pressure. Regulators cannot operate without the vent. If the diaphragm were to rupture or perforate, gas would be vented to the exterior of the device. For this reason, regulators installed indoors must be vented to the outdoors or must comply with the exception to this section. The vent termination outdoors must be protected from blockage and the entry of foreign objects, insects and water, which could cause regulator failure [see commentary Figures 410.3(1) and 410.3(2)].

Photo courtesy of Washington Gas

Figure 410.3(1)
REGULATOR WITH VENT PROTECTION

This section also requires regulators to be vented individually to the outdoors to prevent a potentially hazardous situation that may arise if more than one regulator is served by a single vent. The hazard could occur if the diaphragm of one regulator failed and allowed an increase of pressure within the vent, which would then cause all of the other regulators using the common vent to operate at a higher discharge pressure. ANSI Z223.1 permits engineered vent manifold systems to be used to vent more than one regulator if the manifold is designed to allow unimpeded air movement and to minimize back pressure in the event of a diaphragm failure.

The exception permits installation of regulators indoors without a vent to the outdoors where the regulator is equipped with and labeled for use with a factory-installed vent-limiting device. Vent-limiting devices, as the name implies, are designed to allow the vent to "breathe" under normal operating conditions. In the event of diaphragm failure, the vent-limiting devices will limit gas escapement to a rate near 1 cubic foot per hour (0.0283 m^3/h) for natural gas at 2-psi (13.8 kPa) pressure. A vent-limiting device is a ball check valve with a fixed orifice and is designed to allow air inhalation and escapement while restricting gas leakage to a small rate. Even small leakage rates can be hazardous if inadequate ventilation exists to prevent buildup of explosive concentrations. Although a standard is not referenced in this section, ANSI Z21.18 is appropriate for evaluating such vent-limiting devices. The standard limits the volume of gas that a vent-limiting device is allowed to discharge into a room or space. Vent-limiting devices should be installed in locations that are ventilated sufficiently to dissipate any gas discharged from the device.

Photo courtesy of MAXITROL Company

Figure 410.3(2)
PRESSURE REGULATOR WITH VENT-LIMITING DEVICE

SECTION 411 (IFGC)
APPLIANCE CONNECTIONS

411.1 Connecting appliances. Appliances shall be connected to the piping system by one of the following:

1. Rigid metallic pipe and fittings.

2. Semirigid metallic tubing and metallic fittings. Lengths shall not exceed 6 feet (1829 mm) and shall be located entirely in the same room as the appliance. Semirigid metallic tubing shall not enter a motor-operated appliance through an unprotected knockout opening.

3. Listed and labeled appliance connectors installed in accordance with the manufacturer's installation instructions and located entirely in the same room as the appliance.

4. Listed and labeled quick-disconnect devices used in conjunction with listed and labeled appliance connectors.

5. Listed and labeled convenience outlets used in conjunction with listed and labeled appliance connectors.
6. Listed and labeled appliance connectors complying with ANSI Z21.69 and listed for use with food service equipment having casters, or that is otherwise subject to movement for cleaning, and other large movable equipment.

❖ Connectors are used to connect the appliance or equipment to the gas distribution system outlet. The pipe sizing methods in this chapter size distribution piping up to but not beyond the appliance shutoff valve outlet, and this section sizes the connection between the shutoff valve outlet and the appliance [see commentary Figure 411.1(1)]. The choice of a connection type must take into consideration such factors as appliance movement, vibration, ambient conditions and susceptibility to physical damage. Between the shutoff valve and the appliance, a union fitting or similar arrangement must allow a means of disconnecting the piping. Flared or ground-joint connections that are part of an approved, labeled semirigid ("flexible") connector can serve as a union. [see commentary Figures 411.1(2) and 411.1(3)].

Item 2 refers to copper or aluminum tubing, which is rarely used today as an appliance connector. Such connectors are field fabricated. Item 3 refers to the commonly used, so called "flexible" connectors typically made of corrugated stainless steel or brass. Although CSST is allowed to directly connect to specific appliances (fixed/nonmoveable), CSST is not to be used as a substitute for a connector where one is required. CSST manufacturers' instructions specify where CSST can directly connect to appliances and where it cannot. Appli- ance connectors are listed and labeled with the label typically being a stamped metal ring placed around the connector tubing or the label information is stamped into the fittings on the end of the connector.

Most appliance connectors manufactured today have attached plastic labels that state the installation and sizing instructions. These connectors are not designed for repeated movement that causes bending of the metal and should not be reused after the initial installation. Reuse is typically prohibited by the manufacturers' instructions. Repeated bending and/or vibration can cause metal fatigue, stress cracking and gas leakage [see commentary Figures 411.1(4) and 411.1(5)].

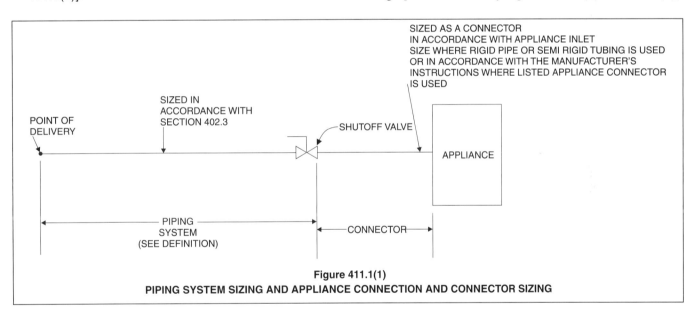

Figure 411.1(1)
PIPING SYSTEM SIZING AND APPLIANCE CONNECTION AND CONNECTOR SIZING

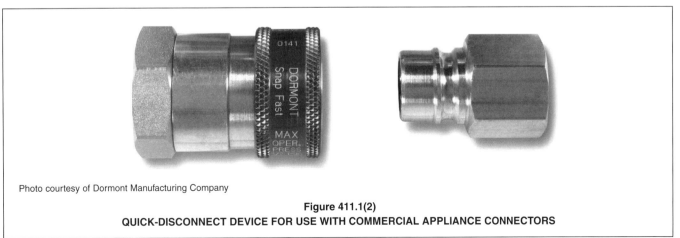

Photo courtesy of Dormont Manufacturing Company

Figure 411.1(2)
QUICK-DISCONNECT DEVICE FOR USE WITH COMMERCIAL APPLIANCE CONNECTORS

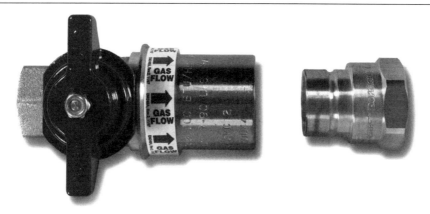

Photo courtesy of Dormont Manufacturing Company

Figure 411.1(3)
QUICK-DISCONNECT DEVICE FOR USE WITH COMMERCIAL APPLIANCE CONNECTORS

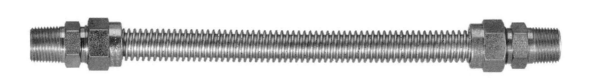

Photo courtesy of Dormont Manufacturing Company

Figure 411.1(4)
SEMI-RIGID (FLEXIBLE) APPLIANCE CONNECTOR

Photo courtesy of Dormont Manufacturing Company

Figure 411.1(5)
OUTDOOR MOBILE HOME CONNECTOR

Item 6 refers to specialized connectors for moveable appliances. These connectors are commonly used for restaurant cooking appliances that are routinely moved for cleaning or equipment line order changes. The appliance must have restraint cables or chains designed to limit movement of the appliance to protect the connector from damage. The restraints limit the travel distance of the appliance to that which can be safely tolerated by the connector [see commentary Figures 411.1(6) and 411.1(7)]. The typical flexible gas appliance connector is not designed to withstand vibration and repeated movement such as might occur for connections to appliances suspended on chains or long hangers. For example, infrared heaters are commonly hung by chains and can move frequently because of air movement in the space. In this application, the heater should be connected with a connector configuration or specialized connector that can tolerate the movement, or the appliance should be stabilized. Infrared tube-type heaters expand and contract during normal operation and can cause connector flexing that can damage the connector if it is not properly installed to accommodate the movement.

The piping system is sized up to the outlet of the appliance shutoff valve. The connector occurs between the shutoff valve outlet and the appliance and is sized by the connector manufacturer or is based on the appliance input rating, gas inlet size and length of the connector. For example, a boiler is supplied by branch piping required to be 1 inch (25.4 mm) based on the sizing methods in Section 402. The branch piping is 1 inch (25.4 mm) up to the boiler shutoff valve, but the boiler inlet opening is only $\frac{1}{2}$ inch (12.7 mm). The piping (connector) downstream of the shutoff valve is sized as a connector because it is 6 feet (1829 mm) or less in length (see commentary Section 409.5).

411.1.1 Protection from damage. Connectors and tubing shall be installed so as to be protected against physical damage.

❖ Appliance connectors, although sound, are not constructed or tested to withstand the same rigors of service as the gas piping system with respect to resisting physical damage. Therefore, they must be installed to minimize the potential for damage. For example, "flexible" and semirigid tubing connectors should not be used where subject to excessive vibration or impact by occupants, vehicles, animals, doors, stored materials, etc. Impact, repeated movement and vibration can cause connector failure.

It is not the intent of this section to require that appliance connectors be hidden from view behind or under an appliance in order to consider them as protected from damage. For typical range and clothes dryer installations, the connectors are physically protected by the appliances themselves. However, for water heaters, furnaces, boilers, unit heaters, infrared heaters, etc., the connector will be located in the open and thus will need protection from damage by proper location, placement, elevation or guards. Protection from damage is commonly afforded by locating the connector out of harm's way such as between a wall and an appliance or 8 feet (2438 mm) or more above the floor.

Photo courtesy of Dormont Manufacturing Company

Figure 411.1(6)
COMMERCIAL COOKING APPLIANCE CONNECTOR
WITH TRAVEL RESTRAINT

411.1.2 Appliance fuel connectors. Connectors shall have an overall length not to exceed 3 feet (914 mm), except for range and domestic clothes dryer connectors, which shall not exceed 6 feet (1829 mm) in length. Connectors shall not be concealed within, or extended through, walls, floors, partitions, ceilings or appliance housings. A shutoff valve not less than the nominal size of the connector shall be installed ahead of the connector in accordance with Section 409.5. Connectors shall be sized to provide the total demand of the connected appliance.

Exception: Fireplace inserts factory equipped with grommets, sleeves, or other means of protection in accordance with the listing of the appliance.

❖ Flexible appliance fuel connectors are primarily intended for use with cooking ranges, clothes dryers and similar appliances where the gas connection is located behind the appliance and some degree of flexibility is necessary to facilitate the hook-up; however, flexible connectors may be used with any appliance where they are not subject to damage (see commentary, Section 411.1.1). Rigid connections are not practical for movable appliances. Flexible connectors are typically constructed of stainless steel and are labeled with tags or metal rings placed over the tubing. The manufacturer's installation instructions dictate the installation requirements that are intended to protect the connector from damage and prevent leakage. Most flexible connectors are intended to be installed only once and are not designed for repeated bending and forming. For other than commercial flexible connections that are designed

Photo courtesy of Dormont Manufacturing Company

Figure 411.1(7)
COMMERCIAL COOKING APPLIANCE
WITH TRAVEL RESTRAINT

for repeated movement, a new connector should be used each time an appliance is reinstalled. Repeated flexing or bending can cause metal fatigue and connector failure. Flexible connectors must be protected from physical damage and are not allowed to pass through any walls, floors, ceilings or appliance housings (see the exception).

Flexible connectors must bear the label of an approved agency and are limited to a maximum of 3 feet (914 mm) in length, except when used in a domestic range or dryer installation where a 6-foot (1829 mm) maximum length is permitted. Longer lengths for range and dryer installations allow the connector to be coiled or otherwise arranged to allow limited movement of the appliance without stressing the connector. Flexible connectors are evaluated in accordance with ANSI Z21.24 or ANSI Z21.69. Multiple connectors may not to be connected together to circumvent the maximum length requirement.

Connectors must also be chosen on the basis that they will provide the total gas demand of the appliance served at the required minimum pressure, as indicated in the manufacturer's instructions. Item 2 of Section 411.1 specifically allows semirigid metallic tubing to be a maximum of 6 feet (1829 mm) long and to enter an appliance housing where the entry opening is designed to protect the tubing.

The exception allows connectors to enter the housing of a fireplace insert appliance where factory-provided means protect the connector from abrasion, corrosion and sharp metal edges. None of the connectors addressed in Section 411.1 are allowed to pass through walls, floors, ceilings or partitions. Connectors are installed downstream of the shutoff valve and that valve must be located in the same room as the appliance. All of the connectors addressed in Section 411.1 must be sized for the connected load.

411.1.3 Movable appliances. Where appliances are equipped with casters or are otherwise subject to periodic movement or relocation for purposes such as routine cleaning and maintenance, such appliances shall be connected to the supply system piping by means of an approved flexible connector designed and labeled for the application. Such flexible connectors shall be installed and protected against physical damage in accordance with the manufacturer's installation instructions.

❖ Commercial appliances such as cooking equipment commonly employ heavy-duty, commercial-type flexible gas connectors. These connectors are designed to tolerate repeated movement to allow appliances on wheels or casters to be moved for cleaning operations or relocation. Such connectors are commonly equipped with quick-disconnect couplings for ease of disconnection and reconnection and restraint cables to limit the amount of appliance movement, thereby preventing stress and damage to the connector.

Flexible connectors designed for specific applications are also used for the connection of equipment and appliances that produce significant vibrations. For example, gas-burning combustion engines in generator sets and HVAC equipment are connected with specially designed vibration dampening flexible connectors. Some pulse-combustion heating appliances employ flexible connectors to control the transmission of noise to gas distribution piping.

SECTION 412 (IFGC)
LIQUEFIED PETROLEUM GAS MOTOR VEHICLE FUEL-DISPENSING STATIONS

412.1 General. Motor fuel-dispensing facilities for LP-gas fuel shall be in accordance with this section and the *International Fire Code*. The operation of LP-gas motor fuel-dispensing facilities shall be regulated by the *International Fire Code*.

❖ Section 2207 of the *International Fire Code®* (IFC®) regulates LP-gas service stations and references Chapter 38 of the IFC.

412.2 Storage and dispensing. Storage vessels and equipment used for the storage or dispensing of LP-gas shall be approved or listed in accordance with Sections 412.3 and 412.4.

❖ Sections 412.3 and 412.4 dictate which components must be listed and which components must be approved by the code official.

412.3 Approved equipment. Containers; pressure-relief devices, including pressure-relief valves; and pressure regulators and piping used for LP-gas shall be approved.

❖ Listing of the specified components could be the basis for approval, but listing is not mandated. Any such components must be designed for the application and recommended for the application by the manufacturer.

412.4 Listed equipment. Hoses, hose connections, vehicle fuel connections, dispensers, LP-gas pumps and electrical equipment used for LP-gas shall be listed.

❖ The specified components must be listed and labeled as complying with the relevant product standards. As with all listed products, the testing/listing agency will apply its seal or mark to the product.

412.5 Attendants. Motor vehicle fueling operations shall be conducted by qualified attendants or in accordance with Section 412.8 by persons trained in the proper handling of LP-gas.

❖ The general public is not allowed to fuel vehicles at LP-gas service stations. The fueling operation must be performed by a service station attendant or trained personnel such as the employees of the company that owns and operates the fuel dispensing system. See Section 412.8.

412.6 Location. In addition to the fuel dispensing requirements of the *International Fire Code*, the point of transfer for dispensing operations shall be 25 feet (7620 mm) or more from buildings having combustible exterior wall surfaces, buildings having noncombustible exterior wall surfaces that are not part of a 1-hour fire-resistance-rated assembly or buildings having combustible overhangs, property which could be built on public streets, or sidewalks and railroads; and at least 10 feet (3048 mm) from driveways and buildings having noncombustible exterior wall surfaces that are part of a fire-resistance-rated assembly having a rating of 1 hour or more.

> **Exception:** The point of transfer for dispensing operations need not be separated from canopies providing weather protection for the dispensing equipment constructed in accordance with the *International Building Code*.

Liquefied petroleum gas containers shall be located in accordance with the *International Fire Code*. Liquefied petroleum gas storage and dispensing equipment shall be located outdoors and in accordance with the *International Fire Code*.

❖ This section specifies a minimum separation distance from buildings, property lines, streets, driveways, sidewalks and railroads to protect them from fire and/or explosion that could occur during a dispensing operation (see Section 2207 and Chapter 38 of the IFC).

412.7 Installation of dispensing devices and equipment. The installation and operation of LP-gas dispensing systems shall be in accordance with this section and the *International Fire Code*. Liquefied petroleum gas dispensers and dispensing stations shall be installed in accordance with manufacturers' specifications and their listing.

❖ Section 412.4 requires dispensers to be listed (see Section 2207 and Chapter 38 of the IFC).

412.7.1 Valves. A manual shutoff valve and an excess flow-control check valve shall be located in the liquid line between the pump and the dispenser inlet where the dispensing device is installed at a remote location and is not part of a complete storage and dispensing unit mounted on a common base.

An excess flow-control check valve or an emergency shutoff valve shall be installed in or on the dispenser at the point at which the dispenser hose is connected to the liquid piping. A differential backpressure valve shall be considered equivalent protection. A listed shutoff valve shall be located at the discharge end of the transfer hose.

❖ Excess flow controls are designed to shut off the gas flow in the event of abnormally high flow rates caused by component failures such as ruptured hoses. Check valves prevent the reverse flow of gas from the vehicle tank into the dispensing system.

412.7.2 Hoses. Hoses and piping for the dispensing of LP-gas shall be provided with hydrostatic relief valves. The hose length shall not exceed 18 feet (5486 mm). An approved method shall be provided to protect the hose against mechanical damage.

❖ Hoses convey liquid fuel; therefore, tremendous hydrostatic pressures could result from pumping-control failures and expansion of the liquid resulting from temperature changes. Hoses are limited in length and must be protected to help prevent hose failure and the subsequent fuel leakage.

412.7.3 Vehicle impact protection. Vehicle impact protection for LP-gas storage containers, pumps and dispensers shall be provided in accordance with the *International Fire Code*.

❖ See Section 2207 of the IFC.

412.8 Private fueling of motor vehicles. Self-service LP-gas dispensing systems, including key, code and card lock dispensing systems, shall not be open to the public and shall be limited to the filling of permanently mounted fuel containers on LP-gas powered vehicles. In addition to the requirements in the *International Fire Code*, self-service LP-gas dispensing systems shall be provided with an emergency shutoff switch located within 100 feet (30 480 mm) of, but not less than 20 feet (6096 mm) from, dispensers and the owner of the dispensing facility shall ensure the safe operation of the system and the training of users.

❖ Self-service dispensers are typically owned and maintained by a company that operates a private fleet of LP-gas vehicles. The dispensers are used by the owners and employees of the company. The emergency dispenser shutoff switch must not be farther than 100 feet (305 m) from and not closer than 20 feet (6.1 m) to the dispenser.

SECTION 413 (IFGC)
COMPRESSED NATURAL GAS MOTOR VEHICLE FUEL-DISPENSING STATIONS

413.1 General. Motor fuel-dispensing facilities for CNG fuel shall be in accordance with this section and the *International Fire Code*. The operation of CNG motor fuel-dispensing facilities shall be regulated by the *International Fire Code*.

❖ Section 2208 of the IFC regulates CNG service stations (see commentary Figure 413.1).

413.2 General. Storage vessels and equipment used for the storage, compression or dispensing of CNG shall be approved or listed in accordance with Sections 413.2.1 and 413.2.2.

❖ This section dictates which components must be listed and which components must be approved by the code official.

413.2.1 Approved equipment. Containers; compressors; pressure-relief devices, including pressure-relief valves; and pressure regulators and piping used for CNG shall be approved.

❖ Listing of the specified components could be the basis for approval, but listing is not mandated. Any such components must be designed for the application and recommended for the application by the manufacturer.

413.2.2 Listed equipment. Hoses, hose connections, dispensers, gas detection systems and electrical equipment used for CNG shall be listed. Vehicle fueling connections shall be listed and labeled.

❖ The specified components must be listed and labeled as complying with the relevant product standards. As with all listed products, the testing/listing agency will apply its seal or mark to the product.

**Figure 413.1
CNG FUEL DISPENSING STATION**

413.3 Location of dispensing operations and equipment. Compression, storage and dispensing equipment shall be located above ground outside.

Exceptions:

1. Compression, storage or dispensing equipment is allowed in buildings of noncombustible construction, as set forth in the *International Building Code*, which are unenclosed for three-quarters or more of the perimeter.

2. Compression, storage and dispensing equipment is allowed to be located indoors in accordance with the *International Fire Code*.

❖ Because of the potential for leakage, the compression, storage and dispensing equipment must be located either outdoors or in a noncombustible, substantially open building except as allowed by Section 2208 and Chapter 30 of the IFC.

413.3.1 Location on property. In addition to the fuel-dispensing requirements of the *International Fire Code*, compression, storage and dispensing equipment shall not be installed:

1. Beneath power lines,
2. Less than 10 feet (3048 mm) from the nearest building or property line which could be built on, public street, sidewalk, or source of ignition.

   Exception: Dispensing equipment need not be separated from canopies providing weather protection for the dispensing equipment constructed in accordance with the *International Building Code*.

3. Less than 25 feet (7620 mm) from the nearest rail of any railroad track.

4. Less than 50 feet (15 240 mm) from the nearest rail of any railroad main track or any railroad or transit line where power for train propulsion is provided by an outside electrical source such as third rail or overhead catenary.

5. Less than 50 feet (15 240 mm) from the vertical plane below the nearest overhead wire of a trolley bus line.

❖ This section lists the limitations on location of CNG dispensing stations with respect to power lines, buildings, property lines, railroad tracks, streets, sidewalks, sources of ignition, electric trains and trolleys.

413.4 Private fueling of motor vehicles. Self-service CNG-dispensing systems, including key, code and card lock dispensing systems, shall be limited to the filling of permanently mounted fuel containers on CNG-powered vehicles.

In addition to the requirements in the *International Fire Code*, the owner of a self-service CNG-dispensing facility shall ensure the safe operation of the system and the training of users.

❖ Unlike LP-gas self-service dispensing stations, CNG self-service dispensing systems can be open to the public. The owner of the system must provide for the training of any users of the system.

413.5 Pressure regulators. Pressure regulators shall be designed, installed or protected so their operation will not be affected by the elements (freezing rain, sleet, snow, ice, mud or debris). This protection is allowed to be integral with the regulator.

❖ Pressure regulator failure could result in dangerous overpressure and the opening of relief valves; therefore, they must be dependable. The regulator vent is susceptible to blockage by debris and ice.

413.6 Valves. Piping to equipment shall be provided with a manual shutoff valve. Such valve shall be provided with ready access.

❖ Shutoff valves allow isolation of components for service, repair, replacement and emergency shutdown. See the definition of "Ready access."

413.7 Emergency shutdown equipment. An emergency shutdown device shall be located within 75 feet (22 860 mm) of, but not less than 25 feet (7620 mm) from, dispensers and shall also be provided in the compressor area. Upon activation, the emergency shutdown shall automatically shut off the power supply to the compressor and close valves between the main gas supply and the compressor and between the storage containers and dispensers.

❖ CNG systems take natural gas from the utility supply line, compress it to very high pressures and store the compressed gas in vessels from which the dispensers draw the gas for transfer to the vehicle onboard containers. The emergency shutdown device must be located not farther than 75 feet (22.9 m) from and not closer than 25 feet (7.6 m) to the dispensers. An additional shutdown device must be located near the compressors. The compressors may be remote from the dispensers. The shutdown device must kill the power to the compressors, must actuate automatic valves that isolate the gas supply from the compressors and isolate the storage vessels from the dispenser, thus limiting accidental gas discharge.

413.8 Discharge of CNG from motor vehicle fuel storage containers. The discharge of CNG from motor vehicle fuel cylinders for the purposes of maintenance, cylinder certification, calibration of dispensers or other activities shall be in accordance with this section. The discharge of CNG from motor vehicle fuel cylinders shall be accomplished through a closed transfer system or an approved method of atmospheric venting in accordance with Section 413.8.1 or 413.8.2.

❖ Gas is intentionally discharged from vehicle containers for several reasons including vehicle repairs, certification of the container integrity and container replacement. The gas must be discharged in a safe manner as dictated by either Section 413.8.1 or 413.8.2. The vehicle storage container is referred to as a "vessel" and as a "cylinder" in the text to follow.

413.8.1 Closed transfer system. A documented procedure which explains the logical sequence for discharging the cylinder shall be provided to the code official for review and approval. The procedure shall include what actions the operator will take in the event of a low-pressure or high-pressure natural gas release during the discharging activity. A drawing illustrating the arrangement of piping, regulators and equipment settings shall be provided to the code official for review and approval. The drawing shall illustrate the piping and regulator arrangement and shall be shown in spatial relation to the location of the compressor, storage vessels and emergency shutdown devices.

❖ A closed transfer system uses the same basic components as a dispensing system and withdraws the gas from the vehicle container and stores it in vessels. Drawings and a description of the sequence of operation of the transfer system must be submitted to the code official for approval.

413.8.2 Atmospheric venting. Atmospheric venting of motor vehicle fuel cylinders shall be in accordance with Sections 413.8.2.1 through 413.8.2.6.

❖ The six subsections that follow list the conditions and requirements under which gas may be discharged to the atmosphere. Such discharge should be avoided where practical because:

1. there is inherent hazard in doing so,
2. methane is an air contaminant, and
3. natural resources should never be wasted.

A closed transfer system is the preferable way to remove gas from vehicle containers.

413.8.2.1 Plans and specifications. A drawing illustrating the location of the vessel support, piping, the method of grounding and bonding, and other requirements specified herein shall be provided to the code official for review and approval.

❖ Plans of the proposed atmospheric venting apparatus and piping system must be reviewed by the code official for approval.

413.8.2.2 Cylinder stability. A method of rigidly supporting the vessel during the venting of CNG shall be provided. The selected method shall provide not less than two points of support and shall prevent the horizontal and lateral movement of the vessel. The system shall be designed to prevent the movement of the vessel based on the highest gas-release velocity through valve orifices at the vessel's rated pressure and volume. The structure or appurtenance shall be constructed of noncombustible materials.

❖ Vehicle CNG containers (vessels) can contain extremely high pressures, which if released quickly can produce large thrust forces that would propel the container like a rocket. Natural gas does not liquify at normal ambient temperatures; therefore, to hold the required amount of fuel on board the vehicle, the gaseous fuel must be compressed to extreme pressures of up to 3600 psia (24.8 mPa). Such pressures are unnecessary for LP-gas because it can be stored as a liquid at normal ambient temperatures.

413.8.2.3 Separation. The structure or appurtenance used for stabilizing the cylinder shall be separated from the site equipment, features and exposures and shall be located in accordance with Table 413.8.2.3.

❖ This section and Table 413.8.2.3 require that the location of the cylinder to be discharged (vessel) be separated from buildings, building openings, lot lines, public ways, vehicles and other CNG equipment. This precaution is reasonable considering the potential hazard of working with highly pressurized containers of a flammable gas.

TABLE 413.8.2.3
SEPARATION DISTANCE FOR ATMOSPHERIC VENTING OF CNG

EQUIPMENT OR FEATURE	MINIMUM SEPARATION (feet)
Buildings	25
Building openings	25
Lot lines	15
Public ways	15
Vehicles	25
CNG compressor and storage vessels	25
CNG dispensers	25

For SI: 1 foot = 304.8 mm.

413.8.2.4 Grounding and bonding. The structure or appurtenance used for supporting the cylinder shall be grounded in accordance with the ICC *Electrical Code*. The cylinder valve shall be bonded prior to the commencement of venting operations.

❖ "Grounding" means to intentionally connect to the earth. This could be accomplished by installing a conductor between the cylinder support and the building grounding electrode system. Bonding means to join metallic parts to form a continuous electrical pathway. Grounding and bonding required by this section are intended to control sparking that could result from current flow produced by voltage differentials across parts of the venting set-up and the building components. Grounding the venting set-up and attaching a bonding jumper to the cylinder valve will put all such components and the building components at the same voltage potential, thereby reducing the possibility of sparks that could ignite flammable vapors and/or harm the cylinder assembly. The provisions of this section will also help prevent the buildup of static electrical charges that could be a source of ignition.

413.8.2.5 Vent tube. A vent tube that will divert the gas flow to the atmosphere shall be installed on the cylinder prior to the commencement of the venting and purging operation. The vent tube shall be constructed of pipe or tubing materials approved for use with CNG in accordance with the *International Fire Code*.

The vent tube shall be capable of dispersing the gas a minimum of 10 feet (3048 mm) above grade level. The vent tube shall not be provided with a rain cap or other feature which would limit or obstruct the gas flow.

At the connection fitting of the vent tube and the CNG cylinder, a listed bidirectional detonation flame arrester shall be provided.

❖ The cylinder must be discharged through a vent tube/pipe that is constructed of a material that is compatible with the gas and that has the required strength to withstand the pressure to which it can be exposed. The pressure that the vent can be exposed to must be calculated based on the size of the vent tube, the length of the vent, the friction loss through the vent and the maximum pressure and discharge rate of the cylinder. Vent failure could cause injury resulting from projectile debris and could cause a severe fire/explosion hazard.

413.8.2.6 Signage. Approved NO SMOKING signs shall be posted within 10 feet (3048 mm) of the cylinder support structure or appurtenance. Approved CYLINDER SHALL BE BONDED signs shall be posted on the cylinder support structure or appurtenance.

❖ Approved signs are intended to remind personnel of the potential danger of cylinder venting. See Section 413.8.2.4.

SECTION 414 (IFGC)
SUPPLEMENTAL AND STANDBY GAS SUPPLY

414.1 Use of air or oxygen under pressure. Where air or oxygen under pressure is used in connection with the gas supply, effective means such as a backpressure regulator and relief valve shall be provided to prevent air or oxygen from passing back into the gas piping. Where oxygen is used, installation shall be in accordance with NFPA 51.

❖ All sources of oxygen must be isolated from the fuel supply to prevent the possibility of creating a flammable/explosive mixture in the piping system. NFPA 51 applies to oxygen and fuel gas piping systems used for welding, cutting and heat treating metals.

414.2 Interconnections for standby fuels. Where supplementary gas for standby use is connected downstream from a meter or a service regulator where a meter is not provided, a device to prevent backflow shall be installed. A three-way valve installed to admit the standby supply and at the same time shut off the regular supply shall be permitted to be used for this purpose.

❖ Some applications require a continuous supply of fuel gas, and a standby gas supply may be provided that will be used in the event that the normal supply is interrupted. Where a standby fuel gas, such as LP-gas, is used as a fuel for a different gas supply system, a three-way valve or other approved safeguard system must be used to prevent the two different supply systems from being interconnected. The three-way valve is analogous to a single-pole, double-throw switch and provides an opening configuration that allows either fuel supply to be connected to the outlet (load), but does not allow both supplies to be connected simultaneously to the outlet or connected to each other. A manual or automatic valve is used to control which fuel gas line is supplying the fuel to the fuel-burning equipment.

SECTION 415 (IFGS)
PIPING SUPPORT INTERVALS

415.1 Interval of support. Piping shall be supported at intervals not exceeding the spacing specified in Table 415.1. Spacing of supports for CSST shall be in accordance with the CSST manufacturer's instructions.

TABLE 415.1
SUPPORT OF PIPING

STEEL PIPE, NOMINAL SIZE OF PIPE (inches)	SPACING OF SUPPORTS (feet)	NOMINAL SIZE OF TUBING (SMOOTH-WALL) (inch O.D.)	SPACING OF SUPPORTS (feet)
$1/2$	6	$1/2$	4
$3/4$ or 1	8	$5/8$ or $3/4$	6
$1\,1/4$ or larger (horizontal)	10	$7/8$ or 1 (Horizontal)	8
$1\,1/4$ or larger (vertical)	Every floor level	1 or Larger (vertical)	Every floor level

For SI: 1 inch = 25.4 mm, 1 foot = 304.8 mm.

❖ This table is two tables side-by-side. The first two columns are for steel pipe, and the second two columns are for tubing materials. See Section 407. Support for CSST is dictated by the manufacturers' installation instructions.

Bibliography

The following resource material is referenced in this chapter or is relevant to the subject matter addressed in this chapter.

ANSI LC1-97, *Gas Piping Systems Using Corrugated Stainless Steel Tubing*. New York, NY: American National Standards Institute, 1997.

ANSI Z21.18-00, *Gas Appliance Pressure Regulators*. New York, NY: American National Standards Institute, 2000.

ANSI Z21.15-97, *Manually Operated Gas Valves for Appliances, Appliance Connector Valves, and Hose End Valves*. New York, NY: American National Standards Institute, 1997.

ASHRAE Handbook of HVAC Systems and Equipment. Atlanta, GA: American Society of Heating, Refrigerating and Air-Conditioning Engineers, 1999.

ASTM D 2513-01A, *Specification for Thermoplastic Gas Pressure Pipe, Tubing, and Fittings*. West Conshohocken, PA: American Society for Testing and Materials, 1998.

AWS A5.8-92, *Filler Metals for Brazing and Braze Welding*. Miami, FL: American Welding Society, 1992.

Design and installation guide for gastite system Titeflex corp. Springfield, MA 01139-0054, 603 Hendee Street.

Crocker, Sabin and R. C. King. *Piping Handbook*. Columbus, OH: McGraw-Hill, 1973

GPSA Engineering Data Book. Tulsa, OK: Gas Processors Suppliers Association, 1998.

NFPA 51-97, *Design and Installation of Oxygen-Fuel Gas Systems for Welding, Cutting, and Allied Processes*. Quincy, MA: National Fire Protection Association, 1997.

NFPA 58-01, *Liquified Petroleum Gases Code*. Quincy, MA: National Fire Protection Association, 1998.

CHAPTER 5
CHIMNEYS AND VENTS

General Comments

This chapter intends to regulate the design, construction, installation, maintenance, repair and approval of chimneys, vents and their connections to gas-fired appliances. A properly designed chimney or vent system is needed to conduct to the outdoors the flue gases produced by a gas-fired appliance. In the case of natural-draft appliances (including Category I), the chimney or vent system serving the appliances is depended on to produce a draft at the appliance connection. Draft refers to a phenomenon created by the buoyancy of lighter (less dense) gases in the presence of heavier (more dense) gases. Draft is measured as pressure in inches of water column (kPa) and is a negative pressure, meaning that it is less than the atmospheric pressure at the point of measurement. Draft is produced by the temperature difference between the combustion gases (flue gases) and the ambient atmosphere. Because hotter gases are less dense, they are buoyant and will rise in the chimney, causing negative pressures to develop that are directly related to the height of the chimney or vent and directly related to the temperature difference between the flue gases and the ambient air. Draft produced by a chimney or vent is the motivating force that conveys combustion products from natural-draft appliances to the outside atmosphere through the chimney or vent passageway. It is also the force that draws combustion air into natural-draft atmospheric-burner appliances.

The type of vent or chimney used in a given installation is generally based on the temperature of the flue gases produced by the appliance and the appliance category. The development of higher efficiency appliances continues to produce lower flue gas temperatures, and the diversity of mechanical venting systems continues to expand.

Many vented gas-fired appliances have been further classified into four categories to simplify the process of selection of the appropriate venting system. The four categories of vented appliances are addressed in the definition of "Vented appliance categories." With the advent of new technology and higher efficiency appliances/equipment, venting means are becoming increasingly diversified and complex. Many of today's appliances are designed to be vented by more than one type of venting system.

Purpose

The provisions in this chapter are intended to minimize the hazards associated with the venting of combustion products produced by fuel-gas-fired appliances and equipment. The requirements of this chapter are intended to achieve the complete removal of the products of combustion from vented fuel-gas-fired appliances and equipment. This chapter includes regulations for the proper selection, design, construction and installation of a chimney or vent, along with measures to minimize the related potential fire hazards. A chimney or vent must be designed for the type of appliance or equipment it serves. Chimneys and vents are designed for specific applications depending on the flue gas temperatures and the pressures at which the chimney or vent operates.

The primary hazards associated with chimneys and vents are high temperatures and the chemical nature of combustion by-products.

SECTION 501 (IFGC)
GENERAL

501.1 Scope. This chapter shall govern the installation, maintenance, repair and approval of factory-built chimneys, chimney liners, vents and connectors and the utilization of masonry chimneys serving gas-fired appliances. The requirements for the installation, maintenance, repair and approval of factory-built chimneys, chimney liners, vents and connectors serving appliances burning fuels other than fuel gas shall be regulated by the *International Mechanical Code*. The construction, repair, maintenance and approval of masonry chimneys shall be regulated by the *International Building Code*.

❖ Chapter 5 contains requirements for the installation, maintenance, repair and approval of residential, commercial and industrial chimney and venting systems that convey the products of combustion from a gas-fired appliance to the outside atmosphere. Venting systems for fuel-fired appliances other than gas-fired appliances, such as oil and solid-fuel appliances, are covered in the *International Mechanical Code®* (IMC®). The construction of masonry chimneys is regulated by the *International Building Code®* (IBC®). See Section 501.3.

501.2 General. Every appliance shall discharge the products of combustion to the outdoors, except for appliances exempted by Section 501.8.

❖ Appliances other than those listed in Section 501.8 must be vented to convey the potentially harmful combustion by-products to the outdoor atmosphere (see Section 501.8).

501.3 Masonry chimneys. Masonry chimneys shall be constructed in accordance with Section 503.5.3 and the *International Building Code*.

❖ A masonry chimney is a field-constructed assembly that can consist of solid masonry units, reinforced concrete, rubble stone, fire-clay liners and mortars. A masonry chimney is permitted to serve low-, medium-, and high-heat appliances. The IBC outlines the general code requirements regarding the construction details for masonry chimneys, including those serving masonry fireplaces.

501.4 Minimum size of chimney or vent. Chimneys and vents shall be sized in accordance with Section 504.

❖ Section 504 contains the requirements for sizing Category I appliance venting systems. Sizing of venting systems for other categories is dictated by the appliance manufacturer.

501.5 Abandoned inlet openings. Abandoned inlet openings in chimneys and vents shall be closed by an approved method.

❖ Abandoned inlet openings result from appliances being disconnected or reconnected at different elevations. Unused openings in chimneys and vents can allow combustion gases to enter the building; can cause loss of draft at other appliances connected to the chimney or vent; can allow conditioned air to escape, resulting in energy losses, and can allow the entry of birds and rodents. For example, consider an appliance that was connected to a chimney and has been replaced with an appliance that vents through an exterior wall. The opening in the chimney left by the disconnection of the appliance must be properly closed to prevent a hazardous condition. Combustion gases from appliances connected to chimneys and vents can escape through openings in the chimney and vent. Those openings can affect the draft of the remaining appliances, causing them to spill combustion gases into the building.

The method used to close an unused opening must not create a protrusion in the flue passageway, which could restrict vent gas flow. It is not uncommon to find multiple openings in chimneys that were used at one time but have since been abandoned and covered or closed in some makeshift fashion, or left open. Inlet openings that are not closed could allow the escape of vent gases and could affect the draft produced by the chimney or vent. Excess dilution air will be admitted through unused openings, which could weaken draft by cooling the vent gases and bypassing the flow through appliance connectors. It is rare to find unused openings in vents because vent openings are easily closed or adapted to different appliance connections; however, the hazard of abandoned openings is the same as for chimneys. Vent and factory-built chimney openings are easily closed by eliminating tees and wyes or by using metal caps. Closure of masonry chimney openings may require the replacement of masonry, especially where the chimney has been structurally weakened by the openings.

501.6 Positive pressure. Where an appliance equipped with a mechanical forced draft system creates a positive pressure in the venting system, the venting system shall be designed for positive pressure applications.

❖ Commonly used chimney and vent systems are not designed for positive pressure (above atmospheric). Chimneys and vents are intended to produce a draft and thus operate with negative internal pressures. Specialized vents and chimneys are available that are designed for positive pressure applications (see commentary Figure 501.6). The misapplication of materials can result in leakage of combustion gases and can cause chimney or vent deterioration from condensate corrosion. Masonry chimneys, Type B vents, Type L vents and most factory-built chimneys are designed for negative pressure applications, meaning that the venting system is producing draft. Negative-pressure venting systems will leak if subjected to positive internal pressures. Leakage can allow combustion gases to enter the building, and condensation can cause rapid deterioration of the chimney or vent.

Photo courtesy of Selkirk L.L.C.

Figure 501.6
VENT DESIGNED FOR CATEGORY III AND IV
(POSITIVE PRESSURE) APPLICATIONS

501.7 Connection to fireplace. Connection of appliances to chimney flues serving fireplaces shall be in accordance with Sections 501.7.1 through 501.7.3.

❖ This section regulates installations where fireplace chimneys are used as the venting means for appliances such as room heaters and fireplace-insert-type heaters.

501.7.1 Closure and access. A noncombustible seal shall be provided below the point of connection to prevent entry of room air into the flue. Means shall be provided for access to the flue for inspection and cleaning.

❖ Without an airtight connection between the chimney flue and the appliance chimney connector, chimney draft will draw room air into the flue, thus weakening the draft drawn through the appliance. The chimney connection must be designed for access for inspection and cleaning of the chimney because the chimney flue is no longer open to the fireplace firebox.

501.7.2 Connection to factory-built fireplace flue. An appliance shall not be connected to a flue serving a factory-built fireplace unless the appliance is specifically listed for such installation. The connection shall be made in accordance with the appliance manufacturer's installation instructions.

❖ Factory-built fireplaces may or may not be tested for any use other than as a traditional fireplace. Installing an appliance or fireplace-insert-type heater could result in abnormally high fireplace or chimney temperatures, component deterioration and hazardous operation. The fireplace manufacturer will provide specific instructions covering what is or is not allowed regarding appliance installations. Typically the manufacturer's instructions will prohibit any modifications and any use of the fireplace other than what it was designed for.

501.7.3 Connection to masonry fireplace flue. A connector shall extend from the appliance to the flue serving a masonry fireplace such that the flue gases are exhausted directly into the flue. The connector shall be accessible or removable for inspection and cleaning of both the connector and the flue. Listed direct connection devices shall be installed in accordance with their listing.

❖ The appliance chimney connector must extend up through the fireplace damper and smoke chamber and terminate in the chimney flue. This helps produce an adequate draft.

501.8 Equipment not required to be vented. The following appliances shall not be required to be vented.

1. Ranges.
2. Built-in domestic cooking units listed and marked for optional venting.
3. Hot plates and laundry stoves.
4. Type 1 clothes dryers (Type 1 clothes dryers shall be exhausted in accordance with the requirements of Section 613).
5. A single booster-type automatic instantaneous water heater, where designed and used solely for the sanitizing rinse requirements of a dishwashing machine, provided that the heater is installed in a commercial kitchen having a mechanical exhaust system. Where installed in this manner, the draft hood, if required, shall be in place and unaltered and the draft hood outlet shall be not less than 36 inches (914 mm) vertically and 6 inches (152 mm) horizontally from any surface other than the heater.
6. Refrigerators.
7. Counter appliances.
8. Room heaters listed for unvented use.
9. Direct-fired make-up air heaters.
10. Other equipment listed for unvented use and not provided with flue collars.
11. Specialized equipment of limited input such as laboratory burners and gas lights.

Where the appliances and equipment listed in Items 5 through 11 above are installed so that the aggregate input rating exceeds 20 British thermal units (Btu) per hour per cubic feet (207 watts per m^3) of volume of the room or space in which such appliances and equipment are installed, one or more shall be provided with venting systems or other approved means for conveying the vent gases to the outdoor atmosphere so that the aggregate input rating of the remaining unvented appliances and equipment does not exceed the 20 Btu per hour per cubic foot (207 watts per m^3) figure. Where the room or space in which the equipment is installed is directly connected to another room or space by a doorway, archway, or other opening of comparable size that cannot be closed, the volume of such adjacent room or space shall be permitted to be included in the calculations.

❖ The specified appliances do not have to be vented unless required by the last paragraph of this section, which addresses items 5 through 11 and not items 1 through 4. Section 301.3 requires all appliances to be listed.

Item 1 recognizes that residential-type cooking appliances are listed for unvented operation. Item 7 refers to gas-fired cooking and food-service appliances designed for use on countertop surfaces.

Item 4 clarifies that clothes dryers are not vented but instead are exhausted. The products of combustion for gas-fired clothes dryers are discharged to the outdoors along with the moisture exhaust; in effect, the dryer is "vented" but not in the sense that other appliances are vented. Clothes dryer exhaust ducts are not referred to as vents because they are exhaust ducts, not a type of vent. Clothes dryers, both gas and electric, must exhaust to the outdoors.

Twenty Btu per hour per cubic foot of room volume (345 W/m^3/min) is the same ratio as 50 cubic feet per 1000 Btu per hour (4832 cm^3/W) on which Section 304.5.1 is based. In other words, this section requires that unvented appliances be located in a space that meets the volume requirement for indoor combustion air specified in Section 304.5.1. If the total appliance input rating exceeds the ratio of 20 Btu/h per cu ft of space volume, one or more appliances would have to be vented or removed to keep the total at or below the limit.

Maintaining a minimum space volume will help control the level of combustion products in the space by dilution with infiltration air.

501.9 Chimney entrance. Connectors shall connect to a masonry chimney flue at a point not less than 12 inches (305 mm) above the lowest portion of the interior of the chimney flue.

❖ This requirement is intended to prevent blockage of the connector opening by debris that has collected at the bottom of the flue passage. The 12-inch deep (305 mm) space below the connector functions as a trap (dirt leg) for debris such as dead animals, leaves, flaking tile, mortar and soot. This section is not intended to apply to factory-built chimneys because those chimneys are protected from the entry of debris by the required cap.

501.10 Connections to exhauster. Appliance connections to a chimney or vent equipped with a power exhauster shall be made on the inlet side of the exhauster. Joints on the positive pressure side of the exhauster shall be sealed to prevent flue-gas leakage as specified by the manufacturer's installation instructions for the exhauster.

❖ When a mechanical draft device or power exhauster is installed, the resulting "forced draft" creates a positive pressure inside the chimney or vent on the discharge or outlet side of the fan. A hazardous condition will result if an appliance is connected on the outlet or discharge side of a power exhauster. The appliance connector provides an alternate path for the flue gases to travel back into the building. Also, the appliance connected on the discharge side (outlet) of the exhausted will not be supplied with draft, and its combustion products will spill into the building interior.

A hazardous condition can also occur when a vent connector serving an appliance vented by natural draft is connected to the vent of an appliance equipped with integral power-venting means. This section does not prohibit Category I fan-assisted appliances from being common-vented with draft-hood (natural draft) appliances as addressed in Section 504.

Any vent, chimney or connector piping into which an exhauster discharges must be rated and approved for positive pressure and properly installed to prevent leakage of combustion gases. For example, Type B vents cannot be used on the outlet side of an exhauster or a power-vented appliance, regardless of any appliance manufacturer's recommendations. Type B vent material cannot be sealed to prevent leakage when exposed to positive pressure. Positive pressure can force combustion products into the annular space between the pipe walls where condensation could occur, causing deterioration of the piping.

501.11 Masonry chimneys. Masonry chimneys utilized to vent appliances shall be located, constructed and sized as specified in the manufacturer's installation instructions for the appliances being vented and Section 503.

❖ Manufacturers' instructions for today's gas-fired appliances specify very limited conditions under which the appliance is allowed to vent to a masonry chimney. The conditions include chimney size, state of condition, location and construction. Because many existing masonry chimneys are oversized, unlined, in poor structural condition or partially exposed to the outdoors, they might not be suitable to serve gas-fired appliances. Masonry chimneys serving to vent fan-assisted gas-fired appliances are becoming rare. In many cases, a masonry chimney must be relined with a retrofit metal liner system, or a vent such as Type B vent will have to be installed within the flueway of the chimney. Masonry chimneys have a large thermal mass that has to be heated to above the dew point of the combustion gases to avoid condensation. The flue gas temperatures discharged from many appliances will not be high enough to sufficiently warm the chimney mass, especially where the chimney is oversized and/or exterior.

501.12 Residential and low-heat appliances flue lining systems. Flue lining systems for use with residential-type and low-heat appliances shall be limited to the following:

1. Clay flue lining complying with the requirements of ASTM C 315 or equivalent. Clay flue lining shall be installed in accordance with the *International Building Code*.
2. Listed chimney lining systems complying with UL 1777.
3. Other approved materials that will resist, without cracking, softening or corrosion, flue gases and condensate at temperatures up to 1,800°F (982°C).

❖ Flue lining systems for masonry chimneys include clay tile, poured-in-place refractory materials and stainless steel pipe.

501.13 Category I appliance flue lining systems. Flue lining systems for use with Category I appliances shall be limited to the following:

1. Flue lining systems complying with Section 501.12.
2. Chimney lining systems listed and labeled for use with gas appliances with draft hoods and other Category I gas appliances listed and labeled for use with Type B vents.

❖ Chimney lining systems for chimneys serving gas-fired appliances include stainless steel and aluminum piping (uninsulated or insulated) designed for both new construction and for relining existing chimneys.

Type B vent is also used as a chimney liner, although it is not considered to be a lining system, and in such cases the chimney is actually a masonry chase containing a vent. Relining of existing chimneys is quite common now because of the incompatibility of mid-efficiency appliances and masonry chimneys. Relining systems allow reuse of existing chimneys, thereby saving expense and retaining desirable architectural features of buildings [see commentary Figures 501.13(1) through 501.13(4)].

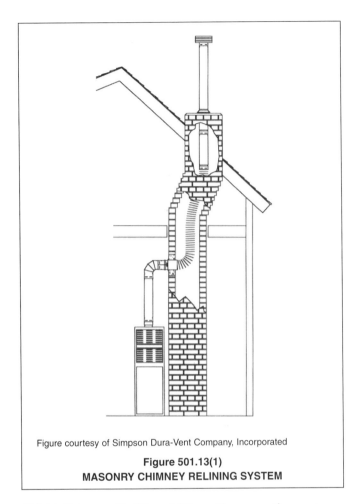

Figure courtesy of Simpson Dura-Vent Company, Incorporated
Figure 501.13(1)
MASONRY CHIMNEY RELINING SYSTEM

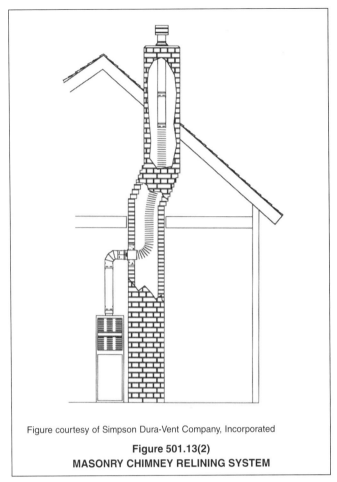

Figure courtesy of Simpson Dura-Vent Company, Incorporated
Figure 501.13(2)
MASONRY CHIMNEY RELINING SYSTEM

501.14 Category II, III and IV appliance venting systems. The design, sizing and installation of vents for Category II, III and IV appliances shall be in accordance with the appliance manufacturer's installation instructions.

❖ Venting systems for these appliances are special vent systems specified by the appliance manufacturer and are not addressed further in the code (see commentary Figure 501.14).

501.15 Existing chimneys and vents. Where an appliance is permanently disconnected from an existing chimney or vent, or where an appliance is connected to an existing chimney or vent during the process of a new installation, the chimney or vent shall comply with Sections 501.15.1 through 501.15.4.

❖ Existing chimneys and vents must be reevaluated for continued suitability whenever the conditions of use change. Size, which is covered in Section 501.15.1, is of primary importance. Other considerations are the presence of liner obstructions, combustible deposits in the liner, the structural condition of the liner, the inclusion of a cleanout and clearances to combustibles, which are addressed in Sections 501.15.2 through 501.15.4. It is the intent of this section that whenever a new appliance is connected or an existing appliance is disconnected from a chimney or vent, the chimney or vent is subject to the requirements in Sections 501.15.2 through 501.15.4, which include requirements for inspection, cleaning, possible repair, the installation of a cleanout if one is not already there and the establishment of clearances to combustibles in accordance with the requirements for new chimneys or vent installations.

Chimneys, vents and the appliances served are all designed to function together as a system. Any change to an existing chimney or vent system will have an impact on the performance of that system. Something as simple as disconnecting an appliance from a chimney can upset the system balance and cause the venting system to fail to produce a draft for the remaining appliances, and could cause the venting system to produce harmful condensation (see commentary Figure 501.15.1).

501.15.1 Size. The chimney or vent shall be resized as necessary to control flue gas condensation in the interior of the chimney or vent and to provide the appliance or appliances served with the required draft. For Category I appliances, the resizing shall be in accordance with Section 502.

❖ The combined input from multiple appliances, especially older lower-efficiency appliances, can maintain sufficiently high chimney or vent temperatures to provide the necessary draft and to avoid condensation. Changing an existing configuration by disconnecting and eliminating an appliance or by substituting a higher-efficiency appliance can cause a decrease in flue gas temperature, resulting in condensation or poor draft. Also, the elimination of one or more draft-hood-equipped appliances will reduce the amount of dilution air in a venting system, thus increasing the likelihood of condensation. Often, it is necessary to resize a chimney or vent by replacing it with a smaller-size chimney or vent or by installing a liner system (see commentary Figure 501.15.1).

A common scenario involves removing chimney-vented furnaces or boilers and leaving a water heater as the only appliance vented to the chimney. In such cases, the chimney would typically be grossly oversized for the water heater, could fail to produce adequate draft and could be subject to continuous condensation. This scenario has received much attention and has created the phrase "orphaned water heaters."

❖ A chimney or flue must be cleaned to examine the flue liner surface, to eliminate fuel for a possible chimney fire, to reduce frictional resistance to flow and to prevent any residual deposits from falling into lower parts of the venting system. Installation or removal of an appliance without an inspection of the existing flue is prohibited. If the flue served a solid- or liquid-fuel-burning appliance or fireplace, cleaning is required in order to make the required inspection and to remove any combustible deposits. In that case, the addition or removal of an appliance without both cleaning and inspection of the existing flue would be a code violation. If inspection shows that the flueway of the vent or chimney is incapable of preventing the leakage of combustion gases, the chimney or vent would have to be repaired or relined.

Figure courtesy of Simpson Dura-Vent Company, Incorporated

Figure 501.13(4)
MASONRY CHIMNEY RELINING SYSTEM

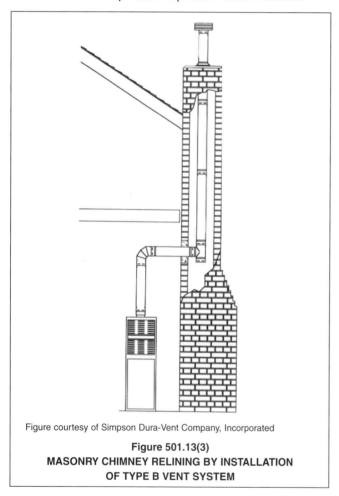

Figure courtesy of Simpson Dura-Vent Company, Incorporated

Figure 501.13(3)
MASONRY CHIMNEY RELINING BY INSTALLATION OF TYPE B VENT SYSTEM

501.15.2 Flue passageways. The flue gas passageway shall be free of obstructions and combustible deposits and shall be cleaned if previously used for venting a solid or liquid fuel-burning appliance or fireplace. The flue liner, chimney inner wall or vent inner wall shall be continuous and shall be free of cracks, gaps, perforations or other damage or deterioration which would allow the escape of combustion products, including gases, moisture and creosote.

501.15.3 Cleanout. Masonry chimney flues shall be provided with a cleanout opening having a minimum height of 6 inches (152 mm). The upper edge of the opening shall be located not less than 6 inches (152 mm) below the lowest chimney inlet opening. The cleanout shall be provided with a tight-fitting, noncombustible cover.

❖ This section requires installation of a cleanout in a chimney to facilitate cleaning and inspection. A fireplace inherently provides access to its chimney through the firebox, throat and smoke chamber. The cleanout cover and opening frame must be of an approved mate-

Photo courtesy of Selkirk L.L.C.

Figure 501.14
SPECIAL VENT SYSTEM FOR CATEGORY II, III, AND IV APPLIANCES

rial such as cast iron, precast cement or other noncombustible material and must be arranged to remain tightly closed. A loose fitting or unsecured cleanout can allow air to flow into the chimney, affecting draft at the appliance or fireplace, and can result in an energy loss in the building. Under certain conditions, combustion products could escape into the building through an ill-fitting cleanout door or cover. The requirement for placing the cleanout at least 6 inches (152 mm) below the lowest connection to the chimney is intended to minimize the possibility of combustion products exiting the chimney through the cleanout. It is the intent of this section that a cleanout opening be installed, if one is not already provided, for existing masonry chimney flues whenever another appliance is installed or removed.

501.15.4 Clearances. Chimneys and vents shall have airspace clearance to combustibles in accordance with the *International Building Code* and the chimney or vent manufacturer's installation instructions. Noncombustible firestopping or fireblocking shall be provided in accordance with the *International Building Code*.

Exception: Masonry chimneys equipped with a chimney lining system tested and listed for installation in chimneys in contact with combustibles in accordance with UL 1777, and installed in accordance with the manufacturer's instructions, shall not be required to have clearance between combustible materials and exterior surfaces of the masonry chimney.

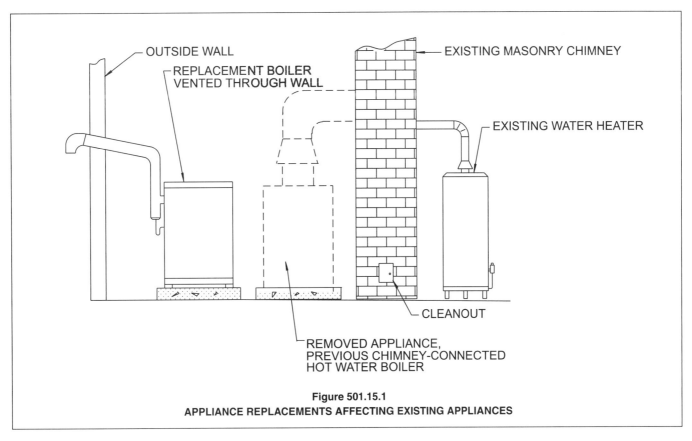

Figure 501.15.1
APPLIANCE REPLACEMENTS AFFECTING EXISTING APPLIANCES

❖ This section requires that the existing chimney or vent be inspected for clearances to combustibles for the entire length of the chimney or vent. Clearances to combustibles for masonry chimneys are found in the IBC. The required clearances for a listed factory-built chimney or vent would be in accordance with the existing chimney or vent manufacturer's installation instructions. The clearances to combustibles for vents and factory-built chimneys are typically indicated on the vent or chimney. If they are not, the manufacturer must be contacted and the original installation instructions consulted.

Existing masonry chimneys commonly do not have the clearances required by code, especially if the structure is fairly old. Listed chimney lining systems, which provide insulation value to compensate for the lack of clearance, are available for use in such chimneys. The exception gives the criteria for these tested lining systems, citing UL 1777 as the test standard.

Fireblocking of any annular spaces where the chimney passes through ceilings or floors must be installed in accordance with Chapters 7 and 21 of the IBC.

SECTION 502 (IFGC)
VENTS

502.1 General. All vents, except as provided in Section 503.7, shall be listed and labeled. Type B and BW vents shall be tested in accordance with UL 441. Type L vents shall be tested in accordance with UL 641. Vents for Category II and III appliances shall be tested in accordance with UL 1738. Plastic vents for Category IV appliances shall not be required to be listed and labeled where such vents are as specified by the appliance manufacturer and are installed in accordance with the appliance manufacturer's installation instructions.

❖ Vents regulated by this section are factory fabricated and must be listed and labeled. The labeling requirement applies to all components of the system such as the sections of pipe, fittings, terminal caps, supports and spacers.

The provision for unlisted plastic vents is necessary to allow for the venting systems commonly specified in the installation instructions for Category IV condensing appliances. Unlisted plastic pipe (commonly PVC or CPVC plumbing pipe) vents are distinct from listed high-temperature plastic special gas vents that were designed primarily for Category III appliances. Unlisted plastic pipe can be used only when specified by the appliance manufacturer and must be installed as directed in the appliance installation instructions. Unlisted plastic pipe is used only for high efficiency Category IV gas-fired appliances. The pipe functions as an exhaust pipe for combustion gases and as a combustion air intake pipe.

Gas appliances (furnaces and boilers) are categorized so that the proper venting system can be selected. Four categories have been developed based on the pressure produced in the vent passageway and whether or not the flue gas temperatures approach the dew point, making condensation likely (see the definition for "Vented gas appliance categories").

Category I appliances operate under a negative or neutral vent pressure, are noncondensing appliances, and may generate vent gases with temperatures up to 550°F (288°C). A Category I appliance can use a draft hood or an integral vent blower or draft inducer (fan assisted) [see commentary Figure 502.1(1)].

Figure 502.1(1)
CATEGORY I FAN-ASSISTED FURNACE

Category II appliances also operate under negative or neutral vent pressure, but the vent gas temperature and percent of CO_2 yield a flue loss of less than 17 percent, meaning that condensation of flue gases can occur. Vent systems serving Category II appliances fall under the special vent provisions of Sections 501.14, 503.4 and 503.6.9.2.

Category III appliances operate with a positive vent pressure and can produce condensation, although they are not considered to be "condensing appliances." The positive pressure in the vent system necessitates air-tight flue passageways to avoid leakage. Vent systems serving Category III appliances fall under the special vent provisions of Sections 501.14, 503.4, and 503.6.9.2. They are typically constructed of high-temperature plastic or stainless steel pipe. The type of stainless steel specified today is typically AL29-4C, a special alloy known for its corrosion resistance.

Category IV appliances operate with a positive vent pressure and condense flue gases. Condensing (high-efficiency) appliances require airtight vent passageways and a method of collecting and draining condensation. The vent system serving this type of appliance is usually constructed of plastic, such as PVC or CPVC. Vent systems serving Category IV appliances also fall under the special vent provisions of Sections 501.14, 503.4 and 503.6.9.2. [see commentary Figure 502.1(2)].

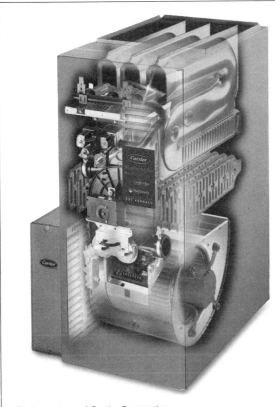

Photo courtesy of Carrier Corporation

**Figure 502.1(2)
CATEGORY IV APPLIANCE (FURNACE)**

A vent system can be used only in conjunction with gas-burning appliances that are listed and labeled for use with a vent. Appliances equipped with draft hoods and Category I (including fan-assisted) appliances both depend on the vent to produce a draft to convey the combustion products to the outdoors. Solid-fuel-burning appliances must not be connected to a vent system because they produce flue-gas temperatures that are much higher than the temperatures for which vents are designed. The burning of solid fuels also produces creosote. Creosote formation leaves a combustible deposit on the surface of the flue passageway that, if ignited, can produce a fire with temperatures exceeding 2,000°F (1093°C), far exceeding the safe temperature limitations of a vent system.

Type B vents are designed for venting noncondensing gas appliances equipped with a draft hood and fan-assisted appliances that operate with a nonpositive vent pressure and that are listed and labeled for use with a Type B vent. Type B vents are capable of withstanding continuous firing at flue temperatures not exceeding 400°F (222°C) above room temperature. This section requires that a Type B vent be tested in accordance with UL 441. Type B vents must not be used with incinerators, appliances readily converted to the use of solid or liquid fuels or combination gas/oil burning appliances. Type B vents can serve multiple gas-fired appliance installations [see commentary Figure 502.1(3)].

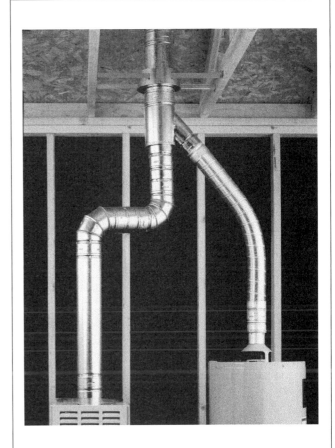

Photo courtesy of Simpson Dura-Vent Company, Incorporated

**Figure 502.1(3)
TYPE B GAS VENT SERVING TWO CATEGORY I APPLIANCES**

Type BW vents must also be tested in accordance with UL 441 and can serve only labeled gas-fired wall furnaces. A Type BW vent is an oval-shaped vent designed and tested for installation in a stud wall cavity. This type of vent is capable of withstanding continuous firing at flue temperatures not exceeding 400°F (222°C) above room temperature. Type BW vents must not be used to vent incinerators, appliances readily converted to the use of solid or liquid fuels or combination gas/oil-burning appliances.

Type L vents must be tested in accordance with UL 641 and are designed to vent oil-burning appliances, which produce a higher flue gas temperature than gas-fired appliances. Type L vents can also be used with appliances listed and labeled for Type B vent systems, meaning that Type L vents can substitute for Type B vents, but Type B vents cannot serve appliances that require Type L vents.

Type B and BW vents are double-walled, air-insulated vent piping systems constructed of galvanized steel outer walls and aluminum inner walls [see commentary Figure 502.1(4)]. Type L vents are typically double-wall, air-insulated vent piping systems constructed of galvanized and stainless steel.

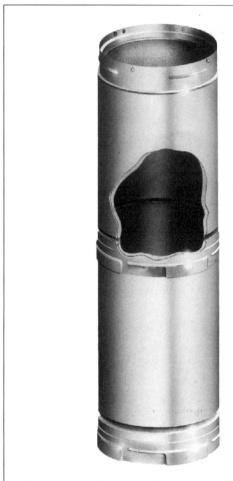

Photo courtesy of Simpson Dura-Vent Company, Incorporated

Figure 502.1(4)
TYPE B VENT SECTIONS WITH AIR SPACE BETWEEN PIPE WALLS

Type B, BW and L vents are designed for natural draft applications only and must not be used to convey combustion gases under positive pressure. For example, a Type B vent must not be used with Category III or IV gas-fired appliances and must not be used on the discharge side of an exhauster or power-vented appliance. Positive pressures will cause a Type B vent to leak combustion gases into the building, and the leakage can cause condensation to form within the vent system pipe, fittings and joints, resulting in corrosion damage to the vent and eventual vent failure. Fan-assisted Category I appliances are not considered power vented and do not produce positive vent pressures when vented to a properly sized draft-producing vent such as a Type B vent system. See the commentary for Section 504. UL 1738 covers factory-built vents for Category II, III and IV appliances. These vents can be constructed of metallic or plastic materials including aluminum alloys, steel (coated, galvanized, stainless or aluminized), PVC and CPVC.

502.2 Connectors required. Connectors shall be used to connect appliances to the vertical chimney or vent, except where the chimney or vent is attached directly to the appliance. Vent connector size, material, construction and installation shall be in accordance with Section 503.

❖ Connectors designed, constructed and installed in accordance with Section 503 are required except where a chimney or vent serves a single appliance and connects directly to that appliance. In such cases, a connector does not exist. The use of single-wall connectors for fan-assisted appliances is extremely limited by the sizing tables because of the high heat input necessary (minimum capacity column) to offset the heat loss of single-wall pipe.

502.3 Vent application. The application of vents shall be in accordance with Table 503.4.

❖ See Section 503.4.

502.4 Insulation shield. Where vents pass through insulated assemblies, an insulation shield constructed of not less than 26 gage sheet (0.016 inch) (0.4 mm) metal shall be installed to provide clearance between the vent and the insulation material. The clearance shall not be less than the clearance to combustibles specified by the vent manufacturer's installation instructions. Where vents pass through attic space, the shield shall terminate not less than 2 inches (51 mm) above the insulation materials and shall be secured in place to prevent displacement. Insulation shields provided as part of a listed vent system shall be installed in accordance with the manufacturer's installation instructions.

❖ Loose insulation in attic floor assemblies or roof assemblies can fall against vents, creating a fire hazard, and can also cause abnormally high vent temperatures. Even though clearances to combustible construction such as wood framing are maintained, the continued heating of combustible insulation materials that have fallen against a vent could be a source of ignition. This section applies regardless of the combustibility of the insulation. Shields constructed in the field must be sufficient to serve their purpose and must be securely attached to building construction because shields that are merely resting in place could be inadvertently dislodged or removed during maintenance activities.

502.5 Installation. Vent systems shall be sized, installed and terminated in accordance with the vent and appliance manufacturer's installation instructions and Section 503.

❖ The standards with which vents and appliances comply require that venting instructions be supplied by the manufacturer of the appliance and the manufacturer of the vent system. These instructions are part of the labeling requirements. Any deviation from them is in violation of the code.

The instructions for Type B vents and for gas appliances approved for use with Type B vents may be specific to a certain appliance or manufacturer's B-vent material. The clearance to combustibles for a vent system is determined by the testing agency and is stated on the component labels and in the manufacturer's instructions. Not all vents have the same required clearances to combustibles. The clearances are determined by the vent's performance during testing in accordance with the applicable standard. Different vent materials and designs impact the vent's ability to control the amount of heat transmitted to surrounding combustibles. It is critical that the vent system be installed in accordance with the clearances listed on the vent's label and in the manufacturer's installation instructions.

The termination of a natural draft vent must comply with the requirements of the manufacturer's installation instructions and Sections 503.6.3, 503.6.6 and 503.6.7. A vent used in conjunction with a mechanical exhauster must meet the termination requirements established in Sections 503.3.3 and 503.8.

Physical protection of the vent system is required to prevent damage to the vent and to prevent combustibles from coming into contact with or being placed too close to the vents. Such protection is typically provided by enclosing the vent in chases, shafts or cavities in the building construction. Physical protection is not required in the room or space where the vent originates (at the appliance connection) and would not be required in such locations as attics that are not occupied or used for storage. For example, assume that a vent is installed in an existing building and that it extends from the basement, through a first-floor closet, through the attic and through the roof. The portion of the vent passing through the closet must be protected from damage. The means of physical damage protection should also be designed to maintain separation between the vent and any combustible storage. To prevent the passage of fire and smoke through the annular space around a vent penetration through a floor or ceiling, the vent must be fireblocked with a noncombustible material in accordance with the IBC. Vent manufacturers provide installation instructions and factory-built components for fireblocking penetrations.

502.6 Support of vents. All portions of vents shall be adequately supported for the design and weight of the materials employed.

❖ Vent manufacturers supply support parts, and manufacturers' instructions contain detailed requirements for support of vent systems. Improper support can cause strain on vent components, appliance connectors and fittings, resulting in vent, appliance or connector damage, as well as loss of required clearance to combustibles and proper slope. Frequently, venting installations suffer from lack of or improperly installed supports, brackets, and hangers.

SECTION 503 (IFGS)
VENTING OF EQUIPMENT

503.1 General. This section recognizes that the choice of venting materials and the methods of installation of venting systems are dependent on the operating characteristics of the equipment being vented. The operating characteristics of vented equipment can be categorized with respect to (1) positive or negative pressure within the venting system; and (2) whether or not the equipment generates flue or vent gases that might condense in the venting system. See Section 202 for the definition of these vented appliance categories.

❖ This section is in itself commentary and does not state any requirements. Section 503 contains general sizing, design and installation requirements for venting systems. Section 504 covers vent system sizing for Category I appliances.

503.2 Venting systems required. Except as permitted in Sections 503.2.1 through 503.2.4 and 501.8, all equipment shall be connected to venting systems.

❖ All appliances must be vented by a venting system except those allowed to be unvented in accordance with Sections 501.8 and 503.2.2. The appliances addressed in Sections 503.2.1, 503.2.3 and 503.2.4 must have a venting system.

503.2.1 Ventilating hoods. Ventilating hoods and exhaust systems shall be permitted to be used to vent equipment installed in commercial applications (see Section 503.3.4) and to vent industrial equipment, such as where the process itself requires fume disposal.

❖ In commercial and industrial applications only, appliances can to be vented by an exhaust hood instead of a vent or chimney system. This is addressed further in Section 503.3.4. The exhaust system will usually exist for some other purpose such as heat, vapor and/or smoke capture and containment. It is convenient and cost effective to allow the exhaust system to also serve as the venting means for the combustion products produced by the process or appliance. This arrangement is common in manufacturing and food processing. The exhaust system must be interlocked with the appliances to prevent appliance operation when the exhaust system is not operating.

503.2.2 Well-ventilated spaces. Where located in a large and well-ventilated space, industrial equipment shall be permitted to be operated by discharging the flue gases directly into the space.

❖ This section should have extremely limited application. The allowance to operate industrial appliances without combustion product venting is intended only for industrial occupancies such as factories, foundries and manufacturing plants. This section states no limit on Btu/h input, does not specify a minimum space volume and does not define "large" or "well-ventilated"; therefore, one must apply this section with extreme caution, mindful of the intent. The space into which the combustion products are discharged must be large enough, open enough and ventilated enough to prevent the accumulation of harmful contaminants. These criteria might be met in a factory structure with very high ceilings, high infiltration rates and high rates of ventilation air movement. See Section 612, which addresses direct-fired industrial air heaters and specifies a mechanical ventilation rate to control contaminants created by the combustion process.

503.2.3 Direct-vent equipment. Listed direct-vent equipment shall be considered properly vented where installed in accordance with the terms of its listing, the manufacturer's instructions, and Section 503.8, Item 3.

❖ Direct-vent appliances are vented by a natural or mechanical draft system that is designed as part of the appliance and that usually requires some assembling on the job site. Except for the vent terminal requirements of Section 503.8, item 3, the code depends on the appliance manufacturer's installation instructions for all aspects of venting direct-vent appliances. Direct-vent appliances include some Category IV condensing furnaces and boilers and power- and gravity-vented direct-vent water heaters, unit heaters, space heaters and packaged terminal units.

Direct-vent appliances have the distinct advantage of providing their own combustion air supply in addition to their own dedicated venting means [see the definition of "Direct-vent appliance," as well as commentary Figures 202(6) and 503.2.3(1) through 503.2.3(6)].

503.2.4 Equipment with integral vents. Equipment incorporating integral venting means shall be considered properly vented when installed in accordance with its listing, the manufacturer's instructions, and Section 503.8, Items 1 and 2.

❖ As it does with direct-vent appliances, the code trusts the manufacturer's installation instructions for the installation of appliances having an integral venting means. Integral-vent appliances such as rooftop HVAC units, outdoor makeup air heaters, and outdoor heaters for swimming pools have a built-in natural or mechanical venting means. The natural-draft integral venting means is simply a short vertical vent with a factory supplied vent cap assembly. The mechanical integral-vent appliances use supervised blowers to discharge the combustion gases directly to the atmosphere. The typical means of blower supervision is one or more pressure switches or a centrifugal switch mounted to the blower motor shaft. Integral-vent appliances must not be confused with unvented appliances and appliances that need not be vented in accordance with Section 501.8. Integral-vent appliances must be vented to the outside atmosphere and thus are always designed for outdoor use only [see Section 503.8, Items 1 and 2, and commentary Figures 503.2.4(1) through 503.2.4(3)].

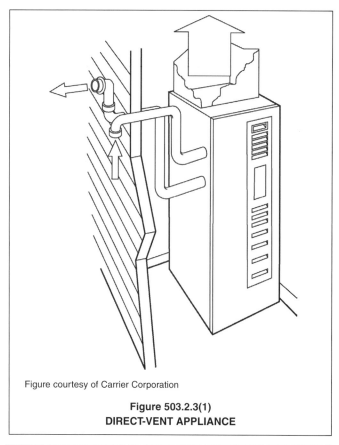

Figure courtesy of Carrier Corporation

Figure 503.2.3(1)
DIRECT-VENT APPLIANCE

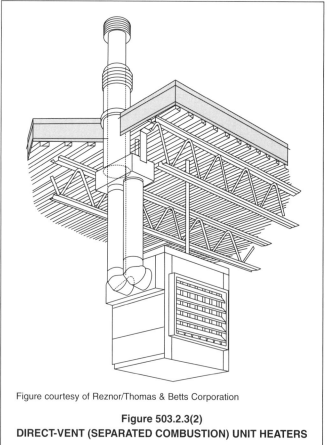

Figure courtesy of Reznor/Thomas & Betts Corporation

Figure 503.2.3(2)
DIRECT-VENT (SEPARATED COMBUSTION) UNIT HEATERS

CHIMNEYS AND VENTS

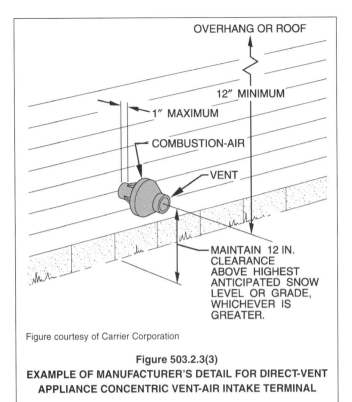

Figure courtesy of Carrier Corporation

Figure 503.2.3(3)
EXAMPLE OF MANUFACTURER'S DETAIL FOR DIRECT-VENT APPLIANCE CONCENTRIC VENT-AIR INTAKE TERMINAL

Photo courtesy of Lochinvar Corporation

Figure 503.2.3(4)
DIRECT-VENT APPLIANCES

503.3 Design and construction. A venting system shall be designed and constructed so as to develop a positive flow adequate to convey flue or vent gases to the outdoor atmosphere.

❖ This section is stated in performance language and requires venting systems to convey the products of combustion to the outdoors. The reference to a "positive flow" does not imply that the venting system is under a positive pressure, but rather implies that the vent produces a dependable flow at its design operating pressure.

503.3.1 Equipment draft requirements. A venting system shall satisfy the draft requirements of the equipment in accordance with the manufacturer's instructions.

❖ This section is stated in performance language and requires venting systems to produce the draft necessary for the operation of the appliance being vented. An appliance manufacturer will specify the required type of venting means and, to assure the required amount of draft, may specify the size and height of a vent or chimney or may specify a mechanical draft system.

503.3.2 Design and construction. Gas utilization equipment required to be vented shall be connected to a venting system designed and installed in accordance with the provisions of Sections 503.4 through 503.15.

Photo courtesy of A.O. Smith Water Products

Figure 503.2.3(5)
GRAVITY DIRECT-VENT WATER HEATER

2003 INTERNATIONAL FUEL GAS CODE® COMMENTARY

❖ As defined in Chapter 2, a venting system includes natural-induced- and forced-draft systems, which are addressed in Sections 503.3.3 through 503.15.

Photo courtesy of Simpson Dura-Vent Company, Incorporated

Figure 503.2.3(6)
DIRECT-VENT SYSTEM COMPONENTS

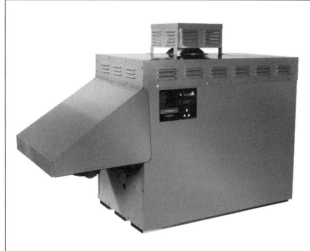

Photo courtesy of A.O. Smith Water Products

Figure 503.2.4(1)
HOT WATER BOILER WITH INTEGRAL VENT FOR OUTDOOR INSTALLATION

503.3.3 Mechanical draft systems. Mechanical draft systems shall comply with the following:

1. Mechanical draft systems shall be listed and shall be installed in accordance with the terms of their listing and both the appliance and the mechanical draft system manufacturer's instructions.

2. Equipment, except incinerators, requiring venting shall be permitted to be vented by means of mechanical draft systems of either forced or induced draft design.

3. Forced draft systems and all portions of induced draft systems under positive pressure during operation shall be designed and installed so as to prevent leakage of flue or vent gases into a building.

4. Vent connectors serving equipment vented by natural draft shall not be connected into any portion of mechanical draft systems operating under positive pressure.

5. When a mechanical draft system is employed, provision shall be made to prevent the flow of gas to the main burners when the draft system is not performing so as to satisfy the operating requirements of the equipment for safe performance.

6. The exit terminals of mechanical draft systems shall be not less than 7 feet (2134 mm) above grade where located adjacent to public walkways and shall be located as specified in Section 503.8, Items 1 and 2.

❖ The appliances and equipment installations addressed in this section employ auxiliary or integral fans and blowers to force the flow of combustion products to the outdoors.

This section applies to externally installed power exhausters, integrally power-exhausted appliances and venting systems equipped with draft inducers. This section does not address fan-assisted Category I appliances and does not address direct-vent appliances. Category III and IV appliances are subject to the requirements of this section if they are not listed as direct-vent appliances. Some Category I appliances are field convertible to Category III. Item 4 also applies to Category I fan-assisted appliances because they are vented by natural draft.

Power exhausters are field-installed pieces of equipment that are independent of, but used in conjunction with, other appliances [see commentary Figures 503.3.3(1) through 503.3.3(4)].

Photo courtesy of Reznor/Thomas & Betts Corporation

Figure 503.2.4(2)
OUTDOOR DUCT FURNACE WITH INTEGRAL POWER VENTING

CHIMNEYS AND VENTS

Photo courtesy of Carrier Corporation

Figure 503.2.4(3)
ROOFTOP HVAC UNIT WITH INTEGRAL POWER VENT

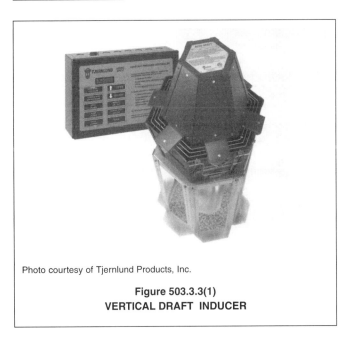

Photo courtesy of Tjernlund Products, Inc.

Figure 503.3.3(1)
VERTICAL DRAFT INDUCER

Power exhausters are typically used where other means of venting are impractical, impossible or uneconomical. Power exhausters are typically designed for use with gas-fired and oil-fired natural draft appliances. Some are also designed for use with fan-assisted appliances and appliances equipped with integral forced-draft (power-vent) fans. Power exhausters produce negative pressures at their inlet connection and positive pressures at their outlet (discharge connection). The most common installation locates the exhauster at the point of termination of the vent or chimney. In such installations, the vent or chimney between the appliance and the exhauster operates under negative pressure [see commentary Figures 503.3.3(1) through 503.3.3(9)].

Photo courtesy of Tjernlund Products, Inc.

Figure 503.3.3(2)
HORIZONTAL POWER EXHAUSTER

Induced draft systems employ separate field-installed units designed to boost draft in a natural (gravity) draft chimney or vent [see commentary Figures 503.3.3(10) and 503.3.3(11)].

Power-vented (self-venting) appliances are equipped with factory-installed integral blowers that force the combustion products through the special venting systems that are addressed in Sections 501.14 and 503.4. Power-vented appliances include mid- to high-efficiency furnaces, boilers, water heaters and heating units that are designed for through-the-wall or through-the-roof venting [see commentary Figure 503.3.3(12)]. Appliances for outdoor installation are typically power vented and discharge directly to the atmosphere. They are addressed in Section 503.2.4.

Although not specifically referred to as an appliance, a power exhauster (mechanical draft system) is considered a mechanical appliance appurtenance and therefore must bear the label of an approved agency in accordance with Item 1. See the definition of "Appliance." A power exhauster is an essential component of the appliance installation it serves and must be installed in accordance with the manufacturer's installation instructions for both the power exhauster and the appliance served. Mechanical draft devices that are an integral part of an appliance are covered by the appliance listing.

Vent or chimney systems installed downstream (discharge side) of a power exhauster must be designed and approved for positive-pressure applications. For example, Type B vent cannot be used on the discharge side of an exhauster because such pipe is not designed for positive pressure applications. The application and installation of exhausters (power venters) must comply

Photo courtesy of Exhausto, Inc.

**Figure 503.3.3(3)
HORIZONTAL POWER EXHAUSTERS**

Photo courtesy of Exhausto, Inc.

**Figure 503.3.3(4)
VERTICAL POWER EXHAUSTER**

with the manufacturers' installation instructions for the exhausters and the appliance(s) served by the exhausters (see commentary, Section 501.6).

There are three distinct variations in the use of blowers at the combustion chamber inlet working in conjunction with the fuel burner.

1. Blowers that supply turbulent combustion air to aid fuel-air mixing in a combustion chamber that is under negative pressure. To obtain proper negative over-fire draft (which optimizes combustion) also requires steady negative (below atmospheric) pressure at the flue outlet. This will be produced by a natural draft chimney and may be controlled by a barometric draft regulator. This is tabulated on line "A" in commentary Figure 503.3.3(13).
2. Blowers that supply sufficient combustion air and pressure to produce flow through the combustion chamber, but the combustion process does not need additional vent or chimney draft. This permits use of gravity or neutral draft venting products such as Type B vent for gas conversion burners. A draft regulator may be used for such equipment to prevent excess draft from affecting combustion efficiency. This is tabulated on line "B" in commentary Figure 503.3.3(13).
3. Blowers with enough power to overcome internal flue passage pressure losses (i.e., in fire-tube boilers) that also produce positive pressure at the outlet. This outlet pressure must be added to gravity draft as the motive force for flow in a chimney. If positive outlet pressure exists, the use of a sealed pressure-tight chimney is required. This is tabulated on line "C" in commentary Figure 503.3.3(13).

These three types of equipment usually have integral blower/burner systems, and all could be considered forced combustion systems. Only those described in the third paragraph truly produce forced chimney draft. When power-venting equipment is installed, it becomes an essential part of the appliance it serves. The appliance relies on the power-venting equipment (exhauster) to provide sufficient draft for proper appliance operation and for venting of the combustion products. Power exhausters are electrically interlocked with the appliance or appliances they serve to assure that the appliances will not operate if there is insufficient draft to vent the products of combustion. If an exhauster fails, the appliances served by the exhauster would discharge the products of combustion into the building unless shut off by controls that monitor draft and draft-hood spillage. An improper or lacking interlock can cause the appliance(s) to malfunction and spill harmful flue gases into the building space. The electrical interlock (interconnection) is typically accomplished with

Photo courtesy of Exhausto, Inc.

**Figure 503.3.3(5)
VERTICAL POWER EXHAUSTER**

controls provided by the exhauster manufacturer [see commentary Figure 503.3.3(14)].

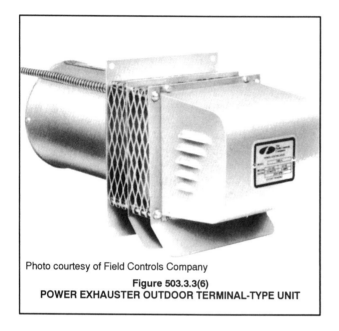

Photo courtesy of Field Controls Company
Figure 503.3.3(6)
POWER EXHAUSTER OUTDOOR TERMINAL-TYPE UNIT

Photo courtesy of Tjernlund Products, Inc.
Figure 503.3.3(8)
TYPICAL POWER EXHAUSTER

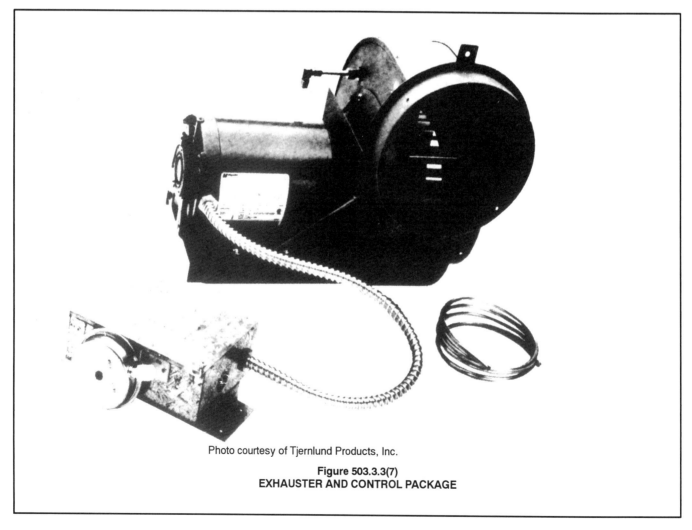

Photo courtesy of Tjernlund Products, Inc.
Figure 503.3.3(7)
EXHAUSTER AND CONTROL PACKAGE

CHIMNEYS AND VENTS

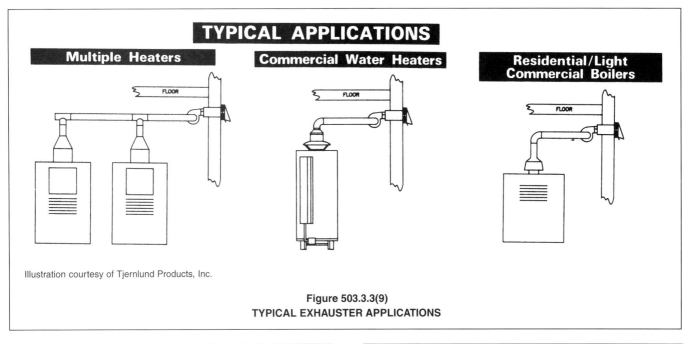

Figure 503.3.3(9)
TYPICAL EXHAUSTER APPLICATIONS

Figure 503.3.3(10)
DRAFT INDUCER

Figure 503.3.3(11)
DRAFT INDUCER

The usual sequence of operation is as follows: The call for heat from the appliance operating control starts the exhauster, and pressure controls start the appliance only after adequate draft has been proven to exist. In some cases, temperature sensors are used in addition to pressure controls to sense draft-hood spillage or unusual vent system temperatures and shut off the appliance.

Item 6 addresses the concern for pedestrian safety relative to exposure to combustion gases (see Section 503.8).

Photo courtesy of A.O. Smith Water Products

**Figure 503.3.3(12)
POWER VENTED WATER HEATER**

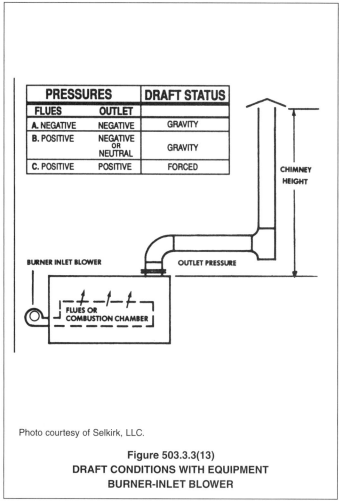

Photo courtesy of Selkirk, LLC.

**Figure 503.3.3(13)
DRAFT CONDITIONS WITH EQUIPMENT
BURNER-INLET BLOWER**

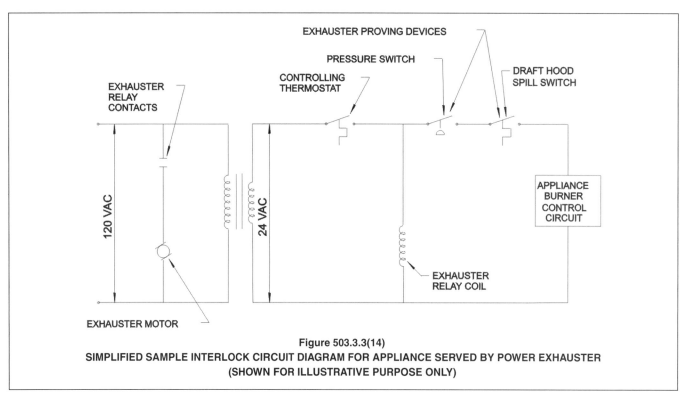

**Figure 503.3.3(14)
SIMPLIFIED SAMPLE INTERLOCK CIRCUIT DIAGRAM FOR APPLIANCE SERVED BY POWER EXHAUSTER
(SHOWN FOR ILLUSTRATIVE PURPOSE ONLY)**

503.3.4 Ventilating hoods and exhaust systems. Ventilating hoods and exhaust systems shall be permitted to be used to vent gas utilization equipment installed in commercial applications. Where automatically operated equipment is vented through a ventilating hood or exhaust system equipped with a damper or with a power means of exhaust, provisions shall be made to allow the flow of gas to the main burners only when the damper is open to a position to properly vent the equipment and when the power means of exhaust is in operation.

❖ This section addresses commercial occupancies such as restaurant kitchens where the kitchen exhaust system is used to vent gas-fired appliances such as booster water heaters for dishwashers. Commercial gas-fired appliances must be vented (see Section 501.8) either by individual venting means such as Type B vents or exhausters, or by venting the products of combustion into the kitchen exhaust hood system that serves the appliances. This section requires the appliance and the ventilating hood or exhaust system to be interlocked to assure simultaneous operation of the hood or exhaust system whenever the appliance is firing. See Sections 505.1.1 and 623.6 for a discussion of this type of interlock.

503.3.5 Circulating air ducts and furnace plenums. No portion of a venting system shall extend into or pass through any circulating air duct or furnace plenum.

❖ If a vent or chimney or a connector passes through or extends into a duct or furnace plenum, it is possible to subject the venting system to negative or positive pressures and/or temperature extremes that could cause the venting system to deteriorate, fail to produce the required draft, produce condensation and/or leak combustion gases. A furnace plenum is a component of the ductwork system constructed as a junction box for multiple duct connections. Consider that pressure differentials can also occur between a space such as an air ceiling plenum and occupied spaces, thus creating a condition similar to what would occur if a venting system component was exposed to the interior of a ductwork plenum.

503.4 Type of venting system to be used. The type of venting system to be used shall be in accordance with Table 503.4.

❖ The table lists various types of vents and the corresponding types of appliances that can be served by the vents. The vent system must be tested and specifically approved for use with the approved appliance. If the vent system is not a tested and labeled component of the appliance, the material must be approved for use with the appliance and installed in accordance with the manufacturer's installation instructions.

Gas-burning appliances (listed as Category II, III or IV) require special vent systems that are specific to the type of appliance. Special vent systems are typically associated with mid- to high-efficiency appliances and include vent materials such as polyvinyl chloride (PVC), chlorinated polyvinyl chloride (CPVC) and special alloys of stainless steel (see commentary Figure 503.4).

A primary consideration for the design of a special vent system that serves a high-efficiency appliance is the selection of a material that is capable of withstanding the corrosive effects of condensate. Flue gas condensate is slightly acidic and corrosive and can deteriorate many vent materials. Sulfur in the fuel and halogens carried in the combustion air can combine with the combustion products to create acids. To prevent the corrosion of a vent and possible escape of flue gas, metallic flue passageways can be protected with appropriate coatings; however, any imperfections, pinholes or discontinuities in the coating can create areas for condensate to collect and accelerate the deterioration of the chimney or vent. The solution to this problem has been to use a material such as PVC, CPVC or high-temperature plastic that is resistant to acidic condensation and compatible with the temperature of the appliance flue gases. Plastic pipe is a suitable material because it is impervious to corrosive condensation and can easily be designed to drain off accumulated condensate. The flue gas temperatures of most condensing-type appliances are not high enough to affect such materials adversely.

Special vent systems must be designed and installed in compliance with the manufacturers' instructions, which will specify installation requirements that are specific to that type of vent and appliance. For example, a maximum developed length, a maximum number of directional fittings, fitting turn radius and specific support requirements will be specified for special vents using plastic pipe. Special vent systems also have special vent termination requirements, and typically incorporate condensate collection and drainage fittings and connections.

Gas- and oil-burning appliances can be used with a vent system if the appliance is so labeled and approved. The type and size of the vent must be as dictated by the manufacturer's installation instructions for the appliance. The design and installation instructions provided by the vent manufacturer must also be consulted when designing a vent system for any particular application. Figure 503.4 shows some of the variables that affect vent system sizing. In addition to Section 504, the vent manufacturers' design and application handbooks and installation instructions contain information and tables for the sizing and application of the vent systems.

503.4.1 Plastic piping. Plastic piping used for venting equipment listed for use with such venting materials shall be approved.

❖ Plastic pipe has been used as a venting material for Category III and IV appliances for many years and has more recently been used to vent specialized power-vented appliances that are equipped with draft hoods and blowers that introduce enough dilution air to keep the vent gases within the service temperature range of the plastic pipe (see Sections 502.1 and 503.4).

TABLE 503.4
TYPE OF VENTING SYSTEM TO BE USED

GAS UTILIZATION EQUIPMENT	TYPE OF VENTING SYSTEM
Listed Category I equipment Listed equipment equipped with draft hood Equipment listed for use with Type B gas vent	Type B gas vent (Section 503.6) Chimney (Section 503.5) Single-wall metal pipe (Section 503.7) Listed chimney lining system for gas venting (Section 503.5.3) Special gas vent listed for this equipment (Section 503.4.2)
Listed vented wall furnaces	Type B-W gas vent (Sections 503.6, 608)
Category II equipment	As specified or furnished by manufacturers of listed equipment (Sections 503.4.1, 503.4.2)
Category III equipment	As specified or furnished by manufacturers of listed equipment (Sections 503.4.1, 503.4.2)
Category IV equipment	As specified or furnished by manufacturers of listed equipment (Sections 503.4.1, 503.4.2)
Incinerators, indoors	Chimney (Section 503.5)
Incinerators, outdoors	Single-wall metal pipe (Sections 503.7, 503.7.6)
Equipment which may be converted to use of solid fuel	Chimney (Section 503.5)
Unlisted combination gas and oil-burning equipment	Chimney (Section 503.5)
Listed combination gas and oil-burning equipment	Type L vent (Section 503.6) or chimney (Section 503.5)
Combination gas and solid fuel-burning equipment	Chimney (Section 503.5)
Equipment listed for use with chimneys only	Chimney (Section 503.5)
Unlisted equipment	Chimney (Section 503.5)
Decorative appliance in vented fireplace	Chimney
Gas-fired toilets	Single-wall metal pipe (Section 625)
Direct vent equipment	See Section 503.2.3
Equipment with integral vent	See Section 503.2.4

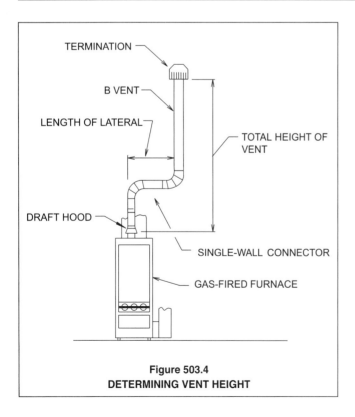

Figure 503.4
DETERMINING VENT HEIGHT

❖ This table lists the currently available types of appliances and the types of venting systems to be used with such appliances. In the first row, the three types of appliances can be matched up with any of the five types of venting systems, if allowed by the appliance manufacturers' installation instructions. The fifth row up from the bottom addresses unlisted appliances that are not allowed by the code except under the alternate approval provisions of Section 105.

503.4.2 Special gas vent. Special gas vent shall be listed and installed in accordance with the terms of the special gas vent listing and the manufacturers' instructions.

❖ Special gas vents include high-temperature plastic pipe, stainless steel alloy pipe and positive-pressure pipe that are specified by the appliance manufacturer for use with a specific appliance. This section does not require listing plastic pipe as a venting system component if the pipe is specified by the appliance manufacturer for venting appliances such as Category IV condensing furnaces and boilers. The installation instructions for Category IV appliances often specify PVC, CPVC and ABS plastic plumbing pipe for venting the appliances, and that pipe is not listed as a component of a venting system (see Section 503.4.1).

503.5 Masonry, metal, and factory-built chimneys. Masonry, metal and factory-built chimneys shall comply with Sections 503.5.1 through 503.5.10.

❖ See Sections 503.5.1 through 503.5.10.

503.5.1 Factory-built chimneys. Factory-built chimneys shall be installed in accordance with their listing and the manufacturers' instructions. Factory-built chimneys used to vent appliances that operate at positive vent pressure shall be listed for such application.

❖ Prefabricated chimney systems must bear the label of an approved agency. A label is required on all components of the chimney system such as the pipe sections, shields, fireblocks, fittings, termination caps and supports. The label states information such as the type of appliance the chimney was tested for use with, a reference to the manufacturer's installation instructions and the minimum required clearances to combustibles [see commentary Figures 202(4) and 202(5)].

Most factory-built chimneys are either of the double-walled fiber-insulated design or the triple-walled air-cooled design. Factory-built chimneys are constructed of stainless steel inner liners with stainless steel outer walls. Factory-built chimneys, like vent systems, are composed of components that must be installed as a complete system. Components from different manufacturers are not designed to be mixed and installed together.

The manufacturer's instructions contain sizing criteria and requirements for every aspect of a factory-built chimney installation. The requirements include component assembly, clearances to combustibles, support, terminations, connections, protection from damage and fireblocking.

Any vent or chimney that conveys vent gases under positive pressure must be factory designed for that application. There is no allowable method of taping or sealing a chimney or vent that will convert a non-positive-pressure chimney or vent into a positive-pressure chimney.

503.5.2 Metal chimneys. Metal chimneys shall be built and installed in accordance with NFPA 211.

❖ This section addresses metal chimneys, which are often referred to as smokestacks. Metal chimneys are used primarily in industrial applications where the vent gases are very high temperature. Discharging high temperature vent gases is a waste of energy and is thus becoming increasingly rare.

503.5.3 Masonry chimneys. Masonry chimneys shall be built and installed in accordance with NFPA 211 and shall be lined with approved clay flue lining, a listed chimney lining system, or other approved material that will resist corrosion, erosion, softening, or cracking from vent gases at temperatures up to 1800°F (982°C).

Exception: Masonry chimney flues serving listed gas appliances with draft hoods, Category I appliances, and other gas appliances listed for use with Type B vent shall be permitted to be lined with a chimney lining system specifically listed for use only with such appliances. The liner shall be installed in accordance with the liner manufacturer's instructions and the terms of the listing. A permanent identifying label shall be attached at the point where the connection is to be made to the liner. The label shall read: "This chimney liner is for appliances that burn gas only. Do not connect to solid or liquid fuel-burning appliances or incinerators."

For information on installation of gas vents in existing masonry chimneys, see Section 503.6.5.

❖ A chimney liner might be listed for use with solid-, liquid- and gas-fuel-fired appliances. The label required by this section is intended to warn the unknowing installer who sees a masonry chimney and thinks that it can serve a solid- or liquid-fuel-fired appliance. The label would be applicable to chimney liner systems that are designed for gas-fired appliances only (see Section 501.3).

503.5.4 Chimney termination. Chimneys for residential-type or low-heat gas utilization equipment shall extend at least 3 feet (914 mm) above the highest point where it passes through a roof of a building and at least 2 feet (610 mm) higher than any portion of a building within a horizontal distance of 10 feet (3048 mm) (see Figure 503.5.4). Chimneys for medium-heat equipment shall extend at least 10 feet (3048 mm) higher than any portion of any building within 25 feet (7620 mm). Chimneys shall extend at least 5 feet (1524 mm) above the highest connected equipment draft hood outlet or flue collar. Decorative shrouds shall not be installed at the termination of factory-built chimneys except where such shrouds are listed and labeled for use with the specific factory-built chimney system and are installed in accordance with the manufacturers' installation instructions.

❖ Low-heat and residential-type chimneys must extend at least 3 feet (914 mm) above the roof, measured from the highest point of the roof penetration. They must also be at least 2 feet (610 mm) higher than any portion of the roof within a 10-foot (3048mm) horizontal distance [see commentary Figure 503.5.4(1)].

The 2-foot (610 mm) termination requirement is intended to prevent wind and pressure zones from reducing the amount of draft produced by the chimney. Wind and wind-induced eddy currents can react with building structural surfaces, thereby creating air pressure zones that can diminish chimney draft and cause reverse flow (backdraft) in the chimney. Loss of draft will cause the appliance or fireplace served by the chimney to discharge combustion products into the building interior. Locating the chimney outlet well into the undisturbed wind stream and away from the cavity and wake (eddy) zones around the building can counteract the adverse effects and also prevent the reentry of vent gases into the building through openings and fresh air intakes. Terminating a chimney in the eddy current area recircu-

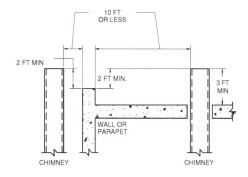
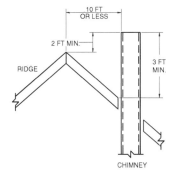

A. TERMINATION 10 FT OR LESS FROM RIDGE, WALL, OR PARAPET

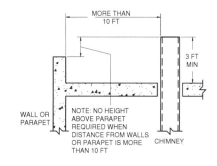
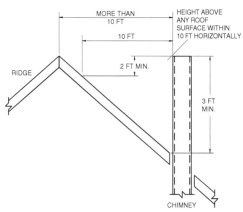

B. TERMINATION MORE THAN 10 FT FROM RIDGE, WALL, OR PARAPET

For SI: 1 foot = 304.5.4 mm.

**FIGURE 503.5.4
TYPICAL TERMINATION LOCATIONS FOR CHIMNEYS AND SINGLE-WALL METAL PIPES SERVING RESIDENTIAL-TYPE AND LOW-HEAT EQUIPMENT**

lates the combustion products and allows them to enter the building via infiltration, wall openings and air intakes. A chimney terminal properly located above the eddy current area allows the wind to carry the combustion products away from the building. Factory-built chimneys have termination caps that are part of the labeled chimney system and that must be installed. Like vent systems, the chimney cap is designed to keep out precipitation and animals and minimize the adverse effects of wind.

Decorative shrouds are designed to be aesthetically pleasing and to hide chimney terminations from view [see commentary Figure 503.5.4(2)]. These shrouds have, in some cases, caused fires resulting from overheated combustible construction and the accumulation of debris and animal nesting. Shrouds can also interfere with the functioning of a vent or chimney. The listing of the shroud must state the specific make and model of chimney with which the shroud is intended to be used. At this time, the author is unaware of any shrouds that are listed for use with a specific make and model of chimney. Most often, shrouds are field-constructed of sheet metal, masonry, or stucco on frame.

503.5.5 Size of chimneys. The effective area of a chimney venting system serving listed appliances with draft hoods, Category I appliances, and other appliances listed for use with Type B vents shall be determined in accordance with one of the following methods:

1. The provisions of Section 504.
2. For sizing an individual chimney venting system for a single appliance with a draft hood, the effective areas of the vent connector and chimney flue shall be not less than the area of the appliance flue collar or draft hood outlet, nor greater than seven times the draft hood outlet area.
3. For sizing a chimney venting system connected to two appliances with draft hoods, the effective area of the chimney flue shall be not less than the area of the larger draft hood outlet plus 50 percent of the area of the smaller draft hood outlet, nor greater than seven times the smallest draft hood outlet area.
4. Chimney venting systems using mechanical draft shall be sized in accordance with approved engineering methods.
5. Other approved engineering methods.

❖ Chimneys can be sized in accordance with Section 504, by Items 2, 3 or 4 of this section or by an engineering method acceptable to the code official. Some appliances are not categorized; however, appliances with draft hoods and appliances listed for use with Type B vents would fit in Category I if they were categorized. The manufacturer's instructions for some Category I appliances, such as fan-assisted furnaces and boilers, may not allow connecting the appliance to be connected to a masonry chimney except under specific limited conditions.

Item 2 applies only to a single appliance that is factory equipped with a draft hood. The "seven times rule" prevents the chimney from being too large, which could

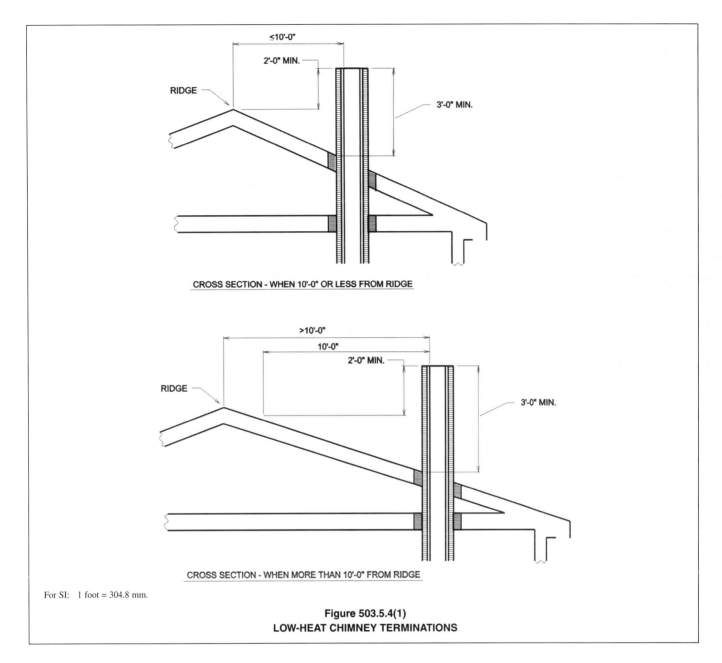

Figure 503.5.4(1)
LOW-HEAT CHIMNEY TERMINATIONS

result in poor draft and condensation problems. Like vents, chimneys that are too large in area can be as problematic as chimneys that are too small in area.

Item 3 is limited to only one venting arrangement. It can only be used for a chimney that serves two appliances, both of which must be factory-equipped with draft hoods. The method is based on the areas of the chimneys and draft hood outlets, not on the diameters.

Item 4 recognizes that some manufacturers of mechanical draft equipment (exhausters) have their own sizing methodology specific to their product. Recall that Item 1 of Section 503.3.3 requires listing of such equipment.

503.5.5.1 Incinerator venting. Where an incinerator is vented by a chimney serving other gas utilization equipment, the gas input to the incinerator shall not be included in calculating chimney size, provided the chimney flue diameter is not less than 1 inch (25 mm) larger in equivalent diameter than the diameter of the incinerator flue outlet.

❖ This section appears to have limited application because Section 503.5.7.1 does not allow gas-fired appliances to be served by a chimney flue that also serves a solid-fuel-fired appliance. An incinerator reduces solid combustibles to ashes and, therefore, could be considered as a solid-fuel appliance. If an incinerator is connected to a dedicated chimney (serving no other appliances), the gas input to the incinerator must be considered in sizing the chimney.

503.5.6 Inspection of chimneys. Before replacing an existing appliance or connecting a vent connector to a chimney, the chimney passageway shall be examined to ascertain that it is

Figure 503.5.4(2)
DECORATIVE SHROUDS

clear and free of obstructions and it shall be cleaned if previously used for venting solid or liquid fuel-burning appliances or fireplaces.

❖ If a chimney is going to serve a new appliance installation or a replacement appliance installation, the chimney must be inspected to determine whether it is still serviceable and free of deposits. Chimneys can become obstructed by debris such as leaves, animal carcasses, loose mortar, and pieces of masonry and liner. Combustible deposits can accumulate on chimney walls used to vent liquid- and solid-fuel-fired appliances, and this section mandates that the chimney be cleaned if it served such appliances in the past. Masonry chimneys are especially susceptible to internal deterioration and should be inspected regularly, in addition to when a new or replacement appliance is connected.

503.5.6.1 Chimney lining. Chimneys shall be lined in accordance with NFPA 211.

Exception: Existing chimneys shall be permitted to have their use continued when an appliance is replaced by an appliance of similar type, input rating, and efficiency.

❖ The liner forms the flue passageway and is the actual conductor of all products of combustion. The chimney liner must be able to withstand exposure to high temperatures and corrosive chemicals. The chimney lining protects the masonry construction of the chimney walls and allows the chimney to be constructed gas tight. This section regulates liners and relining systems. Liners are often used to reline an existing chimney to salvage a masonry chimney or allow connection of higher-efficiency appliances.

Flue lining systems for masonry chimneys include clay tile, poured-in-place refractory materials and stainless steel pipe. Chimney lining systems for chimneys serving gas-fired appliances include stainless steel and aluminum piping (uninsulated or insulated) designed for both new construction and for relining existing chimneys [see commentary Figures 503.5.6.1(1) and 503.5.6.1(2)]. Type B vent is also used as a chimney liner, although it is not considered a lining system, and in such cases the chimney is actually a masonry chase containing a vent.

Relining of existing chimneys is quite common now because of the incompatibility of mid-efficiency appliances and masonry chimneys. Relining systems allow reuse of existing chimneys, thereby saving expense and retaining desirable architectural features of buildings.

The exception is intended to allow an unlined chimney to serve a new appliance that is installed to replace a previously served appliance if the new appliance does not create any different operating conditions in the chimney; that is, the volume, water vapor content and dilution air content of the new appliance flue gases are the same as those of the old appliance.

The exception assumes that the existing chimney and the existing appliances operated as intended by this code. If there is evidence that the previous chimney and appliance system was not working as required by this code, the exception is not intended to be applicable. Because this exception would allow an installation that is in conflict with the appliance manufacturer's installation instructions, it is in effect, negated because the manufacturer's instructions will prevail in this case (see Sections 102.8, 305.1, and 503.5.6.3).

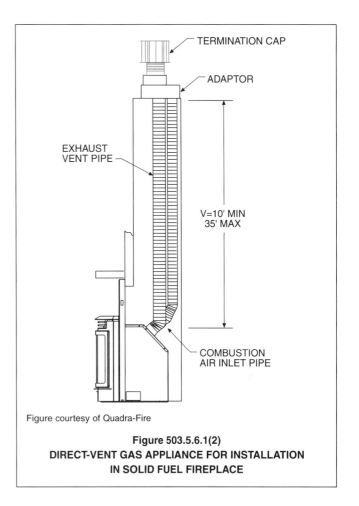

Figure courtesy of Quadra-Fire

**Figure 503.5.6.1(2)
DIRECT-VENT GAS APPLIANCE FOR INSTALLATION IN SOLID FUEL FIREPLACE**

503.5.6.3 Unsafe chimneys. Where inspection reveals that an existing chimney is not safe for the intended application, it shall be repaired, rebuilt, lined, relined, or replaced with a vent or chimney to conform to NFPA 211 and it shall be suitable for the equipment to be vented.

❖ If a chimney is inspected and found to be unsafe for any reason, it must be made safe or be replaced. See Section 501.3. Some chimneys and appliances are not compatible as dictated by the appliance and/or factory-built chimney manufacturer. Masonry chimneys usually become unsafe as a result of deteriorated mortar joints and liners. Other causes include structural instability, cracks, faulty cleanouts, deposit formation and moisture damage.

503.5.7 Chimneys serving equipment burning other fuels. Chimneys serving equipment burning other fuels shall comply with Sections 503.5.7.1 through 503.5.7.4.

❖ See Sections 503.7.1 through 503.7.4.

503.5.7.1 Solid fuel-burning appliances. Gas utilization equipment shall not be connected to a chimney flue serving a separate appliance designed to burn solid fuel.

❖ Each solid-fuel-burning appliance or fireplace must be connected to a dedicated independent chimney, or a

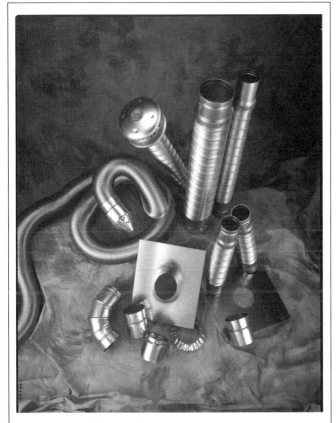

Photo courtesy of Selkirk L.L.C.

**Figure 503.5.6.1(1)
CHIMNEY LINING SYSTEM FOR GAS APPLIANCES**

503.5.6.2 Cleanouts. Cleanouts shall be examined to determine if they will remain tightly closed when not in use.

❖ A cleanout door or panel must be capable of securely sealing the opening to prevent air from entering the chimney, which would reduce draft for the appliances served. Under certain conditions, a faulty cleanout could also cause vent gases to leak from the chimney.

dedicated independent flue in multiple-flue chimney constructions. Solid-fuel-burning appliances and fireplaces cannot share a common chimney or flue-way with any other appliance or fireplace.

Solid-fuel-burning appliances produce creosote deposits on the interior walls of chimney liners. The creosote formation is highly combustible and creates a potential fire hazard. Because of the potential for a chimney fire, other connections to the chimney can allow fire to break out of the chimney into the building.

Also, chimney passageways can become restricted by the creosote formations. If other appliances were vented to such chimneys, combustion products could be discharged into the building and the other appliances might seriously interfere with the draft for the solid-fuel appliance.

This section is also intended to prevent the possibility of creosote leaking out of the chimney through appliance connectors.

Combination (dual fuel) gas- or oil- and solid-fuel-burning appliances are designed to be connected to a single chimney passageway. Such dual-fuel appliances must be listed, labeled and installed in accordance with the manufacturers' instructions.

503.5.7.2 Liquid fuel-burning appliances. Where one chimney flue serves gas utilization equipment and equipment burning liquid fuel, the equipment shall be connected through separate openings or shall be connected through a single opening where joined by a suitable fitting located as close as practical to the chimney. Where two or more openings are provided into one chimney flue, they shall be at different levels. Where the gas utilization equipment is automatically controlled, it shall be equipped with a safety shutoff device.

❖ Gas-fired and oil-fired appliances are allowed to share a chimney or flue. The code does not include a sizing method for such arrangements; therefore, the system would have to be engineered or otherwise approved by the code official. Placing the openings at different levels in the chimney flue will minimize interference between the flow of the chimney connectors. The last sentence is consistent with the design of all automatically controlled appliances listed to today's appliance standards as required by this code.

503.5.7.3 Combination gas and solid fuel-burning appliances. A combination gas- and solid fuel-burning appliance shall be permitted to be connected to a single chimney flue where equipped with a manual reset device to shut off gas to the main burner in the event of sustained backdraft or flue gas spillage. The chimney flue shall be sized to properly vent the appliance.

❖ A dual fuel appliance, gas and solid fuel, can be served by a single chimney or flue if the appliance is equipped with a safety control that monitors chimney spillage. A solid-fuel fire cannot be turned on and off like other fuel fires; thus, it is possible that the gas burner could be operated while the solid-fuel fire is still burning, and the chimney could be overloaded, causing spillage of combustion gases. The chimney size would have to be engineered or would have to comply with the appliance manufacturer's instructions.

503.5.7.4 Combination gas- and oil fuel-burning appliances. A listed combination gas- and oil fuel-burning appliance shall be permitted to be connected to a single chimney flue. The chimney flue shall be sized to properly vent the appliance.

❖ See Section 503.5.7.3.

503.5.8 Support of chimneys. All portions of chimneys shall be supported for the design and weight of the materials employed. Factory-built chimneys shall be supported and spaced in accordance with their listings and the manufacturer's instructions.

❖ See Section 502.6. Chimneys, including factory-built, are very heavy compared to vents and need substantial support to carry their weight and prevent displacement. Factory-built chimneys require special support fittings at offsets to prevent the elbows from being damaged by bearing the weight of the chimney sections above the offset. Chimneys, like piping systems, often suffer from lack of adequate support.

503.5.9 Cleanouts. Where a chimney that formerly carried flue products from liquid or solid fuel-burning appliances is used with an appliance using fuel gas, an accessible cleanout shall be provided. The cleanout shall have a tight-fitting cover and shall be installed so its upper edge is at least 6 inches (152 mm) below the lower edge of the lowest chimney inlet opening.

❖ If an existing chimney has no cleanout and was used for oil or solid-fuel appliances, a cleanout must be added. The cleanout allows access for a person to inspect and clean the chimney and monitor deposits on the interior walls. The cleanout must be located so that vent gases will not be in contact with the cleanout door/cover.

503.5.10 Space surrounding lining or vent. The remaining space surrounding a chimney liner, gas vent, special gas vent, or plastic piping installed within a masonry chimney flue shall not be used to vent another appliance. The insertion of another liner or vent within the chimney as provided in this code and the liner or vent manufacturer's instructions shall not be prohibited.

The remaining space surrounding a chimney liner, gas vent, special gas vent, or plastic piping installed within a masonry, metal or factory-built chimney, shall not be used to supply combustion air. Such space shall not be prohibited from supplying combustion air to direct-vent appliances designed for installation in a solid fuel-burning fireplace and installed in accordance with the listing and the manufacturer's instructions.

❖ If a vent or a chimney liner is installed within a chimney flue, there will be space between the vent or liner and the chimney walls. This space must not be used to convey vent gases because of its irregular size and geometry and because the vent gases could damage the vent or liner installed in the chimney. A metallic vent or liner could be corroded by surrounding vent gases, and plastic pipes could be overheated. Liner systems and vent systems are not listed for use within an atmosphere of vent gases. Installing multiple liners or vents within a chimney is not the same as using the annular space between a vent or liner and the inside chimney walls.

The space between a liner, vent or pipe and the interior chimney walls cannot be used as a duct for conveying combustion air. That space would have poor air-flow characteristics, and moving cold combustion air over a liner or vent can cause condensation to occur in the venting system. An exception is made for direct-vent appliances that are listed for installation in solid-fuel-burning fireplaces.

503.6 Gas vents. Gas vents shall comply with Sections 503.6.1 through 503.6.12 (see Section 202, Definitions).

❖ See Sections 503.6.1 through 503.6.12.

503.6.1 Installation, general. Gas vents shall be installed in accordance with the terms of their listings and the manufacturer's instructions.

❖ Gas vents, such as Type B vents, come with installation instructions available at the place of purchase or enclosed within the boxed items. The manufacturers also will provide, upon request, very detailed and informative design manuals that discuss sizing and venting fundamentals. These design manuals are invaluable for someone who installs, inspects or designs Type B vent systems (see Sections 502.1 and 502.5).

503.6.2 Type B-W vent capacity. A Type B-W gas vent shall have a listed capacity not less than that of the listed vented wall furnace to which it is connected.

❖ A Type B-W vent is an oval shaped version of a Type B vent designed to fit within a nominal 2 x 4 inch framed wall and serve a wall furnace (see Section 608).

503.6.3 Roof penetration. A gas vent passing through a roof shall extend through the roof flashing, roof jack, or roof thimble and shall be terminated by a listed termination cap.

❖ This section duplicates the typical vent manufacturer's instructions and emphasizes that a vent is a system of components that are all necessary for proper functioning. The cap must be as specified by the vent manufacturer and is a listed component. A vent cap not only keeps out moisture, debris and animals, it also serves to prevent wind interference that could negatively affect the vent's ability to produce the required draft. Vent caps are tested for flow resistance and performance in wind. The emphasis here is that the vent system must extend through the roof all the way to the cap using components that are listed as part of the listed venting system.

503.6.4 Offsets. Type B and Type L vents shall extend in a generally vertical direction with offsets not exceeding 45 degrees (0.79 rad), except that a vent system having not more than one 60-degree (1.04 rad) offset shall be permitted. Any angle greater than 45 degrees (0.79 rad) from the vertical is considered horizontal. The total horizontal length of a vent plus the horizontal vent connector length serving draft-hood-equipped appliances shall not be greater than 75 percent of the vertical height of the vent.

Exception: Systems designed and sized as provided in Section 504 or in accordance with other approved engineering methods.

Vents serving Category I fan-assisted appliances shall be installed in accordance with the appliance manufacturer's instructions and Section 504 or other approved engineering methods.

❖ This section addresses vent offsets differently than Section 504 and is in no way related to Section 504. This section would be applied only if Section 504 is not applied; therefore, it is rarely applied, because Section 504 is necessary for sizing the majority of today's Category I venting systems. The only time that a Type B vent system would not be designed in accordance with Section 504 is if it were an engineered system or if it were designed in accordance with the exceptions of Section 503.6.9.1, applicable only to draft-hood-equipped appliances.

This section regulates vent offsets, but without the tables and design requirements of Section 504 there are only very limited design requirements in Section 503 to accompany the offset provisions. In other words, this section would be only one of many parts necessary to design a vent system. The exception to this section will almost always circumvent the main paragraph because in all but the rarest of cases the vent system will either be engineered or designed in accordance with Section 504. This section is not intended for application in conjunction with the requirements of Section 504.

The last paragraph states that this section is not applicable to fan-assisted appliances.

503.6.5 Gas vents installed within masonry chimneys. Gas vents installed within masonry chimneys shall be installed in accordance with the terms of their listing and the manufacturer's installation instructions. Gas vents installed within masonry chimneys shall be identified with a permanent label installed at the point where the vent enters the chimney. The label shall contain the following language: "This gas vent is for appliances that burn gas. Do not connect to solid or liquid fuel-burning appliances or incinerators."

❖ One method of resizing a chimney is to install a Type B vent system within it. When this is done, the chimney is no longer seen as a chimney and becomes nothing more than a chase. This type of installation is practical only for chimneys without offsets. The Type B vent listing conditions and installation instructions will state whether such an installation is allowed and will include installation requirements such as support of the vent, weatherproofing the chimney top and the required condition of the chimney. Special instructions may be applicable for the support of the vent pipe within the chimney. The chimney should be structurally sound to avoid damage to the Type B vent system from falling and/or shifting masonry. There is also a concern for moisture and chimney deposits being corrosive to the Type B vent. The warning label is intended to prevent someone from connecting a solid-fuel- or oil-fired appliance to what appears to be a chimney on the outside but is actually a Type B vent in disguise [see commentary Figure 501.13(3)].

503.6.6 Gas vent terminations. A gas vent shall terminate in accordance with one of the following:

1. Above the roof surface with a listed cap or listed roof assembly. Gas vents 12 inches (305 mm) in size or smaller with listed caps shall be permitted to be terminated in accordance with Figure 503.6.6, provided that such vents are at least 8 feet (2438 mm) from a vertical wall or similar obstruction. All other gas vents shall terminate not less than 2 feet (610 mm) above the highest point where they pass through the roof and at least 2 feet (610 mm) higher than any portion of a building within 10 feet (3048 mm).

2. As provided for industrial equipment in Section 503.2.2.

3. As provided for direct-vent systems in Section 503.2.3.

4. As provided for equipment with integral vents in Section 503.2.4.

5. As provided for mechanical draft systems in Section 503.3.3.

6. As provided for ventilating hoods and exhaust systems in Section 503.3.4.

❖ This section duplicates typical vent manufacturers' instructions and emphasizes that a vent is a system of components that are all necessary for proper functioning. The cap must be as specified by the vent manufacturer and is a listed component. A vent cap not only keeps out moisture, debris and animals, it also serves to prevent wind interference that could negatively affect the vent's ability to produce the required draft. Vent caps are tested for flow resistance and performance in wind.

This section is consistent with typical vent manufacturers' instructions and applies only to vents equipped with listed caps as required by Section 503.6.3. The relationship between vent terminal height and roof pitch is based on mitigating the effects of wind and maintaining a minimum separation between the vent terminal and the roof surface.

See Section 504.2.9 for a discussion of vent terminations that extend above the roof higher than required by the code.

Figure 503.6.6 requires greater vent height above the roof as the roof approaches being a vertical surface. The greater the roof pitch, the greater the effect of wind striking the roof surface and the closer the cap becomes (horizontally) to the roof surface (see Figure 503.6.6).

It is a common misapplication for code users to apply chimney termination height requirements to vents, thereby causing vents to extend above roofs much higher than required in many cases. For example, a 6-inch (152 mm) diameter Type B vent must extend above a simple gabled roof only 1 foot (305 mm) for roofs having up to 6/12 pitch, if the terminal is not within 8 feet (2438mm) of a vertical surface such as an upper-story exterior wall. Consideration should be given to vent location in a building in the design/installation phase so that the vent termination will not be difficult or require long extensions of vent pipe exposed to the outdoors. Vent piping exposed to the outdoors encourages condensation in cold weather, could require guy wires or braces depending on height and is considered aesthetically unattractive. Another good reason to plan for vent location during the design phase of a building is to allow straight vertical runs of vent and avoid offsets. Good planning can also prevent vents from passing through the roof on the street side of the building. Quite often, vent offsets in attics are installed for the sole purpose of penetrating the roof on the back side of a building.

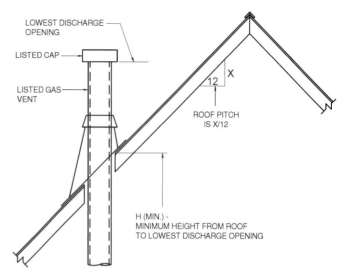

ROOF PITCH	H (min) ft
Flat to 6/12	1.0
Over 6/12 to 7/12	1.25
Over 7/12 to 8/12	1.5
Over 8/12 to 9/12	2.0
Over 9/12 to 10/12	2.5
Over 10/12 to 11/12	3.25
Over 11/12 to 12/12	4.0
Over 12/12 to 14/12	5.0
Over 14/12 to 16/12	6.0
Over 16/12 to 18/12	7.0
Over 18/12 to 20/12	7.5
Over 20/12 to 21/12	8.0

For SI: 1 inch = 25.4 mm, 1 foot = 304.8 mm.

**FIGURE 503.6.6
GAS VENT TERMINATION LOCATIONS FOR
LISTED CAPS 12 INCHES OR LESS IN SIZE AT
LEAST 8 FEET FROM A VERTICAL WALL**

Vent termination heights are affected by the pitch of the roof and the proximity of walls and other vertical or near vertical structural components. Where vent terminal locations within 8 feet (2438mm) of a gable end, upper story or other vertical surface cannot be avoided, vent terminations must be regulated in a manner similar to chimneys [i.e., 2-feet (610 mm) minimum height and not less than 2 feet (610 mm) above any part of building within 10 feet (3048 mm)] (see commentary Figure 503.6.6). Figure 503.6.6 does not apply to vent sizes larger than 12 inches (305 mm). This is consistent with manufacturers' instructions and relates to the testing of gas vents to the referenced standard.

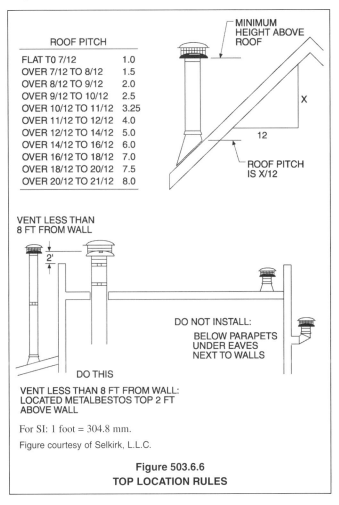

**Figure 503.6.6
TOP LOCATION RULES**

The intent of Section 504.2.9 is for vents exposed to the outdoors below the roof penetration to be sized by a method other than the tables in Section 504. This means that if a vent is within 8 feet of a vertical wall or is larger in size than 12 inches, it must extend 2 feet above any part of the building within 10 feet horizontally, and this could expose a substantial run of vent to the outdoors. In such cases, the tables of Section 504 would not apply, and the vent system would have to be engineered.

503.6.6.1 Decorative shrouds. Decorative shrouds shall not be installed at the termination of gas vents except where such shrouds are listed for use with the specific gas venting system and are installed in accordance with manufacturer's installation instructions.

❖ Decorative shrouds have become a popular architectural feature. They are designed to be aesthetically pleasing and to hide chimney and vent terminations. These shrouds have, in some cases, caused fires resulting from overheated combustible construction and the accumulation of debris and animal nesting. Shrouds can also interfere with the functioning of a chimney or vent system. Shrouds are allowed only where they are listed and labeled for use with the specific factory-built chimney system (see commentary Figure 503.6.6.1).

503.6.7 Minimum height. A Type B or a Type L gas vent shall terminate at least 5 feet (1524 mm) in vertical height above the highest connected equipment draft hood or flue collar. A Type B-W gas vent shall terminate at least 12 feet (3658 mm) in vertical height above the bottom of the wall furnace.

❖ The amount of draft produced by a vent is directly related to the height of the vent. A minimum height must be established to produce the minimum draft necessary for the appliance served. This is made evident by looking at the tables in Section 504 that indicate increasing vent capacity as the vent height increases. This is a result of the increase in draft and vent flow velocity. The tables in Section 504 and some vent manufacturers' instructions indicate a minimum height of 6 feet (1829 mm) because this is the lowest entry for height in the sizing tables.

This section is often violated for appliances installed near the roof such as suspended unit heaters and appliances installed under shed roofs.

503.6.8 Exterior wall penetrations. A gas vent extending through an exterior wall shall not terminate adjacent to the wall or below eaves or parapets, except as provided in Sections 503.2.3 and 503.3.3.

❖ The termination locations prohibited by this section would result in poor draft or back draft because of the effects of wind. Also, termination in such locations could result in a fire hazard, damage to the structure and entry of vent gases into the building. If a gas vent (typically Type B) penetrates an exterior wall, the vent would have to extend vertically above the roof in accordance with Section 503.6.6; however, this exterior extension of vent presents a problem in itself. Gas vents, such as Type B vents, are not designed for outdoor exposure, except for the short section of vent that passes through the roof. Sections 504.2.9 and 504.3.20 reinforce this by stating that the vent sizing tables in Section 504 do not apply to vents exposed to the outdoors below the roof line. Vents exposed to the outdoors will suffer from poor draft and condensation production in most climates because of the heat loss through the exposed vent walls. The bottom line is that vents sized in accordance with Section 504 cannot extend up the exterior side of a building as implied by this section and Section

Figure 503.6.6.1
DECORATIVE SHROUDS

503.6.6. This type of installation could occur only if the vent system was engineered to account for the outdoor exposure. Note that this is also consistent with typical vent manufacturers' design manuals and installation instructions. Even though it is common to see gas vents run up the exterior wall of a building, such installations are not allowed unless they are engineered and allowed by the vent manufacturer, neither of which is likely.

503.6.9 Size of gas vents. Venting systems shall be sized and constructed in accordance with Section 504 or other approved engineering methods and the gas vent and gas equipment manufacturers' instructions.

❖ Gas vents are to be sized by either Section 504 or an engineering method acceptable to the code official and, in all cases, in accordance with the vent manufacturer's instructions (see Sections 503.6.9.1, 503.6.9.2 and 503.6.9.3).

503.6.9.1 Category I appliances. The sizing of natural draft venting systems serving one or more listed appliances equipped with a draft hood or appliances listed for use with Type B gas vent, installed in a single story of a building, shall be in accordance with one of the following methods:

1. The provisions of Section 504.

2. For sizing an individual gas vent for a single, draft-hood-equipped appliance, the effective area of the vent connector and the gas vent shall be not less than the area of the appliance draft hood outlet, nor greater than seven times the draft hood outlet area.

3. For sizing a gas vent connected to two appliances with draft hoods, the effective area of the vent shall be not less than the area of the larger draft hood outlet plus 50 percent of the area of the smaller draft hood outlet, nor greater than seven times the smaller draft hood outlet area.

4. Approved engineering practices.

❖ This section reiterates the intent of the preceding section, is specific to Category I appliances, and applies only to appliances installed within the same story. Multiple-story applications are addressed in Section 503.6.10. Category I appliances are vented by natural draft and, depending upon the design and efficiency, are either equipped with a draft hood or a fan-assisted combustion system. Because of the higher thermal efficiencies and lack of dilution air, vent designs for fan-assisted appliances are different from those for draft hood-equipped appliances.

Item 3 allows a very limited application of the old "fifty-percent rule." Item 2 applies only to a single draft-hood-equipped appliance. Item 3 applies only to a common vent system serving two draft-hood-equipped appliances. Neither Item 2 nor 3 can be applied to the venting of fan-assisted appliances, because the old alternate method (fifty-percent rule) was not created with fan-assisted appliances in mind and will not work with such appliances. Besides being limited to draft hood appliances, Items 2 and 3 should be further limited to simple venting system configurations having short laterals; few, if any, changes in direction (fittings); short connector lengths with maximum vertical rise and tall vent heights. Vent manufacturers emphatically warn against the use of the alterative sizing rule in Item 3 because this method does not account for many factors now known to be important in vent sizing.

503.6.9.2 Category II, III, and IV appliances. The sizing of gas vents for Category II, III, and IV equipment shall be in accordance with the equipment manufacturer's instructions.

❖ Venting systems for category II, III and IV appliances are special vent systems specified by the appliance manufacturer and are not addressed further in the code (see Section 503.4).

503.6.9.3 Mechanical draft. Chimney venting systems using mechanical draft shall be sized in accordance with approved engineering methods.

❖ The manufacturers of exhausters (mechanical draft systems) provide sizing criteria for the vents and chimneys served by such equipment. The code's sizing criteria are based on natural draft chimneys and vents; therefore, engineered sizing methods must be used for other than natural draft (see Section 503.3.3).

503.6.10 Gas vents serving equipment on more than one floor. A single or common gas vent shall be permitted in multistory installations to vent Category I equipment located on more than one floor level, provided the venting system is designed and installed in accordance with this section and approved engineering methods.

❖ With rare exception, venting systems serve only appliances located within the same story. This section addresses those rare exceptions where a common venting system serves appliances that are installed within different stories. For example, a single vent serves two appliances on the first floor of a building and also serves two more on the second floor, for a total of four appliances. Such a system requires a unique design approach as dictated in Sections 503.6.10.1 and 503.6.10.2. The code is silent on the circumstance of appliances in attics common vented with appliances on lower floors. The conservative and safest approach would be to consider the attic a story.

503.6.10.1 Equipment separation. All equipment connected to the common vent shall be located in rooms separated from habitable space. Each of these rooms shall have provisions for an adequate supply of combustion, ventilation, and dilution air that is not supplied from habitable space (see Figure 503.6.10.1).

❖ This section addresses chimneys or vents that serve multiple appliances located on different floor levels. For example, a common vent may be designed to serve one or more appliances per floor of a multistory building, or a vent that rises up through the furnace rooms of single-story apartments in a multi-story apartment building would be a common vent for the "stacked" apartments. At each level there may be a single appliance or several appliances entering into the passageway. Multistory venting involves special design criteria and poses a unique hazard. When a vent is overloaded, blocked or otherwise produces insufficient draft, the flue gases from lower-story appliances could discharge into upper stories through the upper-story appliance connections. The lower-story vent could appear to be functioning normally, while the upper-story occupants would be exposed to a life-threatening hazard.

To avoid this potential hazard, multistory vent systems are designed to isolate the vent system and the appliances from the occupied portions of the building. This isolation is commonly accomplished by locating the common vent and all of the appliances served in mechanical rooms that are accessed from the outdoors only. This is commonly accomplished by placing mechanical rooms on outside walls and making them accessible only from a balcony. The sizing and design of multistory common chimney and vent systems must be in accordance with the vent or chimney manufacturer's installation and design instructions.

This section speaks of "habitable space," but the intent should apply to "occupied/living space" because many spaces such as bathrooms, toilet rooms and halls are not defined as habitable, yet are occupied, even if only for short periods.

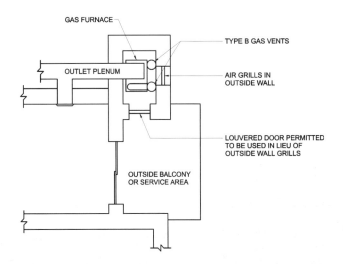

Figure 503.6.10.1
PLAN VIEW OF PRACTICAL SEPARATION METHOD FOR MULTISTORY GAS VENTING

This section is not limited to residential occupancies and would apply to all occupancies. The common vent and the appliances served by the common vent have to be located in a space that does not communicate with occupied spaces. In other words, a door cannot be the means of isolating the common vent and appliances from an occupied space because there is no way to control the position of a door or make it gas-tight. The most common method of isolation is to locate the common vent and appliances in a space that is accessed only from the outdoors [see commentary Figures 503.6.10.1(1) and 503.6.10.1(2)]. This location limitation makes it obvious that only appliances such as furnaces, boilers and water heaters are candidates for this type of venting system. (Room heaters and gas fireplace appliances cannot be separated from the occupied space.) Combustion, dilution and ventilation air for the appliances must not be taken from an occupied space because this would allow communication of atmospheres between the appliance space and the occupied space. The method of supplying combustion air must not violate the separation between the appliances and the occupied space.

503.6.10.2 Sizing. The size of the connectors and common segments of multistory venting systems for equipment listed for use with Type B double-wall gas vent shall be in accordance with Table 504.3(1) and Figures B-13 and B-14 in Appendix B, provided:

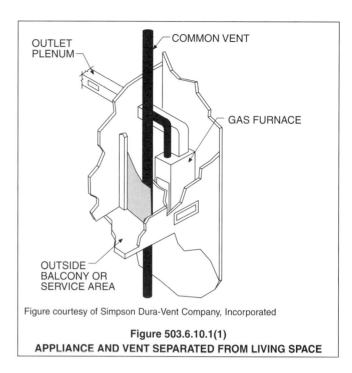

Figure courtesy of Simpson Dura-Vent Company, Incorporated

Figure 503.6.10.1(1)
APPLIANCE AND VENT SEPARATED FROM LIVING SPACE

1. The available total height (H) for each segment of a multistory venting system is the vertical distance between the level of the highest draft hood outlet or flue collar on that floor and the centerline of the next highest interconnection tee (see Figure B-13).

Figure 503.6.10.1(2)
APPLIANCE CLOSETS ACCESSED FROM OUTDOOR BALCONY

2. The size of the connector for a segment is determined from its gas utilization equipment heat input and available connector rise, and shall not be smaller than the draft hood outlet or flue collar size.

3. The size of the common vertical segment, and of the interconnection tee at the base of that segment, shall be based on the total gas utilization equipment heat input entering that segment and its available total height.

❖ Because Table 504.3(1) is referenced, vent connectors within the multistory venting system must be constructed of double-wall gas vent. Manufacturers' instructions may contain additional requirements, such as that the common vent must be entirely vertical with no offsets. The figures in Appendix B clearly depict items 1, 2, and 3 of this section.

503.6.11 Support of gas vents. Gas vents shall be supported and spaced in accordance with their listings and the manufacturer's instructions.

❖ Type B vents must be supported to prevent the weight of the system from bearing on appliances and vent fittings. Fittings such as adjustable elbows are not designed to carry the weight of upper or lower sections of vent pipe, nor are appliance draft hoods and flue collars. Support bases, brackets and spacers are available from the vent manufacturer and are used to support the weight of the vent and maintain alignment and clearance to combustibles. On most job sites, the vents will be supported by various field-constructed brackets and straps made of sheet metal scraps, duct slips and drives and plumber's perforated straps. The installations should be carefully examined to determine that such field-constructed supports are adequate. Sheet metal screws must not penetrate the inner liner of any Type B vent component. Type B vents are designed for installation without the use of any additional fasteners that penetrate the vent walls.

503.6.12 Marking. In those localities where solid and liquid fuels are used extensively, gas vents shall be permanently identified by a label attached to the wall or ceiling at a point where the vent connector enters the gas vent. The determination of where such localities exist shall be made by the code official. The label shall read:

"This gas vent is for appliances that burn gas. Do not connect to solid or liquid fuel-burning appliances or incinerators."

❖ Type B vents installed in localities where oil- and solid-fuel-fired appliances are common must be labeled to warn the installers and occupants that Type B vents cannot be used to vent any appliances other than gas-fired. The label is not placed on the vent; rather, it is to be placed on a wall or ceiling where it would be conspicuous to someone attempting to connect to the vent. The code official determines whether their locality is the intended target for this provision. It is assumed that ignorance of gas vents will be more likely in regions where oil and solid fuel have been commonly used for fuel.

503.7 Single-wall metal pipe. Single-wall metal pipe vents shall comply with Sections 503.7.1 through 503.7.12.

❖ Although discouraged or prohibited by designers, code officials and appliance and vent manufacturers alike, the code recognizes the use of unlisted single-wall metal pipe as a vent. All other types of vents are listed systems as required by Section 502.1. As evidenced by Sections 503.7.1 through 503.7.12, single-wall metal pipe is restricted to very limited applications, and extraordinary installation precautions are necessary. Single-wall metal pipe was once common but is rarely used today. Appliance manufacturers' instructions will typically prohibit the use of a single-wall metal vent (see Section 503.7.8).

503.7.1 Construction. Single-wall metal pipe shall be constructed of galvanized sheet steel not less than 0.0304 inch (0.7 mm) thick, or other approved, noncombustible, corrosion-resistant material.

❖ The galvanized sheet steel thickness specified is equivalent to 22 gage, which is heavier than the pipe commonly used for connectors for most residential solid-fuel-fired appliances.

503.7.2 Cold climate. Uninsulated single-wall metal pipe shall not be used outdoors in cold climates for venting gas utilization equipment.

❖ This section limits the application of single-wall vents to regions such as the southwestern and southeastern parts of the United States, unless the pipe is insulated to compensate for the high heat loss of single-wall pipe. The code contains no guidance for the type of insulation required or what R-value is required. At the minimum, insulation applied directly to the pipe would have to be a noncombustible material, the R-value would have to exceed that of Type B vent, and the insulation would have to be permanently attached and protected from the weather. Insulating single-wall pipe could be considerably more difficult and expensive than using a venting system designed for outdoor use. Bear in mind that air insulated double-wall vent pipe (Type B) is not designed for use outdoors in cold climates. When applying Section 504, single-wall vents are also prohibited from being exposed to the outdoors below the point of roof penetration.

503.7.3 Termination. Single-wall metal pipe shall terminate at least 5 feet (1524 mm) in vertical height above the highest connected equipment draft hood outlet or flue collar. Single-wall metal pipe shall extend at least 2 feet (610 mm) above the highest point where it passes through a roof of a building and at least 2 feet (610 mm) higher than any portion of a building within a horizontal distance of 10 feet (3048 mm) (see Figure 503.5.4). An approved cap or roof assembly shall be attached to the terminus of a single-wall metal pipe (see also Section 503.7.8, Item 3).

❖ Single-wall metal vents are not allowed to terminate as allowed by Section 503.6.6 and must comply with the termination requirements for chimneys, except that the minimum height above the roof penetration is 2 feet, whereas it is 3 feet for chimneys [see Figure 503.5.4(1)]. The minimum height and vent cap requirements are consistent with Sections 503.6.6 and 503.6.7.

503.7.4 Limitations of use. Single-wall metal pipe shall be used only for runs directly from the space in which the equipment is located through the roof or exterior wall to the outdoor atmosphere.

❖ Single-wall metal pipe is further restricted by this section to only two applications. It can run from the appliance directly to the outdoors through an exterior wall or it can run from the appliance directly to the outdoors through a roof. Single-wall metal pipe cannot pass through any floors, interior walls or partitions and cannot pass through attics or concealed spaces. This means that the appliance served must be in a space with an exterior wall or in the same story as the roof.

503.7.5 Roof penetrations. A pipe passing through a roof shall extend without interruption through the roof flashing, roof jack, or roof thimble. Where a single-wall metal pipe passes through a roof constructed of combustible material, a noncombustible, nonventilating thimble shall be used at the point of passage. The thimble shall extend at least 18 inches (457 mm) above and 6 inches (152 mm) below the roof with the annular space open at the bottom and closed only at the top. The thimble shall be sized in accordance with Section 503.10.16.

❖ A thimble is a sheet metal assembly designed to provide clearance between a vent and combustible materials, analogous to a thimble that protects one's fingers from a sewing needle. It is essentially a spacer that provides an annular space around the vent. Annular space is usually ventilated by punched holes in the faces of the thimble assembly. Some thimbles are constructed with insulation in addition to an annular space.

A roof jack is an assembly that passes through the roof and contains a thimble and weather protection all in one. Roof jacks are flashing/thimble combinations and typically mount to a roof curb.

503.7.6 Installation. Single-wall metal pipe shall not originate in any unoccupied attic or concealed space and shall not pass through any attic, inside wall, concealed space, or floor. The installation of a single-wall metal pipe through an exterior combustible wall shall comply with Section 503.10.15. Single-wall metal pipe used for venting an incinerator shall be exposed and readily examinable for its full length and shall have suitable clearances maintained.

❖ A single-wall metal vent must be exposed for its entire length, except for the section of pipe that is within a thimble or roof jack assembly. Single-wall vents can penetrate only exterior walls and roofs. A roof/ceiling assembly cannot be penetrated if there is an attic or a concealed space between the ceiling and the roof (see Section 503.7.4).

503.7.7 Clearances. Minimum clearances from single-wall metal pipe to combustible material shall be in accordance with Table 503.7.7. The clearance from single-wall metal pipe to combustible material shall be permitted to be reduced where the combustible material is protected as specified for vent connectors in Table 308.2.

❖ Consideration should be given to the fact that single-wall vents will have high surface temperatures compared to insulated vents such as Type B vents. High surface temperatures could be an ignition or burn injury hazard.

TABLE 503.7.7[a]
CLEARANCES FOR CONNECTORS

EQUIPMENT	MINIMUM DISTANCE FROM COMBUSTIBLE MATERIAL			
	Listed Type B gas vent material	Listed Type L vent material	Single-wall metal pipe	Factory-built chimney sections
Listed equipment with draft hoods and equipment listed for use with Type B gas vents	As listed	As listed	6 inches	As listed
Residential boilers and furnaces with listed gas conversion burner and with draft hood	6 inches	6 inches	9 inches	As listed
Residential appliances listed for use with Type L vents	Not permitted	As listed	9 inches	As listed
Listed gas-fired toilets	Not permitted	As listed	As listed	As listed
Unlisted residential appliances with draft hood	Not permitted	6 inches	9 inches	As listed
Residential and low-heat equipment other than above	Not permitted	9 inches	18 inches	As listed
Medium-heat equipment	Not permitted	Not permitted	36 inches	As listed

For SI: 1 inch = 25.4 mm.
a. These clearances shall apply unless the listing of an appliance or connector specifies different clearances, in which case the listed clearances shall apply.

❖ This table is a vent connector clearance table. The second column from the right in the table applies to single-wall vents as well as single-wall vent connectors.

503.7.8 Size of single-wall metal pipe. A venting system constructed of single-wall metal pipe shall be sized in accordance with one of the following methods and the equipment manufacturer's instructions:

1. For a draft-hood-equipped appliance, in accordance with Section 504.
2. For a venting system for a single appliance with a draft hood, the areas of the connector and the pipe each shall be not less than the area of the appliance flue collar or draft hood outlet, whichever is smaller. The vent area shall not be greater than seven times the draft hood outlet area.
3. Other approved engineering methods.

❖ As can be seen from items 1, 2, and 3 and Tables 504.2(5) and 504.3(5), single-wall metal pipe vents are limited to use with draft-hood-equipped appliances, except where the system is engineered.

503.7.9 Pipe geometry. Any shaped single-wall metal pipe shall be permitted to be used, provided that its equivalent effective area is equal to the effective area of the round pipe for which it is substituted, and provided that the minimum internal dimension of the pipe is not less than 2 inches (51 mm).

❖ Single-wall metal pipe does not have to be round; it could be oval, square, or rectangular. Round pipe has the best flow characteristics as evidenced by the fact that a round chimney liner will have a greater capacity than a rectangular liner of equal area.

503.7.10 Termination capacity. The vent cap or a roof assembly shall have a venting capacity not less than that of the pipe to which it is attached.

❖ Section 503.7.3 requires an approved vent cap or roof termination assembly, and this section requires that the terminus be sized and designed to not reduce the capacity of the vent. An unlisted, untested cap or roof assembly will have an unknown capacity unless it is engineered.

503.7.11 Support of single-wall metal pipe. All portions of single-wall metal pipe shall be supported for the design and weight of the material employed.

❖ Single-wall vents, like all vents and chimneys, must be supported to prevent structural failure and joint separation and to maintain the required clearances.

503.7.12 Marking. Single-wall metal pipe shall comply with the marking provisions of Section 503.6.12.

❖ See the commentary for Section 503.6.12.

503.8 Venting system termination location. The location of venting system terminations shall comply with the following (see Appendix C):

1. A mechanical draft venting system shall terminate at least 3 feet (914 mm) above any forced-air inlet located within 10 feet (3048 mm).

Exceptions:

1. This provision shall not apply to the combustion air intake of a direct-vent appliance.
2. This provision shall not apply to the separation of the integral outdoor air inlet and flue gas discharge of listed outdoor appliances.

2. A mechanical draft venting system, excluding direct-vent appliances, shall terminate at least 4 feet (1219 mm) below, 4 feet (1219 mm) horizontally from, or 1 foot (305 mm) above any door, operable window, or gravity air inlet into any building. The bottom of the vent terminal shall be located at least 12 inches (305 mm) above grade.

3. The vent terminal of a direct-vent appliance with an input of 10,000 Btu per hour (3 kW) or less shall be located at least 6 inches (152 mm) from any air opening into a building, and such an appliance with an input over 10,000 Btu per hour (3 kW) but not over 50,000 Btu per hour (14.7 kW) shall be installed with a 9-inch (230 mm) vent termination clearance, and an appliance with an input over 50,000 Btu/h (14.7 kw) shall have at least a 12-inch (305 mm) vent termination clearance. The bottom of the vent terminal and the air intake shall be located at least 12 inches (305 mm) above grade.

4. Through-the-wall vents for Category II and IV appliances and noncategorized condensing appliances shall not terminate over public walkways or over an area where condensate or vapor could create a nuisance or hazard or could be detrimental to the operation of regulators, relief valves, or other equipment. Where local experience indicates that condensate is a problem with Category I and III appliances, this provision shall also apply.

❖ This section addresses the terminations of mechanical drafts systems and direct-vent appliances. To prevent vent gases from entering the building, item 1 requires forced air (mechanical) intakes to be at least 10 feet away from a mechanical draft termination, or the termination must be at least 3 feet above the air intake. The natural buoyancy of vent gases is the justification for allowing a closer distance where the terminal is above the air intake. A combustion air intake for a direct-vent appliance is not what item 1 intends to address. Direct-vent appliance exhaust and intake opening locations are dictated by the appliance listing and manufacturer's instructions. Commentary Figures 503.8(1) and 503.8(2) are extracted from Z21.47b-2002/CSA 2.3b-2002, the standard for gas-fired central furnaces, and dictate the content of appliance installation instructions. Note the distinction between direct-vent and other than direct-vent in the figures. Item 1 also does not apply to appliances such as rooftop units that have built-in outdoor air intakes and combustion gas exhaust outlets. The code assumes that the design and listing of the outdoor appliance accounts for the location of those openings.

Item 2 addresses gravity air inlets in the exterior envelope of a building, including doors, operable (openable) windows and intake louvers and grilles. The appliances addressed by this item use auxiliary or integral fans and blowers to force the flow of combustion

FIGURE 503.8(1)　　CHIMNEYS AND VENTS

		Canadian Installations[1]	US Installations[2]			Canadian Installations[1]	US Installations[2]
A=	Clearance above grade, veranda, porch, deck or balcony	12 inches (30 cm)	12 inches (30 cm)	J=	Clearance to nonmechanical air supply inlet to building or the combustion air inlet to any other appliance	6 inches (15 cm) for appliances ≤ 10,000 Btuh (3 kW), 12 inches (30 cm) for appliances > 10,000 Btuh (3 kW) and ≤ 100,000 Btuh (30 kW), 36 inches (91 cm) for appliances > 100,000 Btuh (30 kW)	6 inches (15 cm) for appliances ≤ 10,000 Btuh (3 kW), 9 inches (23 cm) for appliances > 10,000 Btuh (3 kW) and ≤ 50,000 Btuh (15 kW), 12 inches (30 cm) for appliances > 50,000 Btuh (15 kW)
B=	Clearance to window or door that may be opened	6 inches (15 cm) for appliances ≤ 10,000 Btuh (3 kW), 12 inches (30 cm) for appliances > 10,000 Btuh (3 kW) and ≤ 100,000 Btuh (30 kW), 36 inches (91 cm) for appliances >100,000 Btuh (30 kW)	6 inches (15 cm) for appliances ≤ 10,000 Btuh (3 kW), 9 inches (23 cm) for appliances > 10,000 Btuh (3 kW) and ≤ 50,000 Btuh (15 kW), 12 inches (30 cm) for appliances >50,000 Btuh (15 kW)	K=	Clearance to a mechanical air supply inlet	6 feet (1.83 m)	3 feet (91 cm) above if within 10 feet (3 m) horizontally
C=	Clearance to permanently closed window	*	*	L=	Clearance above paved sidewalk or paved driveway located on public property	7 feet (2.13 m) †	*
D=	Vertical clearance to ventilated soffit located above the terminal within a horizontal distance of 2 feet (61 cm) from the center line of the terminal	*	*	M=	Clearance under veranda, porch deck or balcony	12 inches (30 cm) ‡	*
E=	Clearance to unventilated soffit	*	*				
F=	Clearance to outside corner	*	*				
G=	Clearance to inside corner	*	*				
H=	Clearance to each side of center line extended above meter/regulator assembly	3 feet (91 cm) within a height 15 feet (4.6 m) above the meter/regulator assembly	*				
I=	Clearance to service regulator vent outlet	3 feet (91 cm)	*				

[1] in accordance with the current CSA B149.1, *Natural Gas and Propane Installation Code.*
[2] in accordance with the current ANSI Z223.1/NFPA 54, *National Fuel Gas Code*
† A vent shall not terminate directly above a sidewalk or paved driveway that is located between two single family dwellings and serves both dwellings.
‡ Permitted only if veranda, porch, deck or balcony is fully open on a minimum of two sides beneath the floor.
* For clearances not specified in Ansi Z223.1/NFPA 54 or CSA B149.1, the following statement shall be included:
"Clearance in accordance with local installation codes, the requirements of the gas supplier and the manufacturer's installation instructions."

With the permission of Canadian Standards Association, material is reproduced from CSA Standard ANSI Z21.47 b-2002/CSA 2.3 b-2002, *Gas-Fired Central Furnaces*, which is copyrighted by Canadian Standards Association, 178 Rexdale Blvd., Toronto, Ontario, M9W 1R3, www.csa.ca. Although use of this material has been authorized, CSA shall not be responsible for the manner in which the information is presented, or for any interpretations thereof.

Figure 503.8(1)
DIRECT-VENT TERMINAL CLEARANCES

CHIMNEYS AND VENTS

FIGURE 503.8(2)

	Canadian Installations[1]	US Installations[2]		Canadian Installations[1]	US Installations[2]
A= Clearance above grade, veranda, porch, deck or balcony	12 inches (30 cm)	12 inches (30 cm)	J= Clearance to nonmechanical air supply inlet to building or the combustion air inlet to any other appliance	6 inches (15 cm) for appliances ≤ 10,000 Btuh (3 kW), 12 inches (30 cm) for appliances > 10,000 Btuh (3 kW) and ≤ 100,000 Btuh (30 kW), 36 inches (91 cm) for appliances > 100,000 Btuh (30 kW)	4 feet (1.2 m) below or to side of opening; 1 foot (30 cm) above opening
B= Clearance to window or door that may be opened	6 inches (15 cm) for appliances ≤ 10,000 Btuh (3 kW), 12 inches (30 cm) for appliances > 10,000 Btuh (3 kW) and ≤ 100,000 Btuh (30 kW), 36 inches (91 cm) for appliances >100,000 Btuh (30 kW)	4 feet (1.2 m) below or to side of opening; 1foot (30 cm) above opening	K= Clearance to a mechanical air supply inlet	6 feet (1.83 m)	3 feet (91 cm) above if within 10 feet (3 m) horizontally
C= Clearance to permanently closed window	*	*	L= Clearance above paved sidewalk or paved driveway located on public property	7 feet (2.13 m) †	7 feet (2.13 m)
D= Vertical clearance to ventilated soffit located above the terminal within a horizontal distance of 2 feet (61 cm) from the center line of the terminal	*	*	M= Clearance under veranda, porch deck or balcony	12 inches (30 cm) ‡	*
E= Clearance to unventilated soffit	*	*			
F= Clearance to outside corner	*	*			
G= Clearance to inside corner	*	*			
H= Clearance to each side of center line extended above meter/regulator assembly	3 feet (91 cm) within a height 15 feet (4.6 m) above the meter/regulator assembly	*			
I= Clearance to service regulator vent outlet	3 feet (91 cm)	*			

[1] in accordance with the current CSA B149.1, *Natural Gas and Propane Installation Code.*
[2] in accordance with the current ANSI Z223.1/NFPA 54, *National Fuel Gas Code*
† A vent shall not terminate directly above a sidewalk or paved driveway that is located between two single family dwellings and serves both dwellings.
‡ Permitted only if veranda, porch, deck or balcony is fully open on a minimum of two sides beneath the floor.
* For clearances not specified in Ansi Z223.1/NFPA 54 or CSA B149.1, the following statement shall be included:
"Clearance in accordance with local installation codes, the requirements of the gas supplier and the manufacturer's installation instructions."

With the permission of Canadian Standards Association, material is reproduced from CSA Standard ANSI Z21.47 b-2002/CSA 2.3 b-2002, *Gas-Fired Central Furnaces*, which is copyrighted by Canadian Standards Association, 178 Rexdale Blvd., Toronto, Ontario, M9W 1R3, www.csa.ca. Although use of this material has been authorized, CSA shall not be responsible for the manner in which the information is presented, or for any interpretations thereof.

Figure 503.8(2)
OTHER THAN DIRECT-VENT TERMINAL CLEARANCES

products to the outdoors. This item applies to externally installed power exhausters, integrally power-exhausted appliances and venting systems equipped with draft inducers. This item does not address fan-assisted Category I appliances and does not address direct-vent appliances. Category III and IV appliances are subject to the requirements of this item if those appliances are not listed as direct-vent appliances. Some Category I appliances are field convertible to Category III.

Power exhausters are field-installed pieces of equipment that are independent of, but used in conjunction with, other appliances [see commentary Figure 503.8(3)].

This item is consistent with appliance installation instructions that will contain additional details for the location of the venting system terminals, including but not limited to clearance to grade, inside corners, outside corners, decks, porches, balconies, soffits, roof eaves, utility meters, combustion air intakes, other appliance vents, service regulators, sidewalks and driveways [see commentary Figures 503.8(1) and 503.8(2)].

There are significant differences between the requirements of items 2 and 3 as they relate to nondirect-vent and direct-vent appliances, respectively. A Category IV, 90-percent efficient furnace or boiler would be subject to item 2 if installed without a combustion air intake pipe to the outdoors but would be subject to item 3, not item 2, if installed with a combustion air intake pipe to the outdoors. This is because such a furnace can be listed as both a direct-vent and nondirect-vent appliance, and as such can be installed either way. Less restrictive clearances, improved operation efficiency, longer appliance life, immunity from the effects of building exhaust systems, and ease of providing combustion air are some of the advantages of direct-vent appliances. For example, according to item 2, a sidewall vent terminal must be 4 feet or more to either side of an openable window, or must be 4 feet or more below the bottom of the window or must be 1 foot or more above the top of the window.

Item 3 is consistent with appliance manufacturers' instructions. Research has shown that the distances associated with the input ranges are necessary to allow the combustion gases to dissipate in the atmosphere, thus avoiding reentry into the building through openings. This item is applicable only to direct-vent appliances. For example, a Category IV, 90-percent efficient furnace installed without an outdoor air intake pipe for combustion air would be subject to item 2, not 3 [see commentary Figures 503.8(4), 503.8(5) and 503.8(6)].

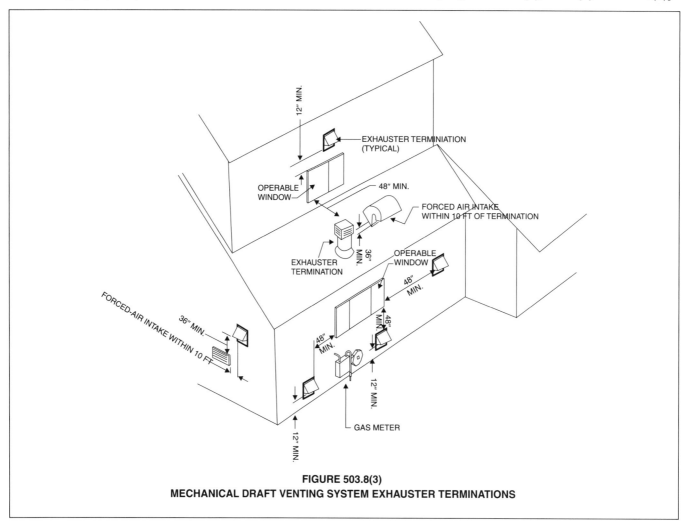

FIGURE 503.8(3)
MECHANICAL DRAFT VENTING SYSTEM EXHAUSTER TERMINATIONS

CHIMNEYS AND VENTS

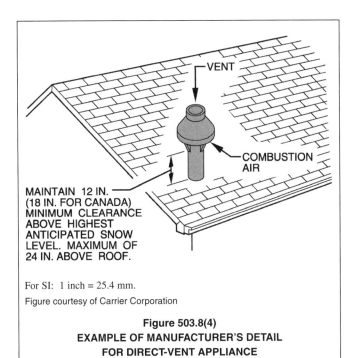

Figure 503.8(4)
EXAMPLE OF MANUFACTURER'S DETAIL FOR DIRECT-VENT APPLIANCE

Item 4 addresses the potential problem of condensation relative to its possible effect on pedestrians, service equipment, etc. Condensate is corrosive and could create a hazard when it freezes. Condensate ice could be a slip hazard and could obstruct regulator and relief valve vents. Category I and III appliances do not produce as much condensate as appliances in the other two categories, but they can also cause condensate problems in cold climates. It is common to see large masses of ice build up on or near the vent terminals of Category I and III vents. Such ice masses can fall or slide from roofs and injure someone or cause damage to a building, vehicle or equipment.

503.9 Condensation drainage. Provision shall be made to collect and dispose of condensate from venting systems serving Category II and IV equipment and noncategorized condensing appliances in accordance with Section 503.8, Item 4. Where local experience indicates that condensation is a problem, provision shall be made to drain off and dispose of condensate from venting systems serving Category I and III equipment in accordance with Section 503.8, Item 4.

❖ See Section 503.8, item 4.

503.10 Vent connectors for Category I equipment. Vent connectors for Category I equipment shall comply with Sections 503.10.1 through 503.10.16.

❖ See Sections 503.10.1 through 503.10.17.

503.10.1 Where required. A vent connector shall be used to connect equipment to a gas vent, chimney, or single-wall metal pipe, except where the gas vent, chimney, or single-wall metal pipe is directly connected to the equipment.

❖ Unless the chimney or vent is connected directly to the appliance, a connector, defined in Chapter 2 as pipe used to connect an approved fuel-burning appliance to a chimney or vent, is necessary. This includes the fittings necessary to make a connection or change in di-

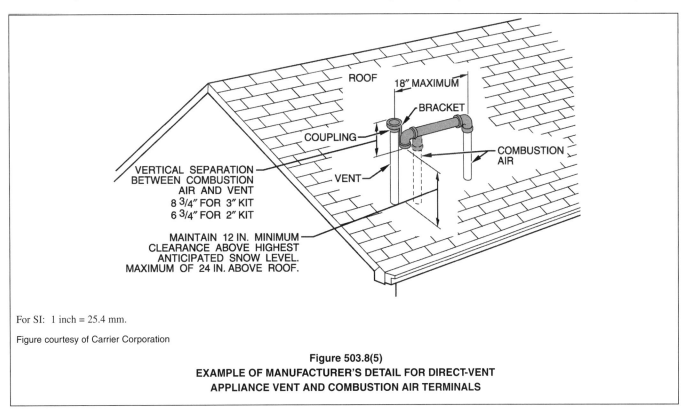

Figure 503.8(5)
EXAMPLE OF MANUFACTURER'S DETAIL FOR DIRECT-VENT APPLIANCE VENT AND COMBUSTION AIR TERMINALS

Figure 503.8(6)
DIRECT-VENT HOT WATER BOILER

rection. This is usually accomplished with a single-wall metal pipe, but it is also common practice to use listed and labeled chimney and vent pipe or listed factory-built single-wall or double-wall bendable connectors.

Many factors affect the design and configuration of a connector. The most important of these is the appliance location with respect to the chimney or vent system. This impacts the connector size, length and rise. Another important factor is the number of appliances being vented. The appliance manufacturer's installation instructions may prohibit the use of single-wall connectors. For example, a single-wall vent connector may not be appropriate for fan-assisted appliances because of the possibility of condensation, corrosion and leakage. They are also not permitted in attic or crawl space installations that are subject to cold temperatures. The cold temperatures increase the possibility of condensation, which leads to accelerated material failure. If the vent or chimney extends all the way to a single appliance vent outlet, no connector is needed.

503.10.2 Materials. Vent connectors shall be constructed in accordance with Sections 503.10.2.1 through 503.10.2.5.

❖ See Sections 503.10.2.1 through 503.10.2.5.

503.10.2.1 General. A vent connector shall be made of noncombustible corrosion-resistant material capable of withstanding the vent gas temperature produced by the equipment and of sufficient thickness to withstand physical damage.

❖ See Sections 503.10.2.2 through 503.10.2.5.

503.10.2.2 Vent connectors located in unconditioned areas. Where the vent connector used for equipment having a draft hood or a Category I appliance is located in or passes through attics, crawl spaces or other unconditioned spaces, that portion of the vent connector shall be listed Type B or Type L or listed vent material or listed material having equivalent insulation properties.

Exception: Single-wall metal pipe located within the exterior walls of the building in areas having a local 99 percent winter design temperature of 5°F (-15°C) or higher shall be permitted to be used in unconditioned spaces other than attics and crawl spaces.

❖ Compared to double-wall pipe, single-wall pipe has a much higher heat loss, and when installed in an unconditioned space, condensation can occur, deteriorating the pipe. A typical scenario is an attic, garage or crawl space appliance installation where a connector would be exposed to cold temperatures, practically ensuring the formation of condensation in the connector. Except as allowed in the exception, this section effectively prohibits the use of single-wall connectors in garages, attics and crawl spaces when the spaces are unheated. The term "unconditioned" is not defined and the debate over the intent has created new text in the exception, which is climate related. Attics and crawl spaces are unconditioned spaces without doubt; however, the uncertainty has been with residential basements and garages, which are common locations for gas-fired appliances. In many climates, the garage winter temperatures will be cold enough to cause the combustion gases in a connector to reach their dew point, meaning that condensate will form and linger in the connector, causing its rapid destruction. Most residential garages and some basements are considered unconditioned insofar as they are not directly or indirectly heated as is the living space.

The typical residential garage is not heated (conditioned) and, in most climates, will certainly have low ambient temperatures. The exception allows single-wall metal pipe in unconditioned spaces (other than attics and crawl spaces) if the climate is as described. The winter design temperatures can be found in the ASHRAE *Handbook of Fundamentals*. To put the climate criterion in perspective, Philadelphia, Pennsylvania, Louisville, Kentucky, Fall River, Massachusetts, and Providence, Rhode Island all have 99 percent winter design temperatures of 5°F (-15°C) or higher. The exception is applicable only to vent connectors that are inside a building (i.e. located within the exterior walls). (see commentary Figure 503.10.2.2).

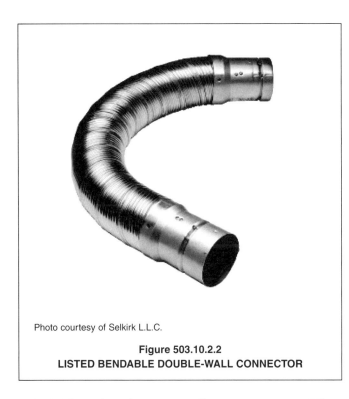

Photo courtesy of Selkirk L.L.C.

**Figure 503.10.2.2
LISTED BENDABLE DOUBLE-WALL CONNECTOR**

503.10.2.3 Residential-type appliance connectors. Where vent connectors for residential-type appliances are not installed in attics or other unconditioned spaces, connectors for listed appliances having draft hoods and for appliances having draft hoods and equipped with listed conversion burners shall be one of the following:

1. Type B or Type L vent material;
2. Galvanized sheet steel not less than 0.018 inch (0.46 mm) thick;
3. Aluminum (1100 or 3003 alloy or equivalent) sheet not less than 0.027 inch (0.69 mm) thick;
4. Stainless steel sheet not less than 0.012 inch (0.31 mm) thick;
5. Smooth interior wall metal pipe having resistance to heat and corrosion equal to or greater than that of Item 2, 3 or 4 above; or
6. A listed vent connector.

Vent connectors shall not be covered with insulation.

Exception: Listed insulated vent connectors shall be installed according to the terms of their listing.

❖ This section applies to draft-hood-equipped appliances only and lists the allowable materials, including sheet metals, listed connectors and listed vent system pipe. The minimum thickness for single-wall galvanized steel pipe is equivalent to 26 gage. Installers have field-installed insulation on vent connectors in an attempt to control heat loss and the resultant condensation; however, this is prohibited because of the combustibility of most insulation materials and the possibility of excessive connector temperatures. Listed double-wall (air insulated) Type B connectors are available as an option to using Type B vent pipe.

503.10.2.4 Low-heat equipment. A vent connector for low-heat equipment shall be a factory-built chimney section or steel pipe having resistance to heat and corrosion equivalent to that for the appropriate galvanized pipe as specified in Table 503.10.2.4. Factory-built chimney sections shall be joined together in accordance with the chimney manufacturers' instructions.

❖ Residential appliances are also low-heat by definition, and the provisions of Section 503.10.2.3 apply as well. The table is intended for non-residential-type appliances, which is apparent when compared to the materials allowed by section 503.10.2.3.

**TABLE 503.10.2.4
MINIMUM THICKNESS FOR GALVANIZED STEEL VENT CONNECTORS FOR LOW-HEAT APPLIANCES**

DIAMETER OF CONNECTOR (inches)	MINIMUM THICKNESS (inch)
Less than 6	0.019
6 to less than 10	0.023
10 to 12 inclusive	0.029
14 to 16 inclusive	0.034
Over 16	0.056

For SI: 1 inch = 25.4 mm.

❖ See the commentary for Section 503.10.2.4.

503.10.2.5 Medium-heat appliances. Vent connectors for medium-heat equipment and commercial and industrial incinerators shall be constructed of factory-built medium-heat chimney sections or steel of a thickness not less than that specified in Table 503.10.2.5 and shall comply with the following:

1. A steel vent connector for equipment with a vent gas temperature in excess of 1000°F (538°C), measured at the entrance to the connector shall be lined with medium-duty fire brick (ASTM C 64, Type F), or the equivalent.
2. The lining shall be at least $2^1/_2$ inches (64 mm) thick for a vent connector having a diameter or greatest cross-sectional dimension of 18 inches (457 mm) or less.
3. The lining shall be at least $4^1/_2$ inches (114 mm) thick laid on the $4^1/_2$-inch (114 mm) bed for a vent connector having a diameter or greatest cross-sectional dimension greater than 18 inches (457 mm).
4. Factory-built chimney sections, if employed, shall be joined together in accordance with the chimney manufacturers' instructions.

❖ Medium-heat appliances have very high temperature flue gases [(1000°F) (538°C)], and because of the waste of enormous amounts of energy are often equipped with secondary heat exchangers to reclaim heat energy and substantially reduce the vent gas temperatures.

TABLE 503.10.2.5
MINIMUM THICKNESS FOR STEEL VENT CONNECTORS FOR MEDIUM-HEAT EQUIPMENT AND COMMERCIAL AND INDUSTRIAL INCINERATORS VENT CONNECTOR SIZE

DIAMETER (inches)	AREA (square inches)	MINIMUM THICKNESS (inch)
Up to 14	Up to 154	0.053
Over 14 to 16	154 to 201	0.067
Over 16 to 18	201 to 254	0.093
Over 18	Larger than 254	0.123

For SI: 1 inch = 25.4 mm, 1 square inch = 645.16 mm².

503.10.3 Size of vent connector. Vent connectors shall be sized in accordance with Sections 503.10.3.1 through 503.10.3.5.

❖ See the commentary for Sections 503.10.3.1 through 503.10.3.6.

503.10.3.1 Single draft hood and fan-assisted. A vent connector for equipment with a single draft hood or for a Category I fan-assisted combustion system appliance shall be sized and installed in accordance with Section 504 or other approved engineering methods.

❖ The sizing tables in Section 504 will determine connector size with the fundamental requirement that a connector not be smaller than the appliance vent connection, except as allowed by Sections 504.2.2 and 504.3.21. The tables will often require a connector to be larger than the appliance vent outlet (see Section 504).

503.10.3.2 Multiple draft hood. For a single appliance having more than one draft hood outlet or flue collar, the manifold shall be constructed according to the instructions of the appliance manufacturer. Where there are no instructions, the manifold shall be designed and constructed in accordance with approved engineering practices. As an alternate method, the effective area of the manifold shall equal the combined area of the flue collars or draft hood outlets and the vent connectors shall have a minimum 1-foot (305 mm) rise.

❖ Some appliances are designed with more than one venting (flue) outlet. For example, some large furnaces were built with dual heat exchangers and burner sections, and thus have two draft hoods or flue collar outlets. The manufacturer's instructions must be followed. However, in the event that the manufacturer's instructions are not available, this section specifies required sizing criteria. Section 305.1 requires the manufacturer's instructions to be on the job site.

503.10.3.3 Multiple appliances. Where two or more appliances are connected to a common vent or chimney, each vent connector shall be sized in accordance with Section 504 or other approved engineering methods.

As an alternative method applicable only when all of the appliances are draft hood equipped, each vent connector shall have an effective area not less than the area of the draft hood outlet of the appliance to which it is connected.

❖ The sizing tables in Section 504 will determine connector size with the fundamental requirement that a connector not be smaller than the appliance vent connection, except as allowed by Sections 504.2.2 and 504.3.21. The tables will often require a connector that is larger than the appliance vent outlet. See Section 504. The alternative method paragraph of this section is incompatible with Section 504 and is intended only for use with the alternative sizing method of Section 503.6.9.1.

503.10.3.4 Common connector/manifold. Where two or more gas appliances are vented through a common vent connector or vent manifold, the common vent connector or vent manifold shall be located at the highest level consistent with available headroom and the required clearance to combustible materials and shall be sized in accordance with Section 504 or other approved engineering methods.

As an alternate method applicable only where there are two draft hood equipped appliances, the effective area of the common vent connector or vent manifold and all junction fittings shall be not less than the area of the larger vent connector plus 50 percent of the area of the smaller flue collar outlet.

❖ Vent manufacturers' instructions contain detailed design criteria for manifolds. Connectors serving gas-fired appliances are allowed to join to form a common connector manifold rather than connecting independently to the vertical common vent [see commentary Figures 503.10.3.4(1) and 503.10.3.4(2)]. Various manufacturers' instructions require a 10-percent to 20-percent reduction in the capacity of a common vent where a manifold is used (see commentary, Section 504.3.4). A manifold is defined as a vent or connector that is a lateral (horizontal) extension of the lower end of a common vent. Because a manifold serves multiple appliances and thus conveys flue gases from two or more appliances, manifolds must be sized by the common vent tables. This section requires that manifolds be installed as high as possible, respective of ceiling height and clearance to combustibles requirements. This will result in the most appliance connector rise, which is always beneficial for venting performance. Connector rise takes advantage of the energy of hot gases discharging directly from the appliance and develops flow velocity and draft in addition to that produced by the common vent [see commentary Figures 503.10.3.4(1) and 503.10.3.4(2)]. The alternative method in the second paragraph is intended for use only with the alternative sizing method of Section 503.6.9.1 and is not compatible with Section 504.

503.10.3.5 Size increase. Where the size of a vent connector is increased to overcome installation limitations and obtain connector capacity equal to the equipment input, the size increase shall be made at the equipment draft hood outlet.

❖ If a connector is required to be larger than the appliance vent outlet, the size increase must occur at the appliance connection by use of an appropriate increaser fit-

CHIMNEYS AND VENTS

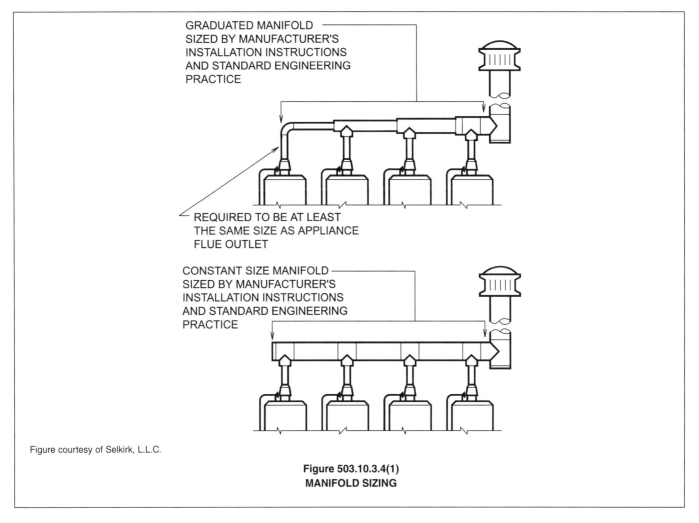

Figure 503.10.3.4(1)
MANIFOLD SIZING

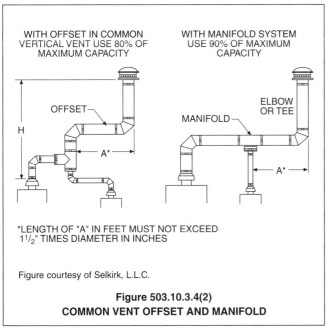

Figure 503.10.3.4(2)
COMMON VENT OFFSET AND MANIFOLD

ting. For example, it is common for a water heater with a 3-inch draft-hood outlet to be connected to a 4-inch connector by a 3 x 4 inch increaser fitting, as dictated by the vent tables in Section 504.

503.10.4 Two or more appliances connected to a single vent. Where two or more vent connectors enter a common gas vent, chimney flue, or single-wall metal pipe, the smaller connector shall enter at the highest level consistent with the available headroom or clearance to combustible material. Vent connectors serving Category I appliances shall not be connected to any portion of a mechanical draft system operating under positive static pressure, such as those serving Category III or IV appliances.

❖ A common vent is somewhat oversized when only a single appliance is operating; therefore, every effort is made to achieve proper vent/chimney operation during all possible operating circumstances. Placing the smallest appliance connection at the highest elevation in the common vent/chimney will allow the greatest connector rise for the smaller appliance connector and will take advantage of any draft priming effect caused by the lower connector. For example, it has been common practice to connect a domestic water heater connector above the furnace or boiler connector in a chimney or vent system. Category I appliances require a vent that produces a draft and therefore cannot be connected to any venting system that produces a positive pressure.

For example, a Category I appliance and a Category III appliance cannot share the same vent unless they both connect to the negative pressure (inlet) side of a mechanical draft system in accordance with the appliance and draft system installation instructions (see Section 501.10).

503.10.5 Clearance. Minimum clearances from vent connectors to combustible material shall be in accordance with Table 503.7.7.

Exception: The clearance between a vent connector and combustible material shall be permitted to be reduced where the combustible material is protected as specified for vent connectors in Table 308.2.

❖ Flue gas passageways must have minimum clearances to ignitable materials. The clearance requirements in Table 503.7.7 apply to unlisted single-wall connectors. Connectors that are listed and labeled for this use must be installed with a clearance to combustibles as required by the connector manufacturer's installation instructions. If the appliance or connector manufacturer's installation instructions specify larger clearances than those prescribed by this section, the manufacturer's installation instructions govern. Table 503.7.7 dictates the required air-space clearances between vent connectors and combustible materials and assemblies. The provisions of Section 308 would allow reduction of connector clearances where the prescribed clearance cannot be provided or is impractical. Lack of vent connector clearance is a common code violation. Connectors usually lack the required clearances to wood joists, plastic plumbing piping, building and piping insulation materials and gypsum board. A major disadvantage of single-wall connectors is the large clearance required to combustibles. Listed connectors usually require far less clearance, in some cases only 1 inch (25.4 mm).

503.10.6 Flow resistance. A vent connector shall be installed so as to avoid turns or other construction features that create excessive resistance to flow of vent gases.

❖ A vent connector must have no more bends or fittings than absolutely necessary to accomplish the installation because changes in direction add to flow resistance. Excessive resistance to flow could be defined as any resistance in excess of what would be caused by the two changes in direction anticipated by the vent sizing tables in Section 504. This means that the installer and the designer will have to plan the installations of vents and appliances so that the connectors can be as short and direct as possible. It is quite common to see installations with three, four or more changes of direction in a connector because the appliance and vent location were not planned well. Section 504 will penalize connector installations that contain more than two changes in direction (fittings). Listed bendable connectors offer the advantage of long sweeping changes in direction, which can have better flow characteristics than short sweep fittings (see commentary Figure 503.10.2.2). If a connector could have been installed with fewer fittings than the installer used, the installation could be rejected under this section (see Section 503.10.9 and commentary, Section 504.3.2).

503.10.7 Joints. Joints between sections of connector piping and connections to flue collars and hood outlets shall be fastened by one of the following methods:

1. Sheet metal screws.
2. Vent connectors of listed vent material assembled and connected to flue collars or draft hood outlets in accordance with the manufacturers' instructions.
3. Other approved means.

❖ Single-wall connectors (unlisted) have been traditionally fastened with sheet metal screws and rivets. All connector joints must be fastened, including the joint at the appliance draft hood or flue collar. A displaced connector could result in a life-threatening condition; therefore, connectors must be fastened and supported well. Appliance connectors are often located where they can be impacted or otherwise disturbed by building occupants, making proper fastening even more important. Item 2 speaks of listed connectors such as Type B vent material and factory-built corrugated (bendable) connectors, which employ adapter fittings or integral collars that are mechanically fastened to the appliance with screws or rivets (see commentary Figure 503.10.2.2).

503.10.8 Slope. A vent connector shall be installed without dips or sags and shall slope upward toward the vent or chimney at least $^1/_4$ inch per foot (21 mm/m).

Exception: Vent connectors attached to a mechanical draft system installed in accordance with the manufacturers' instructions.

❖ The ideal chimney or vent configuration is a totally vertical system, even though it is not always practical. This section requires all portions of a chimney or vent connector to rise vertically a minimum of a 1/4 inch per each foot (0.02 mm/m) of its horizontal length. The connector slope is intended to induce the flow of flue gases using the natural buoyancy of the hot gases. Connector slope can promote the priming of a cold venting system and can partially compensate for short connector vertical rise. Low points, dips and sags could also trap condensate and accelerate corrosion of the connector. The exception allows connectors without slope where connected to mechanical draft systems because the connector is on the negative pressure (inlet) side of the exhauster and slope would provide no benefit.

503.10.9 Length of vent connector. A vent connector shall be as short as practical and the equipment located as close as practical to the chimney or vent. Except as provided for in Section 503.10.3, the maximum horizontal length of a single-wall connector shall be

75 percent of the height of the chimney or vent. Except as provided for in Section 503.10.3, the maximum horizontal length of a Type B double-wall connector shall be 100 percent of the height of the chimney or vent. For a chimney or vent system serving multiple appliances, the maximum length of an individual connector, from the appliance outlet to the junction with the common vent or another connector, shall be 100 percent of the height of the chimney or vent.

❖ This section is applied in conjunction with Section 504.3.2. The appliance and chimney or vent must be located to keep the connector length as short as practicable. This section establishes the maximum allowable length for uninsulated chimney and vent connectors, insulated chimney and vent connectors and individual connectors for a chimney or vent system serving multiple appliances. These requirements are based on the heat loss of the connector and the ability of the vent or chimney system to produce a draft.

An insulated connector (double wall) reduces the amount of heat transfer through the connector pipe; thus, the flue gas is maintained at a higher temperature (see commentary, Section 504.3.2). The amount of draft is directly related to the chimney or vent height and the difference in temperature between the flue gases and the ambient air. An uninsulated connector is a run of single-wall pipe. Commentary Figure 503.10.9 shows that the total chimney or vent height is measured from the top of the highest appliance flue outlet connection to the termination point of the chimney or vent. The length limitations of this section are based on the total vertical rise of the chimney or vent and on the developed length of horizontal connectors within the chimney or vent system. The venting tables will often prohibit the use of single-wall connectors with fan-assisted appliances, and most appliance and vent manufacturers recommend against the use of single-wall (uninsulated) connectors.

Connectors are limited in length because of the flow resistance of the connector pipe and because heat loss through the connector is directly related to its length. For a venting system to work, the draft produced by the vertical vent or chimney must be able to overcome the resistance to flow created by the connector. The longer the connector, the longer it takes to prime a cold venting system and develop draft. As stated in Section 504.3.2, connectors should always be as short as the installation conditions will permit. The sizing methodology and tables in Section 504 for connectors in single-appliance installations might allow connector lengths greater than the general limits given in this section or might require a shorter connector length than would be allowed by these limits, depending on the combination of appliance type and input, connector size and type and vent height.

Because the sizing tables in Section 504 are based on specific computer modeling of the installation configuration, the limitations or allowances in the single-appliance tables supersede what would be required in this section. For multiple-appliance systems, the sizing methods and tables do not contain the same limitations on connector length. This section sets a reasonable length limitation for connectors in multiple-appliance systems (100 percent of the height of the chimney or vent), which still provides liberal installation flexibility without permitting excessively long connectors. This section functions as an absolute cap for the connector length provisions of Sections 504.3.2 and 504.3.3.

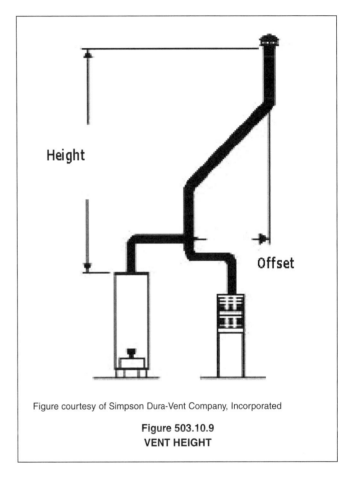

Figure courtesy of Simpson Dura-Vent Company, Incorporated

**Figure 503.10.9
VENT HEIGHT**

As discussed in the commentary to Section 503.10.6, the appliance and vent location must be well planned to allow the shortest horizontal run of connector piping. The code official must exercise good judgment in determining what is short as practical and close as practical because this section could require relocation of an appliance, vent or chimney to comply with the intent of the first sentence (see Sections 504.3.2 and 504.3.3).

This section refers to Section 503.10.3 as if the section contains an exception to the length limits. However, Section 503.10.3 is devoted to connector sizing only and does not address length. The intent of the reference is to allow connector length to be determined by design as part of an engineered venting system.

503.10.10 Support. A vent connector shall be supported for the design and weight of the material employed to maintain clearances and prevent physical damage and separation of joints.

❖ A connector must be supported for the design and weight of the material used. Proper support is necessary to maintain the clearances required by Section 503.10.5, to maintain the required slope and to prevent physical damage and separation of joints. The joints between connectors and appliance draft-hood outlets and flue collars must be secured with screws, rivets or other approved means. The joints between pipe sections also must be secured with screws, rivets or other approved means. Some connectors are available with a proprietary-type fastening method, which does not require screws, rivets or other fasteners if it provides equivalent resistance to disengagement or displacement. This section addresses connectors, not chimneys and vents, and most chimney and vent manufacturers do not require or might even prohibit the use of screws or rivets for securing joints for chimney and vent pipe sections and fittings.

503.10.11 Chimney connection. Where entering a flue in a masonry or metal chimney, the vent connector shall be installed above the extreme bottom to avoid stoppage. Where a thimble or slip joint is used to facilitate removal of the connector, the connector shall be firmly attached to or inserted into the thimble or slip joint to prevent the connector from falling out. Means shall be employed to prevent the connector from entering so far as to restrict the space between its end and the opposite wall of the chimney flue (see Section 501.9).

❖ All of the precautions taken to provide adequate chimney design may be ineffective if the connection between the appliance and the chimney is not properly accomplished. Improper connections can lead to appliance or connector failure and the leakage of vent gases into the building. This section contains requirements for the connection between a chimney and the appliance connector. The connection to a vent is properly accomplished by vent fittings and likewise for factory-built chimneys. Chimney connectors are required to pass through a masonry chimney wall to the inner face of the liner, but not beyond. A connector that extends into a chimney passageway can restrict the flow of vent gases and provide a ledge on which debris can accumulate. The joint between the connector and the chimney must be fastened in a manner that will prevent separation. If the connector enters a masonry chimney, it must be cemented in place with an approved material such as refractory mortar or other heat-resistant cement.

This section also permits the use of thimbles at a masonry chimney opening to provide for easy removal of the connector to facilitate cleaning. When a thimble is installed, it is to be permanently cemented in place with an approved high-temperature cement. A connector must be attached to a thimble in an approved manner to prevent displacement. A collar on the connector or a similar arrangement must be provided to limit the penetration of the connector into the chimney. Otherwise, it is quite possible that a connector will be shoved blindly into a chimney, reaching to the back wall of the chimney, and thus be obstructed.

503.10.12 Inspection. The entire length of a vent connector shall be provided with ready access for inspection, cleaning, and replacement.

❖ A connector cannot be concealed by any form of building construction, whether or not such construction is removable. Connectors must be inspected periodically for corrosion damage, structural integrity, joint separation and clearances to combustibles (see Section 503.10.14).

503.10.13 Fireplaces. A vent connector shall not be connected to a chimney flue serving a fireplace unless the fireplace flue opening is permanently sealed.

❖ A fireplace chimney cannot serve any appliance unless the fireplace is retired by permanently sealing off the chimney from the fireplace. If this is not done, the fireplace can affect draft for the connected appliance, and/or vent gases could escape into the building through the fireplace opening.

503.10.14 Passage through ceilings, floors, or walls. A vent connector shall not pass through any ceiling, floor or fire-resistance-rated wall . A single-wall metal pipe connector shall not pass through any interior wall.

Exception: Vent connectors made of listed Type B or Type L vent material and serving listed equipment with draft hoods and other equipment listed for use with Type B gas vents shall be permitted to pass through walls or partitions constructed of combustible material if the connectors are installed with not less than the listed clearance to combustible material.

❖ A single-wall connector has more than double the heat loss of a listed metal vent or chimney system. This produces high temperatures on the outside surface and cooling of the flue gases on the inside of the connector. The cooling of flue gases produces condensation, which can cause connector deterioration, while the excessive surface temperatures pose a potential fire hazard. For these reasons, a single-wall connector can be located only within the room or space in which the appliance is located. Connectors are prohibited from passing through any ceiling, floor or fire-resistance-rated wall. Connectors that pass through walls, floors or ceilings might not be readily observable, and a potential fire hazard or connector deterioration could go undetected. Also, pass-throughs increase the likelihood that the connector will be excessively long, exposed to low ambient temperatures or subject to contact with combustibles. Connectors that pass through a wall, floor or ceiling would be out of sight and therefore out of mind, meaning that they would not get the attention needed to avoid hazardous conditions.

The exception allows connectors made of Type B or L vent to pass through walls because these vents may be concealed within construction and pass-through assemblies. Clearances to vents must be observed, regardless of their use as vents or as connectors. Although the exception appears to allow connectors to penetrate only combustible walls, it is the intent to allow penetration of noncombustible walls also (see Section 503.10.15 for exterior wall penetrations).

503.10.15 Single-wall connector penetrations of combustible walls. A vent connector made of a single-wall metal pipe shall not pass through a combustible exterior wall unless guarded at the point of passage by a ventilated metal thimble not smaller than the following:

1. For listed appliances equipped with draft hoods and appliances listed for use with Type B gas vents, the thimble shall be not less than 4 inches (102 mm) larger in diameter than the vent connector. Where there is a run of not less than 6 feet (1829 mm) of vent connector in the open between the draft hood outlet and the thimble, the thimble shall be permitted to be not less than 2 inches (51 mm) larger in diameter than the vent connector.

2. For unlisted appliances having draft hoods, the thimble shall be not less than 6 inches (152 mm) larger in diameter than the vent connector.

3. For residential and low-heat appliances, the thimble shall be not less than 12 inches (305 mm) larger in diameter than the vent connector.

Exception: In lieu of thimble protection, all combustible material in the wall shall be removed from the vent connector a sufficient distance to provide the specified clearance from such vent connector to combustible material. Any material used to close up such opening shall be noncombustible.

❖ This section applies to exterior (outside) walls. Section 503.10.2.2 does not allow a single-wall connector to extend outdoors; thus, this section has no apparent application for vent connectors. The only application for the provisions of this section would be for single-wall vents under Section 503.7.6 (see also Sections 503.10.14, 503.7.2 and 503.7.6). Item 1 dictates the construction of thimbles used to protect the combustible wall. Ventilated thimbles have holes in the faces of the thimble to allow air to cool the thimble by convection. Item 2 would rarely apply because unlisted appliances are not allowed by this code unless specifically approved by the code official under Section 105.2. Item 3 would rarely apply because residential incinerators are believed to be extinct, and gas-fired appliances without draft hoods and appliances not listed for use with Type B vent are either factory/industrial types or Category II, III or IV. The exception requires that the combustible wall be made noncombustible at the point of penetration. The noncombustible portion of the wall must extend from the connector in all directions a distance equal to the required airspace clearance for the connector.

503.10.16 Medium-heat connectors. Vent connectors for medium-heat equipment shall not pass through walls or partitions constructed of combustible material.

❖ When the vent gas temperatures are 1000°F (538°C) and higher, penetration of combustible assemblies is an unacceptable risk.

503.11 Vent connectors for Category II, III, and IV appliances. Vent connectors for Category II, III and IV appliances shall be as specified for the venting systems in accordance with Section 503.4.

❖ See Sections 503.4, 503.4.1, and 503.4.2.

503.12 Draft hoods and draft controls. The installation of draft hoods and draft controls shall comply with Sections 503.12.1 through 503.12.7.

❖ See Sections 503.12.1 through 503.12.7.

503.12.1 Equipment requiring draft hoods. Vented equipment shall be installed with draft hoods.

Exception: Dual oven-type combination ranges, incinerators, direct-vent equipment, fan-assisted combustion system appliances, equipment requiring chimney draft for operation, single firebox boilers equipped with conversion burners with inputs greater than 400,000 Btu per hour (117 kw), equipment equipped with blast, power, or pressure burners that are not listed for use with draft hoods, and equipment designed for forced venting.

❖ Draft hoods are integral to or are supplied with natural draft atmospheric-burner gas-fired appliances (see commentary Figure 503.12.1). With the development of higher appliance efficiencies, fan-assisted combustion, direct-vent appliances and Category III and IV appliances, fewer draft-hood appliances will be used. Draft hoods are still common on conventional tank-type and tankless water heaters and 80 percent efficient boilers. A draft-hood-equipped appliance will allow a continuous flow of air into the venting system, which represents an energy loss except where an automatic vent damper is installed to close off the vent in the off cycle. Without an automatic vent damper, conditioned air is taken from the space in which the appliance is located and residual heat is taken away from the appliance itself, both of which contribute to energy loss. Not all appliances are categorized, but if they were, draft-hood-equipped appliances would fall under Category I. Draft hoods have multiple functions. They serve as a relief opening in the event of vent or chimney backdraft, they stabilize/regulate the amount of draft that occurs in the appliance combustion chamber and they allow the introduction of dilution air into the chimney or vent.

503.12.2 Installation. A draft hood supplied with or forming a part of listed vented equipment shall be installed without alteration, exactly as furnished and specified by the equipment manufacturer.

❖ Draft hoods are often shipped with appliances as a part that must be field installed. Installers have been known to shorten or otherwise modify the draft hoods to compensate for installation constraints and have also installed them in an orientation other than vertical.

Figure 503.12.1
DRAFT HOOD EQUIPPED APPLIANCE

503.12.2.1 Draft hood required. If a draft hood is not supplied by the equipment manufacturer where one is required, a draft hood shall be installed, shall be of a listed or approved type and, in the absence of other instructions, shall be of the same size as the equipment flue collar. Where a draft hood is required with a conversion burner, it shall be of a listed or approved type.

❖ In the unlikely event that an appliance manufacturer does not supply the required draft hood, a listed or approved draft hood must be obtained and installed.

503.12.2.2 Special design draft hood. Where it is determined that a draft hood of special design is needed or preferable for a particular installation, the installation shall be in accordance with the recommendations of the equipment manufacturer and shall be approved.

❖ There could be unusual circumstances, such as where a horizontal type draft hood is needed, because of low ceilings and/or insufficient connector rise.

503.12.3 Draft control devices. Where a draft control device is part of the equipment or is supplied by the equipment manufacturer, it shall be installed in accordance with the manufacturers' instructions. In the absence of manufacturers' instructions, the device shall be attached to the flue collar of the equipment or as near to the equipment as practical.

❖ Draft hoods also serve to control draft; however, this section would be redundant with Section 503.12.2 if it were not addressing draft controls such as barometric dampers. Two types of draft controls are installed in the appliance vent outlet or vent connector: the passive draft hood and the active barometric damper. Barometric dampers consist of a damper plate mounted on an axle that rotates in low friction bearing points such as needle or knife-edge bearings. The device is mounted in a tee fitting arrangement and connects with the appliance vent connector. It admits air in varying amounts into the vent connector to control the amount of draft through the appliance combustion chamber. These devices respond to varying chimney or vent draft caused by changes in indoor and outdoor temperatures, changes in atmospheric (barometric) pressure and changes in wind speed and direction. Barometric dampers are tuned by adjustable balancing weights attached to the damper plate. The damper manufacturer will provide instructions on how and where to install the device [see commentary Figure 202(2)].

503.12.4 Additional devices. Equipment (except incinerators) requiring controlled chimney draft shall be permitted to be equipped with a listed double-acting barometric-draft regulator installed and adjusted in accordance with the manufacturers' instructions.

❖ Like Section 503.12.3, this section addresses barometric dampers, specifically, the double-acting type. These dampers open inward to admit air into the vent system like all other barometric dampers, but can also open outward in the event of backdraft to allow vent gases to be spilled into the room or space in which the appliance is located, just as a draft hood would do. The manufacturer's installation instructions may require or recommend that a sensor (spill switch) be installed that will sense prolonged vent-gas spillage from the damper and shut down the appliance. In such cases, a manufacturer's recommendation should be interpreted as an enforceable requirement.

Some Category I fan-assisted furnaces can be fitted with a factory-supplied conversion part that is, in effect, a retrofit draft hood. This will lower the efficiency of the appliance by converting it to a draft-hood-equipped appliance that will now introduce dilution air (conditioned room air) into the vent. This is done solely to allow fan-assisted furnaces to connect to existing masonry chimneys that would otherwise not be allowed to serve such appliances. These conversion parts use a sensor (spill switch) in the relief/air inlet opening. The manufacturer of the appliance and the draft hood conversion kit will provide specific installation instructions that must be followed for the conversion.

Vent gas spillage must not occur from an incinerator venting system; therefore, a draft regulating device must be of the single-acting type.

503.12.5 Location. Draft hoods and barometric draft regulators shall be installed in the same room or enclosure as the equipment in such a manner as to prevent any difference in pressure between the hood or regulator and the combustion air supply.

❖ Draft-regulation devices are designed to maintain a constant draft through the appliance combustion chamber and flue passages for the purpose of maintaining proper combustion conditions and achieving the highest possible efficiency. If the draft control sees a different ambient pressure than the appliance, it could allow too much or too little draft through the appliance. The positive, negative or neutral pressures within a building must act equally on both the appliance and the draft control to allow the draft control to be adjusted and perform its intended function.

503.12.6 Positioning. Draft hoods and draft regulators shall be installed in the position for which they were designed with reference to the horizontal and vertical planes and shall be located so that the relief opening is not obstructed by any part of the equipment or adjacent construction. The equipment and its draft hood shall be located so that the relief opening is accessible for checking vent operation.

❖ Draft hoods and barometric dampers are both affected by their orientation. There are two basic types of draft hoods: vertical draft hoods and horizontal draft hoods. These types of draft hoods are not interchangeable. Barometric dampers can be adjusted to connect to either vertical or horizontal vent connectors; however, they are always positioned so that the closed damper plate is in a vertical plane. The manufacturer's instructions must always be followed. The relief opening on draft controls is also the dilution air inlet, and it must be unobstructed to allow air to enter under normal conditions, allow vent gases to spill in the event of backdraft and allow testing of the vent system for proper draft. For example, for many years, appliance venting/draft has been checked by observing the movement of smoke introduced near the relief/inlet opening.

503.12.7 Clearance. A draft hood shall be located so its relief opening is not less than 6 inches (152 mm) from any surface except that of the equipment it serves and the venting system to which the draft hood is connected. Where a greater or lesser clearance is indicated on the equipment label, the clearance shall be not less than that specified on the label. Such clearances shall not be reduced.

❖ The specified clearance is necessary to assure the unimpeded flow of air into the draft hood and flow of vent gas spillage from the hood. The clearance also protects materials from the hot gases that can spill from the draft hood and allows access for testing and observation.

503.13 Manually operated dampers. A manually operated damper shall not be placed in the vent connector for any equipment. Fixed baffles shall not be classified as manually operated dampers.

❖ Manual dampers are associated with solid-fuel appliances but not with gas-fired appliances. A manual damper requires manual operation by a human, which cannot be relied on because humans forget. If a damper is left closed or partially closed during appliance operation, a severe hazard could result from vent-gas spillage and/or appliance malfunction (see Section 503.14).

503.14 Automatically operated vent dampers. An automatically operated vent damper shall be of a listed type.

❖ Automatic vent dampers are intended for use with gas-fired natural-draft appliances. An automatic vent damper must be installed in strict compliance with the manufacturer's installation instructions. Because automatic vent-damper failure can result in a hazardous condition, automatic dampers must be listed and labeled. An automatic vent damper is installed on the draft hood outlet of an individual gas-fired appliance (see commentary Figure 503.14). These dampers must not serve more than one appliance. The manufacturer's installation instructions require the damper to be installed by a qualified installer in accordance with the terms of the listing and the manufacturer's instructions.

Because the purpose of a vent damper is to close or restrict the flue passageway of an appliance, it is imperative that the device be properly installed to minimize the possibility of failure. A malfunctioning or improperly installed vent-damper device can cause the appliance to malfunction and could cause the discharge of products of combustion directly into the building interior. Automatic vent dampers are energy-saving devices designed to close off or restrict an appliance flue passageway when the appliance is not operating and is in its "off" cycle. These devices save energy by trapping residual heat in a heat exchanger after the burners shut off and by preventing the escape of conditioned room air up the chimney or vent. Thus, appliance efficiency can be boosted, and building air infiltration can be reduced. Such devices can be field-installed additions to existing equipment and can be a factory-supplied component of an appliance (see commentary Figure 504.2.1).

A common application for vent dampers today is for midefficiency (80 to 83 percent) hot water boilers, which are draft-hood equipped. There are three types of automatic vent dampers: electrical-, mechanical- and thermal-actuating (see commentary Figure 503.14). Electrical- and mechanical-actuating automatic dampers are required to open automatically prior to main burner ignition and must remain open until the completion of the burner cycle. They must also be interlocked electrically with the appliance control circuitry to prevent the operation of the appliance in the event of damper failure. Thermal-actuating dampers open in response to flue gas temperature after burner ignition and close after burner shutdown. Thermal-actuated dampers use bimetal components that convert heat energy into mechanical energy. Some thermal-actuated damp-

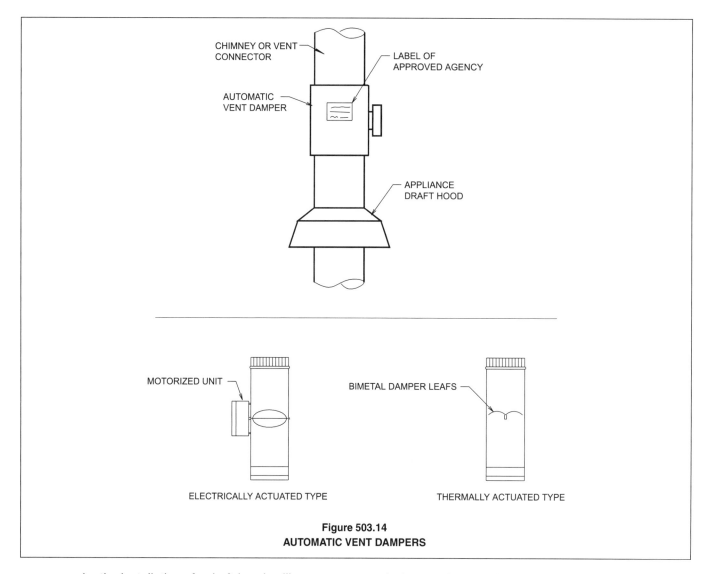

Figure 503.14
AUTOMATIC VENT DAMPERS

ers require the installation of a draft-hood spillage sensor that would shut off the appliance if the damper failed to open. The most common type of vent damper uses an electric motor to rotate a damper blade. The device uses switches that prove the damper position and allow the appliance to start only after the damper is opened and verified. The sequence of operation is the same as that described in the commentary to Section 503.3.3.

503.15 Obstructions. Devices that retard the flow of vent gases shall not be installed in a vent connector, chimney, or vent. The following shall not be considered as obstructions:

1. Draft regulators and safety controls specifically listed for installation in venting systems and installed in accordance with the terms of their listing.
2. Approved draft regulators and safety controls that are designed and installed in accordance with approved engineering methods.
3. Listed heat reclaimers and automatically operated vent dampers installed in accordance with the terms of their listing.
4. Approved economizers, heat reclaimers, and recuperators installed in venting systems of equipment not required to be equipped with draft hoods, provided that the gas utilization equipment manufacturer's instructions cover the installation of such a device in the venting system and performance in accordance with Sections 503.3 and 503.3.1 is obtained.
5. Vent dampers serving listed appliances installed in accordance with Sections 504.2.1 and 504.3.1 or other approved engineering methods.

❖ Manual dampers, flow restricting orifices and similar devices are not accounted for in the vent design methods of this code. The tables in Section 504 assume that no such devices are installed in the vent. Item 3 addresses the automatic vent dampers covered in Section 503.14. These dampers open fully before the appliance is allowed to fire and offer negligible flow resistance because the damper plate is parallel to the vent gas flow. Item 4 addresses devices that extract heat from the vent gases to save energy that would otherwise be lost to the outside atmosphere. Appliances without draft

CHIMNEYS AND VENTS

hoods, excluding Category I fan-assisted, Category II, III and IV appliances, usually have higher vent gas temperatures than those with draft hoods; therefore, waste heat can be reclaimed with less chance of diminishing draft (see Section 504.2.1).

SECTION 504 (IFGS)
SIZING OF CATEGORY I APPLIANCE VENTING SYSTEMS

504.1 Definitions. The following definitions apply to the tables in this section.

❖ The definitions in this section are necessary for application of the vent sizing tables and are exclusively related to Section 504. As stated in the main section title, this section applies to Category I appliances. See the definition of "Vented Appliance Categories." However, it also applies to draft-hood-equipped appliances, which may or may not be categorized at all, even though they could be Category I because they meet the definition of Category I appliances. For example, a draft-hood water heater need not be categorized by its governing standard, but if it were categorized it would be a Category I appliance. A gas-fired appliance with a draft hood depends on gravity flow for supply of combustion air, for flow of combustion products through the heat exchanger and for proper gas venting. The draft hood is designed to maintain proper combustion in the event of draft fluctuations. Its relief opening serves as a flue product exit in the event of a blocked vent or downdraft. The relief opening, however, also allows dilution air to enter the vent during normal appliance/vent operation. When this dilution air is obtained from within the heated space, there is a loss of seasonal efficiency, which can be reduced by installation of a vent damper designed to reduce the flow of air through the vent when the burner is not firing.

To improve annual fuel utilization efficiency (AFUE), many appliance manufacturers have designed fan-assisted combustion using mechanical means (blowers or fans) to obtain either induced or forced flow of combustion air and combustion products. Greater efficiency results from three major effects. First, heat exchange improves because of higher internal flow velocity, which enhances heat transfer by creating turbulent flow and the "scrubbing" of heat exchanger surfaces. Second, induced draft through the appliance combustion chambers and heat exchangers allows the heat exchangers to be constructed with longer passes, which results in greater surface area and thus longer flue gas retention time. This allows extraction of more heat from the combustion gases. Third, there is no longer a draft hood to cause heated air loss up the vent, both when the appliance is operating and when it is not [see commentary Figures 504.1(1) through 504.1(4)].

A fan-assisted combustion appliance uses pressure-sensing and heat-sensing controls to monitor proper venting. The controls must prove that the appliance flue outlet is at neutral or negative pressure and that adequate flow of flue gases exists. Without a draft hood and dilution air to dilute the combustion products, gases entering the vent will have a higher water vapor content than those from a draft-hood appliance. [see commentary Figure 504.1(1)]. There will likely be a longer period of condensation in the vent (wet time), particularly in the upper and/or colder portions of the venting

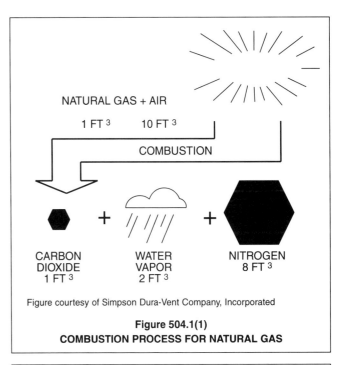

Figure courtesy of Simpson Dura-Vent Company, Incorporated

Figure 504.1(1)
COMBUSTION PROCESS FOR NATURAL GAS

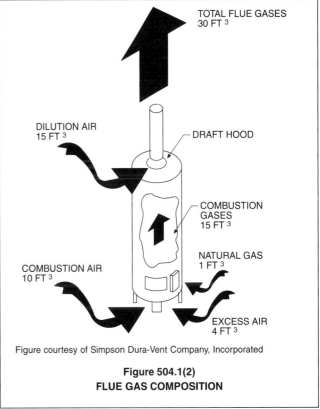

Figure courtesy of Simpson Dura-Vent Company, Incorporated

Figure 504.1(2)
FLUE GAS COMPOSITION

system. These differences between fan-assisted and draft-hood appliances are reflected in the tables by the lack of minimum capacities for draft hood appliances. The minimums for fan-assisted appliances are based on the heat input needed to control the duration of wet time or condensation during appliance start-up and operation.

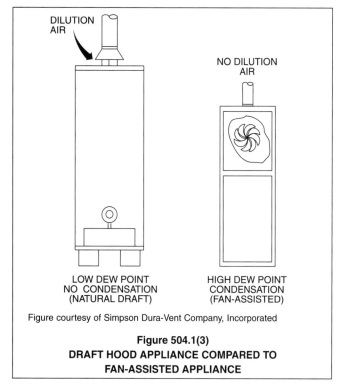

Figure 504.1(3)
DRAFT HOOD APPLIANCE COMPARED TO FAN-ASSISTED APPLIANCE

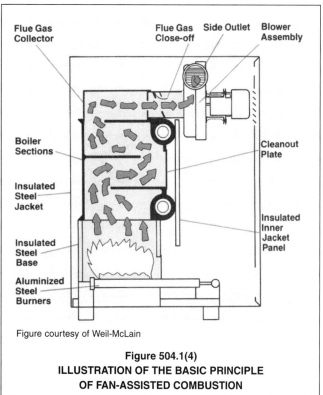

Figure 504.1(4)
ILLUSTRATION OF THE BASIC PRINCIPLE OF FAN-ASSISTED COMBUSTION

The tables do not apply to:

1. Wall furnaces (recessed heaters), which require Type BW vents.
2. Decorative gas appliances, which generally require a specific vent size and are best individually vented.
3. Category II, III or IV gas appliances. If permitted by the appliance manufacturer, some Category III appliances can also be vented in accordance with Category I conditions. Specifically, if the vent for a Category III appliance is sized and configured (connector rise, total height, etc.) for the input, the vent will operate under nonpositive pressure. This allows the use of a Type B gas vent. For a Category III appliance to operate as a Category I, appliance manufacturers' instructions must be followed, and the vent size may need to be larger than the appliance flue collar.
4. Gas-fired appliances listed for use only with chimneys or dual-fuel appliances such as oil/gas, wood/gas, or coal/gas. Dual-fuel appliances require chimneys sized in accordance with the appliance manufacturer's instructions and in accordance with the chimney manufacturer's instructions.

Capacities in the tables assume cold starts such as with an intermittent ignition system (no standing pilot). The type of ignition system does not affect maximum capacities, but it does affect minimums. A standing pilot on an appliance will keep vent gas temperatures slightly above outdoor ambient temperature, whereas a tank of hot water plus water-heater pilot operation will maintain flue gas temperature ahead of the draft hood at approximately the same temperature as the stored water. This flow of heat into a vent aids in priming as well as reducing wet time. In some cases, the fan-assisted appliance manufacturer and/or the sizing tables in the code will require connection of a draft-hood-type appliance to the venting system because the relatively inefficient draft-hood appliance contributes heat to the venting system necessary to make it function.

The tables also assume that there are no adverse or building depressurization effects. A strong wind or mechanical ventilation/exhaust might cause a downdraft with draft-hood appliances that will prevent the vent from priming properly.

The input capacity values in the tables are computed for typical natural gas. They can be used for LP gases, such as propane and butane, and mixtures of these with air. With LP gases, the maximum capacity remains the same, but because these have less hydrogen and produce less water vapor, the possibility of condensation is somewhat lower. To simplify matters, assume that minimum capacities are the same regardless of fuel type.

Oversized or excessively long vents can cause condensation and should be avoided. The tables define limits that minimize wet time in the indoor portions as well as in the upper exposed end of the vent.

The maximum vent lateral lengths for FAN appliances are calculated on the basis of several assumed conditions, including flue gas temperature and composition, and an outdoor ambient temperature of 42°F (6°C), chosen as a representative value. Colder assumed temperatures or greater outdoor exposure of the vent would lead to shorter maximum allowable lengths or to greater possibilities of condensation. To minimize condensation, it is essential to operate closer to maximum than minimum capacity and also to use the smallest allowable vent size.

APPLIANCE CATEGORIZED VENT DIAMETER/AREA. The minimum vent area/diameter permissible for Category I appliances to maintain a nonpositive vent static pressure when tested in accordance with nationally recognized standards.

❖ The minimum Category I appliance vent area/diameter necessary to maintain a nonpositive vent static pressure when tested in accordance with the applicable standard. The appliance categorized vent diameter/area is determined by the appliance manufacturer and is typically the vent outlet collar installed on the appliance.

FAN-ASSISTED COMBUSTION SYSTEM. An appliance equipped with an integral mechanical means to either draw or force products of combustion through the combustion chamber or heat exchanger.

❖ An appliance equipped with an integral mechanical means to either draw or force products of combustion through the combustion chamber or heat exchanger [see commentary Figures 504.1(4) and (5)].

FAN Min. The minimum input rating of a Category I fan-assisted appliance attached to a vent or connector.

❖ The minimum appliance input rating of a Category I appliance with a fan-assisted combustion system that is capable of being attached to a vent.

FAN Max. The maximum input rating of a Category I fan-assisted appliance attached to a vent or connector.

❖ The maximum appliance input rating of a Category I appliance with a fan-assisted combustion system that is capable of being attached to a vent.

NAT Max. The maximum input rating of a Category I draft-hood-equipped appliance attached to a vent or connector.

❖ The maximum input rating of a Category I appliance equipped with a draft hood that is capable of being attached to a vent. There are no minimum appliance input ratings for draft-hood-equipped appliances.

FAN + FAN. The maximum combined appliance input rating of two or more Category I fan-assisted appliances attached to the common vent.

❖ The maximum combined input rating of two or more fan-assisted appliances attached to a common vent.

Photo courtesy of Weil-McLain

Figure 504.1(5)
FAN-ASSISTED HOT WATER BOILER

FAN + NAT. The maximum combined appliance input rating of one or more Category I fan-assisted appliances and one or more Category I draft-hood-equipped appliances attached to the common vent.

❖ The maximum combined input rating of one or more fan-assisted appliances and one or more draft-hood-equipped appliances attached to a common vent.

NA. Vent configuration is not allowed due to potential for condensate formation or pressurization of the venting system, or not applicable due to physical or geometric restraints.

❖ Vent configuration is prohibited.

NAT + NAT. The maximum combined appliance input rating of two or more Category I draft-hood-equipped appliances attached to the common vent.

❖ The maximum combined input rating of two or more draft-hood-equipped appliances attached to a common vent.

504.2 Application of single-appliance vent Tables 504.2(1) through 504.2(5). The application of Tables 504.2(1) through 504.2(5) shall be subject to the requirements of Sections 504.2.1 through 504.2.15.

❖ Note that all of the table titles have been reformatted to make them easier to use and to lessen the possibility for error in choosing the appropriate table. The tables do not apply to Type B-W vents; vents for decorative gas appliances; vents for Category II, III or IV appliances; vents for dual-fuel appliances; vents for appliances not listed for use with Type B vent and appliances listed only for connection to chimneys. Category I appliances can be a fan-assisted design or can be a draft-hood-equipped design, and each design has different vent system design considerations. A noteworthy difference between fan-assisted and draft-hood-equipped designs is that the vents for fan-assisted appliances have both a maximum and a minimum vent capacity. In the past, lower-efficiency draft-hood-type appliances were dominant, and vent sizing was based only on the maximum venting capacity of the vent. In other words, the only concern was to make sure that the vent was big enough. This is no longer the case with the advent of mid-efficiency fan-assisted-type appliances. Today, it is possible for a vent to be either too small or too large. An undersized vent will have insufficient capacity and can allow flue gas spillage or cause positive pressure to occur within the vent. An oversized vent can fail to produce sufficient draft and can be subject to the continuous formation of water vapor condensation on the interior vent surfaces. A typical individual vent is shown in Commentary Figure 504.2(1) for a FAN furnace with an input rating of 150,000 Btu/h (44 kW) and a 5-inch (127mm) outlet connection.

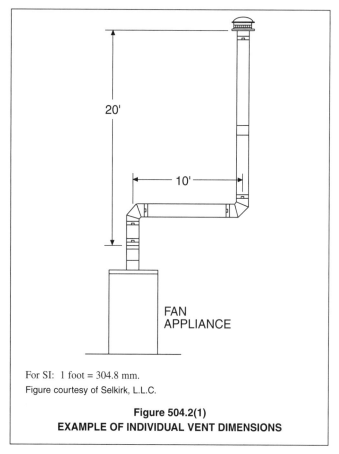

For SI: 1 foot = 304.8 mm.
Figure courtesy of Selkirk, L.L.C.

Figure 504.2(1)
EXAMPLE OF INDIVIDUAL VENT DIMENSIONS

Procedure for a Type B vent, Table 503.2(1): Go down the height column to 20 feet (6096 mm) and across on the 10-foot (3048 mm) lateral line.

For this furnace, the MAX capacity under the 5-inch (127mm) size heading is 229,000 Btu/h (67.1 kW) and the MIN capacity is 50,000 Btu/h (14.7 kW). A 5-inch (127 mm) size Type B vent will be correct. See Commentary Figure 504.2(1). In applying the tables, extreme care must be taken to observe the title of the table and to apply the conditions and instructions set forth in Sections 504.2.1 through 504.2.15.

To determine the proper size for an individual vent, apply the table as follows:

1. Determine the total vent height and length of lateral, based on the appliance and vent location and the height to the top of the vent, as indicated in Commentary Figure 504.2(1). If gas appliances, such as a furnace, boiler or water heater, have not been chosen or installed, estimate the height beginning at 6 feet (1829 mm) above the floor. For attic or horizontal furnaces, floor furnaces, room heaters and small boilers, the height location of the draft hood outlet or vent collar should be known.

2. Read down the height column to a height equal to or less than the estimated total height.

3. Select the horizontal row for the appropriate lateral (L) length.

4. Read across to the first column under the type of appliance (FAN or NAT) that shows an appliance input rating equal to or greater than the name plate sea level input rating of the appliance to be vented.

Commentary Figure 504.2(2) illustrates the measurement of height and laterals for draft hood appliances.

Some vent manufacturers include special requirements regarding certain types of appliances in their installation/design instructions. The following example is taken from the 1992 *Chimney and Gas Vent Sizing Handbook* by Selkirk Metalbestos:

1. Room Heaters, Floor Furnaces, Wall Furnaces: If the appliance has a draft hood, assume an adjusted input 40 percent greater than the name plate value and design the vent for this increased input.

   Example: The vent for a 50,000 Btu/h room heater will be 10 feet high with a 2-foot lateral. Adjusted input is 1.4 x 50,000 = 70,000. The NAT column in the tables shows that a 4-inch vent with 81,000 Btu/h capacity will be adequate. If the appliance has a fan-assisted burner, use the FAN column without adjustment.

2. Central Heating Boilers, Gravity and Forced-Air Furnaces, Duct Furnaces and Unit Heaters, Conversion Burners with Draft Hoods: Use full tabulated capacities, depending on the type of appliance (NAT or FAN).

3. Cooking Equipment–Domestic, Restaurant, and Commercial: Use the outlet size or five times the input to determine the individual vent size.
4. Decorative Appliances: Use the outlet size or five times the input to determine the individual vent size. Appendix B contains vent sizing examples.

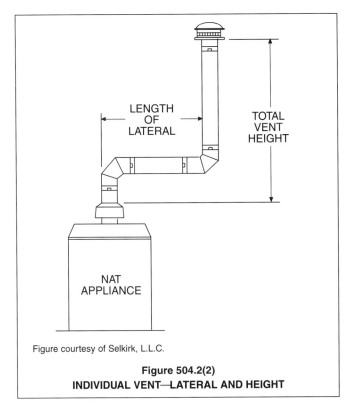

Figure courtesy of Selkirk, L.L.C.

Figure 504.2(2)
INDIVIDUAL VENT—LATERAL AND HEIGHT

504.2.1 Vent obstructions. These venting tables shall not be used where obstructions, as described in Section 503.15, are installed in the venting system. The installation of vents serving listed appliances with vent dampers shall be in accordance with the appliance manufacturer's instructions or in accordance with the following:

1. The maximum capacity of the vent system shall be determined using the "NAT Max" column.

2. The minimum capacity shall be determined as if the appliance were a fan-assisted appliance, using the "FAN Min" column to determine the minimum capacity of the vent system. Where the corresponding "FAN Min" is "NA," the vent configuration shall not be permitted and an alternative venting configuration shall be utilized.

❖ An engineered-vent sizing method or appliance and vent/chimney manufacturer's instructions must be used to size vents where devices (obstructions) are installed in a venting system. The tables do not account for the flow resistance caused by such devices, nor do they account for the reduction of flue gas temperatures resulting from heat reclaim exchangers. Section 503.15 lists several items that are not to be viewed as obstructions, including automatic vent dampers. This section recognizes the use of the vent sizing tables with appliances equipped with automatic vent-damper devices. By referring to the appliance manufacturer's installation instructions, this section implies that the appliances are factory equipped with automatic vent dampers. Item 1 recognizes that during the firing cycle, an appliance with an automatic vent damper affects a venting system no differently than a draft-hood appliance without one. Item 2 addresses the effect that appliances equipped with automatic vent dampers have on venting systems during the appliance "off" cycle. The appliance acts as if it were an idle fan-assisted appliance because a closed vent damper will allow little or no air to enter the vent system, just as an idle fan-assisted appliance allows no air to enter the vent system. The lack of dilution air entering the venting system is a primary reason that the minimum input columns were created in the vent sizing tables. Boilers and water heaters with draft hoods commonly use automatic vent dampers to boost the efficiency of the appliance by trapping residual heat of the firing cycle, reducing standby losses and reducing the amount of conditioned air that escapes up the vent through the draft hood (see commentary Figure 504.2.1).

Photo courtesy of Weil-McLain

Figure 504.2.1
CATEGORY I NATURAL DRAFT HOT WATER BOILER WITH AUTOMATIC VENT DAMPER

TABLE 504.2(1)
TYPE B DOUBLE-WALL GAS VENT

Number of Appliances	Single
Appliance Type	Category I
Appliance Vent Connection	Connected directly to vent

HEIGHT (H) (feet)	LATERAL (L) (feet)	VENT DIAMETER—(D) inches																			
		3			4			5			6			7			8			9		
		FAN		NAT	FAN		NAT	FAN		NAT	FAN		NAT	FAN		NAT	FAN		NAT	FAN		NAT
		Min	Max	Max	Min	Max	Max	Min	Max	Max	Min	Max	Max	Min	Max	Max	Min	Max	Max	Min	Max	Max
6	0	0	78	46	0	152	86	0	251	141	0	375	205	0	524	285	0	698	370	0	897	470
	2	13	51	36	18	97	67	27	157	105	32	232	157	44	321	217	53	425	285	63	543	370
	4	21	49	34	30	94	64	39	153	103	50	227	153	66	316	211	79	419	279	93	536	362
	6	25	46	32	36	91	61	47	149	100	59	223	149	78	310	205	93	413	273	110	530	354
8	0	0	84	50	0	165	94	0	276	155	0	415	235	0	583	320	0	780	415	0	1,006	537
	2	12	57	40	16	109	75	25	178	120	28	263	180	42	365	247	50	483	322	60	619	418
	5	23	53	38	32	103	71	42	171	115	53	255	173	70	356	237	83	473	313	99	607	407
	8	28	49	35	39	98	66	51	164	109	64	247	165	84	347	227	99	463	303	117	596	396
10	0	0	88	53	0	175	100	0	295	166	0	447	255	0	631	345	0	847	450	0	1,096	585
	2	12	61	42	17	118	81	23	194	129	26	289	195	40	402	273	48	533	355	57	684	457
	5	23	57	40	32	113	77	41	187	124	52	280	188	68	392	263	81	522	346	95	671	446
	10	30	51	36	41	104	70	54	176	115	67	267	175	88	376	245	104	504	330	122	651	427
15	0	0	94	58	0	191	112	0	327	187	0	502	285	0	716	390	0	970	525	0	1,263	682
	2	11	69	48	15	136	93	20	226	150	22	339	225	38	475	316	45	633	414	53	815	544
	5	22	65	45	30	130	87	39	219	142	49	330	217	64	463	300	76	620	403	90	800	529
	10	29	59	41	40	121	82	51	206	135	64	315	208	84	445	288	99	600	386	116	777	507
	15	35	53	37	48	112	76	61	195	128	76	301	198	98	429	275	115	580	373	134	755	491
20	0	0	97	61	0	202	119	0	349	202	0	540	307	0	776	430	0	1,057	575	0	1,384	752
	2	10	75	51	14	149	100	18	250	166	20	377	249	33	531	346	41	711	470	50	917	612
	5	21	71	48	29	143	96	38	242	160	47	367	241	62	519	337	73	697	460	86	902	599
	10	28	64	44	38	133	89	50	229	150	62	351	228	81	499	321	95	675	443	112	877	576
	15	34	58	40	46	124	84	59	217	142	73	337	217	94	481	308	111	654	427	129	853	557
	20	48	52	35	55	116	78	69	206	134	84	322	206	107	464	295	125	634	410	145	830	537

(continued)

CHIMNEYS AND VENTS

TABLE 504.2(1)—continued
TYPE B DOUBLE-WALL GAS VENT

Number of Appliances	Single
Appliance Type	Category I
Appliance Vent Connection	Connected directly to vent

HEIGHT (H) (feet)	LATERAL (L) (feet)	VENT DIAMETER—(D) inches																	
		3			4			5			6			7		8		9	
		FAN		NAT	FAN		NAT	FAN		NAT	FAN		NAT	FAN	NAT	FAN	NAT	FAN	NAT
		Min	Max	Max	Min	Max	Max	Min	Max	Max	Min	Max	Max	Min	Max	Max	Min	Max	Max	Min	Max	Max
30	0	0	100	64	0	213	128	0	374	220	0	587	336	0	853	475	0	1,173	650	0	1,548	855
	2	9	81	56	13	166	112	14	283	185	18	432	280	27	613	394	33	826	535	42	1,072	700
	5	21	77	54	28	160	108	36	275	176	45	421	273	58	600	385	69	811	524	82	1,055	688
	10	27	70	50	37	150	102	48	262	171	59	405	261	77	580	371	91	788	507	107	1,028	668
	15	33	64	NA	44	141	96	57	249	163	70	389	249	90	560	357	105	765	490	124	1,002	648
	20	56	58	NA	53	132	90	66	237	154	80	374	237	102	542	343	119	743	473	139	977	628
	30	NA	NA	NA	73	113	NA	88	214	NA	104	346	219	131	507	321	149	702	444	171	929	594
50	0	0	101	67	0	216	134	0	397	232	0	633	363	0	932	518	0	1,297	708	0	1,730	952
	2	8	86	61	11	183	122	14	320	206	15	497	314	22	715	445	26	975	615	33	1,276	813
	5	20	82	NA	27	177	119	35	312	200	43	487	308	55	702	438	65	960	605	77	1,259	798
	10	26	76	NA	35	168	114	45	299	190	56	471	298	73	681	426	86	935	589	101	1,230	773
	15	59	70	NA	42	158	NA	54	287	180	66	455	288	85	662	413	100	911	572	117	1,203	747
	20	NA	NA	NA	50	149	NA	63	275	169	76	440	278	97	642	401	113	888	556	131	1,176	722
	30	NA	NA	NA	69	131	NA	84	250	NA	99	410	259	123	605	376	141	844	522	161	1,125	670
100	0	NA	NA	NA	0	218	NA	0	407	NA	0	665	400	0	997	560	0	1,411	770	0	1,908	1,040
	2	NA	NA	NA	10	194	NA	12	354	NA	13	566	375	18	831	510	21	1,155	700	25	1,536	935
	5	NA	NA	NA	26	189	NA	33	347	NA	40	557	369	52	820	504	60	1,141	692	71	1,519	926
	10	NA	NA	NA	33	182	NA	43	335	NA	53	542	361	68	801	493	80	1,118	679	94	1,492	910
	15	NA	NA	NA	40	174	NA	50	321	NA	62	528	353	80	782	482	93	1,095	666	109	1,465	895
	20	NA	NA	NA	47	166	NA	59	311	NA	71	513	344	90	763	471	105	1,073	653	122	1,438	880
	30	NA	NA	NA	NA	NA	NA	78	290	NA	92	483	NA	115	726	449	131	1,029	627	149	1,387	849
	50	NA	NA	NA	NA	NA	NA	NA	NA	NA	147	428	NA	180	651	405	197	944	575	217	1,288	787

(continued)

TABLE 504.2(1)—continued
TYPE B DOUBLE-WALL GAS VENT

													Number of Appliances	Single		
													Appliance Type	Category I		
													Appliance Vent Connection	Connected directly to vent		

		VENT DIAMETER—(D) inches																			
		10			12			14			16			18			20			22			24		
		FAN		NAT	FAN		NAT	FAN		NAT	FAN		NAT	FAN		NAT	FAN		NAT	FAN		NAT	FAN		NAT
HEIGHT (H) (feet)	LATERAL (L) (feet)	Min	Max	Max	Min	Max	Max	Min	Max	Max	Min	Max	Max	Min	Max	Max	Min	Max	Max	Min	Max	Max	Min	Max	Max

APPLIANCE INPUT RATING IN THOUSANDS OF BTU/H

HEIGHT (H) (feet)	LATERAL (L) (feet)	Min	Max	Max	Min	Max	Max	Min	Max	Max	Min	Max	Max	Min	Max	Max	Min	Max	Max	Min	Max	Max	Min	Max	Max
6	0	0	1,121	570	0	1,645	850	0	2,267	1,170	0	2,983	1,530	0	3,802	1,960	0	4,721	2,430	0	5,737	2,950	0	6,853	3,520
	2	75	675	455	103	982	650	138	1,346	890	178	1,769	1,170	225	2,250	1,480	296	2,782	1,850	360	3,377	2,220	426	4,030	2,670
	4	110	668	445	147	975	640	191	1,338	880	242	1,761	1,160	300	2,242	1,475	390	2,774	1,835	469	3,370	2,215	555	4,023	2,660
	6	128	661	435	171	967	630	219	1,330	870	276	1,753	1,150	341	2,235	1,470	437	2,767	1,820	523	3,363	2,210	618	4,017	2,650
8	0	0	1,261	660	0	1,858	970	0	2,571	1,320	0	3,399	1,740	0	4,333	2,220	0	5,387	2,750	0	6,555	3,360	0	7,838	4,010
	2	71	770	515	98	1,124	745	130	1,543	1,020	168	2,030	1,340	212	2,584	1,700	278	3,196	2,110	336	3,882	2,560	401	4,634	3,050
	5	115	758	503	154	1,110	733	199	1,528	1,010	251	2,013	1,330	311	2,563	1,685	398	3,180	2,090	476	3,863	2,545	562	4,612	3,040
	8	137	746	490	180	1,097	720	231	1,514	1,000	289	2,000	1,320	354	2,552	1,670	450	3,163	2,070	537	3,850	2,530	630	4,602	3,030
10	0	0	1,377	720	0	2,036	1,060	0	2,825	1,450	0	3,742	1,925	0	4,782	2,450	0	5,955	3,050	0	7,254	3,710	0	8,682	4,450
	2	68	852	560	93	1,244	850	124	1,713	1,130	161	2,256	1,480	202	2,868	1,890	264	3,556	2,340	319	4,322	2,840	378	5,153	3,390
	5	112	839	547	149	1,229	829	192	1,696	1,105	243	2,238	1,461	300	2,849	1,871	382	3,536	2,318	458	4,301	2,818	540	5,132	3,371
	10	142	817	525	187	1,204	795	238	1,669	1,080	298	2,209	1,430	364	2,818	1,840	459	3,504	2,280	546	4,268	2,780	641	5,099	3,340
15	0	0	1,596	840	0	2,380	1,240	0	3,323	1,720	0	4,423	2,270	0	5,678	2,900	0	7,099	3,620	0	8,665	4,410	0	10,393	5,300
	2	63	1,019	675	86	1,495	985	114	2,062	1,350	147	2,719	1,770	186	3,467	2,260	239	4,304	2,800	290	5,232	3,410	346	6,251	4,080
	5	105	1,003	660	140	1,476	967	182	2,041	1,327	229	2,696	1,748	283	3,442	2,235	355	4,278	2,777	426	5,204	3,385	501	6,222	4,057
	10	135	977	635	177	1,446	936	227	2,009	1,289	283	2,659	1,712	346	3,402	2,193	432	4,234	2,739	510	5,159	3,343	599	6,175	4,019
	15	155	953	610	202	1,418	905	257	1,976	1,250	318	2,623	1,675	385	3,363	2,150	479	4,192	2,700	564	5,115	3,300	665	6,129	3,980
20	0	0	1,756	930	0	2,637	1,350	0	3,701	1,900	0	4,948	2,520	0	6,376	3,250	0	7,988	4,060	0	9,785	4,980	0	11,753	6,000
	2	59	1,150	755	81	1,694	1,100	107	2,343	1,520	139	3,097	2,000	175	3,955	2,570	220	4,916	3,200	269	5,983	3,910	321	7,154	4,700
	5	101	1,133	738	135	1,674	1,079	174	2,320	1,498	219	3,071	1,978	270	3,926	2,544	337	4,885	3,174	403	5,950	3,880	475	7,119	4,662
	10	130	1,105	710	172	1,641	1,045	220	2,282	1,460	273	3,029	1,940	334	3,880	2,500	413	4,835	3,130	489	5,896	3,830	573	7,063	4,600
	15	150	1,078	688	195	1,609	1,018	248	2,245	1,425	306	2,988	1,910	372	3,835	2,465	459	4,786	3,090	541	5,844	3,795	631	7,007	4,575
	20	167	1,052	665	217	1,578	990	273	2,210	1,390	335	2,948	1,880	404	3,791	2,430	495	4,737	3,050	585	5,792	3,760	689	6,953	4,550

(continued)

CHIMNEYS AND VENTS

TABLE 504.2(1)—continued
TYPE B DOUBLE-WALL GAS VENT

Number of Appliances	Single
Appliance Type	Category I
Appliance Vent Connection	Connected directly to vent

HEIGHT (H) (feet)	LATERAL (L) (feet)	VENT DIAMETER—(D) inches																					
		10			12			14			16			18			20			22			24		
		FAN		NAT	FAN		NAT	FAN		NAT	FAN		NAT	FAN		NAT	FAN		NAT	FAN		NAT	FAN		NAT
		Min	Max	Max	Min	Max	Max	Min	Max	Max	Min	Max	Max	Min	Max	Max	Min	Max	Max	Min	Max	Max	Min	Max	Max
		APPLIANCE INPUT RATING IN THOUSANDS OF BTU/H																					
30	0	0	1,977	1,060	0	3,004	1,550	0	4,252	2,170	0	5,725	2,920	0	7,420	3,770	0	9,341	4,750	0	11,483	5,850	0	13,848	7,060
	2	54	1,351	865	74	2,004	1,310	98	2,786	1,800	127	3,696	2,380	159	4,734	3,050	199	5,900	3,810	241	7,194	4,650	285	8,617	5,600
	5	96	1,332	851	127	1,981	1,289	164	2,759	1,775	206	3,666	2,350	252	4,701	3,020	312	5,863	3,783	373	7,155	4,622	439	8,574	5,552
	10	125	1,301	829	164	1,944	1,254	209	2,716	1,733	259	3,617	2,300	316	4,647	2,970	386	5,803	3,739	456	7,090	4,574	535	8,505	5,471
	15	143	1,272	807	187	1,908	1,220	237	2,674	1,692	292	3,570	2,250	354	4,594	2,920	431	5,744	3,695	507	7,026	4,527	590	8,437	5,391
	20	160	1,243	784	207	1,873	1,185	260	2,633	1,650	319	3,523	2,200	384	4,542	2,870	467	5,686	3,650	548	6,964	4,480	639	8,370	5,310
	30	195	1,189	745	246	1,807	1,130	305	2,555	1,585	369	3,433	2,130	440	4,442	2,785	540	5,574	3,565	635	6,842	4,375	739	8,239	5,225
50	0	0	2,231	1,195	0	3,441	1,825	0	4,934	2,550	0	6,711	3,440	0	8,774	4,460	0	11,129	5,635	0	13,767	6,940	0	16,694	8,430
	2	41	1,620	1,010	66	2,431	1,513	86	3,409	2,125	113	4,554	2,840	141	5,864	3,670	171	7,339	4,630	209	8,980	5,695	251	10,788	6,860
	5	90	1,600	996	118	2,406	1,495	151	3,380	2,102	191	4,520	2,813	234	5,826	3,639	283	7,295	4,597	336	8,933	5,654	394	10,737	6,818
	10	118	1,567	972	154	2,366	1,466	196	3,332	2,064	243	4,464	2,767	295	5,763	3,585	355	7,224	4,542	419	8,855	5,585	491	10,652	6,749
	15	136	1,536	948	177	2,327	1,437	222	3,285	2,026	274	4,409	2,721	330	5,701	3,534	396	7,155	4,511	465	8,779	5,546	542	10,570	6,710
	20	151	1,505	924	195	2,288	1,408	244	3,239	1,987	300	4,356	2,675	361	5,641	3,481	433	7,086	4,479	506	8,704	5,506	586	10,488	6,670
	30	183	1,446	876	232	2,214	1,349	287	3,150	1,910	347	4,253	2,631	412	5,523	3,431	494	6,953	4,421	577	8,557	5,444	672	10,328	6,603
100	0	0	2,491	1,310	0	3,925	2,050	0	5,729	2,950	0	7,914	4,050	0	10,485	5,300	0	13,454	6,700	0	16,817	8,600	0	20,578	10,300
	2	30	1,975	1,170	44	3,027	1,820	72	4,313	2,550	95	5,834	3,500	120	7,591	4,600	138	9,577	5,800	169	11,803	7,200	204	14,264	8,800
	5	82	1,955	1,159	107	3,002	1,803	136	4,282	2,531	172	5,797	3,475	208	7,548	4,566	245	9,528	5,769	293	11,748	7,162	341	14,204	8,756
	10	108	1,923	1,142	142	2,961	1,775	180	4,231	2,500	223	5,737	3,434	268	7,478	4,509	318	9,447	5,717	374	11,658	7,100	436	14,105	8,683
	15	126	1,892	1,124	163	2,920	1,747	206	4,182	2,469	252	5,678	3,392	304	7,409	4,451	358	9,367	5,665	418	11,569	7,037	487	14,007	8,610
	20	141	1,861	1,107	181	2,880	1,719	226	4,133	2,438	277	5,619	3,351	330	7,341	4,394	387	9,289	5,613	452	11,482	6,975	523	13,910	8,537
	30	170	1,802	1,071	215	2,803	1,663	265	4,037	2,375	319	5,505	3,267	378	7,209	4,279	446	9,136	5,509	514	11,310	6,850	592	13,720	8,391
	50	241	1,688	1,000	292	2,657	1,550	350	3,856	2,250	415	5,289	3,100	486	6,956	4,050	572	8,841	5,300	659	10,979	6,600	752	13,354	8,100

For SI: 1 inch = 25.4 mm, 1 foot = 304.8 mm, 1 British thermal unit per hour = 0.2931 W.

❖ For Tables 504.2(1) through 504.2(5), see the commentary for Section 504.2.

TABLE 504.2(2)

CHIMNEYS AND VENTS

TABLE 504.2(2)
TYPE B DOUBLE-WALL GAS VENT

Number of Appliances	Single
Appliance Type	Category I
Appliance Vent Connection	Single-wall metal connector

VENT DIAMETER—(D) inches

APPLIANCE INPUT RATING IN THOUSANDS OF BTU/H

HEIGHT (H) (feet)	LATERAL (L) (feet)	3 FAN Min	3 FAN Max	3 NAT Max	4 FAN Min	4 FAN Max	4 NAT Max	5 FAN Min	5 FAN Max	5 NAT Max	6 FAN Min	6 FAN Max	6 NAT Max	7 FAN Min	7 FAN Max	7 NAT Max	8 FAN Min	8 FAN Max	8 NAT Max	9 FAN Min	9 FAN Max	9 NAT Max	10 FAN Min	10 FAN Max	10 NAT Max	12 FAN Min	12 FAN Max	12 NAT Max
6	0	38	77	45	59	151	85	85	249	140	126	373	204	165	522	284	211	695	369	267	894	469	371	1,118	569	537	1,639	849
6	2	39	51	36	60	96	66	85	156	104	123	231	156	159	320	213	201	423	284	251	541	368	347	673	453	498	979	648
6	4	NA	NA	NA	74	92	63	102	152	102	146	225	152	187	313	208	237	416	277	295	533	360	409	664	443	584	971	638
6	6	NA	NA	NA	83	89	60	114	147	99	163	220	148	207	307	203	263	409	271	327	526	352	449	656	433	638	962	627
8	0	37	83	50	58	164	93	83	273	154	123	412	234	161	580	319	206	777	414	258	1,002	536	360	1,257	658	521	1,852	967
8	2	39	56	39	59	108	75	83	176	119	121	261	179	155	363	246	197	482	321	246	617	417	339	768	513	486	1,120	743
8	5	NA	NA	NA	77	102	69	107	168	114	151	252	171	193	352	235	245	470	311	305	604	404	418	754	500	598	1,104	730
8	8	NA	NA	NA	90	95	64	122	161	107	175	243	163	223	342	225	280	458	300	344	591	392	470	740	486	665	1,089	715
10	0	37	87	53	57	174	99	82	293	165	120	444	254	158	628	344	202	844	449	253	1,093	584	351	1,373	718	507	2,031	1,057
10	2	39	61	41	59	117	80	82	193	128	119	287	194	153	400	272	193	531	354	242	681	456	332	849	559	475	1,242	848
10	5	52	56	39	76	111	76	105	185	122	148	277	186	190	388	261	241	518	344	299	667	443	409	834	544	584	1,224	825
10	10	NA	NA	NA	97	100	68	132	171	112	188	261	171	237	369	241	296	497	325	363	643	423	492	808	520	688	1,194	788
15	0	36	93	57	56	190	111	80	325	186	116	499	283	153	713	388	195	966	523	244	1,259	681	336	1,591	838	488	2,374	1,237
15	2	38	69	47	57	136	93	80	225	149	115	337	224	148	473	314	187	631	413	232	812	543	319	1,015	673	457	1,491	983
15	5	51	63	44	75	128	86	102	216	140	144	326	217	182	459	298	231	616	400	287	795	526	392	997	657	562	1,469	963
15	10	NA	NA	NA	95	116	79	128	201	131	182	308	203	228	438	284	284	592	381	349	768	501	470	966	628	664	1,433	928
15	15	NA	NA	NA	NA	NA	72	158	186	124	220	290	192	272	418	269	334	568	367	404	742	484	540	937	601	750	1,399	894
20	0	35	96	60	54	200	118	78	346	201	114	537	306	149	772	428	190	1,053	573	238	1,379	750	326	1,751	927	473	2,631	1,346
20	2	37	74	50	56	148	99	78	248	165	113	375	248	144	528	344	182	708	468	227	914	611	309	1,146	754	443	1,689	1,098
20	5	50	68	47	73	140	94	100	239	158	141	363	239	178	514	334	224	692	457	279	896	596	381	1,126	734	547	1,665	1,074
20	10	NA	NA	NA	93	129	86	125	223	146	177	344	224	222	491	316	277	666	437	339	866	570	457	1,092	702	646	1,626	1,037
20	15	NA	NA	NA	NA	NA	80	155	208	136	216	325	210	264	469	301	325	640	419	393	838	549	526	1,060	677	730	1,587	1,005
20	20	NA	NA	NA	NA	NA	NA	186	192	126	254	306	196	309	448	285	374	616	400	448	810	526	592	1,028	651	808	1,550	973

(continued)

CHIMNEYS AND VENTS

TABLE 504.2(2)

Number of Appliances	Single
Appliance Type	Category I
Appliance Vent Connection	Single-wall metal connector

TABLE 504.2(2)—continued
TYPE B DOUBLE-WALL GAS VENT

HEIGHT (H) (feet)	LATERAL (L) (feet)	VENT DIAMETER—(D) inches																													
		3			4				5				6				7				8				9				10				12			
		FAN		NAT	FAN		NAT	FAN		NAT	FAN		NAT	FAN		NAT	FAN		NAT	FAN		NAT	FAN		NAT	FAN		NAT
		Min	Max	Max	Min	Max	Max	Min	Max	Max	Min	Max	Max	Min	Max	Max	Min	Max	Max	Min	Max	Max	Min	Max	Max	Min	Max	Max
		APPLIANCE INPUT RATING IN THOUSANDS OF BTU/H																												
30	0	34	99	63	53	211	127	76	372	219	110	584	334	144	849	472	184	1,168	647	229	1,542	852	312	1,971	1,056	454	2,996	1,545
	2	37	80	56	55	164	111	76	281	183	109	429	392	139	610	392	175	823	533	219	1,069	698	296	1,346	863	424	1,999	1,308
	5	49	74	52	72	157	106	98	271	173	136	417	279	171	595	382	215	806	521	269	1,049	684	366	1,324	846	524	1,971	1,283
	10	NA	NA	NA	91	144	98	122	255	168	171	397	271	213	570	367	265	777	501	327	1,017	662	440	1,287	821	620	1,927	1,234
	15	NA	NA	NA	115	131	NA	151	239	157	208	377	257	255	547	349	312	750	481	379	985	638	507	1,251	794	702	1,884	1,205
	20	NA	NA	NA	NA	NA	NA	181	223	NA	246	357	242	298	524	333	360	723	461	433	955	615	570	1,216	768	780	1,841	1,166
	30	NA	NA	NA	NA	NA	NA	NA	NA	NA	NA	NA	228	389	477	305	461	670	426	541	895	574	704	1,147	720	937	1,759	1,101
50	0	33	99	66	51	213	133	73	394	230	105	629	361	138	928	515	176	1,292	704	220	1,724	948	295	2,223	1,189	428	3,432	1,818
	2	36	84	61	53	181	121	73	318	205	104	495	312	133	712	443	168	971	613	209	1,273	811	280	1,615	1,007	401	2,426	1,509
	5	48	80	NA	70	174	117	94	308	198	131	482	305	164	696	435	204	953	602	257	1,252	795	347	1,591	991	496	2,396	1,490
	10	NA	NA	NA	89	160	NA	118	292	186	162	461	292	203	671	420	253	923	583	313	1,217	765	418	1,551	963	589	2,347	1,455
	15	NA	NA	NA	112	148	NA	145	275	174	199	441	280	244	646	405	299	894	562	363	1,183	736	481	1,512	934	668	2,299	1,421
	20	NA	NA	NA	NA	NA	NA	176	257	NA	236	420	267	285	622	389	345	866	543	415	1,150	708	544	1,473	906	741	2,251	1,387
	30	NA	NA	NA	NA	NA	NA	NA	NA	NA	315	376	NA	373	573	NA	442	809	502	521	1,086	649	674	1,399	848	892	2,159	1,318
100	0	NA	NA	NA	49	214	NA	69	403	NA	100	659	395	131	991	555	166	1,404	765	207	1,900	1,033	273	2,479	1,300	395	3,912	2,042
	2	NA	NA	NA	51	192	NA	70	351	NA	98	563	373	125	828	508	158	1,152	698	196	1,532	933	259	1,970	1,168	371	3,021	1,817
	5	NA	NA	NA	67	186	NA	90	342	NA	125	551	366	156	813	501	194	1,134	688	240	1,511	921	322	1,945	1,153	460	2,990	1,796
	10	NA	NA	NA	85	175	NA	113	324	NA	153	532	354	191	789	486	238	1,104	672	293	1,477	902	389	1,905	1,133	547	2,938	1,763
	15	NA	NA	NA	132	162	NA	138	310	NA	188	511	343	230	764	473	281	1,075	656	342	1,443	884	447	1,865	1,110	618	2,888	1,730
	20	NA	NA	NA	NA	NA	NA	168	295	NA	224	487	NA	270	739	458	325	1,046	639	391	1,410	864	507	1,825	1,087	690	2,838	1,696
	30	NA	NA	NA	NA	NA	NA	231	264	NA	301	448	NA	355	685	NA	418	988	NA	491	1,343	824	631	1,747	1,041	834	2,739	1,627
	50	NA	NA	NA	NA	NA	NA	NA	NA	NA	NA	NA	NA	540	584	NA	617	866	NA	711	1,205	NA	895	1,591	NA	1,138	2,547	1,489

For SI: 1 inch = 25.4 mm, 1 foot = 304.8 mm, 1 British thermal unit per hour = 0.2931 W.

TABLE 504.2(3)
MASONRY CHIMNEY

							Number of Appliances	Single		
							Appliance Type	Category I		
							Appliance Vent Connection	Type B double-wall connector		

TYPE B DOUBLE-WALL CONNECTOR DIAMETER—(D) inches
to be used with chimney areas within the size limits at bottom

APPLIANCE INPUT RATING IN THOUSANDS OF BTU/H

HEIGHT (H) (feet)	LATERAL (L) (feet)	3 FAN Min	3 FAN Max	3 NAT Max	4 FAN Min	4 FAN Max	4 NAT Max	5 FAN Min	5 FAN Max	5 NAT Max	6 FAN Min	6 FAN Max	6 NAT Max	7 FAN Min	7 FAN Max	7 NAT Max	8 FAN Min	8 FAN Max	8 NAT Max	9 FAN Min	9 FAN Max	9 NAT Max	10 FAN Min	10 FAN Max	10 NAT Max	12 FAN Min	12 FAN Max	12 NAT Max
6	2	NA	NA	28	NA	NA	52	NA	NA	86	NA	NA	130	NA	NA	180	NA	NA	247	NA	NA	320	NA	NA	401	NA	NA	581
6	5	NA	NA	25	NA	NA	49	NA	NA	82	NA	NA	117	NA	NA	165	NA	NA	231	NA	NA	298	NA	NA	376	NA	NA	561
8	2	NA	NA	29	NA	NA	55	NA	NA	93	NA	NA	145	NA	NA	198	NA	NA	266	84	590	350	100	728	446	139	1,024	651
8	5	NA	NA	26	NA	NA	52	NA	NA	88	NA	NA	134	NA	NA	183	NA	NA	247	NA	NA	328	149	711	423	201	1,007	640
8	8	NA	NA	24	NA	NA	48	NA	NA	83	NA	NA	127	NA	NA	175	NA	NA	239	NA	NA	318	173	695	410	231	990	623
10	2	NA	NA	31	NA	NA	61	NA	NA	103	NA	NA	162	NA	NA	221	68	519	298	82	655	388	98	810	491	136	1,144	724
10	5	NA	NA	28	NA	NA	57	NA	NA	96	NA	NA	148	NA	NA	204	NA	NA	277	124	638	365	146	791	466	196	1,124	712
10	10	NA	NA	25	NA	NA	50	NA	NA	87	NA	NA	139	NA	NA	191	NA	NA	263	155	610	347	182	762	444	240	1,093	668
15	2	NA	NA	35	NA	NA	67	NA	NA	114	NA	NA	179	53	475	250	64	613	336	77	779	441	92	968	562	127	1,376	841
15	5	NA	NA	35	NA	NA	62	NA	NA	107	NA	NA	164	NA	NA	231	99	594	313	118	759	416	139	946	533	186	1,352	828
15	10	NA	NA	28	NA	NA	55	NA	NA	97	NA	NA	153	NA	NA	216	126	565	296	148	727	394	173	912	567	229	1,315	777
15	15	NA	NA	NA	NA	NA	48	NA	NA	89	NA	NA	141	NA	NA	201	NA	NA	281	171	698	375	198	880	485	259	1,280	742
20	2	NA	NA	38	NA	NA	74	NA	NA	124	NA	NA	201	51	522	274	61	678	375	73	867	491	87	1,083	627	121	1,548	953
20	5	NA	NA	36	NA	NA	68	NA	NA	116	NA	NA	184	80	503	254	95	658	350	113	845	463	133	1,059	597	179	1,523	933
20	10	NA	NA	NA	NA	NA	60	NA	NA	107	NA	NA	172	NA	NA	237	122	627	332	143	811	440	167	1,022	566	221	1,482	879
20	15	NA	NA	NA	NA	NA	NA	NA	NA	97	NA	NA	159	NA	NA	220	NA	NA	314	165	780	418	191	987	541	251	1,443	840
20	20	NA	NA	NA	NA	NA	NA	NA	NA	83	NA	NA	148	NA	NA	206	NA	NA	296	186	750	397	214	955	513	277	1,406	807

(continued)

CHIMNEYS AND VENTS

TABLE 504.2(3)

TABLE 504.2(3)—continued
MASONRY CHIMNEY

		Number of Appliances	Single
		Appliance Type	Category I
		Appliance Vent Connection	Type B double-wall connector

TYPE B DOUBLE-WALL CONNECTOR DIAMETER—(D) inches
to be used with chimney areas within the size limits at bottom

APPLIANCE INPUT RATING IN THOUSANDS OF BTU/H

HEIGHT (H) (feet)	LATERAL (L) (feet)	3 FAN Min	3 FAN Max	3 NAT Max	4 FAN Min	4 FAN Max	4 NAT Max	5 FAN Min	5 FAN Max	5 NAT Max	6 FAN Min	6 FAN Max	6 NAT Max	7 FAN Min	7 FAN Max	7 NAT Max	8 FAN Min	8 FAN Max	8 NAT Max	9 FAN Min	9 FAN Max	9 NAT Max	10 FAN Min	10 FAN Max	10 NAT Max	12 FAN Min	12 FAN Max	12 NAT Max
30	2	NA	NA	41	NA	NA	82	NA	NA	137	NA	NA	216	47	581	303	57	762	421	68	985	558	81	1,240	717	111	1,793	1,112
30	5	NA	NA	NA	NA	NA	76	NA	NA	128	NA	NA	198	75	561	281	90	741	393	106	962	526	125	1,216	683	169	1,766	1,094
30	10	NA	NA	NA	NA	NA	67	NA	NA	115	NA	NA	184	NA	NA	263	115	709	373	135	927	500	158	1,176	648	210	1,721	1,025
30	15	NA	NA	NA	NA	NA	NA	NA	NA	107	NA	NA	171	NA	NA	243	NA	NA	353	156	893	476	181	1,139	621	239	1,679	981
30	20	NA	NA	NA	NA	NA	NA	NA	NA	91	NA	NA	159	NA	NA	227	NA	NA	332	176	860	450	203	1,103	592	264	1,638	940
30	30	NA	NA	NA	NA	NA	NA	NA	NA	NA	NA	NA	188	NA	NA	188	NA	NA	288	NA	NA	416	249	1,035	555	318	1,560	877
50	2	NA	NA	NA	NA	NA	92	NA	NA	161	NA	NA	251	NA	NA	351	51	840	477	61	1,106	633	72	1,413	812	99	2,080	1,243
50	5	NA	NA	NA	NA	NA	NA	NA	NA	151	NA	NA	230	NA	NA	323	83	819	445	98	1,083	596	116	1,387	774	155	2,052	1,225
50	10	NA	NA	NA	NA	NA	NA	NA	NA	138	NA	NA	215	NA	NA	304	NA	NA	424	126	1,047	567	147	1,347	733	195	2,006	1,147
50	15	NA	NA	NA	NA	NA	NA	NA	NA	127	NA	NA	199	NA	NA	282	NA	NA	400	146	1,010	539	170	1,307	702	222	1,961	1,099
50	20	NA	NA	NA	NA	NA	NA	NA	NA	NA	NA	NA	185	NA	NA	264	NA	NA	376	165	977	511	190	1,269	669	246	1,916	1,050
50	30	NA	NA	NA	NA	NA	NA	NA	NA	NA	NA	NA	NA	NA	NA	NA	NA	NA	327	NA	NA	468	233	1,196	623	295	1,832	984
Minimum Internal Area of Chimney (square inches)			12			19			28			38			50			63			78			95			132	
Maximum Internal Area of Chimney (square inches)			49			88			137			198			269			352			445			550			792	

For SI: 1 inch = 25.4 mm, 1 square inch = 645.16 mm², 1 foot = 304.8 mm, 1 British thermal unit per hour = 0.2931 W.

TABLE 504.2(4)
MASONRY CHIMNEY

Number of Appliances	Single
Appliance Type	Category I
Appliance Vent Connection	Single-wall metal connector

TYPE B DOUBLE-WALL CONNECTOR DIAMETER—(D) inches
to be used with chimney areas within the size limits at bottom

APPLIANCE INPUT RATING IN THOUSANDS OF BTU/H

HEIGHT (H) (feet)	LATERAL (L) (feet)	3 FAN Min	3 FAN Max	3 NAT Max	4 FAN Min	4 FAN Max	4 NAT Max	5 FAN Min	5 FAN Max	5 NAT Max	6 FAN Min	6 FAN Max	6 NAT Max	7 FAN Min	7 FAN Max	7 NAT Max	8 FAN Min	8 FAN Max	8 NAT Max	9 FAN Min	9 FAN Max	9 NAT Max	10 FAN Min	10 FAN Max	10 NAT Max	12 FAN Min	12 FAN Max	12 NAT Max
6	2	NA	NA	28	NA	NA	52	NA	NA	86	NA	NA	130	NA	NA	180	NA	NA	247	NA	NA	319	NA	NA	400	NA	NA	580
6	5	NA	NA	25	NA	NA	48	NA	NA	81	NA	NA	116	NA	NA	164	NA	NA	230	NA	NA	297	NA	NA	375	NA	NA	560
8	2	NA	NA	29	NA	NA	55	NA	NA	93	NA	NA	145	NA	NA	197	NA	NA	265	NA	NA	349	382	725	445	549	1,021	650
8	5	NA	NA	26	NA	NA	51	NA	NA	87	NA	NA	133	NA	NA	182	NA	NA	246	NA	NA	327	NA	NA	422	673	1,003	638
8	8	NA	NA	23	NA	NA	47	NA	NA	82	NA	NA	126	NA	NA	174	NA	NA	237	NA	NA	317	NA	NA	408	747	985	621
10	2	NA	NA	31	NA	NA	61	NA	NA	102	NA	NA	161	NA	NA	220	216	518	297	271	654	387	373	808	490	536	1,142	722
10	5	NA	NA	28	NA	NA	56	NA	NA	95	NA	NA	147	NA	NA	203	NA	NA	276	334	635	364	459	789	465	657	1,121	710
10	10	NA	NA	24	NA	NA	49	NA	NA	86	NA	NA	137	NA	NA	189	NA	NA	261	NA	NA	345	547	758	441	771	1,088	665
15	2	NA	NA	35	NA	NA	67	NA	NA	113	NA	NA	178	166	473	249	211	611	335	264	776	440	362	965	560	520	1,373	840
15	5	NA	NA	32	NA	NA	61	NA	NA	106	NA	NA	163	NA	NA	230	261	591	312	325	775	414	444	942	531	637	1,348	825
15	10	NA	NA	27	NA	NA	54	NA	NA	96	NA	NA	151	NA	NA	214	NA	NA	294	392	722	392	531	907	504	749	1,309	774
15	15	NA	NA	NA	NA	NA	46	NA	NA	87	NA	NA	138	NA	NA	198	NA	NA	278	452	692	372	606	873	481	841	1,272	738
20	2	NA	NA	38	NA	NA	73	NA	NA	123	NA	NA	200	163	520	273	206	675	374	258	864	490	252	1,079	625	508	1,544	950
20	5	NA	NA	35	NA	NA	67	NA	NA	115	NA	NA	183	80	NA	252	255	655	348	317	842	461	433	1,055	594	623	1,518	930
20	10	NA	NA	NA	NA	NA	59	NA	NA	105	NA	NA	170	NA	NA	235	312	622	330	382	806	437	517	1,016	562	733	1,475	875
20	15	NA	NA	NA	NA	NA	NA	NA	NA	95	NA	NA	156	NA	NA	217	NA	NA	311	442	773	414	591	979	539	823	1,434	835
20	20	NA	NA	NA	NA	NA	NA	NA	NA	80	NA	NA	144	NA	NA	202	NA	NA	292	NA	NA	392	663	944	510	911	1,394	800

(continued)

CHIMNEYS AND VENTS TABLE 504.2(4)

TABLE 504.2(4)—continued
MASONRY CHIMNEY

Number of Appliances	Single
Appliance Type	Category I
Appliance Vent Connection	Single-wall metal connector

TYPE B DOUBLE-WALL CONNECTOR DIAMETER—(D) inches
to be used with chimney areas within the size limits at bottom

APPLIANCE INPUT RATING IN THOUSANDS OF BTU/H

HEIGHT (H) (feet)	LATERAL (L) (feet)	3 FAN Min	3 FAN Max	3 NAT Max	4 FAN Min	4 FAN Max	4 NAT Max	5 FAN Min	5 FAN Max	5 NAT Max	6 FAN Min	6 FAN Max	6 NAT Max	7 FAN Min	7 FAN Max	7 NAT Max	8 FAN Min	8 FAN Max	8 NAT Max	9 FAN Min	9 FAN Max	9 NAT Max	10 FAN Min	10 FAN Max	10 NAT Max	12 FAN Min	12 FAN Max	12 NAT Max
30	2	NA	NA	41	NA	NA	81	NA	NA	136	NA	NA	215	158	578	302	200	759	420	249	982	556	340	1,237	715	489	1,789	1,110
30	5	NA	NA	NA	NA	NA	75	NA	NA	127	NA	NA	196	NA	NA	279	245	737	391	306	958	524	417	1,210	680	600	1,760	1,090
30	10	NA	NA	NA	NA	NA	66	NA	NA	113	NA	NA	182	NA	NA	260	300	703	370	370	920	496	500	1,168	644	708	1,713	1,020
30	15	NA	NA	NA	NA	NA	NA	NA	NA	105	NA	NA	168	NA	NA	240	NA	NA	349	428	884	471	572	1,128	615	798	1,668	975
30	20	NA	NA	NA	NA	NA	NA	NA	NA	88	NA	NA	155	NA	NA	223	NA	NA	327	NA	NA	445	643	1,089	585	883	1,624	932
30	30	NA	NA	NA	NA	NA	NA	NA	NA	NA	NA	NA	NA	NA	NA	182	NA	NA	281	NA	NA	408	NA	NA	544	1,055	1,539	865
50	2	NA	NA	NA	NA	NA	91	NA	NA	160	NA	NA	250	NA	NA	350	191	837	475	238	1,103	631	323	1,408	810	463	2,076	1,240
50	5	NA	NA	NA	NA	NA	NA	NA	NA	149	NA	NA	228	NA	NA	321	NA	NA	442	293	1,078	593	398	1,381	770	571	2,044	1,220
50	10	NA	NA	NA	NA	NA	NA	NA	NA	136	NA	NA	212	NA	NA	301	NA	NA	420	355	1,038	562	447	1,337	728	674	1,994	1,140
50	15	NA	NA	NA	NA	NA	NA	NA	NA	124	NA	NA	195	NA	NA	278	NA	NA	395	NA	NA	533	546	1,294	695	761	1,945	1,090
50	20	NA	NA	NA	NA	NA	NA	NA	NA	NA	NA	NA	180	NA	NA	258	NA	NA	370	NA	NA	504	616	1,251	660	844	1,898	1,040
50	30	NA	NA	NA	NA	NA	48	NA	NA	NA	NA	NA	NA	NA	NA	318	NA	NA	318	NA	NA	458	NA	NA	610	1,009	1,805	970
Minimum Internal Area of Chimney (square inches)		12			19			28			38			50			63			78			95			132		
Maximum Internal Area of Chimney (square inches)		49			88			137			198			269			352			445			550			792		

For SI: 1 inch = 25.4 mm, 1 square inch = 645.16 mm², 1 foot = 304.8 mm, 1 British thermal unit per hour = 0.2931 W.

TABLE 504.2(5)
SINGLE-WALL METAL PIPE OR TYPE B ASBESTOS CEMENT VENT

Number of Appliances	Single
Appliance Type	Draft hood equipped
Appliance Vent Connection	Connected directly to pipe or vent

HEIGHT (H) (feet)	LATERAL (L) (feet)	VENT DIAMETER—(D) inches							
		3	4	5	6	7	8	10	12
		MAXIMUM APPLIANCE INPUT RATING IN THOUSANDS OF BTU/H							
6	0	39	70	116	170	232	312	500	750
	2	31	55	94	141	194	260	415	620
	5	28	51	88	128	177	242	390	600
8	0	42	76	126	185	252	340	542	815
	2	32	61	102	154	210	284	451	680
	5	29	56	95	141	194	264	430	648
	10	24	49	86	131	180	250	406	625
10	0	45	84	138	202	279	372	606	912
	2	35	67	111	168	233	311	505	760
	5	32	61	104	153	215	289	480	724
	10	27	54	94	143	200	274	455	700
	15	NA	46	84	130	186	258	432	666
15	0	49	91	151	223	312	420	684	1,040
	2	39	72	122	186	260	350	570	865
	5	35	67	110	170	240	325	540	825
	10	30	58	103	158	223	308	514	795
	15	NA	50	93	144	207	291	488	760
	20	NA	NA	82	132	195	273	466	726
20	0	53	101	163	252	342	470	770	1,190
	2	42	80	136	210	286	392	641	990
	5	38	74	123	192	264	364	610	945
	10	32	65	115	178	246	345	571	910
	15	NA	55	104	163	228	326	550	870
	20	NA	NA	91	149	214	306	525	832
30	0	56	108	183	276	384	529	878	1,370
	2	44	84	148	230	320	441	730	1,140
	5	NA	78	137	210	296	410	694	1,080
	10	NA	68	125	196	274	388	656	1,050
	15	NA	NA	113	177	258	366	625	1,000
	20	NA	NA	99	163	240	344	596	960
	30	NA	NA	NA	NA	192	295	540	890
50	0	NA	120	210	310	443	590	980	1,550
	2	NA	95	171	260	370	492	820	1,290
	5	NA	NA	159	234	342	474	780	1,230
	10	NA	NA	146	221	318	456	730	1,190
	15	NA	NA	NA	200	292	407	705	1,130
	20	NA	NA	NA	185	276	384	670	1,080
	30	NA	NA	NA	NA	222	330	605	1,010

For SI: 1 inch = 25.4 mm, 1 foot = 304.8 mm, 1 British thermal unit per hour = 0.2931 W.

504.2.2 Minimum size. Where the vent size determined from the tables is smaller than the appliance draft hood outlet or flue collar, the smaller size shall be permitted to be used provided that all of the following requirements are met:

1. The total vent height (H) is at least 10 feet (3048 mm).
2. Vents for appliance draft hood outlets or flue collars 12 inches (305 mm) in diameter or smaller are not reduced more than one table size.
3. Vents for appliance draft hood outlets or flue collars larger than 12 inches (305 mm) in diameter are not reduced more than two table sizes.
4. The maximum capacity listed in the tables for a fan-assisted appliance is reduced by 10 percent (0.90 × maximum table capacity).
5. The draft hood outlet is greater than 4 inches (102 mm) in diameter. Do not connect a 3-inch-diameter (76 mm) vent to a 4-inch-diameter (102 mm) draft hood outlet. This provision shall not apply to fan-assisted appliances.

❖ If a vent complies with all of items 1 through 5 and is sized in accordance with the appropriate table, it is allowed to be smaller than the appliance connection, primarily because of the increase in capacity of the vent resulting from its height and because vent capacity is based on the Btu/h input rating of the appliance and not the size of the appliance flue collar. This code section suggests that flue collars are sized for worst case vent designs and are thus somewhat oversized for vents with higher venting capacities. Item 4 is necessary to prevent positive pressure from developing in the vent serving fan-assisted appliances. A 6-inch draft hood connected to a 5-inch Type B vent, for example, might look like an installation error, but this section could allow it in some cases, and operating a vent nearer its maximum capacity is always desirable (see Section 504.2.10).

504.2.3 Vent offsets. Single-appliance venting configurations with zero (0) lateral lengths in Tables 504.2(1), 504.2(2), and 504.2(5) shall not have elbows in the venting system. For vent configurations with lateral lengths, the venting tables include allowance for two 90-degree (1.57 rad) turns. For each elbow up to and including 45 degrees (0.79 rad), the maximum capacity listed in the venting tables shall be reduced by 5 percent. For each elbow greater than 45 degrees (0.79 rad) up to and including 90 degrees (1.57 rad), the maximum capacity listed in the venting tables shall be reduced by 10 percent.

❖ A vent system with a zero lateral length is an ideal system because it is entirely vertical and thus has the least flow resistance and greatest draft. To create a lateral run of vent or vent connector, one or two elbows (adjustable fittings) are necessary [see commentary Figure 504.2(2)]. The tables account for two 90-degree (1.57 rad) changes in direction, which will offer considerable resistance to the flow of flue gases. Depending on the appliance flue collar or draft hood outlet orientation, a single elbow can create a lateral, but the tables assume that a vertical outlet orientation and two elbows are used. Two 45-degree (0.79 rad) fittings joined together, for example, are the equivalent of a single 90-degree (1.57 rad) fitting but would offer slightly less resistance to flow because of the larger turning radius. The capacity reduction penalty for each change in direction in excess of the two "free" 90-degree (1.57 rad) fittings accounts for the increased flow resistance caused by the additional fittings (elbows). For example, a vent employing two 90-degree (1.57 rad) fittings and two 45-degree (0.79 rad) fittings would have a 10 percent capacity reduction penalty. A vent employing two 90-degree (1.57 rad) fittings and two 30-degree (0.52 rad) fittings would also have a 10 percent capacity penalty. All changes of direction, regardless of the angle, (fittings) that are in addition to the two "free" 90-degree fittings are accounted for by a capacity reduction penalty. A fitting adjusted to 45-degrees (0.79 rad) or less will cause a 5 percent penalty and a fitting adjusted to greater than 45-degrees (0.79 rad) and up to 90 degrees (1.57 rad) will cause a 10 percent penalty.

Whenever possible, an individual vent should be located directly over the appliance outlet. If the appliance flue outlet is horizontal, one 90-degree (1.57 rad) elbow should be used with the vent directly over it. A straight vertical vent is more easily supported and has less flow resistance.

There is no need to offset the vertical vent to include a tee and bracket. The use of a tee for cleanout or inspection purposes is quite unnecessary for three reasons. First, using a B-vent cap keeps debris and birds out. Second, clean-burning gas does not produce any deposits needing removal. Third, Type B vent joints are easily opened for inspection of the inside of the piping.

Should an offset be needed, the use of two elbows will provide somewhat greater capacity than an elbow and a tee. A tee has greater flow resistance than an elbow. The two 90-degree turns that are accounted for by the tables for laterals can occur anywhere within the vent system. This means that if the turns are used where the appliance connects to the system, the rest of the vent must be entirely vertical, or else the penalties of this section must be applied. The two "free" 90-degree (1.57 rad) turns can be composed of any number and angle of fittings that total not more than 180 degrees (3.1 rad). Simply put, there are free turns up to a maximum of 180 degrees (3.1 rad), and after that each additional turn will cause a capacity penalty to apply.

Vent offsets should always be avoided where possible, especially offsets in the upper portion of the vent and in cold spaces such as attics and truss spaces. Offsets can be avoided with simple planning of the vent system and appliance locations. The attic offset is prevalent because the vent and appliance location is often not planned, and the building owners and designers do not want the vent to terminate on the street side of a roof for aesthetic reasons. This type of offset causes the most problems for several reasons. The typical attic will have winter temperatures close to the outdoor temperature; therefore, offsets in attics will cause more vent pipe to be exposed to the cold attic. See the commentary for Section 504.2.9. This can increase the production of condensate in the vent, and the fittings used will be points at which the condensate can leak from the inner vent lining. Most condensate leakage from vents can be traced to a fitting used in an offset. The code (Sections 504.2.9 and 504.3.20) assumes that a vent will not be exposed to outdoor temperatures except for the portion that extends above the roof. The temperature in a typical attic today is nearly equivalent to the outdoors.

Studies have shown that offsets in the upper portion of vent systems can produce slight positive pressures because of the flow resistance, and this encourages water vapor to escape from the vent and condense between the vent walls. Offsets in attics usually involve long runs of pipe that increase the surface area of vent pipe exposed to cold temperature. It is quite common to find that attic offsets exceed the horizontal length allowed by Section 504.3.5.

504.2.4 Zero lateral. Zero (0) lateral (L) shall apply only to a straight vertical vent attached to a top outlet draft hood or flue collar.

❖ The zero lateral rows in the tables refer only to an entirely vertical vent run without offsets, lateral runs and directional fittings. This would be the ideal vent. Unfortunately, it is seldom seen because of lack of planning and coordination of vent and appliance locations. Architectural features of buildings usually dictate that vents take less than desirable paths to get from the appliances to the outdoors.

504.2.5 High-altitude installations. Sea-level input ratings shall be used when determining maximum capacity for high altitude installation. Actual input (derated for altitude) shall be used for determining minimum capacity for high altitude installation.

❖ Appliances installed at elevations above sea level must have their Btu/h (W) input reduced (derated) because the atmosphere is less dense and oxygen levels decrease as altitude increases. The actual volume of flue gases entering a vent (i.e., primary and secondary combustion air, excess air and combustion by-products) does not significantly change for different elevations; therefore, the maximum vent capacity is based on the sea-level input rating. The minimum capacity of a vent is dependent upon heat input to the vent to control condensation and maintain the required draft. Therefore, the actual heat input (as derated) is used to determine the minimum capacity of a vent.

504.2.6 Multiple input rate appliances. For appliances with more than one input rate, the minimum vent capacity (FAN Min) determined from the tables shall be less than the lowest appliance input rating, and the maximum vent capacity (FAN Max/NAT Max) determined from the tables shall be greater than the highest appliance rating input.

❖ To improve efficiency, many appliances are equipped with modulating input or high-low two-stage input. The venting system must be capable of operating properly throughout the entire firing range of the appliance. The venting system must be capable of handling the maximum input of the appliance and also must be compatible with the appliance when the appliance is operating at its lowest input. Such appliances will usually operate at the "low-fire" input most of their total operating time. To satisfy the minimum input requirements for fan-assisted appliance vents, the vent design must be based on the lowest Btu/h rating of the appliance because this is what the vent senses as the connected load. The lowest appliance input rate is used for minimum vent capacity, and the highest input rate must be used for maximum vent capacity. This requirement can be easily overlooked if the appliance label is not carefully studied because the appliance will usually be described by its maximum input, and any minimum input will have to obtained from the label.

504.2.7 Liner system sizing. Listed corrugated metallic chimney liner systems in masonry chimneys shall be sized by using Table 504.2(1) or 504.2(2) for Type B vents with the maximum capacity reduced by 20 percent (0.80 × maximum capacity) and the minimum capacity as shown in Table 504.2(1) or 504.2(2). Corrugated metallic liner systems installed with bends or offsets shall have their maximum capacity further reduced in accordance with Section 504.2.3. The 20-percent reduction for corrugated metallic chimney liner systems includes an allowance for one long-radius 90-degree (157 rad) turn at the bottom of the liner.

❖ A liner system installed within a masonry chimney is sized and designed as if the chimney/liner system combination were a Type B vent. Corrugated metal liners are constructed of semi-rigid piping having a higher flow resistance than smooth-wall rigid pipe. A capacity reduction penalty compensates for the poorer flow characteristics of corrugated pipe. Bends in semirigid liners count as elbows and are regulated by Section 504.2.3. Note that the sweeping 90-degree (1.57 rad) bend in the liner at the point where it exits the bottom of the chimney and connects to the appliance(s) is already accounted for in the 20-percent capacity penalty and no further capacity adjustment is required.

504.2.8 Vent area and diameter. Where the vertical vent has a larger diameter than the vent connector, the vertical vent diameter shall be used to determine the minimum vent capacity, and the connector diameter shall be used to determine the maximum vent capacity. The flow area of the vertical vent shall not exceed seven times the flow area of the listed appliance categorized vent area, flue collar area, or draft hood outlet area unless designed in accordance with approved engineering methods.

❖ The maximum capacity of a venting system must be based on the element of the system having the least capacity (i.e., the weakest link in the chain). Likewise, the minimum Btu/h (W) input required for proper vent system operation must be based on the largest element of the system because the largest element will naturally have the highest minimum input requirement. A connector that is smaller than the vent will set the maximum capacity limit for the system because it has the greater flow resistance and smaller cross-sectional area. The minimum Btu/h input is based on the larger vent size because the vent will require more heat to produce adequate draft and avoid condensation. The ratio of appliance vent-connection area to vertical vent area is intended to prevent oversizing of the vent, which can result in poor draft and condensation problems. Because the tables will allow vents to be larger than needed, this section serves as a limit to prevent gross oversizing. For example, in Table 504.2(1) it can be seen that a fan-assisted appliance having an input rating near the maximum end of the minimum/maximum range for any 4-inch venting system configuration can be served by not only a 4-inch vent but also a 5-inch and sometimes a 6-inch or 7-inch vent. Consistent with vent manufacturers' instructions, a venting system should always be designed to use the smallest vent allowed by the tables. Vents that are larger than they need to be usually will not perform as well as accurately sized vents and are more likely to suffer from poor draft and condensate problems.

504.2.9 Chimney and vent locations. Tables 504.2(1), 504.2(2), 504.2(3), 504.2(4) and 504.2(5) shall be used for chimneys and vents not exposed to the outdoors below the roof line. A Type B vent or listed chimney lining system passing through an unused masonry chimney flue shall not be considered to be exposed to the outdoors. Table 504.2(3) in combination with Table 504.3(6) shall be used for clay-tile-lined exterior masonry chimneys, provided all of the following are met:

1. Vent connector is Type B double-wall.
2. Vent connector length is limited to $1^1/_2$ feet for each inch (18 mm per mm) of vent connector diameter.
3. The appliance is draft hood equipped.
4. The input rating is less than the maximum capacity given by Table 504.2(3).
5. For a water heater, the outdoor design temperature is not less than 5°F (-15°C).
6. For a space-heating appliance, the input rating is greater than the minimum capacity given by Table 504.3(6).

Where these conditions cannot be met, an alternative venting design shall be used, such as a listed chimney lining system.

Exception: The installation of vents serving listed appliances shall be permitted to be in accordance with the appliance manufacturer's instructions and the terms of the listing.

❖ Chimneys that have one or more walls exposed to the outdoors and vents exposed to the outdoors can develop condensation and weak draft problems in cold weather. Commentary Figure 504.2.9 illustrates why masonry chimneys, including interior chimneys, are more prone to draft and condensation problems. The tables assume that the only portion of the chimney or vent that is exposed to the outdoors is the portion that extends above the roof penetration. Even though an attic is technically not outdoors, the temperature in an attic can be very close to the outdoor temperature. In some climates, the actual attic temperature can be far lower than the assumed temperature used in the computer modeling on which the vent sizing tables are based. Therefore, the portion of a venting system that passes through an attic should be kept to an absolute minimum. If at all possible, the installation should be planned to avoid the need for vent offsets in an attic. This section states that an exterior chimney lined with a listed liner or Type B vent is to be sized as if it were a Type B vent. The walls of an exterior chimney will provide some insulating effect for a properly installed Type B vent or lining system installed within the chimney. Thus, such installations can be sized as interior chimneys or vents. A chimney lined with a Type B vent is not considered a chimney, but is treated as a Type B vent installed within a masonry chase. See Section 504.2.3.

This section intends that the listed tables be applicable only to vents and chimneys that are not exposed to the outdoors below the roof penetration. If the vent or chimney is exposed to the outdoors below the roof penetration, the listed sizing tables do not apply. The intent is that vents and chimneys be enclosed within the building envelope to lessen exposure to low ambient temperatures and wind. It is these conditions that were considered in the development of the vent sizing tables. The sizing tables are based on an assumed outdoor temperature of 42°F (6° C) and the assumption that all portions of a vent or chimney below the roof penetration are exposed to an ambient temperature of 60°F (16°C). This makes it obvious that long runs of vent piping through attics and long extensions above roofs should be avoided in cold climates. See Section 504.3.5.

This section prohibits the use of the vent sizing tables for the installation of a Type B vent run up the outside of an exterior wall of a building. This is consistent with vent manufacturers' instructions. Outdoor vents obviously suffer from poorer draft and increased condensation formation because they are exposed to the elements, low temperatures being the main factor. "Stack effect" in the building interior can cause back drafts (reverse flow) to occur in any vent or chimney, particularly vents exposed to the outdoors. For the same reasons, vents should extend above the roof no more than required by the code because the more vent pipe that extends above the roof, the more vent pipe will be exposed to the cold and wind. In some cases, designers and installers will intentionally raise a vent above a roof more than Section 503.6.6 requires just to increase the vent capacity. In doing so, additional vent pipe is exposed to the outdoors, thus inviting condensation and possibly diminishing draft. Section 503.6.6 could also require a vent to extend well above a lower roof penetration, thus causing the sizing tables of this section to be inapplicable to such a vent.

The tables for exterior masonry chimneys obviously apply to chimneys that are exposed to the outdoors below the roof line, and the list of conditions intends to compensate for the cold masonry. Conditions such as requiring double-wall connectors and draft-hood-equipped appliances are intended to put as much heat into the chimney as possible because it will be needed to produce draft and minimize condensation in a less-than-ideal chimney. Exterior chimneys require greater heat input to make them work. An alternative to the continued use of an exterior chimney would be relining the chimney or abandoning the chimney and using high-efficiency appliances.

The exception addresses the possibility of an appliance that was tested and listed for venting applications other than what is allowed by this section. An example might be an appliance that is field-converted to a draft-hood appliance for the purpose of using a masonry chimney (see commentary, Section 503.12.4).

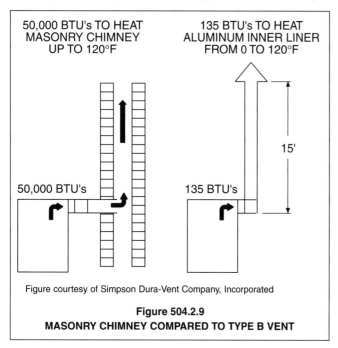

Figure courtesy of Simpson Dura-Vent Company, Incorporated

Figure 504.2.9
MASONRY CHIMNEY COMPARED TO TYPE B VENT

504.2.10 Corrugated vent connector size. Corrugated vent connectors shall be not smaller than the listed appliance categorized vent diameter, flue collar diameter, or draft hood outlet diameter.

❖ Corrugated vent connectors are becoming more common because they can save installation time and avoid the use of fittings. This section is consistent with Section 504.3.21 and applies to both draft-hood-equipped and fan-assisted appliances. Corrugated materials have higher resistance to flow than smooth wall materials; therefore, to assure the required capacity, full size connectors must be used.

504.2.11 Vent connector size limitation. Vent connectors shall not be increased in size more than two sizes greater than the listed appliance categorized vent diameter, flue collar diameter, or draft hood outlet diameter.

❖ The sizing tables of this section will often require that a connector be larger than the appliance vent connection. The appliance manufacturer's instructions may also require that a connector be increased in size. The increase in size compensates for flow resistance in the connector and prevents positive pressurization. A grossly oversized connector will allow the vent gases to expand and cool too much, thus negatively affecting draft and contributing to the formation of condensation. Connectors are sometimes increased for the sole purpose of gaining more allowable length (Section 504.3.2), and this section will limit that practice.

504.2.12 Component commingling. In a single run of vent or vent connector, different diameters and types of vent and connector components shall be permitted to be used, provided that all such sizes and types are permitted by the tables.

❖ In the unlikely event that an installer wants to mix, for example, Type B vent with single-wall pipe in a single connector (for example, Type B vent from a draft hood, switching to single wall and connecting to Type B vent again), the mixed system must comply with the code based on each type of material as if the run were constructed entirely of that material. There may be some unique situations that would justify commingling of different types and/or sizes of components, but clearances, heat loss, ambient temperatures, joining methods and flow characteristics must be considered.

504.2.13 Table interpolation. Interpolation shall be permitted in calculating capacities for vent dimensions that fall between the table entries (see Example 3, Appendix B).

❖ Because Tables 504.2(1) through 504.2(5) do not contain values for every possible height of a vent or length of a lateral, interpolation could be necessary. The maximum and minimum capacity of vents having heights not listed in the tables can be determined using the following equations.

Interpolated Maximum Vent Capacity = $(A/B \times C) + D$
Interpolated Minimum Vent Capacity = $E - (A/B \times F)$

where:

A = The difference between the design height entry and the next lower height entry in the applicable table.

B = The difference between the closest consecutive height entries in the applicable table.

C = The difference between the maximum capacity column consecutive entries in the applicable table.

D = The lower maximum capacity entry in the applicable table.

E = The higher minimum capacity entry in the applicable table.

F = The difference between the minimum capacity column consecutive entries in the applicable table.

For SI: 1 inch = 25.4 mm, 1 foot = 304.8 mm,
1 Btu/h = 0.2931 W.

Example 1:

Determine the minimum and maximum vent capacity for a single fan-assisted appliance given the following:

Vertical design height of vent	17 feet
Diameter of vent	6 inches
Vent type	Type B
Length of lateral	10 feet

Using Table 504.2(1):

Interpolated Maximum Vent Capacity =
[(17 ft − 15 ft)/(20 ft − 15 ft) × (351,000 Btu/h − 315,000 Btu/h)] + 315,000 Btu/h
= (2/5 × 36,000 Btu/h) + 315,000 Btu/h)]
= 329,000 Btu/h

Interpolated Minimum Vent Capacity =
64,000 Btu/h − [(17 ft − 15 ft)/(20 ft − 15 ft)
× (64,000 Btu/h − 62,000 Btu/h)]
= 64,000 Btu/h − (2.5 × 2,000 Btu/h)
= 63,200 Btu/h

Example 2:

Determine the maximum vent capacity for a single fan assisted furnace given the following:

Input rating of furnace	210,000 Btu/h
Vertical design height of vent	$12^1/_2$ feet
Diameter of vent	6 inches
Vent type	Type B
Length of lateral	2 feet

Using Table 504.2(1):

Step 1: Find the maximum vent capacity (MAX) at the first height entry greater than $12^1/_2$ feet, i.e., 15 feet.

MAX_{15} = 226,000 Btu/h

Step 2: Find the maximum vent capacity (MAX) at the next height entry lower than $12^1/_2$ feet, i.e., 10 feet.

MAX_{10} = 194,000 Btu/h

Step 3: Determine the difference between the two maximum vent capacities.

$MAX_{15} - MAX_{10}$ = 226,000 Btu/h − 194,000 Btu/h
= 32,000 Btu/h

Step 4: Determine the maximum vent capacity for a $12^1/_2$-foot high vent using:

$MAX_{12^1/_2} = [(A/B) \times (MAX_{15} - MAX_{10})] + MAX_{10}$

where:

A = The difference between the next higher table vent height and the design vent height.

B = The difference between the consecutive closest table vent heights.

$MAX_{12^1/_2}$ = [(15 ft − $12^1/_2$ ft)/(15 ft − 10 ft)
× 32,000 Btu/h] + 194,000 Btu/h
= [($2^1/_2$ ft)/(5 ft) × 32,000 Btu/h] + 194,000 Btu/h
= 16,000 Btu/h + 194,000/h
= 210,000 Btu/h

When a vent height falls between height entries in a table, the code user can choose to interpolate or use the closest table height entry that is lower than the actual height of the vent. These are the only allowable options.

This example also illustrates how the tables may be used to reduce vent size if there is adequate height and capacity to do so (see commentary, Section 504.2.2).

It is important to consider interpolation before assuming that an "in between" height vent installation is not in compliance with the code. For example, a vent with a 25-foot height could fail to comply with a table when it is viewed as a 20-foot vent, but that same vent could quite possibly comply with the table when viewed as what it actually is, a 25-foot high vent.

This interpolation process can also be used to estimate in-between capacities for intermediate lengths of laterals, as well as in-between minimum capacities.

There is no way to estimate in-between capacity between a zero lateral (straight vertical vent) and a 2-foot lateral (which has two elbows). If the vent has just one 90-degree turn, 2-foot lateral capacity applies.

504.2.14 Extrapolation prohibited. Extrapolation beyond the table entries shall not be permitted.

❖ Projecting capacities below the lowest or above the highest boundary entries of the tables is not allowed because the tables are based on the known operational limits of venting systems. The tables cannot predict the operating characteristics of venting systems having heights shorter or taller than the table limits; therefore, any such system must be an engineered system.

504.2.15 Engineering calculations. For vent heights less than 6 feet (1829 mm) and greater than shown in the tables, engineering methods shall be used to calculate vent capacities.

❖ This section parallels the previous section by stating that vents having heights not within the tables must be designed by an engineered method. In other words, extrapolation above or below the table height entries is not allowed (see Section 504.2.14).

504.3 Application of multiple appliance vent Tables 504.3(1) through 504.3(8). The application of Tables 504.3(1) through 504.3(8) shall be subject to the requirements of Sections 504.3.1 through 504.3.26.

❖ The application of Tables 504.3(1) through 504.3(8) shall be subject to the requirements of Sections 504.3.1 through 504.3.26. This part of Section 504 regulates vents and chimneys that serve more than one appliance. These venting systems can be described as combined vents and include multiple connectors. A "combined vent" is a vent for two or more appliances at one level served by a common vent. "Least total height" is the vertical distance from the highest appliance outlet (draft hood or flue collar) to the lowest discharge opening of the vent top. This height dimension is illustrated in commentary Figure 504.3(1) for a typical (FAN + NAT) system. The same height measurement applies to FAN+FAN and NAT+NAT systems. Least total height is used for vent sizing of all connected appliances on one level. "Connector rise" for any appliance is the vertical distance from its outlet connection to the level at which it joins the common vent, as shown in commentary Figure 504.3(1).

A "connector" for purposes of designing a combined vent is that part of the vent piping between the appliance outlet and its junction with or interconnection to the rest of the system. A chimney is referred to as a venting means, although a chimney and a vent are distinct. An improperly sized common vent or chimney could fail to vent the combustion products of the appliances served. Many factors influence the sizing of a vent or chimney, including total height, offsets, lateral lengths, the type of appliance, the appliance energy input and the appliance connector configurations. The "common vent" is that portion of the system serving two or more connected appliances. If connectors are joined before reaching the vertical vent, the run between the last entering connector and the vertical portion is also treated as part of the common vent. In commentary Figure 504.3(1), the vertical common vent is cross-hatched beginning at the interconnection tee.

For each connector, the correct size must be found from the applicable tables based on its appliance input, rise and least total vent height. For draft-hood (NAT) appliances, the outlet size may be too small if there is not enough rise; therefore, connector design involves choosing the correct size and verifying that the use of a connector the same size as the outlet is within input rating limits.

For the common vent, the capacity table shows maximum combined ratings only. The size of the common vent is thus based on least total height and the combination of attached appliances: FAN+FAN, FAN+NAT or NAT+NAT. To find each connector size, use Table 504.3(1) for Type B vent connectors or Table 504.3(2) for single-wall metal connectors and proceed as follows:

1. Determine "least total height" for the system.
2. Determine connector rise for each appliance.
3. Enter the applicable vent connector table at the least total height. Continue across on the line for appliance connector rise to a MAX input rating equal to or greater than that of the appliance. For a FAN appliance, this input rating should also be greater than shown in the MIN column. Read the connector size at the top of the column. If Table 504.3(2) for single wall shows NA, use a double-wall connector in accordance with Table 504.3(1). In some cases, the table will dictate a connector size that is larger than the appliance vent outlet (flue collar).

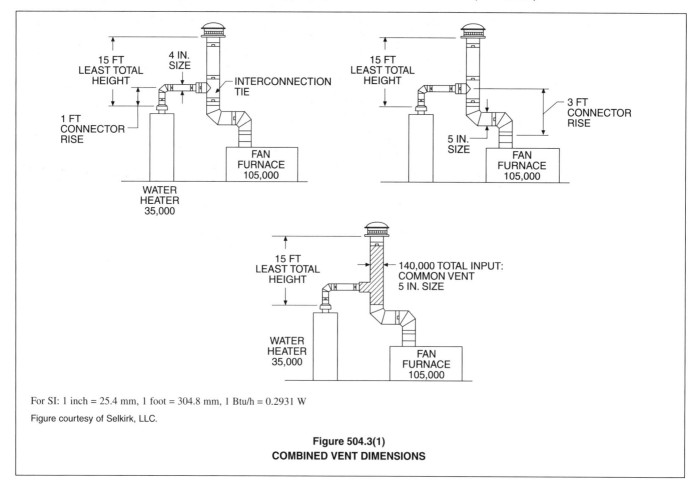

For SI: 1 inch = 25.4 mm, 1 foot = 304.8 mm, 1 Btu/h = 0.2931 W

Figure courtesy of Selkirk, LLC.

**Figure 504.3(1)
COMBINED VENT DIMENSIONS**

To find the common vent size, proceed as follows:

1. Add all appliances' Btu input ratings to get the total Btu input.
2. If one or both connectors are single-wall metal, use Table 504.3(2) for the common vent.
3. Enter the common vent table at the same least total height used for connectors.
4. Continue across and stop at the first applicable column of combined appliance input rating equal to or greater than the total. If Table 504.3(2) for the common vent shows NA, the entire system must be double-wall Type B vent.
5. Read the size of the common vent at the top of the applicable column (NAT+NAT, FAN+NAT or FAN+FAN).

    (a) Regardless of table results for size, the common vent must be at least as large as the largest connector (see Section 504.3.8).

    (b) In accordance with vent manufacturer's instructions, for NAT+NAT appliance combinations, if both connectors are the same size, the common vent must be at least one size larger.

Example 1

Commentary Figure 504.3(1) shows a two- appliance FAN+NAT system, combining a draft-hood water heater (NAT appliance) with a fan-assisted-combustion Category I (FAN) furnace. The system is designed in the following steps using Table 504.3(1) for double-wall connectors.

1. For the water heater, enter the vent connector table at a least total height of 15 feet (4572 mm) and a connector rise of 1 foot (305 mm). Read across to the MAX Btu/h rating for a NAT appliance vent higher than 35,000 Btu/h (10.3 kW). The table shows that a 4-inch (102mm) connector is needed. This size must be used beginning at the draft hood, regardless of draft hood size, which might be 3 inches (76 mm) [see commentary Figure 504.3(1)].
2. For the 105,000 Btu/h (30.8 kW) furnace, enter the vent connector table at the same least total height [15 feet (4572)] at a connector rise of 3 feet (914 mm). Read across to 163,000 under the FAN MAX column. The MIN is 51,000. Therefore, 5 inches (127 mm) is the correct connector size [see commentary Figure 504.3(1)].
3. For the common vent, the sum of the two ratings is 140,000 Btu/h (41.0 kW). Enter the common vent table at 15 foot (4572 mm) least total height.

For a FAN+NAT combination, the maximum input of 5-inch (127 mm) vent is 164,000 (48.1 kW), so 5 inches (127 mm) is the proper size, as shown in commentary Figure 504.3(1). Both draft-hood and fan-assisted appliances may be common vented in any combination, as indicated by the headings in the common vent table. Appliance types include the following:

- Central heating furnaces;
- Central heating boilers (hot water and steam);
- Water heaters;
- Unit heaters;
- Duct furnaces;
- Room heaters*; and
- Floor furnaces*.

*If these have draft hoods, a design input increase of 40 percent is recommended by some vent manufacturers in order to use the tables (see commentary, Section 504.2).

The common vent tables do not apply to the following:

1. Gas cooking appliances, which should be vented into an exhaust hood or vented individually in accordance with the appliance manufacturer's instructions.
2. Forced draft and commercial or industrial hot water or steam boilers without draft hoods. For this equipment, see the boiler and vent/chimney manufacturers' instructions.
3. Categories II, III and IV gas-burning equipment, for which the equipment manufacturer's venting instructions must be used.
4. Gas-fired incinerators.

The manufacturers of chimney and vent systems provide installation and design information as part of their installation instructions. The common chimney or vent must be designed to properly vent the products of combustion when any one, any combination or all of the connected appliances operate. [see commentary Figure 504.3(2)]. Solid-fuel-burning appliances must not connect to common chimneys (see Section 503.5.7.1).

Combined vent systems for two or more gas appliances of either type (FAN or NAT) must be designed to prevent draft-hood spillage for natural draft (NAT) appliances and to avoid positive pressure for fan-assisted (FAN) appliances. The connector and common vent tables have been computed by examining the most critical situations for any operating combination. A common vent must function properly when all, one or any combination of the connected appliances are firing. It is particularly demanding of a vent system to operate properly when only a single appliance having the lowest input rating is firing, thus creating the greatest potential for poor draft and condensation. Oversizing of venting systems should be avoided.

The connector tables are based on the most critical condition for a particular appliance when operating by itself, whereas the common vent tables show sizes that assure adequate capacity and draft, whether one or all appliances are operating simultaneously.

All parts of a combined vent must be checked for capacity. For connectors, the size must be determined from the tables, particularly for low-height vents or where headroom restricts available connector rise.

The connector in a combined-vent system is defined as the piping from a draft hood or flue collar to the junction of the common vent or to a junction in a vent manifold. Proper connector design is vital to obtaining adequate capacity. The connector must produce its share of the total draft for its NAT or FAN appliance and must deliver enough heat to the common vent so that it can contribute the balance of draft needed.

For application of the code's installation provisions, the Type B gas vent connector is a "gas vent". It is essential, however, for system design purposes to use the word "connector" so that its rise and configuration may be explained and tabulated. The connector tables for combined vents show MIN and MAX capacities only for FAN appliances because no minimums are needed for NAT appliances with Type B gas-vent connectors. Important factors in connector design include the following:

- Connector material;
- Connector length;
- Connector rise;
- Number;
- Appliance location as it affects the piping arrangement;
- Number of attached appliances or different connector sizes; and
- Connection to an offset or manifold rather than directly to the vertical common vent.

Example 2

This example is a common scenario [see commentary Figure 504.3(2)].

Given:

- Single-family house with Type B vent serving two appliances;
- Category I fan-assisted furnace, 80 percent efficient, 75,000 Btu/h input (4-inch vent collar);
- Furnace connector single-wall pipe with 3-foot rise;
- Draft-hood-equipped water heater, 40,000 Btu/h (3-inch draft hood outlet);
- Water heater connector single-wall with 2-foot rise;
- Vent height: 25 feet;
- A 45-degree offset occurs in the common vent in the attic.

Step 1. Choose the applicable table. Table 504.3(2) addresses two or more Category I appliances with single-wall connectors.

Step 2. Using the top portion of the table, size the appliance connectors. Because the table has no

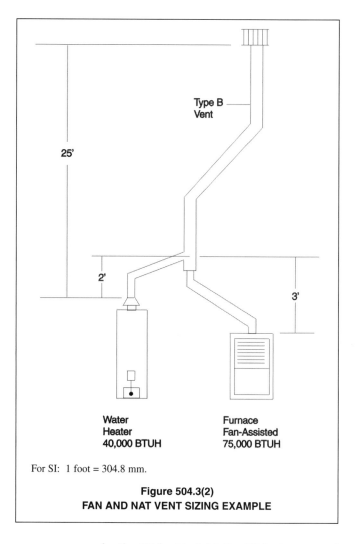

For SI: 1 foot = 304.8 mm.

**Figure 504.3(2)
FAN AND NAT VENT SIZING EXAMPLE**

row for the 25 foot height, the 20 foot row must be used, or interpolation must be used to calculate the capacities at 25 feet. To simplify the procedure, first proceed using the 20 foot row until interpolation is necessary.

A. Looking in the "NAT" columns, see that a 4-inch connector is required for the water heater because the water heater input of 40,000 Btu/h exceeds the maximum capacity (36,000) of a 3-inch connector with a 2-foot rise.

B. Looking in the "FAN" columns, see that the furnace input of 75,000 Btu/h is less than the minimum capacity of 4-inch and larger connectors with a 3-foot rise. This means that single-wall pipe cannot be used for the furnace connector. The furnace connector must be double-wall (Type B or equivalent) and must be sized using Table 504.3(1). In Table 504.3(1), see that a 4-inch double-wall connector is suitable for the furnace because the 75,000 Btu/h input fits within the range of 35,000 to 110,000 Btu/h for a 3-foot rise.

The furnace connector has been determined to be 4-inch double-wall pipe, and the water heater connector has been determined to be 4-inch single-wall pipe.

Step 3. Using the bottom portion of Table 504.3(2), the common vent is sized. (Because both single- and double-wall connectors are used within the same system, Section 504.3.22 requires use of the single-wall table.) The water heater and furnace inputs total 115,000 Btu/h. Under the 4-inch vent heading, FAN + NAT column, see that 4-inch Type B vent has a capacity of 118,000 Btu/h, which exceeds the appliance total of 115,000. However, the system contains an offset, which will reduce this capacity by 20 percent in accordance with Section 504.3.5 (0.80 X 118,000) = 94,400 Btu/h.

The actual capacity of a 4-inch common vent is 94,400 Btu/h, which is less than the 115,000 appliance total; therefore, a 5-inch common vent is required.

Verify 5-inch capacity: (0.80 X 177) = 141,600 Btu/h.

From the 30-foot height rows, it is apparent that interpolation for the actual height of 25 feet would not have changed the outcome of this system design.

Appendix B contains additional sizing examples.

504.3.1 Vent obstructions. These venting tables shall not be used where obstructions, as described in Section 503.15, are installed in the venting system. The installation of vents serving listed appliances with vent dampers shall be in accordance with the appliance manufacturer's instructions or in accordance with the following:

1. The maximum capacity of the vent connector shall be determined using the NAT Max column.
2. The maximum capacity of the vertical vent or chimney shall be determined using the FAN+NAT column when the second appliance is a fan-assisted appliance, or the NAT+NAT column when the second appliance is equipped with a draft hood.
3. The minimum capacity shall be determined as if the appliance were a fan-assisted appliance.
   3.1. The minimum capacity of the vent connector shall be determined using the FAN Min column.
   3.2. The FAN+FAN column shall be used where the second appliance is a fan-assisted appliance, and the FAN+NAT column shall be used where the second appliance is equipped with a draft hood, to determine whether the vertical vent or chimney configuration is not permitted (NA). Where the vent configuration is NA, the vent configuration shall not be permitted and an alternative venting configuration shall be utilized.

❖ See the commentary for Section 504.2.1. This section is a bit more complex than Section 504.2.1 because the tables for multiple appliances are two-part tables. The difference lies in items 2 and 3.2 dealing with the common vent portion of the sizing tables. Item 2 determines the maximum capacity of the common vent by treating the vent-damper-equipped appliance as a draft-hood appliance coupled with either another draft-hood appliance or a fan-assisted appliance. Item 3.2 is a test wherein the vent-damper-equipped appliance is treated as a fan-assisted appliance. If the table says NA, the vent system must be redesigned, regardless of the fact that it may have been allowed by the table under item 2.

504.3.2 Connector length limit. The vent connector shall be routed to the vent utilizing the shortest possible route. Except as provided in Section 504.3.3, the maximum vent connector horizontal length shall be $1^1/_2$ feet for each inch (457 mm per mm) of connector diameter as shown in Table 504.3.2.

❖ Connectors are necessary for chimney and vent systems that serve multiple appliances. To reduce friction loss and flue-gas heat loss and to conserve available draft, connectors must be kept to a minimum length. Thoughtful placement of appliances and chimneys or vents can reduce connector lengths and achieve optimum venting performance. Unlike Tables 504.2(1) through 504.2(5), Tables 504.3(1) through 504.3(8) do not provide a range of permissible lateral lengths. Therefore, Section 503.10.9 places an overall maximum limit on connector length. Section 503.10.9 provides a necessary limiting function because Tables 504.3(1) through 504.3(8) would appear to allow (unintentionally) a connector to be considerably longer than the height of the vent.

TABLE 504.3.2
MAXIMUM VENT CONNECTOR LENGTH

CONNECTOR DIAMETER MAXIMUM (inches)	CONNECTOR HORIZONTAL LENGTH (feet)
3	$4^1/_2$
4	6
5	$7^1/_2$
6	9
7	$10^1/_2$
8	12
9	$13^1/_2$
10	15
12	18
14	21
16	24
18	27
20	30
22	33
24	36

For SI: 1 inch = 25.4 mm, 1 foot = 304.8 mm.

❖ See the commentary for Section 504.3.2.

Increasing the length of the connector increases the required minimum capacity of a vent and decreases its maximum capacity. Some vent manufacturers appear to measure only the horizontal (lateral) run of a connector in their design manuals. The more conservative approach would be to consider the friction loss and heat loss for the entire run of a connector throughout its developed length (see commentary, Sections 503.10.6 and 503.10.9). The vent sizing tables were developed with the assumption that vent connectors are no longer than 18 inches per inch of connector diameter. This section establishes the basic rule, which is modified by Sections 504.3.3 and 503.10.9. Simply put, connector length is limited by this section to 18 inches per inch of diameter without capacity penalties or is limited in length by Section 503.10.9 if the capacity penalties of Section 504.3.3 are applied and the minimum capacity is met for fan-assisted appliances. Section 504.3.3 allows longer connector lengths, but this section and the others mentioned still require connectors to take the shortest route to the vent.

504.3.3 Connectors with longer lengths. Connectors with longer horizontal lengths than those listed in Section 504.3.2 are permitted under the following conditions:

1. The maximum capacity (FAN Max or NAT Max) of the vent connector shall be reduced 10 percent for each additional multiple of the length listed above. For example, the maximum length listed above for a 4-inch (102 mm) connector is 6 feet (1829 mm). With a connector length greater than 6 feet (1829 mm) but not exceeding 12 feet (3658 mm), the maximum capacity must be reduced by 10 percent (0.90 × maximum vent connector capacity). With a connector length greater than 12 feet (3658 mm) but not exceeding 18 feet (5486 mm), the maximum capacity must be reduced by 20 percent (0.80 × maximum vent capacity).

2. For a connector serving a fan-assisted appliance, the minimum capacity (FAN Min) of the connector shall be determined by referring to the corresponding single appliance table. For Type B double-wall connectors, Table 504.2(1) shall be used. For single-wall connectors, Table 504.2(2) shall be used. The height (H) and lateral (L) shall be measured according to the procedures for a single-appliance vent, as if the other appliances were not present.

❖ Section 503.10.9 still applies and serves to limit the application of this section and Section 504.3.2. For fan-assisted appliances, both items 1 and 2 apply. Item 1 allows the maximum length permitted by Section 504.3.2 to be increased (doubled, tripled, etc.) if the vent connector capacity is reduced by 10 percent when the maximum connector length is doubled, by 20 percent when tripled, by 30 percent when quadrupled, etc. Regardless of the length permitted, connectors should always be as short as practicable.

Item 2 applies to fan-assisted appliances and allows connector lengths to be as long as the lateral entries in Table 504.2(1) or 504.2(2) if the limits of Section 503.10.9 are not exceeded and the appliance complies with the minimum input requirement of the respective table.

When horizontal lengths in excess of those stated in Section 504.3.2 are necessary, the minimum capacity of the system is determined by referring to the corresponding single-appliance table. In this case, for each appliance the entire vent connector and common vent from appliance to vent termination is treated as a single-appliance vent (of the same size as the common vent), as if the others were not present. Any appliance failing to meet the Min input may be prone to creating excessive condensation and/or insufficient draft within the vent system if the appliance is operated by itself. In that case, options may include relocation of appliances, selective sequential or simultaneous operation of appliances or separate vent installations. This section is intended to be used only where a longer connector cannot be avoided as required by Sections 503.10.6 and 503.10.9 and the connector takes the shortest possible route to the vent. The longer the connector, the greater the flow resistance and the greater the heat loss, both of which negatively affect vent system operation. The capacity penalties account for the increased flow resistance and item 2 accounts for the increased heat loss. Item 2 addresses the worst case condition when a single appliance is operating and treats a connector serving a fan-assisted appliance as if the appliance were being independently vented, hence, the reference to the single appliance tables. Recall that the single appliance tables limit lateral length, which can be, in effect, a connector length limitation.

In many cases, the sizing tables will prevent someone from taking advantage of the allowances of this section for connectors. For example, a single-wall connector serving a fan-assisted appliance may not be able to meet the minimum input requirement (FAN Min) if the capacity penalty of this section forces increasing the connector size. Note that Section 503.10.9 will not allow a connector length to exceed the vent or chimney height in any case.

504.3.4 Vent connector manifold. Where the vent connectors are combined prior to entering the vertical portion of the common vent to form a common vent manifold, the size of the common vent manifold and the common vent shall be determined by applying a 10-percent reduction (0.90 × maximum common vent capacity) to the common vent capacity part of the common vent tables. The length of the common vent connector manifold (L_m) shall not exceed $1^1/_2$ feet for each inch (457 mm per mm) of common vent connector manifold diameter (D) (see Figure B-11).

❖ This section inflicts a 10-percent capacity penalty on the common vent as a result of the lateral (offset) effect of a manifold. A connector manifold is a horizontal extension of the lower end of a common vent [see commentary Figure 503.10.3.4(2)]. The capacity reduction accounts for the additional turn in the common vent. The manifold portion of a vent is sized based on the common vent portion of the tables. The length of a connector manifold is measured from the vertical common vent to the most upstream interconnection fitting.

CHIMNEYS AND VENTS   TABLE 504.3(1)

TABLE 504.3(1)
TYPE B DOUBLE-WALL VENT

Number of Appliances	Two or more
Appliance Type	Category I
Appliance Vent Connection	Type B double-wall connector

VENT CONNECTOR CAPACITY

		\multicolumn{16}{c}{TYPE B DOUBLE-WALL VENT AND CONNECTOR DIAMETER—(D) inches}

VENT HEIGHT (H) (feet)	CONNECTOR RISE (R) (feet)	3 FAN Min	3 FAN Max	3 NAT Max	4 FAN Min	4 FAN Max	4 NAT Max	5 FAN Min	5 FAN Max	5 NAT Max	6 FAN Min	6 FAN Max	6 NAT Max	7 FAN Min	7 FAN Max	7 NAT Max	8 FAN Min	8 FAN Max	8 NAT Max	9 FAN Min	9 FAN Max	9 NAT Max	10 FAN Min	10 FAN Max	10 NAT Max
6	1	22	37	26	35	66	46	46	106	72	58	164	104	77	225	142	92	296	185	109	376	237	128	466	289
6	2	23	41	31	37	75	55	48	121	86	60	183	124	79	253	168	95	333	220	112	424	282	131	526	345
6	3	24	44	35	38	81	62	49	132	96	62	199	139	82	275	189	97	363	248	114	463	317	134	575	386
8	1	22	40	27	35	72	48	49	114	76	64	176	109	84	243	148	100	320	194	118	408	248	138	507	303
8	2	23	44	32	36	80	57	51	128	90	66	195	129	86	269	175	103	356	230	121	454	294	141	564	358
8	3	24	47	36	37	87	64	53	139	101	67	210	145	88	290	198	105	384	258	123	492	330	143	612	402
10	1	22	43	28	34	78	50	49	123	78	65	189	113	89	257	154	106	341	200	125	436	257	146	542	314
10	2	23	47	33	36	86	59	51	136	93	67	206	134	91	282	182	109	374	238	128	479	305	149	596	372
10	3	24	50	37	37	92	67	52	146	104	69	220	150	94	303	205	111	402	268	131	515	342	152	642	417
15	1	21	50	30	33	89	53	47	142	83	64	220	120	88	298	163	110	389	214	134	493	273	162	609	333
15	2	22	53	35	35	96	63	49	153	99	66	235	142	91	320	193	112	419	253	137	532	323	165	658	394
15	3	24	55	40	36	102	71	51	163	111	68	248	160	93	339	218	115	445	286	140	565	365	167	700	444
20	1	21	54	31	33	99	56	46	157	87	62	246	125	86	334	171	107	436	224	131	552	285	158	681	347
20	2	22	57	37	34	105	66	48	167	104	64	259	149	89	354	202	110	463	265	134	587	339	161	725	414
20	3	23	60	42	35	110	74	50	176	116	66	271	168	91	371	228	113	486	300	137	618	383	164	764	466
30	1	20	62	33	31	113	59	45	181	93	60	288	134	83	391	182	103	512	238	125	649	305	151	802	372
30	2	21	64	39	33	118	70	47	190	110	62	299	158	85	408	215	105	535	282	129	679	360	155	840	439
30	3	22	66	44	34	123	79	48	198	124	64	309	178	88	423	242	108	555	317	132	706	405	158	874	494
50	1	19	71	36	30	133	64	43	216	101	57	349	145	78	477	197	97	627	257	120	797	330	144	984	403
50	2	21	73	43	32	137	76	45	223	119	59	358	172	81	490	234	100	645	306	123	820	392	148	1,014	478
50	3	22	75	48	33	141	86	46	229	134	61	366	194	83	502	263	103	661	343	126	842	441	151	1,043	538
100	1	18	82	37	28	158	66	40	262	104	53	442	150	73	611	204	91	810	266	112	1,038	341	135	1,285	417
100	2	19	83	44	30	161	79	42	267	123	55	447	178	75	619	242	94	822	316	115	1,054	405	139	1,306	494
100	3	20	84	50	31	163	89	44	272	138	57	452	109	78	627	272	97	834	355	118	1,069	455	142	1,327	555

COMMON VENT CAPACITY

| | \multicolumn{21}{c}{TYPE B DOUBLE-WALL COMMON VENT DIAMETER (D)—inches} |

VENT HEIGHT (H) (feet)	4 FAN+FAN	4 FAN+NAT	4 NAT+NAT	5 FAN+FAN	5 FAN+NAT	5 NAT+NAT	6 FAN+FAN	6 FAN+NAT	6 NAT+NAT	7 FAN+FAN	7 FAN+NAT	7 NAT+NAT	8 FAN+FAN	8 FAN+NAT	8 NAT+NAT	9 FAN+FAN	9 FAN+NAT	9 NAT+NAT	10 FAN+FAN	10 FAN+NAT	10 NAT+NAT
6	92	81	65	140	116	103	204	161	147	309	248	200	404	314	260	547	434	335	672	520	410
8	101	90	73	155	129	114	224	178	163	339	275	223	444	348	290	602	480	378	740	577	465
10	110	97	79	169	141	124	243	194	178	367	299	242	477	377	315	649	522	405	800	627	495
15	125	112	91	195	164	144	283	228	206	427	352	280	556	444	365	753	612	465	924	733	565
20	136	123	102	215	183	160	314	255	229	475	394	310	621	499	405	842	688	523	1,035	826	640
30	152	138	118	244	210	185	361	297	266	547	459	360	720	585	470	979	808	605	1,209	975	740
50	167	153	134	279	244	214	421	353	310	641	547	423	854	706	550	1,164	977	705	1,451	1,188	860
100	175	163	NA	311	277	NA	489	421	NA	751	658	479	1,025	873	625	1,408	1,215	800	1,784	1,502	975

(continued)

TABLE 504.3(1)—continued
TYPE B DOUBLE-WALL VENT

Number of Appliances	Two or more
Appliance Type	Category I
Appliance Vent Connection	Type B double-wall connector

VENT CONNECTOR CAPACITY

VENT HEIGHT (H) (feet)	CONNECTOR RISE (R) (feet)	TYPE B DOUBLE-WALL VENT AND DIAMETER—(D) inches																			
		12			14			16			18			20			22			24		
		APPLIANCE INPUT RATING LIMITS IN THOUSANDS OF BTU/H																			
		FAN		NAT	FAN		NAT	FAN		NAT	FAN		NAT	FAN		NAT	FAN		NAT	FAN		NAT
		Min	Max	Max	Min	Max	Max	Min	Max	Max	Min	Max	Max	Min	Max	Max	Min	Max	Max	Min	Max	Max
6	2	174	764	496	223	1,046	653	281	1,371	853	346	1,772	1,080	NA	NA	NA	NA	NA	NA	NA	NA	NA
	4	180	897	616	230	1,231	827	287	1,617	1,081	352	2,069	1,370	NA	NA	NA	NA	NA	NA	NA	NA	NA
	6	NA	NA	NA	NA	NA	NA	NA	NA	NA	NA	NA	NA	NA	NA	NA	NA	NA	NA	NA	NA	NA
8	2	186	822	516	238	1,126	696	298	1,478	910	365	1,920	1,150	NA	NA	NA	NA	NA	NA	NA	NA	NA
	4	192	952	644	244	1,307	884	305	1,719	1,150	372	2,211	1,460	471	2,737	1,800	560	3,319	2,180	662	3,957	2,590
	6	198	1,050	772	252	1,445	1,072	313	1,902	1,390	380	2,434	1,770	478	3,018	2,180	568	3,665	2,640	669	4,373	3,130
10	2	196	870	536	249	1,195	730	311	1,570	955	379	2,049	1,205	NA	NA	NA	NA	NA	NA	NA	NA	NA
	4	201	997	664	256	1,371	924	318	1,804	1,205	387	2,332	1,535	486	2,887	1,890	581	3,502	2,280	686	4,175	2,710
	6	207	1,095	792	263	1,509	1,118	325	1,989	1,455	395	2,556	1,865	494	3,169	2,290	589	3,849	2,760	694	4,593	3,270
15	2	214	967	568	272	1,334	790	336	1,760	1,030	408	2,317	1,305	NA	NA	NA	NA	NA	NA	NA	NA	NA
	4	221	1,085	712	279	1,499	1,006	344	1,978	1,320	416	2,579	1,665	523	3,197	2,060	624	3,881	2,490	734	4,631	2,960
	6	228	1,181	856	286	1,632	1,222	351	2,157	1,610	424	2,796	2,025	533	3,470	2,510	634	4,216	3,030	743	5,035	3,600
20	2	223	1,051	596	291	1,443	840	357	1,911	1,095	430	2,533	1,385	NA	NA	NA	NA	NA	NA	NA	NA	NA
	4	230	1,162	748	298	1,597	1,064	365	2,116	1,395	438	2,778	1,765	554	3,447	2,180	661	4,190	2,630	772	5,005	3,130
	6	237	1,253	900	307	1,726	1,288	373	2,287	1,695	450	2,984	2,145	567	3,708	2,650	671	4,511	3,190	785	5,392	3,790
30	2	216	1,217	632	286	1,664	910	367	2,183	1,190	461	2,891	1,540	NA	NA	NA	NA	NA	NA	NA	NA	NA
	4	223	1,316	792	294	1,802	1,160	376	2,366	1,510	474	3,110	1,920	619	3,840	2,365	728	4,861	2,860	847	5,606	3,410
	6	231	1,400	952	303	1,920	1,410	384	2,524	1,830	485	3,299	2,340	632	4,080	2,875	741	4,976	3,480	860	5,961	4,150
50	2	206	1,479	689	273	2,023	1,007	350	2,659	1,315	435	3,548	1,665	NA	NA	NA	NA	NA	NA	NA	NA	NA
	4	213	1,561	860	281	2,139	1,291	359	2,814	1,685	447	3,730	2,135	580	4,601	2,633	709	5,569	3,185	851	6,633	3,790
	6	221	1,631	1,031	290	2,242	1,575	369	2,951	2,055	461	3,893	2,605	594	4,808	3,208	724	5,826	3,885	867	6,943	4,620
100	2	192	1,923	712	254	2,644	1,050	326	3,490	1,370	402	4,707	1,740	NA	NA	NA	NA	NA	NA	NA	NA	NA
	4	200	1,984	888	263	2,731	1,346	336	3,606	1,760	414	4,842	2,220	523	5,982	2,750	639	7,254	3,330	769	8,650	3,950
	6	208	2,035	1,064	272	2,811	1,642	346	3,714	2,150	426	4,968	2,700	539	6,143	3,350	654	7,453	4,070	786	8,892	4,810

COMMON VENT CAPACITY

VENT HEIGHT (H) (feet)	TYPE B DOUBLE-WALL COMMON VENT DIAMETER—(D) inches																				
	12			14			16			18			20			22			24		
	COMBINED APPLIANCE INPUT RATING IN THOUSANDS OF BTU/H																				
	FAN +FAN	FAN +NAT	NAT +NAT	FAN +FAN	FAN +NAT	NAT +NAT	FAN +FAN	FAN +NAT	NAT +NAT	FAN +FAN	FAN +NAT	NAT +NAT	FAN +FAN	FAN +NAT	NAT +NAT	FAN +FAN	FAN +NAT	NAT +NAT	FAN +FAN	FAN +NAT	NAT +NAT
6	900	696	588	1,284	990	815	1,735	1,336	1,065	2,253	1,732	1,345	2,838	2,180	1,660	3,488	2,677	1970	4,206	3,226	2,390
8	994	773	652	1,423	1,103	912	1,927	1,491	1,190	2,507	1,936	1,510	3,162	2,439	1,860	3,890	2,998	2,200	4,695	3,616	2,680
10	1,076	841	712	1,542	1,200	995	2,093	1,625	1,300	2,727	2,113	1645	3,444	2,665	2,030	4,241	3,278	2,400	5,123	3,957	2,920
15	1,247	986	825	1,794	1,410	1,158	2,440	1,910	1,510	3,184	2,484	1,910	4,026	3,133	2,360	4,971	3,862	2,790	6,016	4,670	3,400
20	1,405	1,116	916	2,006	1,588	1,290	2,722	2,147	1,690	3,561	2,798	2,140	4,548	3,552	2,640	5,573	4,352	3,120	6,749	5,261	3,800
30	1,658	1,327	1,025	2,373	1,892	1,525	3,220	2,558	1,990	4,197	3,326	2,520	5,303	4,193	3,110	6,539	5,157	3,680	7,940	6,247	4,480
50	2,024	1,640	1,280	2,911	2,347	1,863	3,964	3,183	2,430	5,184	4,149	3,075	6,567	5,240	3,800	8,116	6,458	4,500	9,837	7,813	5,475
100	2,569	2,131	1,670	3,732	3,076	2,450	5,125	4,202	3,200	6,749	5,509	4,050	8,597	6,986	5,000	10,681	8,648	5,920	13,004	10,499	7,200

For SI: 1 inch = 25.4 mm, 1 foot = 304.8 mm, 1 British thermal unit per hour = 0.2931 W.

❖ For tables 504.3(1) through 504.3(8), see the commentary for Section 504.3.

CHIMNEYS AND VENTS

TABLE 504.3(2)
TYPE B DOUBLE-WALL VENT

Number of Appliances	Two or more
Appliance Type	Category I
Appliance Vent Connection	Single-wall metal connector

VENT CONNECTOR CAPACITY

VENT HEIGHT (H) (feet)	CONNECTOR RISE (R) (feet)	3 FAN Min	3 FAN Max	3 NAT Max	4 FAN Min	4 FAN Max	4 NAT Max	5 FAN Min	5 FAN Max	5 NAT Max	6 FAN Min	6 FAN Max	6 NAT Max	7 FAN Min	7 FAN Max	7 NAT Max	8 FAN Min	8 FAN Max	8 NAT Max	9 FAN Min	9 FAN Max	9 NAT Max	10 FAN Min	10 FAN Max	10 NAT Max
6	1	NA	NA	26	NA	NA	46	NA	NA	71	NA	NA	102	207	223	140	262	293	183	325	373	234	447	463	286
6	2	NA	NA	31	NA	NA	55	NA	NA	85	168	182	123	215	251	167	271	331	219	334	422	281	458	524	344
6	3	NA	NA	34	NA	NA	62	121	131	95	175	198	138	222	273	188	279	361	247	344	462	316	468	574	385
8	1	NA	NA	27	NA	NA	48	NA	NA	75	NA	NA	106	226	240	145	285	316	191	352	403	244	481	502	299
8	2	NA	NA	32	NA	NA	57	125	126	89	184	193	127	234	266	173	293	353	228	360	450	292	492	560	355
8	3	NA	NA	35	NA	NA	64	130	138	100	191	208	144	241	287	197	302	381	256	370	489	328	501	609	400
10	1	NA	NA	28	NA	NA	50	119	121	77	182	186	110	240	253	150	302	335	196	372	429	252	506	534	308
10	2	NA	NA	33	84	85	59	124	134	91	189	203	132	248	278	183	311	369	235	381	473	302	517	589	368
10	3	NA	NA	36	89	91	67	129	144	102	197	217	148	257	299	203	320	398	265	391	511	339	528	637	413
15	1	NA	NA	29	79	87	52	116	138	81	177	214	116	238	291	158	312	380	208	397	482	266	556	596	324
15	2	NA	NA	34	83	94	62	121	150	97	185	230	138	246	314	189	321	411	248	407	522	317	568	646	387
15	3	NA	NA	39	87	100	70	127	160	109	193	243	157	255	333	215	331	438	281	418	557	360	579	690	437
20	1	49	56	30	78	97	54	115	152	84	175	238	120	233	325	165	306	425	217	390	538	276	546	664	336
20	2	52	59	36	82	103	64	120	163	101	182	252	144	243	346	197	317	453	259	400	574	331	558	709	403
20	3	55	62	40	87	107	72	125	172	113	190	264	164	252	363	223	326	476	294	412	607	375	570	750	457
30	1	47	60	31	77	110	57	112	175	89	169	278	129	226	380	175	296	497	230	378	630	294	528	779	358
30	2	51	62	37	81	115	67	117	185	106	177	290	152	236	397	208	307	521	274	389	662	349	541	819	425
30	3	54	64	42	85	119	76	122	193	120	185	300	172	244	412	235	316	542	309	400	690	394	555	855	482
50	1	46	69	34	75	128	60	109	207	96	162	336	137	217	460	188	284	604	245	364	768	314	507	951	384
50	2	49	71	40	79	132	72	114	215	113	170	345	164	226	473	223	294	623	293	376	793	375	520	983	458
50	3	52	72	45	83	136	82	119	221	123	178	353	186	235	486	252	304	640	331	387	816	423	535	1,013	518
100	1	45	79	34	71	150	61	104	249	98	153	424	140	205	585	192	269	774	249	345	993	321	476	1,236	393
100	2	48	80	41	75	153	73	110	255	115	160	428	167	212	593	228	279	788	299	358	1,011	383	490	1,259	469
100	3	51	81	46	79	157	85	114	260	129	168	433	190	222	603	256	289	801	339	368	1,027	431	506	1,280	527

COMMON VENT CAPACITY

VENT HEIGHT (H) (feet)	4 FAN+FAN	4 FAN+NAT	4 NAT+NAT	5 FAN+FAN	5 FAN+NAT	5 NAT+NAT	6 FAN+FAN	6 FAN+NAT	6 NAT+NAT	7 FAN+FAN	7 FAN+NAT	7 NAT+NAT	8 FAN+FAN	8 FAN+NAT	8 NAT+NAT	9 FAN+FAN	9 FAN+NAT	9 NAT+NAT	10 FAN+FAN	10 FAN+NAT	10 NAT+NAT
6	NA	78	64	NA	113	99	200	158	144	304	244	196	398	310	257	541	429	332	665	515	407
8	NA	87	71	NA	126	111	218	173	159	331	269	218	436	342	285	592	473	373	730	569	460
10	NA	94	76	163	137	120	237	189	174	357	292	236	467	369	309	638	512	398	787	617	487
15	121	108	88	189	159	140	275	221	200	416	343	274	544	434	357	738	599	456	905	718	553
20	131	118	98	208	177	156	305	247	223	463	383	302	606	487	395	824	673	512	1,013	808	626
30	145	132	113	236	202	180	350	286	257	533	446	349	703	570	459	958	790	593	1,183	952	723
50	159	145	128	268	233	208	406	337	296	622	529	410	833	686	535	1,139	954	689	1,418	1,157	838
100	166	153	NA	297	263	NA	469	398	NA	726	633	464	999	846	606	1,378	1,185	780	1,741	1,459	948

For SI: 1 inch = 25.4 mm, 1 foot = 304.8 mm, 1 British thermal unit per hour = 0.2931 W.

TABLE 504.3(3) MASONRY CHIMNEY

Number of Appliances	Two or more
Appliance Type	Category I
Appliance Vent Connection	Type B double-wall connector

VENT CONNECTOR CAPACITY

VENT HEIGHT (H) (feet)	CONNECTOR RISE (R) (feet)	TYPE B DOUBLE-WALL VENT CONNECTOR DIAMETER—(D) inches																			
		3			4			5			6			7			8			9			10		
		_____APPLIANCE INPUT RATING LIMITS IN THOUSANDS OF BTU/H_____																			
		FAN		NAT	FAN		NAT	FAN		NAT	FAN		NAT	FAN		NAT	FAN		NAT	FAN		NAT	FAN		NAT
		Min	Max	Max	Min	Max	Max	Min	Max	Max	Min	Max	Max	Min	Max	Max	Min	Max	Max	Min	Max	Max	Min	Max	Max
6	1	24	33	21	39	62	40	52	106	67	65	194	101	87	274	141	104	370	201	124	479	253	145	599	319
6	2	26	43	28	41	79	52	53	133	85	67	230	124	89	324	173	107	436	232	127	562	300	148	694	378
6	3	27	49	34	42	92	61	55	155	97	69	262	143	91	369	203	109	491	270	129	633	349	151	795	439
8	1	24	39	22	39	72	41	55	117	69	71	213	105	94	304	148	113	414	210	134	539	267	156	682	335
8	2	26	47	29	40	87	53	57	140	86	73	246	127	97	350	179	116	473	240	137	615	311	160	776	394
8	3	27	52	34	42	97	62	59	159	98	75	269	145	99	383	206	119	517	276	139	672	358	163	848	452
10	1	24	42	22	38	80	42	55	130	71	74	232	108	101	324	153	120	444	216	142	582	277	165	739	348
10	2	26	50	29	40	93	54	57	153	87	76	261	129	103	366	184	123	498	247	145	652	321	168	825	407
10	3	27	55	35	41	105	63	58	170	100	78	284	148	106	397	209	126	540	281	147	705	366	171	893	463
15	1	24	48	23	38	93	44	54	154	74	72	277	114	100	384	164	125	511	229	153	658	297	184	824	375
15	2	25	55	31	39	105	55	56	174	89	74	299	134	103	419	192	128	558	260	156	718	339	187	900	432
15	3	26	59	35	41	115	64	57	189	102	76	319	153	105	448	215	131	597	292	159	760	382	190	960	486
20	1	24	52	24	37	102	46	53	172	77	71	313	119	98	437	173	123	584	239	150	752	312	180	943	397
20	2	25	58	31	39	114	56	55	190	91	73	335	138	101	467	199	126	625	270	153	805	354	184	1,011	452
20	3	26	63	35	40	123	65	57	204	104	75	353	157	104	493	222	129	661	301	156	851	396	187	1,067	505
30	1	24	54	25	37	111	48	52	192	82	69	357	127	96	504	187	119	680	255	145	883	337	175	1,115	432
30	2	25	60	32	38	122	58	54	208	95	72	376	145	99	531	209	122	715	287	149	928	378	179	1,171	484
30	3	26	64	36	40	131	66	56	221	107	74	392	163	101	554	233	125	746	317	152	968	418	182	1,220	535
50	1	23	51	25	36	116	51	51	209	89	67	405	143	92	582	213	115	798	294	140	1,049	392	168	1,334	506
50	2	24	59	32	37	127	61	53	225	102	70	421	161	95	604	235	118	827	326	143	1,085	433	172	1,379	558
50	3	26	64	36	39	135	69	55	237	115	72	435	80	98	624	260	121	854	357	147	1,118	474	176	1,421	611
100	1	23	46	24	35	108	50	49	208	92	65	428	155	88	640	237	109	907	334	134	1,222	454	161	1,589	596
100	2	24	53	31	37	120	60	51	224	105	67	444	174	92	660	260	113	933	368	138	1,253	497	165	1,626	651
100	3	25	59	35	38	130	68	53	237	118	69	458	193	94	679	285	116	956	399	141	1,282	540	169	1,661	705

COMMON VENT CAPACITY

VENT HEIGHT (H) (feet)	MINIMUM INTERNAL AREA OF MASONRY CHIMNEY FLUE (square inches)																					
	12			19			28			38			50			63			78			113		
	COMBINED APPLIANCE INPUT RATING IN THOUSANDS OF BTU/H																					
	FAN +FAN	FAN +NAT	NAT +NAT	FAN +FAN	FAN +NAT	NAT +NAT	FAN +FAN	FAN +NAT	NAT +NAT	FAN +FAN	FAN +NAT	NAT +NAT	FAN +FAN	FAN +NAT	NAT +NAT	FAN +FAN	FAN +NAT	NAT +NAT	FAN +FAN	FAN +NAT	NAT +NAT	FAN +FAN	FAN +NAT	NAT +NAT
6	NA	74	25	NA	119	46	NA	178	71	NA	257	103	NA	351	143	NA	458	188	NA	582	246	1,041	853	NA
8	NA	80	28	NA	130	53	NA	193	82	NA	279	119	NA	384	163	NA	501	218	724	636	278	1,144	937	408
10	NA	84	31	NA	138	56	NA	207	90	NA	299	131	NA	409	177	606	538	236	776	686	302	1,226	1,010	454
15	NA	NA	36	NA	152	67	NA	233	106	NA	334	152	523	467	212	682	611	283	874	781	365	1,374	1,156	546
20	NA	NA	41	NA	NA	75	NA	250	122	NA	368	172	565	508	243	742	668	325	955	858	419	1,513	1,286	648
30	NA	NA	NA	NA	NA	NA	NA	270	137	NA	404	198	615	564	278	816	747	381	1,062	969	496	1,702	1,473	749
50	NA	NA	NA	NA	NA	NA	NA	NA	NA	NA	NA	NA	NA	620	328	879	831	461	1,165	1,089	606	1,905	1,692	922
100	NA	NA	NA	NA	NA	NA	NA	NA	NA	NA	NA	NA	NA	348	NA	NA	499	NA	NA	669	2,053	1,921	1,058

For SI: 1 inch = 25.4 mm, 1 square inch = 645.16 mm², 1 foot = 304.8 mm, 1 British thermal unit per hour = 0.2931 W.

CHIMNEYS AND VENTS | TABLE 504.3(4)

TABLE 504.3(4)
MASONRY CHIMNEY

Number of Appliances	Two or more
Appliance Type	Category I
Appliance Vent Connection	Single-wall metal connector

VENT CONNECTOR CAPACITY

		\multicolumn{24}{c}{SINGLE-WALL METAL VENT CONNECTOR DIAMETER (D)—inches}
		3			4			5			6			7			8			9			10		
VENT HEIGHT (H) (feet)	CONNECTOR RISE (R) (feet)	\multicolumn{24}{c}{APPLIANCE INPUT RATING LIMITS IN THOUSANDS OF BTU/H}
		FAN		NAT	FAN		NAT	FAN		NAT	FAN		NAT	FAN		NAT	FAN		NAT	FAN		NAT	FAN		NAT
		Min	Max	Max	Min	Max	Max	Min	Max	Max	Min	Max	Max	Min	Max	Max	Min	Max	Max	Min	Max	Max	Min	Max	Max
6	1	NA	NA	21	NA	NA	39	NA	NA	66	179	191	100	231	271	140	292	366	200	362	474	252	499	594	316
6	2	NA	NA	28	NA	NA	52	NA	NA	84	186	227	123	239	321	172	301	432	231	373	557	299	509	696	376
6	3	NA	NA	34	NA	NA	61	134	153	97	193	258	142	247	365	202	309	491	269	381	634	348	519	793	437
8	1	NA	NA	21	NA	NA	40	NA	NA	68	195	208	103	250	298	146	313	407	207	387	530	263	529	672	331
8	2	NA	NA	28	NA	NA	52	137	139	85	202	240	125	258	343	177	323	465	238	397	607	309	540	766	391
8	3	NA	NA	34	NA	NA	62	143	156	98	210	264	145	266	376	205	332	509	274	407	663	356	551	838	450
10	1	NA	NA	22	NA	NA	41	130	151	70	202	225	106	267	316	151	333	434	213	410	571	273	558	727	343
10	2	NA	NA	29	NA	NA	53	136	150	86	210	255	128	276	358	181	343	489	244	420	640	317	569	813	403
10	3	NA	NA	34	97	102	62	143	166	99	217	277	147	284	389	207	352	530	279	430	694	363	580	880	459
15	1	NA	NA	23	NA	NA	43	129	151	73	199	271	112	268	376	161	349	502	225	445	646	291	623	808	366
15	2	NA	NA	30	92	103	54	135	170	88	207	295	132	277	411	189	359	548	256	456	706	334	634	884	424
15	3	NA	NA	34	96	112	63	141	185	101	215	315	151	286	439	213	368	586	289	466	755	378	646	945	479
20	1	NA	NA	23	87	99	45	128	167	76	197	303	117	265	425	169	345	569	235	439	734	306	614	921	347
20	2	NA	NA	30	91	111	55	134	185	90	205	325	136	274	455	195	355	610	266	450	787	348	627	986	443
20	3	NA	NA	35	96	119	64	140	199	103	213	343	154	282	481	219	365	644	298	461	831	391	639	1,042	496
30	1	NA	NA	24	86	108	47	126	187	80	193	347	124	259	492	183	338	665	250	430	864	330	600	1,089	421
30	2	NA	NA	31	91	119	57	132	203	93	201	366	142	269	518	205	348	699	282	442	908	372	613	1,145	473
30	3	NA	NA	35	95	127	65	138	216	105	209	381	160	277	540	229	358	729	312	452	946	412	626	1,193	524
50	1	NA	NA	24	85	113	50	124	204	87	188	392	139	252	567	208	328	778	287	417	1,022	383	582	1,302	492
50	2	NA	NA	31	89	123	60	130	218	100	196	408	158	262	588	230	339	806	320	429	1,058	425	596	1,346	545
50	3	NA	NA	35	94	131	68	136	231	112	205	422	176	271	607	255	349	831	351	440	1,090	466	610	1,386	597
100	1	NA	NA	23	84	104	49	122	200	89	182	410	151	243	617	232	315	875	328	402	1,181	444	560	1,537	580
100	2	NA	NA	30	88	115	59	127	215	102	190	425	169	253	636	254	326	899	361	415	1,210	488	575	1,570	634
100	3	NA	NA	34	93	124	67	133	228	115	199	438	188	262	654	279	337	921	392	427	1,238	529	589	1,604	687

COMMON VENT CAPACITY

	\multicolumn{24}{c}{MINIMUM INTERNAL AREA OF MASONRY CHIMNEY FLUE (square inches)}
	12			19			28			38			50			63			78			113		
VENT HEIGHT (H) (feet)	\multicolumn{24}{c}{COMBINED APPLIANCE INPUT RATING IN THOUSANDS OF BTU/H}
	FAN+FAN	FAN+NAT	NAT+NAT	FAN+FAN	FAN+NAT	NAT+NAT	FAN+FAN	FAN+NAT	NAT+NAT	FAN+FAN	FAN+NAT	NAT+NAT	FAN+FAN	FAN+NAT	NAT+NAT	FAN+FAN	FAN+NAT	NAT+NAT	FAN+FAN	FAN+NAT	NAT+NAT	FAN+FAN	FAN+NAT	NAT+NAT
6	NA	NA	25	NA	118	45	NA	176	71	NA	255	102	NA	348	142	NA	455	187	NA	579	245	NA	846	NA
8	NA	NA	28	NA	128	52	NA	190	81	NA	276	118	NA	380	162	NA	497	217	NA	633	277	1,136	928	405
10	NA	NA	31	NA	136	56	NA	205	89	NA	295	129	NA	405	175	NA	532	234	171	680	300	1,216	1,000	450
15	NA	NA	36	NA	NA	66	NA	230	105	NA	335	150	NA	400	210	677	602	280	866	772	360	1,359	1,139	540
20	NA	NA	NA	NA	NA	74	NA	247	120	NA	362	170	NA	503	240	765	661	321	947	849	415	1,495	1,264	640
30	NA	NA	NA	NA	NA	NA	NA	135	NA	398	195	NA	558	275	808	739	377	1,052	957	490	1,682	1,447	740
50	NA	NA	NA	NA	NA	NA	NA	NA	NA	NA	NA	NA	612	325	NA	821	456	1,152	1,076	600	1,879	1,672	910
100	NA	NA	NA	NA	NA	NA	NA	NA	NA	NA	NA	NA	NA	NA	494	NA	NA	663	2,006	1,885	1,046

For SI: 1 inch = 25.4 mm, 1 square inch = 645.16 mm^2, 1 foot = 304.8 mm, 1 British thermal unit per hour = 0.2931 W.

TABLE 504.3(5)
SINGLE-WALL METAL PIPE OR TYPE ASBESTOS CEMENT VENT

Number of Appliances	Two or more
Appliance Type	Draft hood-equipped
Appliance Vent Connection	Direct to pipe or vent

VENT CONNECTOR CAPACITY

TOTAL VENT HEIGHT (H) (feet)	CONNECTOR RISE (R) (feet)	VENT CONNECTOR DIAMETER—(D) inches					
		3	4	5	6	7	8
		MAXIMUM APPLIANCE INPUT RATING IN THOUSANDS OF BTU/H					
6-8	1	21	40	68	102	146	205
6-8	2	28	53	86	124	178	235
6-8	3	34	61	98	147	204	275
15	1	23	44	77	117	179	240
15	2	30	56	92	134	194	265
15	3	35	64	102	155	216	298
30 and up	1	25	49	84	129	190	270
30 and up	2	31	58	97	145	211	295
30 and up	3	36	68	107	164	232	321

COMMON VENT CAPACITY

TOTAL VENT HEIGHT (H) (feet)	COMMON VENT DIAMETER—(D) inches						
	4	5	6	7	8	10	12
	COMBINED APPLIANCE INPUT RATING IN THOUSANDS OF BTU/H						
6	48	78	111	155	205	320	NA
8	55	89	128	175	234	365	505
10	59	95	136	190	250	395	560
15	71	115	168	228	305	480	690
20	80	129	186	260	340	550	790
30	NA	147	215	300	400	650	940
50	NA	NA	NA	360	490	810	1,190

For SI: 1 inch = 25.4 mm, 1 foot = 304.8 mm, 1 British thermal unit per hour = 0.2931 W.

CHIMNEYS AND VENTS

TABLE 504.3(6)

TABLE 504.3(6)
EXTERIOR MASONRY CHIMNEY

Number of Appliances	One
Appliance Type	NAT
Appliance Vent Connection	Type B double-wall connector

VENT HEIGHT (feet)	MINIMUM ALLOWABLE INPUT RATING OF SPACE-HEATING APPLIANCE IN THOUSANDS OF BTU PER HOUR							
	Internal area of chimney (square inches)							
	12	19	28	38	50	63	78	113
37°F or Greater	Local 99% Winter Design Temperature: 37°F or Greater							
6	0	0	0	0	0	0	0	0
8	0	0	0	0	0	0	0	0
10	0	0	0	0	0	0	0	0
15	NA	0	0	0	0	0	0	0
20	NA	NA	123	190	249	184	0	0
30	NA	NA	NA	NA	NA	393	334	0
50	NA	NA	NA	NA	NA	NA	NA	579
27 to 36°F	Local 99% Winter Design Temperature: 27 to 36°F							
6	0	0	68	116	156	180	212	266
8	0	0	82	127	167	187	214	263
10	0	51	97	141	183	201	225	265
15	NA	NA	NA	NA	233	253	274	305
20	NA	NA	NA	NA	NA	307	330	362
30	NA	NA	NA	NA	NA	419	445	485
50	NA	NA	NA	NA	NA	NA	NA	763
17 to 26°F	Local 99% Winter Design Temperature: 17 to 26°F							
6	NA	NA	NA	NA	NA	215	259	349
8	NA	NA	NA	NA	197	226	264	352
10	NA	NA	NA	NA	214	245	278	358
15	NA	NA	NA	NA	NA	296	331	398
20	NA	NA	NA	NA	NA	352	387	457
30	NA	NA	NA	NA	NA	NA	507	581
50	NA	NA	NA	NA	NA	NA	NA	NA
5 to 16°F	Local 99% Winter Design Temperature: 5 to 16°F							
6	NA	NA	NA	NA	NA	NA	NA	416
8	NA	NA	NA	NA	NA	NA	312	423
10	NA	NA	NA	NA	NA	289	331	430
15	NA	NA	NA	NA	NA	NA	393	485
20	NA	NA	NA	NA	NA	NA	450	547
30	NA	NA	NA	NA	NA	NA	NA	682
50	NA	NA	NA	NA	NA	NA	NA	972
-10 to 4°F	Local 99% Winter Design Temperature: -10 to 4°F							
6	NA	NA	NA	NA	NA	NA	NA	484
8	NA	NA	NA	NA	NA	NA	NA	494
10	NA	NA	NA	NA	NA	NA	NA	513
15	NA	NA	NA	NA	NA	NA	NA	586
20	NA	NA	NA	NA	NA	NA	NA	650
30	NA	NA	NA	NA	NA	NA	NA	805
50	NA	NA	NA	NA	NA	NA	NA	1,003
-11°F or Lower	Local 99% Winter Design Temperature: -11°F or Lower							
	Not recommended for any vent configurations							

Note: See Figure B-19 in Appendix B for a map showing local 99 percent winter design temperatures in the United States.

For SI: °C = [(°F - 32)]/1.8, 1 inch = 25.4 mm, 1 foot = 304.8 mm, 1 British thermal unit per hour = 0.2931 W.

TABLE 504.3(7a)
EXTERIOR MASONRY CHIMNEY

Number of Appliances	Two or more
Appliance Type	NAT + NAT
Appliance Vent Connection	Type B double-wall connector

Combined Appliance Maximum Input Rating in Thousands of Btu per Hour

VENT HEIGHT (feet)	INTERNAL AREA OF CHIMNEY (square inches)							
	12	19	28	38	50	63	78	113
6	25	46	71	103	143	188	246	NA
8	28	53	82	119	163	218	278	408
10	31	56	90	131	177	236	302	454
15	NA	67	106	152	212	283	365	546
20	NA	NA	NA	NA	NA	325	419	648
30	NA	NA	NA	NA	NA	NA	496	749
50	NA	NA	NA	NA	NA	NA	NA	922
100	NA	NA	NA	NA	NA	NA	NA	NA

TABLE 504.3(7b)
EXTERIOR MASONRY CHIMNEY

Number of Appliances	Two or more
Appliance Type	NAT + NAT
Appliance Vent Connection	Type B double-wall connector

Minimum Allowable Input Rating of Space-Heating Appliance in Thousands of Btu per Hour

VENT HEIGHT (feet)	INTERNAL AREA OF CHIMNEY (square inches)							
	12	19	28	38	50	63	78	113
37°F or Greater	Local 99% Winter Design Temperature: 37°F or Greater							
6	0	0	0	0	0	0	0	NA
8	0	0	0	0	0	0	0	0
10	0	0	0	0	0	0	0	0
15	NA	0	0	0	0	0	0	0
20	NA	NA	NA	NA	NA	184	0	0
30	NA	NA	NA	NA	NA	393	334	0
50	NA	NA	NA	NA	NA	NA	NA	579
100	NA	NA	NA	NA	NA	NA	NA	NA
27 to 36°F	Local 99% Winter Design Temperature: 27 to 36°F							
6	0	0	68	NA	NA	180	212	NA
8	0	0	82	NA	NA	187	214	263
10	0	51	NA	NA	NA	201	225	265
15	NA	NA	NA	NA	NA	253	274	305
20	NA	NA	NA	NA	NA	307	330	362
30	NA	NA	NA	NA	NA	NA	445	485
50	NA	NA	NA	NA	NA	NA	NA	763
100	NA	NA	NA	NA	NA	NA	NA	NA
17 to 26°F	Local 99% Winter Design Temperature: 17 to 26°F							
6	NA	NA	NA	NA	NA	NA	NA	NA
8	NA	NA	NA	NA	NA	NA	264	352
10	NA	NA	NA	NA	NA	NA	278	358
15	NA	NA	NA	NA	NA	NA	331	398
20	NA	NA	NA	NA	NA	NA	387	457
30	NA	NA	NA	NA	NA	NA	NA	581
50	NA	NA	NA	NA	NA	NA	NA	862
100	NA	NA	NA	NA	NA	NA	NA	NA
5 to 16°F	Local 99% Winter Design Temperature: 5 to 16°F							
6	NA	NA	NA	NA	NA	NA	NA	NA
8	NA	NA	NA	NA	NA	NA	NA	NA
10	NA	NA	NA	NA	NA	NA	NA	430
15	NA	NA	NA	NA	NA	NA	NA	485
20	NA	NA	NA	NA	NA	NA	NA	547
30	NA	NA	NA	NA	NA	NA	NA	682
50	NA	NA	NA	NA	NA	NA	NA	NA
100	NA	NA	NA	NA	NA	NA	NA	NA
4°F or Lower	Local 99% Winter Design Temperature: 4°F or Lower							
	Not recommended for any vent configurations							

Note: See Figure B-19 in Appendix B for a map showing local 99 percent winter design temperatures in the United States.

For SI: °C = [(°F - 32)/1.8, 1 inch = 25.4 mm, 1 square inch = 645.16 mm^2, 1 foot = 304.8 mm, 1 British thermal unit per hour = 0.2931 W.

TABLE 504.3(8a) EXTERIOR MASONRY CHIMNEY

Number of Appliances	Two or more
Appliance Type	FAN + NAT
Appliance Vent Connection	Type B double-wall connector

Combined Appliance Maximum Input Rating in Thousands of Btu per Hour

VENT HEIGHT (feet)	INTERNAL AREA OF CHIMNEY (square inches)							
	12	19	28	38	50	63	78	113
6	74	119	178	257	351	458	582	853
8	80	130	193	279	384	501	636	937
10	84	138	207	299	409	538	686	1,010
15	NA	152	233	334	467	611	781	1,156
20	NA	NA	250	368	508	668	858	1,286
30	NA	NA	NA	404	564	747	969	1,473
50	NA	NA	NA	NA	NA	831	1,089	1,692
100	NA	NA	NA	NA	NA	NA	NA	1,921

TABLE 504.3(8b) EXTERIOR MASONRY CHIMNEY

Number of Appliances	Two or more
Appliance Type	FAN + NAT
Appliance Vent Connection	Type B double-wall connector

Minimum Allowable Input Rating of Space-Heating Appliance in Thousands of Btu per Hour

VENT HEIGHT (feet)	INTERNAL AREA OF CHIMNEY (square inches)							
	12	19	28	38	50	63	78	113
37°F or Greater	Local 99% Winter Design Temperature: 37°F or Greater							
6	0	0	0	0	0	0	0	0
8	0	0	0	0	0	0	0	0
10	0	0	0	0	0	0	0	0
15	NA	0	0	0	0	0	0	0
20	NA	NA	123	190	249	184	0	0
30	NA	NA	NA	334	398	393	334	0
50	NA	NA	NA	NA	NA	714	707	579
100	NA	NA	NA	NA	NA	NA	NA	1,600
27 to 36°F	Local 99% Winter Design Temperature: 27 to 36°F							
6	0	0	68	116	156	180	212	266
8	0	0	82	127	167	187	214	263
10	0	51	97	141	183	210	225	265
15	NA	111	142	183	233	253	274	305
20	NA	NA	187	230	284	307	330	362
30	NA	NA	NA	330	319	419	445	485
50	NA	NA	NA	NA	NA	672	705	763
100	NA	NA	NA	NA	NA	NA	NA	1,554
17 to 26°F	Local 99% Winter Design Temperature: 17 to 26°F							
6	0	55	99	141	182	215	259	349
8	52	74	111	154	197	226	264	352
10	NA	90	125	169	214	245	278	358
15	NA	NA	167	212	263	296	331	398
20	NA	NA	212	258	316	352	387	457
30	NA	NA	NA	362	429	470	507	581
50	NA	NA	NA	NA	NA	723	766	862
100	NA	NA	NA	NA	NA	NA	NA	1,669
5 to 16°F	Local 99% Winter Design Temperature: 5 to 16°F							
6	NA	78	121	166	214	252	301	416
8	NA	94	135	182	230	269	312	423
10	NA	111	149	198	250	289	331	430
15	NA	NA	193	247	305	346	393	485
20	NA	NA	NA	293	360	408	450	547
30	NA	NA	NA	377	450	531	580	682
50	NA	NA	NA	NA	NA	797	853	972
100	NA	NA	NA	NA	NA	NA	NA	1,833
-10 to 4°F	Local 99% Winter Design Temperature: -10 to 4°F							
6	NA	NA	145	196	249	296	349	484
8	NA	NA	159	213	269	320	371	494
10	NA	NA	175	231	292	339	397	513
15	NA	NA	NA	283	351	404	457	586
20	NA	NA	NA	333	408	468	528	650
30	NA	NA	NA	NA	NA	603	667	805
50	NA	NA	NA	NA	NA	NA	955	1,003
100	NA	NA	NA	NA	NA	NA	NA	NA
-11°F or Lower	Local 99% Winter Design Temperature: -11°F or Lower							
	Not recommended for any vent configurations							

Note: See Figure B-19 in Appendix B for a map showing local 99 percent winter design temperatures in the United States.

For SI: °C = [(°F - 32)/1.8, 1 inch = 25.4 mm, 1 square inch = 645.16 mm^2, 1 foot = 304.8 mm, 1 British thermal unit per hour = 0.2931 W.

504.3.5 Common vertical vent offset. Where the common vertical vent is offset, the maximum capacity of the common vent shall be reduced in accordance with Section 504.3.6. The horizontal length of the common vent offset (L_o) shall not exceed $1^1/_2$ feet for each inch (457 mm per mm) of common vent diameter.

❖ Offsets, as opposed to a single change in direction, require two fittings (elbows) [see commentary Figures 202(14) and 503.10.3.4(2)]. An offset offers resistance to flow as does any change of direction in a vent chimney or connector, and the capacity reduction penalty compensates for this flow resistance. Common vent locations should be well planned in a building to avoid the need for an offset. Offsets also introduce the need for additional supports to prevent stress on fittings (elbows). The term Offset (Vent) is defined in Chapter 2 [see commentary, Sections 504.2.3 and 504 2.9].

Offsets should be avoided because of the flow resistance, capacity reduction, additional support requirements and possibility of condensation leakage. The length limitation of this section is absolute with no allowance for greater lengths. This provision is commonly violated in attic space offsets with some violations resulting in offsets that are as long or longer than the vent height. The length of an offset is measured in a horizontal plane, not along the developed length of the vent pipe. For example, a 45-degree offset with 7 feet of pipe between the fittings will have an offset length of 5 feet for application of this section. (For a 45-degree offset, actual pipe developed length is approximately 1.4 times the horizontal distance between the parallel vertical sections of pipe.) (See commentary Figure 504.3.5.)

504.3.6 Elbows in vents. For each elbow up to and including 45 degrees (0.79 rad) in the common vent, the maximum common vent capacity listed in the venting tables shall be reduced by 5 percent. For each elbow greater than 45 degrees (0.79 rad) up to

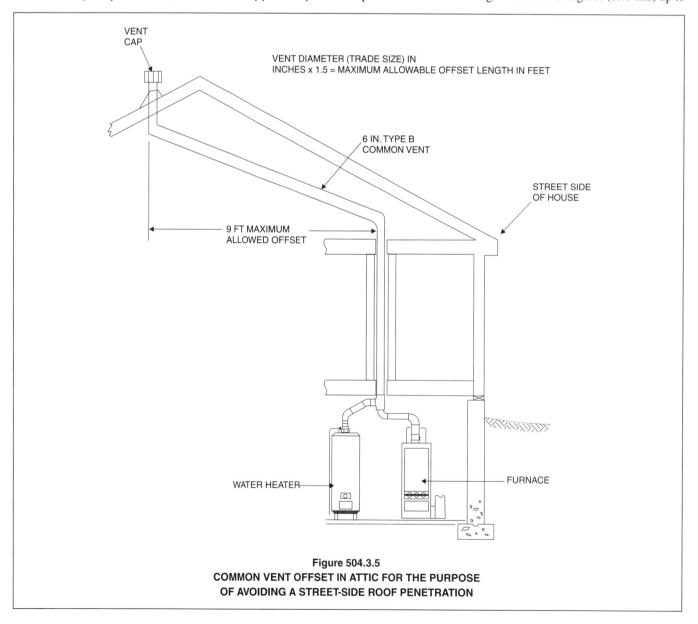

Figure 504.3.5
COMMON VENT OFFSET IN ATTIC FOR THE PURPOSE
OF AVOIDING A STREET-SIDE ROOF PENETRATION

and including 90 degrees (1.57 rad), the maximum common vent capacity listed in the venting tables shall be reduced by 10 percent.

❖ This section applies to the common vent portion of a vent system (see commentary, Section 504.2.3). For example, a common vent may have two 30-degree (0.52 rad) changes of direction at its base and might also have an offset constructed with 45-degree (0.79 rad) fittings in an attic space, thus having a total capacity reduction of 20 percent. If that same vent system had two 45-degree (0.79 rad) changes of direction at the base and an offset constructed with 60-degree (1.05 rad) fittings in the attic, the total capacity penalty would be 30 percent. In Figure 504.3.6(1), an offset is created in the common vent by the use of two fittings (elbows) adjusted to 45 degree (0.79 rad) angles. The common vent capacity would be reduced by 10 percent total (5 percent for each fitting/elbow). If the fittings were adjusted to angles greater than 45 degrees (0.79 rad), the total capacity reduction would be 10 percent per fitting for a total of 20 percent (see commentary Figures 504.3.6(2) and 504.3.6(3)].

504.3.7 Elbows in connectors. The vent connector capacities listed in the common vent sizing tables include allowance for two 90-degree (1.57 rad) elbows. For each additional elbow up to and including 45 degrees (0.79 rad), the maximum vent connector capacity listed in the venting tables shall be reduced by 5 percent. For each elbow greater than 45 degrees (0.79 rad) up to

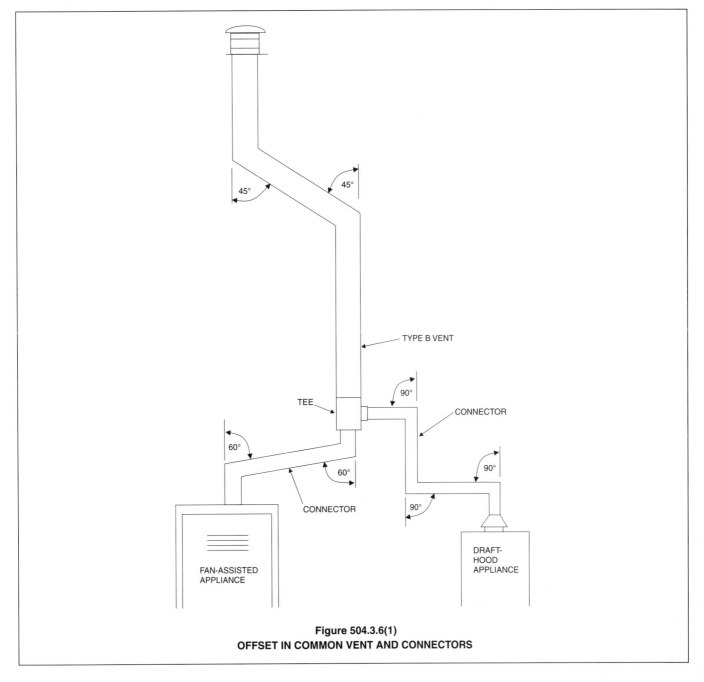

Figure 504.3.6(1)
OFFSET IN COMMON VENT AND CONNECTORS

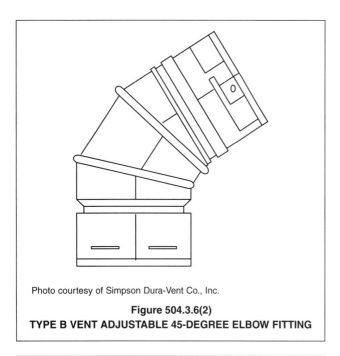

Photo courtesy of Simpson Dura-Vent Co., Inc.

Figure 504.3.6(2)
TYPE B VENT ADJUSTABLE 45-DEGREE ELBOW FITTING

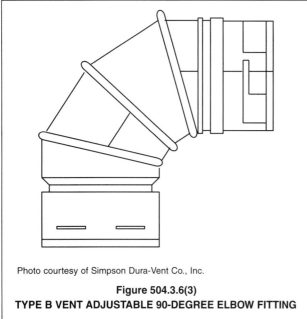

Photo courtesy of Simpson Dura-Vent Co., Inc.

Figure 504.3.6(3)
TYPE B VENT ADJUSTABLE 90-DEGREE ELBOW FITTING

and including 90 degrees (1.57 rad), the maximum vent connector capacity listed in the venting tables shall be reduced by 10 percent.

❖ In Figure 504.3.6(1), the connector for the fan-assisted appliance is offset with two 60-degree (1.05 rad) fittings (elbows). Because the tables include an allowance for two "free" 90-degree (1.57 rad) fittings, no capacity reduction would be required for this connector. The draft-hood-equipped appliance has a connector with a total of four 90-degree (1.57 rad) fittings, including the tee in the common vent. This would require a total capacity reduction of 20 percent. Note that Sections 503.10.6 and 503.10.9 both intend to prohibit connectors with excessive length and fittings/elbows. Excess fittings and length are often the result of poor planning, improper appliance location and obstructions, such as ducts, structural members and other appliances. Capacity reduction penalties can force an increase in connector size, and in the case of fan-assisted appliances, increasing the connector size might raise the minimum required input rating above the appliance rating, thus forcing redesign of the system.

504.3.8 Common vent minimum size. The cross-sectional area of the common vent shall be equal to or greater than the cross-sectional area of the largest connector.

❖ This requirement is consistent with vent manufacturers' instructions and is necessary to prevent overloading of the common vent. It is apparent that a vent's maximum capacity could be exceeded where multiple connectors are involved, especially if one of the connectors is larger than the common vent itself. Vent manufacturers may also require that the common vent be increased one size where more than one appliance connector and the common vent are the same size.

504.3.9 Common vent fittings. At the point where tee or wye fittings connect to a common vent, the opening size of the fitting shall be equal to the size of the common vent. Such fittings shall not be prohibited from having reduced- size openings at the point of connection of appliance vent connectors.

❖ This section specifically allows use of tee and wye fittings with reduced-size openings to join appliance vent connectors to the common vent. The fittings must be the same size as the common vent at the point where they join, but are allowed to have smaller branch openings and smaller openings on the opposite end of the run of the fitting. A common question regarding interconnection fittings is whether or not the code intends to allow single-wall interconnection fittings such as tees and wyes. These fittings have commonly been used with single-wall connectors to join two appliances to a common Type B vent (5 x 4 x 3-inch wyes, 6 x 6 x 4-inch tees, etc.). Vent manufacturers' opinion on this issue is that because the interconnection fitting is part of the common vent, it should be constructed as required for the common vent, meaning Type B vent tees and wyes. Certainly, double-wall vent connectors are not allowed to discharge into a single-wall fitting. However, the code is silent with regard to single-wall connectors discharging into single-wall fittings. Double-wall tees and wyes are preferable because they have far less heat loss and less required clearance to combustibles [see commentary Figures 504.3.9(1) and (2)].

504.3.10 High-altitude installations. Sea-level input ratings shall be used when determining maximum capacity for high-altitude installation. Actual input (derated for altitude) shall be used for determining minimum capacity for high-altitude installation.

❖ See the commentary for Section 504.2.5.

CHIMNEYS AND VENTS

Figure 504.3.9(1)
TYPE B VENT REDUCING TEE

Figure 504.3.9(2)
TYPE B VENT REDUCING WYE

504.3.11 Connector rise measurement. Connector rise (R) for each appliance connector shall be measured from the draft hood outlet or flue collar to the centerline where the vent gas streams come together.

❖ It is desirable to achieve the greatest connector rise that conditions will permit [see commentary Figures 504.3.11 and 504.3(2)].

504.3.12 Vent height measurement. For multiple units of equipment all located on one floor, available total height (H) shall be measured from the highest draft hood outlet or flue collar up to the level of the outlet of the common vent.

❖ The vent design is based on the "worst case" (most conservative) height so that a vent is not given credit for being taller than it actually is [see commentary Figure 504.3(2)].

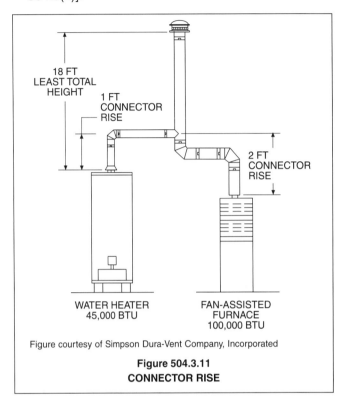

Figure 504.3.11
CONNECTOR RISE

504.3.13 Multistory height measurement. For multistory installations, available total height (H) for each segment of the system shall be the vertical distance between the highest draft hood outlet or flue collar entering that segment and the centerline of the next higher interconnection tee (see Figure B-13).

❖ See Appendix B and Section 503.6.10.

504.3.14 Multistory lowest portion sizing. The size of the lowest connector and of the vertical vent leading to the lowest interconnection of a multistory system shall be in accordance with Table 504.2(1) or 504.2(2) for available total height (H) up to the lowest interconnection (see Figure B-14).

❖ See Appendix B and Section 503.6.10. The lowest appliance vent is sized as an individual vent.

504.3.15 Multistory common vents. Where used in mulistory systems, vertical common vents shall be Type B double wall and shall be installed with a listed vent cap.

❖ The design and sizing criteria for multistory venting systems are based on the known performance of listed Type B vents with the listed cap (See Sections 502.1 and 503.6.6).

504.3.16 Multistory common vent offsets. Offsets in multistory common vent systems shall be limited to a single offset in each system, and systems with an offset shall comply with all of the following:

1. The offset angle shall not exceed 45 degrees (0.79 rad) from vertical.

2. The horizontal length of the offset shall not exceed $1^1/_2$ feet for each inch (457 mm per mm) of common vent diameter of the segment in which the offset is located.

3. For the segment of the common vertical vent containing the offset, the common vent capacity listed in the common venting tables shall be reduced by 20 percent (0.80 × maximum common vent capacity).

4. A multistory common vent shall not be reduced in size above the offset.

❖ This requirement is consistent with vent manufacturers' instructions and is intended to eliminate flow restrictions and pressure differentials in the venting system. Because appliances connect to the multistory common vent at different floor levels, flow disturbances must be avoided in the entire extent of the common vent to prevent having an effect on lower-floor appliances. The criteria for a single offset are based on research conducted by the Gas Research Institute, which used a specialized computer program to investigate the performance of multistory venting systems with offsets. The research indicated that multistory Type B vents with a single offset can perform as well as a nonoffset vent performs if the adjustments and limitations in this section are applied. Those limitations involve a reduction of the vent capacity found in the tables and a limitation on the horizontal length of the offset based on vent size (see commentary Section 504.3.5).

A multistory common vertical vent is allowed to have a single offset if all of the following requirements are met:

1. The offset angle does not exceed 45 degrees (0.79 rad).

2. The horizontal length of the offset does not exceed $1^1/_2$ feet for each inch (457 mm per mm) of common vent diameter of the segment in which the offset is located.

3. For the segment of the common vertical vent containing the offset, the common vent capacity listed in the common venting tables is reduced by 20 percent (0.80 x maximum common vent capacity).

4. A multistory common vent must not be reduced in size above the offset.

504.3.17 Vertical vent maximum size. Where two or more appliances are connected to a vertical vent or chimney, the flow area of the largest section of vertical vent or chimney shall not exceed seven times the smallest listed appliance categorized vent areas, flue collar area, or draft hood outlet area unless designed in accordance with approved engineering methods.

❖ Vent manufacturers' instructions address ratios of areas of appliance connectors connecting to a common vent. This section relates the area ratio limitation to the smallest appliance connection and the area of the vertical common vent. The concern is that the smaller or smallest appliance may not be able to provide enough heat to the common vent to develop sufficient draft and prevent a condensation problem. It can be seen that the common vent tables would allow gross oversizing if it were not for the limit of this section. For example, if the table permits a 5-inch common vent, it would also permit a 6, 7, 8, 9, and 10-inch common vent. This section serves to limit the size of the common vent, whereas the table does not. The larger the common vent, the more difficult it becomes to maintain draft and avoid condensation, especially when the smallest appliance is firing by itself.

504.3.18 Multiple input rate appliances. For appliances with more than one input rate, the minimum vent connector capacity (FAN Min) determined from the tables shall be less than the lowest appliance input rating, and the maximum vent connector capacity (FAN Max or NAT Max) determined from the tables shall be greater than the highest appliance input rating.

❖ See the commentary for Section 504.2.6.

504.3.19 Liner system sizing. Listed, corrugated metallic chimney liner systems in masonry chimneys shall be sized by using Table 504.3(1) or 504.3(2) for Type B vents, with the maximum capacity reduced by 20 percent (0.80 × maximum capacity) and the minimum capacity as shown in Table 504.3(1) or 504.3(2). Corrugated metallic liner systems installed with bends or offsets shall have their maximum capacity further reduced in accordance with Sections 504.3.5 and 504.3.6. The 20-percent reduction for corrugated metallic chimney liner systems includes an allowance for one long-radius 90-degree (1.57 rad) turn at the bottom of the liner.

❖ See the commentary for Section 504.2.7.

504.3.20 Chimney and vent location. Tables 504.3(1), 504.3(2), 504.3(3), 504.3(4), and 504.3(5) shall be used for chimneys and vents not exposed to the outdoors below the roof line. A Type B vent or listed chimney lining system passing through an unused masonry chimney flue shall not be considered to be exposed to the outdoors. Tables 504.3(7) and 504.3(8) shall be used for clay-tile-lined exterior masonry chimneys, provided all of the following conditions are met:

1. Vent connector is Type B double-wall.

2. At least one appliance is draft hood equipped.

3. The combined appliance input rating is less than the maximum capacity given by Table 504.3(7a) for NAT+NAT or Table 504.3(8a) for FAN+NAT.

4. The input rating of each space-heating appliance is greater than the minimum input rating given by Table 504.3(7b) for NAT+NAT or Table 504.3(8b) for FAN+NAT.

5. The vent connector sizing is in accordance with Table 504.3(3).

Where these conditions cannot be met, an alternative venting design shall be used, such as a listed chimney lining system.

Exception: Vents serving listed appliances installed in accordance with the appliance manufacturer's instructions and the terms of the listing.

❖ See the commentary for Section 504.2.9.

504.3.21 Connector maximum and minimum size. Vent connectors shall not be increased in size more than two sizes greater than the listed appliance categorized vent diameter, flue collar diameter, or draft hood outlet diameter. Vent connectors for draft hood-equipped appliances shall not be smaller than the draft hood outlet diameter. Where a vent connector size(s) determined from the tables for a fan-assisted appliance(s) is smaller than the flue collar diameter, the use of the smaller size(s) shall be permitted provided that the installation complies with all of the following conditions:

1. Vent connectors for fan-assisted appliance flue collars 12 inches (305 mm) in diameter or smaller are not reduced by more than one table size [e.g., 12 inches to 10 inches (305 mm to 254 mm) is a one-size reduction] and those larger than 12 inches (305 mm) in diameter are not reduced more than two table sizes [e.g., 24 inches to 20 inches (610 mm to 508 mm) is a two-size reduction].
2. The fan-assisted appliance(s) is common vented with a draft-hood-equipped appliances(s).
3. The vent connector has a smooth interior wall.

❖ Using a connector that is larger than the appliance outlet connection may sometimes be necessary to comply with the requirements of the sizing tables for multiple-appliance systems. However, having a connector significantly larger than the appliance outlet connection could cause reduced flow velocity, excessive loss of heat and could result in the allowance of excessively long connectors.

In accordance with Section 504.2.2, a venting system serving a single appliance can, under certain conditions, be smaller than the appliance outlet connection. However, this is prohibited by this section for common venting systems serving more than one appliance (see commentary, Section 504.2.10). The reason for this prohibition is that a common vent or chimney, being large enough for multiple appliances, may not be able to produce adequate draft or avoid condensation where a single appliance is operating with a reduced-size vent connector.

Recent research involving computerized modeling has shown that connectors for fan-assisted appliances in multiple-appliance systems can be reduced under conditions similar to those for single-appliance installations. Therefore, this section makes an exception for systems with fan-assisted appliances commonly vented with appliances equipped with a draft hood (see commentary, Section 504.2.2). The allowance for a connector to be smaller than an appliance flue collar is applicable only to fan-assisted appliances. A reduction in size is not allowed for corrugated metal connectors.

504.3.22 Component commingling. All combinations of pipe sizes, single-wall, and double-wall metal pipe shall be allowed within any connector run(s) or within the common vent, provided all of the appropriate tables permit all of the desired sizes and types of pipe, as if they were used for the entire length of the subject connector or vent. Where single-wall and Type B double-wall metal pipes are used for vent connectors within the same venting system, the common vent must be sized using Table 504.3(2) or 504.3(4), as appropriate.

❖ Because this section parallels Section 504.2.12, see the commentary for that section. This section addresses the installation that uses both single-wall connectors and double-wall connectors in the same venting system. The penalty for mixing is that the common vent must be sized from the more restrictive single-wall connector tables. Double-wall pipe is not designed to be interchangeable with single-wall pipe, and all such transitions would have to made with fittings designed for the transition.

504.3.23 Multiple sizes permitted. Where a table permits more than one diameter of pipe to be used for a connector or vent, all the permitted sizes shall be permitted to be used.

❖ This section broadly states that any size pipe permitted by the sizing tables may be used. However, do not overlook Sections 504.3.8, 504.3.17 and 504.3.21, which contain provisions that can override this section. Vent manufacturers stress that the smallest size pipe allowed by the code should be used rather than any larger sizes that may be permitted. Operating a vent or chimney at nearer to its maximum capacity will produce stronger draft and result in better protection against the formation of condensation.

504.3.24 Table interpolation. Interpolation shall be permitted in calculating capacities for vent dimensions that fall between table entries (see Appendix B, Example 3).

❖ See Section 504.2.13.

504.3.25 Extrapolation prohibited. Extrapolation beyond the table entries shall not be permitted.

❖ See Section 504.2.14.

504.3.26 Engineering calculations. For vent heights less than 6 feet (1829 mm) and greater than shown in the tables, engineering methods shall be used to calculate vent capacities.

❖ See Section 504.2.15.

SECTION 505 (IFGC)
DIRECT-VENT, INTEGRAL VENT, MECHANICAL VENT AND VENTILATION/EXHAUST HOOD VENTING

505.1 General. The installation of direct-vent and integral vent appliances shall be in accordance with Section 503. Mechanical venting systems and exhaust hood venting systems shall be designed and installed in accordance with Section 503.

❖ See Sections 503.2.3, 503.2.4, 503.3.3, 503.3.4, and 503.8.

505.1.1 Commercial cooking appliances vented by exhaust hoods. Where commercial cooking appliances are vented by means of the Type I or Type II kitchen exhaust hood system that serves such appliances, the exhaust system shall be fan powered and the appliances shall be interlocked with the exhaust hood system to prevent appliance operation when the exhaust hood system is not operating. Dampers shall not be installed in the exhaust system.

❖ This section is an appliance-specific relative of section 503.3.4. Commercial cooking appliances must be vented (see Section 501.2) but are rarely vented by means of direct connection to a vent. Instead, such appliances are typically vented by the Type I or II exhaust system that is required by the IMC. A gravity (nonpowered) exhaust system is not allowed since its ability to vent combustion products is not consistently dependable and there is no practical method of interlocking the appliances with a gravity exhaust system (see commentary Figure 505.1.1).

The appliance control system must be interlocked with the kitchen exhaust hood system to permit appliance operation only when the power means of exhaust is in operation. This requirement is similar to that of Section 503.3.3. The appliances that depend on the exhaust system must be shut down in the event of exhaust system failure or halt operation. This interlock must not allow appliance operation until the exhaust-hood system has been started. Such an interlock is easily accomplished with electrically heated cooking appliances and gas-fired appliances having electronic ignition; however, it is more difficult in the case of gas-fired appliances having standing pilot ignition systems. If the interlock method uses an electrically actuated (solenoid) main fuel valve, that valve will close when the exhaust fan system is shut down, thus extinguishing any standing pilots and necessitating the relighting of pilots each day or each time the kitchen is "off line." Although some attempts have been made to circumvent this problem, a practical and safe solution to the standing pilot problem is not known at this time.

SECTION 506 (IFGC)
FACTORY-BUILT CHIMNEYS

506.1 Building heating appliances. Factory-built chimneys for building heating appliances producing flue gases having a temperature not greater than 1,000°F (538°C), measured at the entrance to the chimney, shall be listed and labeled in accordance with UL 103 and shall be installed and terminated in accordance with the manufacturer's installation instructions.

❖ Factory-built chimneys can be used with gas-fired appliances if allowed by the chimney manufacturer. The sizing tables inSection 504 do not apply to factory-built chimneys; therefore, sizing must be engineered or as specified by the chimney and appliance manufacturers' instructions. Factory-built chimneys will be too large for many residential appliances [see commentary Figures 506.1(1) and 506.1(2)].

Figure 505.1.1
EXHAUST HOOD INTERLOCK WITH
COOKING APPLIANCES (SOLENOID GAS VALVE)

Photo courtesy of Selkirk L.L.C

**Figure 506.1(1)
FACTORY-BUILT CHIMNEY**

Photo courtesy of Simpson Dura-Vent Company, Incorporated

**Figure 506.1(2)
AIR-COOLED FACTORY-BUILT CHIMNEY SYSTEM**

506.2 Support. Where factory-built chimneys are supported by structural members, such as joists and rafters, such members shall be designed to support the additional load.

❖ Factory-built chimneys are supported by the building structure and can impose considerable weight on structural components. Joists, rafters and other structural members must be designed to support the additional loading. Structural evaluation may be necessary, especially where framing for a chimney requires cutting and heading of joists and rafters.

506.3 Medium-heat appliances. Factory-built chimneys for medium-heat appliances producing flue gases having a temperature above 1,000°F (538°C), measured at the entrance to the chimney, shall be listed and labeled in accordance with UL 959 and shall be installed and terminated in accordance with the manufacturer's installation instructions.

❖ Factory-built chimneys serving medium-heat appliances in which the maximum continuous flue gas temperature does not exceed 2,000°F (1093°C) must be tested in accordance with UL 959. Such chimneys are associated with factory and industrial equipment and incinerators. Chimneys serving medium-heat appliances may require greater termination heights and clearances to combustibles.

Bibliography

The following resource material is referenced in this chapter or is relevant to the subject matter addressed in this chapter.

ANSI Z21.47-00, *Gas-Fired Central Furnaces with Addendum Z21.47a-00*. New York: American National Standards Institute, 2001.

Chimney and Gas Vent Sizing Handbook. Dallas: Selkirk Metalbestos, 1992.

NFPA 211-00, *Chimneys, Fireplaces, Vents and Solid Fuel-Burning Appliances*. Quincy, MA: National Fire Protection Association, 2000.

UL 641-95, *Type L Low Temperature Venting Systems with Revisions through April 1999*. Northbrook, IL: Underwriters Laboratories, 1995.

UL 1738-93, *Venting Systems for Gas Burning Appliances: Categories II, III and IV with Revisions through December 2000*. Northbrook, IL: Underwriters Laboratories, 1993.

UL 1777-98, *Chimney Liners with Revisions through July 1998*. Northbrook, IL: Underwriters Laboratories, 1998.

CHAPTER 6
SPECIFIC APPLIANCES

General Comments

Chapter 6 regulates the design and installation of specifically named gas-fired appliances such as furnaces, boilers, water heaters, heaters, cooking and lighting appliances and clothes dryers. This chapter lists all of the appliances that the code regulates within its scope. The chapters of the code, with the exceptions of Chapter 3 and this chapter, are dedicated to single subjects. This chapter, however, is a collection of requirements for various appliances and equipment. The only commonality in this chapter is that all subjects use fuel gas to perform some task or function. Because none of the subjects in this chapter have the volume of text to warrant an individual chapter, they have been combined into a single chapter.

Purpose

This chapter contains requirements for specific appliances and their installation. In addition to being listed and labeled, many appliances have special installation or location requirements. Unlike Chapter 3, titled General Regulations, this chapter contains and organizes the specific regulations applicable to certain appliances.

SECTION 601 (IFGC)
GENERAL

601.1 Scope. This chapter shall govern the approval, design, installation, construction, maintenance, alteration and repair of the appliances and equipment specifically identified herein.

❖ This chapter regulates all aspects of appliances and equipment to the extent found in each of Sections 602 through 634.

SECTION 602 (IFGC)
DECORATIVE APPLIANCES
FOR INSTALLATION IN FIREPLACES

602.1 General. Decorative appliances for installation in approved solid fuel-burning fireplaces shall be tested in accordance with ANSI Z21.60 and shall be installed in accordance with the manufacturer's installation instructions. Manually lighted natural gas decorative appliances shall be tested in accordance with ANSI Z21.84.

❖ These appliances include gas log sets that are designed to simulate wood fires (see commentary Figure 602.1). Chapter 3 addresses the requirements for testing, labeling and installing mechanical equipment and appliances. The gas-burning appliance must be tested to the standard or standards appropriate for the equipment. The testing agency is responsible for determining the standard to be used to test the equipment. In the case of decorative gas log sets, ANSI Z21.60 or Z21.84 is the applicable test standard.

Labeling is the code official's assurance that the subject product is a representative duplication of the product that the testing agency tested in the laboratory. The label indicates that an independent agency has conducted inspections at the plant to verify that all units conform to the specifications that the quality control manual sets forth for fabricating the gas appliances. In-

Photography by D. F. Noyes Studio, courtesy of DESA International

**Figure 602.1
VENTED GAS LOG SET**

formation that must be contained on the label is described in Section 301.5 and includes the manufacturer's identification, the third-party inspection agency's identification, the model number, the serial number, the input ratings and the type of fuel the appliance is designed to burn.

The code requires that the appliance be installed in accordance with the manufacturer's installation instructions. This requirement is also linked to the laboratory testing of the appliance because the laboratory used the same installation instructions to install the prototype appliance being tested. When the appliance has been tested and evaluated for code compliance and judged to meet the performance and construction requirements of the applicable standard, the installation instructions become an integral part of the labeling requirements and must be strictly adhered to.

The intent of this section is to regulate gas-burning appliances that are accessory to, and designed for installation in, vented solid-fuel-burning fireplaces. Gas-fired decorative log sets and log lighters are examples of accessory appliances that are designed for installation in solid-fuel-burning fireplaces. Gas log sets provide some radiant heat; however, their primary function is to create an aesthetically pleasing simulation of a wood log fire. This section addresses vented appliances and does not address "unvented gas log sets" (room heaters).

Decorative gas-burning appliances, such as some gas log sets, are designed to simulate wood fires by intentionally causing incomplete combustion as necessary to yield yellow or yellow-tipped flames. This incomplete combustion results in an increase in the amount of carbon monoxide produced. The highly toxic carbon monoxide is odorless and colorless, and the accompanying products of combustion produced by a gas-burning appliance do not have a strong odor or an odor that is readily recognized by an untrained person.

If a decorative gas appliance were operated with the fireplace damper closed, the carbon monoxide levels could be dangerously high before the building occupants became aware of the hazard. To prevent harm and the possible asphyxiation of the occupants, it is imperative that the fireplace damper be open whenever the appliance is burning. The manufacturer's installation instructions will specify the minimum free area of damper opening required to vent the appliance combustion products. The damper area is proportional to the appliance's input rating. The fireplace damper plate must be removed or permanently fixed in a position that provides the opening area required by the appliance manufacturer's installation instructions.

602.2 Flame safeguard device. Decorative appliances for installation in approved solid fuel-burning fireplaces, with the exception of those tested in accordance with ANSI Z21.84, shall utilize a direct ignition device, an ignitor or a pilot flame to ignite the fuel at the main burner, and shall be equipped with a flame safeguard device. The flame safeguard device shall automatically shut off the fuel supply to a main burner or group of burners when the means of ignition of such burners becomes inoperative.

❖ To eliminate the hazards associated with manually igniting a decorative gas appliance, the code requires that the main burner be ignited by a supervised means of direct ignition, such as a standing pilot. The exception to this requirement is for manually lighted appliances listed to ANSI Z21.84. The use of automatic ignition controls prevents the operator of the appliance from being exposed to the potential hazard of trying to ignite the main burner with a match or other manual lighting device. The flame safeguard device consists of a control valve assembly with an integral means of pilot or igniter supervision and is designed to prevent or shut off the flow of gas to the appliance main burner in the event that the source of ignition is extinguished or otherwise fails.

The typical flame safeguard device used with gas log set appliances is a combination manual control valve, pressure regulator, pilot feed and magnetic pilot safety mechanism with a thermocouple generator. If the pilot flame is extinguished, the drop in thermocouple output voltage will cause the control valve to "lock out" in the closed position, thereby preventing the flow of gas to the main burner and the pilot burner. It is not the intent of this section to require flame safeguard devices for manually operated log lighter appliances used to kindle wood fires.

A flame safeguard, also known as a safety shutoff device, functions to automatically shut off the fuel supply to the main burner or burners of an appliance when the source of ignition becomes inoperative. Gas-burning equipment may operate using standing pilot ignition, which is a small flame kept constantly burning for the sole purpose of igniting the main burner when the thermostat calls for heat or a manual gas valve is opened by the operator. If the pilot is not lit, for whatever reason, the flame safeguard will automatically activate to prevent the flow of fuel to the main burner or burners of the appliance. Appliances that use automatic sparking devices or other means for ignition are subject to the same requirements for providing flame safeguards, and the method of operation of the flame safeguard will differ depending on the type of ignition system in use.

Automatically operated gas-burning equipment and appliances are equipped with some means of monitoring the means of ignition or the main burner flame because fuel introduced into a combustion chamber and not ignited and burned could cause a dangerous condition.

The most commonly used type of flame safeguard device consists of a thermocouple and an electro-mechanical device that function together to supervise a standing pilot flame. In the event of pilot failure, the thermocouple and flame safeguard device will function to prevent the flow of fuel to the main burners, and if the safeguard control is of the 100-percent shutoff type, it will also shut off the flow of fuel to the pilot burner.

Modern appliances seldom have standing pilots and instead use electric-spark-ignited intermittent pilots or direct ignition devices such as hot surface igniters and glow coils. The methods of detecting the presence of a pilot flame include thermocouples, millivolt (power pile) generators, liquid-filled capillary tubes, bimetal mechanisms and flame rectification circuits. Where only the main burner is monitored to verify ignition and continued combustion, the methods of detection include flame rectification, infrared detectors, ultraviolet detectors and bimetal stack sensors. The majority of today's appliances rely on electronic flame rectification or radiant energy detectors (infrared or ultraviolet) as the means of proving ignition of a pilot burner or the main burner(s). The flame rectification method of sensing the presence of flames is very common today and involves two electrodes, a flame rod and the burner itself. An AC voltage is applied between the flame rod and the burner. If a flame is present and impinges on both of these electrodes, a few microamps of current will flow through the gases in the flame, but almost entirely in one direction because of the relative size of the surface areas of the two electrodes. This unidirectional flow rectifies the AC current to DC current, which is then recognized by electronic circuitry, thereby proving the presence of a flame. Manually lighted natural gas decorative appliances tested and listed to Z21.84 do not have to be equipped with a flame safeguard device.

Flame safeguard devices are commonly integral with a combination gas control. These controls typically incorporate multiple solenoid valves, a manual valve, an appliance gas pressure regulator, a pressure operated valve and flame safeguard devices all in one device body. It is also common for flame safeguard devices to be incorporated into an electronic integrated circuit control module that is separate from the combination gas control. For example, spark ignition and hot surface ignition systems incorporate the means of ignition and ignition supervision in an electronic control module that may also contain components and circuitry for control of combustion fans, furnace blowers, combustion chamber purging and monitoring of multiple system pressure and temperature sensors located throughout an appliance. Commentary Figure 602.2(1) illustrates a type of combination gas control designed for use with an outboard intermittent ignition control module. Commentary Figure 602.2(2) illustrates the internal workings that are typical of combination gas controls. These controls have redundant valve mechanisms as an additional safety feature.

Photo courtesy of Emerson Climate Technologies, White Rodgers

Figure 602.2(1)
COMBINATION GAS CONTROL

602.3 Prohibited installations. Decorative appliances for installation in fireplaces shall not be installed where prohibited by Section 303.3.

❖ Section 303.3 lists various locations where gas appliances cannot be installed, such as bedrooms and bathrooms. Those prohibitions also apply to decorative appliances.

SECTION 603 (IFGC)
LOG LIGHTERS

603.1 General. Log lighters shall be tested in accordance with CSA 8 and installed in accordance with the manufacturer's installation instructions.

❖ Log lighters are simple manually operated burners used to start wood fires. Log lighters are functional rather than decorative appliances. The heat produced by the log lighter flames raises the temperature of the wood fuel to its ignition temperature. Log lighters are designed to be turned off manually after the wood fire is capable of sustaining combustion. The code, rather than providing prescriptive installation requirements, requires that log lighters be installed in accordance with the manufacturer's installation instructions.

SECTION 604 (IFGC)
VENTED GAS FIREPLACES
(DECORATIVE APPLIANCES)

604.1 General. Vented gas fireplaces shall be tested in accordance with ANSI Z21.50, shall be installed in accordance with the manufacturer's installation instructions and shall be designed and equipped as specified in Section 602.2.

❖ Unlike the appliances addressed in Section 602, these appliances are self-contained and do not rely on a fireplace to contain or vent them. Such appliances are referred to as gas fireplaces because they are designed to simulate a solid-fuel-burning fireplace. The standard, Z21.50, that regulates these appliances has recently been retitled as "Vented Gas Fireplaces." Therefore, these appliances will be referred to as gas fireplaces even though they do not fall under the definition of "fireplace" [see commentary Figures 604.1(1), 604.1(2) and 604.1(3)]. These appliances are designed for various methods of venting, including direct-venting through the wall or roof and conventional venting with Type B vent or factory-supplied vent material. Section 303.3 controls the type of appliance that is allowed in bathrooms, toilet rooms and bedrooms. Direct-vent appliances have the advantage of a closed combustion chamber that does not communicate with the room in which they are installed, as well as the advantage of an outdoor combustion air supply.

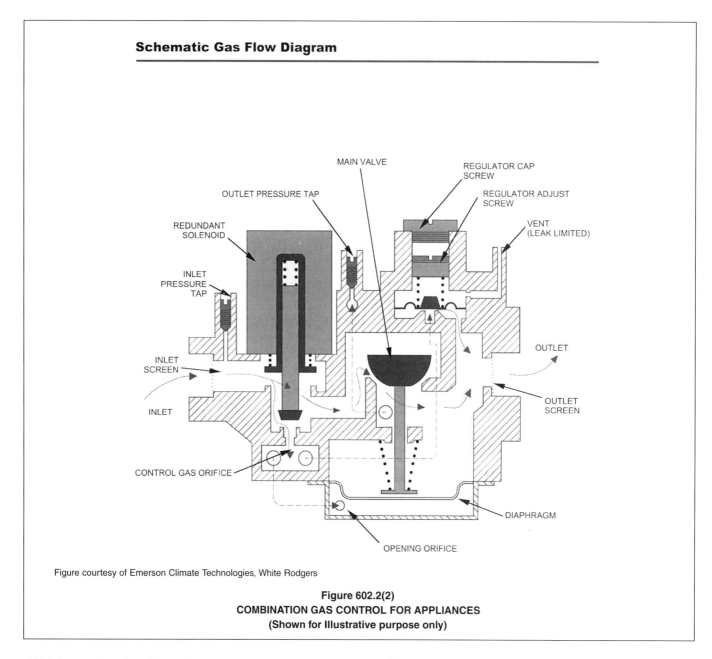

Figure 602.2(2)
COMBINATION GAS CONTROL FOR APPLIANCES
(Shown for Illustrative purpose only)

604.2 Access. Panels, grilles and access doors that are required to be removed for normal servicing operations shall not be attached to the building.

❖ Access to the appliance for maintenance and repair may be through panels, grilles or doors, but such panels must be removable as intended by the manufacturer of the appliance and not a permanent part of the structure (see the definition of "Access" in Section 202).

SECTION 605 (IFGC)
VENTED GAS FIREPLACE HEATERS

605.1 General. Vented gas fireplace heaters shall be installed in accordance with the manufacturer's installation instructions, shall be tested in accordance with ANSI Z21.88 and shall be designed and equipped as specified in Section 602.2.

❖ These appliances are similar to those addressed in Section 604, the main difference being that vented gas fireplace heaters are designed with more emphasis on space heating while maintaining the decorative features. These heaters must comply with minimum thermal efficiency requirements.

SECTION 606 (IFGC)
INCINERATORS AND CREMATORIES

606.1 General. Incinerators and crematories shall be installed in accordance with the manufacturer's installation instructions.

❖ Incinerators used today are installed in hospitals, universities and other relatively large institutions that must dispose of medical wastes, toxic and contaminated substances, pathological waste and human and animal tissues. A crematory is a furnace that is used exclusively to reduce human and animal cadavers to ashes.

Figure 604.1(1)
VENTED DECORATIVE APPLIANCE

Figure 604.1(2)
DIRECT-VENT DECORATIVE APPLIANCE

3 for equipment approval and installation. The incinerator or crematory must be tested to the standard that governs its use, and evidence must be provided indicating that the manufacturer has arranged for a third-party inspection of the manufacturer's facility to determine that the units being produced are the same as those that were tested. As required by Section 301.4.1, the testing agency is responsible for determining the appropriate test protocol, subject ultimately to approval by the code official. When tested in the laboratory, the equipment is installed in accordance with the manufacturer's installation instructions. For the equipment to perform as designed and tested, the equipment must be installed in the same manner as the tested installation. Therefore, the manufacturer's installation instructions become an integral part of the approval process.

Figure 604.1(3)
DIRECT-VENT GAS FIREPLACE

Modern incinerator installations take advantage of the heat produced by incineration by passing the flue gases through boilers or other heat exchangers for space heating, cooling or processing operations. The reclamation of heat produced from incinerators can substantially reduce the energy costs of the facility served by the incinerator.

This section requires factory-built incinerators and crematories to conform to the requirements of Chapter

SECTION 607 (IFGC)
COMMERCIAL-INDUSTRIAL INCINERATORS

607.1 Incinerators, commercial-industrial. Commercial-industrial-type incinerators shall be constructed and installed in accordance with NFPA 82.

❖ Commercial-industrial incinerators are large versions of the incinerators described in Section 606. These units burn large amounts of refuse that reduce the load on land fills and, in many instances, capture the heat of combustion for use in another process such as commercial steam heating, generation of electricity or heat for a manufacturing process. The reclamation of heat from incinerators can substantially reduce the energy costs of the facility served by the incinerator.

Commercial-industrial incinerators used today are installed in hospitals, universities and other relatively large institutions that must dispose of medical wastes, toxic and contaminated substances, pathological waste and human and animal tissues. To achieve a more complete incineration and thereby reduce the amount of pollutants released into the atmosphere, multiple-chamber incinerators are used rather than single-chamber incinerators. There are two basic designs of multiple-chamber incinerators: the in-line type and the retort type. The in-line type is characterized by the flow of combustion gases always traveling straight through the incinerator, with 90-degree (1.57 rad) turns only in the vertical direction. The retort type is characterized by a design that causes the combustion gases to flow through 90-degree (1.57 rad) turns in both lateral and vertical directions.

SECTION 608 (IFGC)
VENTED WALL FURNACES

608.1 General. Vented wall furnaces shall be tested in accordance with ANSI Z21.49 or Z21.86/CSA 2.32 and shall be installed in accordance with the manufacturer's installation instructions.

❖ Wall furnaces are a type of room heater usually designed to be installed within a 2- by 4-inch (51 mm by 102 mm) stud cavity in frame construction. They are typically used in cottages, room additions and homes in mild climates. Some units are designed to serve a single room, and others are designed as through-the-wall units to serve adjacent rooms. Wall furnaces are ductless; however, some units are listed for use with a surface-mounted supply outlet extension. Wall furnaces can be either gravity or forced air type.

608.2 Venting. Vented wall furnaces shall be vented in accordance with Section 503.

❖ Because wall furnaces are designed to fit within a wall cavity, a special type of oval-shaped vent is required to fit within the same wall. Type BW vent is specially designed for wall furnace applications (see commentary, Section 503).

608.3 Location. Vented wall furnaces shall be located so as not to cause a fire hazard to walls, floors, combustible furnishings or doors. Vented wall furnaces installed between bathrooms and adjoining rooms shall not circulate air from bathrooms to other parts of the building.

❖ Wall furnaces, like all room heaters, can present a fire hazard if improperly located. The heat discharged or directly radiated from these units can ignite nearby wall or floor surfaces, furniture, trim items, window treatments and doors (see commentary, Section 608.4). A through-the-wall unit serving both a bathroom and an adjacent room must not be capable of recirculating air between those spaces.

608.4 Door swing. Vented wall furnaces shall be located so that a door cannot swing within 12 inches (305 mm) of an air inlet or air outlet of such furnace measured at right angles to the opening. Doorstops or door closers shall not be installed to obtain this clearance.

❖ A combustible door that swings close to a wall furnace could be a fire hazard if at some position the door would be within 12 inches (305 mm) of an air inlet or outlet. Because door closers and door stops are easily defeated, they must not be depended upon to secure the required clearance. If the door swing cannot comply with this section, the door would have to be removed or rehung to change its swing direction.

608.5 Ducts prohibited. Ducts shall not be attached to wall furnaces. Casing extension boots shall not be installed unless listed as part of the appliance.

❖ Most wall furnaces are not designed to force air through ducts, especially those gravity types that do not use fans. Attachment of ducts would add resistance to the flow of air through the furnace, thereby causing an abnormally high temperature rise across the heat exchanger and abnormally high temperatures on furnace surfaces. Some wall furnaces are listed and designed for use with supply duct extensions intended for wall mounting and intended to improve heat distribution. Such duct extensions are factory-built and supplied only by the furnace manufacturer.

608.6 Access. Vented wall furnaces shall be provided with access for cleaning of heating surfaces, removal of burners, replacement of sections, motors, controls, filters and other working parts, and for adjustments and lubrication of parts requiring such attention. Panels, grilles and access doors that are required to be removed for normal servicing operations shall not be attached to the building construction.

❖ All access panels and doors must be removable as intended by the manufacturer of the furnace.

SECTION 609 (IFGC)
FLOOR FURNACES

609.1 General. Floor furnaces shall be tested in accordance with ANSI Z21.48 or Z21.86/CSA 2.32 and shall be installed in accordance with the manufacturer's installation instructions.

❖ Floor furnaces are vented appliances that are installed in an opening in the floor. These units heat the room by gravity convection and direct radiation and usually serve as the sole source of space heating. Such furnaces are common in cottages, small homes, seasonally occupied structures and rural homes. Because the floor grille can become hot, extreme care must be exercised to prevent occupants, especially children, from contacting the grille by walking or falling on it. Also, care must be taken to avoid a fire hazard caused by placement of materials or furnishings on or near the furnace floor grille (see commentary Figure 609.1).

Photo courtesy of Empire Comfort Systems Inc.

**Figure 609.1
FLOOR FURNACE**

609.2 Placement. The following provisions apply to floor furnaces.

1. Floors. Floor furnaces shall not be installed in the floor of any doorway, stairway landing, aisle or passageway of any enclosure, public or private, or in an exitway from any such room or space.

2. Walls and corners. The register of a floor furnace with a horizontal warm-air outlet shall not be placed closer than 6 inches (152 mm) to the nearest wall. A distance of at least 18 inches (457 mm) from two adjoining sides of the floor furnace register to walls shall be provided to eliminate the necessity of occupants walking over the warm-air discharge. The remaining sides shall be permitted to be placed not closer than 6 inches (152 mm) to a wall. Wall-register models shall not be placed closer than 6 inches (152 mm) to a corner.

3. Draperies. The furnace shall be placed so that a door, drapery or similar object cannot be nearer than 12 inches (305 mm) to any portion of the register of the furnace.

4. Floor construction. Floor furnaces shall not be installed in concrete floor construction built on grade.

5. Thermostat. The controlling thermostat for a floor furnace shall be located within the same room or space as the floor furnace or shall be located in an adjacent room or space that is permanently open to the room or space containing the floor furnace.

❖ 1. Floor furnaces must not be located where they would interfere with or impede egress. In an emergency egress situation, a floor furnace could be a tripping hazard and could collapse under the live load of many occupants.

2. The floor furnace register must be at least 6 inches from walls to avoid creating a fire hazard by raising the wall temperature to a combustible level. The same reasoning applies to the 6-inch distance from wall registers to a corner. The 18-inch clearance from adjoining sides of a register to walls provides space for the occupants to walk around the grille, which can reach temperatures high enough to cause burns.

3. As stated in Section 608.3, the furnace location must not create a fire hazard by being too close to wall surfaces, trim items, furnishings and window treatments.

4. A floor furnace installed in a slab on grade would have to be in a pit, would be subject to flooding and corrosion, and would be inaccessible for service and inspection.

5. If the controlling thermostat does not sense the air temperature in the room in which the furnace is installed, dangerous overheating could result. A thermostat isolated from the source of heat it controls would not respond to the condition in the space served by the furnace.

609.3 Bracing. The floor around the furnace shall be braced and headed with a support framework designed in accordance with the *International Building Code*.

❖ The framing around the floor opening must be capable of supporting the floor system, the anticipated floor loads, and the weight of the furnace. The structural requirements of the *International Building Code®* (IBC®) must be complied with.

609.4 Clearance. The lowest portion of the floor furnace shall have not less than a 6-inch (152 mm) clearance from the grade level; except where the lower 6-inch (152 mm) portion of the floor furnace is sealed by the manufacturer to prevent entrance of water, the minimum clearance shall be not less than 2 inches (51 mm). Where such clearances cannot be provided, the ground below and to the sides shall be excavated to form a pit under the furnace so that the required clearance is provided beneath the lowest portion of the furnace. A 12-inch (305 mm) minimum clearance shall be provided on all sides except the control side, which shall have an 18-inch (457 mm) minimum clearance.

❖ This section specifies clearances between the ground and the furnace, which apply to crawl space installations. The clearances allow access for service and inspection and help prevent corrosion of the furnace assembly.

609.5 First floor installation. Where the basement story level below the floor in which a floor furnace is installed is utilized as habitable space, such floor furnaces shall be enclosed as specified in Section 608.6 and shall project into a nonhabitable space.

❖ Where a floor furnace is installed above a habitable basement space, the furnace must project into a nonhabitable space and be separated from that space by noncombustible construction (see commentary, Section 609.6).

609.6 Upper floor installations. Floor furnaces installed in upper stories of buildings shall project below into nonhabitable space and shall be separated from the nonhabitable space by an enclosure constructed of noncombustible materials. The floor furnace shall be provided with access, clearance to all sides and bottom of not less than 6 inches (152 mm) and combustion air in accordance with Section 304.

❖ Where the floor furnace is installed in an upper floor of a building, the furnace must project into a nonhabitable space, similar to the installation in Section 609.5, and must be separated from that space by noncombustible construction. Access and clearance must be provided for maintenance and servicing. An adequate source of combustion air must also supply to the furnace.

SECTION 610 (IFGC)
DUCT FURNACES

610.1 General. Duct furnaces shall be tested in accordance with ANSI Z83.9 or UL 795 and shall be installed in accordance with the manufacturer's installation instructions.

❖ Duct furnaces are designed to be installed "in-line" in an air duct served by an external blower. Such units are basically a heat exchanger and burner assembly in a cabinet and rely upon an independent air handler [see commentary Figures 610.1(1) through 610.1(4)].

610.2 Access panels. Ducts connected to duct furnaces shall have removable access panels on both the upstream and downstream sides of the furnace.

❖ Access panels in the ducts on both sides of the furnace provide full access to the furnace for inspection, repair and regular maintenance of the heat exchanger.

610.3 Location of draft hood and controls. The controls, combustion air inlets and draft hoods for duct furnaces shall be located outside of the ducts. The draft hood shall be located in the same enclosure from which combustion air is taken.

❖ For access to the controls without the necessity of opening the duct access panels on a regular basis, the furnace controls must be located outside the ducts. The combustion air inlets and draft hood cannot operate properly inside the ducts when exposed to the pressurized air flow from the circulating fan.

610. 4 Circulating air. Where a duct furnace is installed so that supply ducts convey air to areas outside the space containing the furnace, the return air shall also be conveyed by a duct(s) sealed

Photo courtesy of Reznor/Thomas & Betts Corporation

Figure 610.1(1)
INTEGRAL POWER-VENTED OUTDOOR DUCT FURNACE

Photo courtesy of Reznor/Thomas & Betts Corporation

Figure 610.1(2)
INTEGRAL GRAVITY-VENTED OUTDOOR DUCT FURNACE

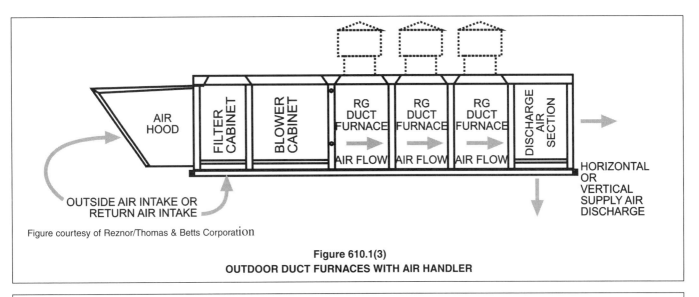

Figure 610.1(3)
OUTDOOR DUCT FURNACES WITH AIR HANDLER

Photo courtesy of Reznor/Thomas & Betts Corporation

Figure 610.1(4)
DUCT FURNACE

to the furnace casing and terminating outside the space containing the furnace.

The duct furnace shall be installed on the positive pressure side of the circulating air blower.

❖ Where the furnace supplies air to areas outside the space containing the furnace, return air must also be supplied from the space outside the furnace area. This will prevent the creation of negative pressure in the space containing the furnace.

SECTION 611 (IFGC)
NONRECIRCULATING DIRECT-FIRED INDUSTRIAL AIR HEATERS

611.1 General. Nonrecirculating direct-fired industrial air heaters shall be listed to ANSI Z83.4/CSA 3.7 and shall be installed in accordance with the manufacturer's instructions.

❖ Direct-fired air-heating equipment is unique in that there is no heat exchanger to separate the burners from the air being heated. As the name implies, the heat of combustion and all by-products are directly introduced into the airstream. Because there is no heat exchanger, chimney or vent, direct-fired heaters are close to 100-percent energy efficient (see commentary Figure 611.1). The term "nonrecirculating" refers to the fact that the same air is not passed through the combustion zone more than once. In other words, these units heat 100 percent outdoor air and are what used to be referred to as make-up air heaters. The air that passes across the burner must be 100 percent outdoor air.

611.2 Installation. Nonrecirculating direct-fired industrial air heaters shall not be used to supply any area containing sleeping quarters. Nonrecirculating direct-fired industrial air heaters shall be installed only in industrial or commercial occupancies. Nonrecirculating direct-fired industrial air heaters shall be permitted to provide ventilation air.

❖ Direct-fired make-up air heaters are used to preheat outdoor air that is supplied as make-up air for large exhaust systems such as those in a commercial kitchen or manufacturing facility. Direct-fired make-up air heaters must not be used to supply any area containing sleeping quarters. The intent of this section is to prohibit the use of such equipment to serve sleeping quarters because direct-fired heaters are unvented and discharge all combustion by-products into the air being heated. There is a health concern because of the possible buildup of carbon monoxide, nitrogen oxides and other combustion gases.

Direct-fired heaters are well suited for large heating loads where thermal efficiency is the primary concern. The heating of outdoor areas, make-up air to be exhausted, and buildings with high infiltration rates is considered energy wasteful; therefore, the highest possible combustion and thermal efficiency is sought for such applications.

611.3 Clearance from combustible materials. Nonrecirculating direct-fired industrial air heaters shall be installed with a clearance from combustible materials of not less than that shown on the rating plate and in the manufacturer's instructions.

❖ Like all listed appliances, the manufacturer's instructions and nameplate (label) will specify the minimum required clearances to combustibles.

611.4 Supply air. All air handled by a nonrecirculating direct-fired industrial air heater, including combustion air, shall be ducted directly from the outdoors.

❖ See the commentary for Section 611.1.

611.5 Outdoor air louvers. If outdoor air louvers of either the manual or automatic type are used, such devices shall be proven to be in the open position prior to allowing the main burners to operate.

❖ Nonrecirculating direct-fired industrial (make-up) air heaters convey 100 percent outdoor air; therefore, it is imperative that any air intake dampers be proven open prior to operation of the heater. As with all interlocks required by the code, the intent is to require a proof-of-operation sequence as opposed to a simple parallel operation arrangement. In other words, the dampers must be electrically proven to be open before heater operation begins.

611.6 Atmospheric vents and gas reliefs or bleeds. Non-recirculating direct-fired industrial air heaters with valve train com-

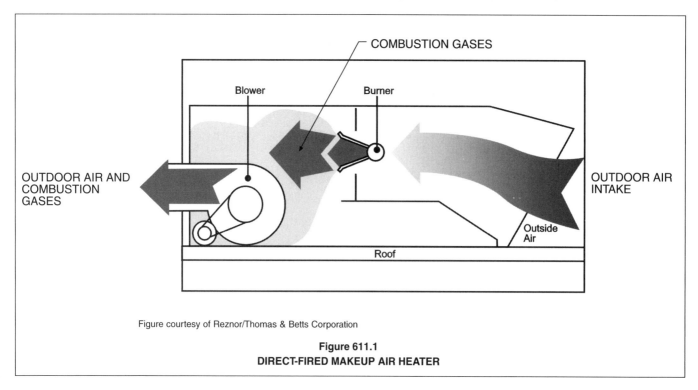

Figure courtesy of Reznor/Thomas & Betts Corporation

Figure 611.1
DIRECT-FIRED MAKEUP AIR HEATER

ponents equipped with atmospheric vents or gas reliefs or bleeds shall have their atmospheric vent lines or gas reliefs or bleeds lead to the outdoors. Means shall be employed on these lines to prevent water from entering and to prevent blockage by insects and foreign matter. An atmospheric vent line shall not be required to be provided on a valve train component equipped with a listed vent limiter.

❖ Large gas-fired equipment is commonly equipped with a complicated gas control arrangement (fuel/valve train components) that can include high- and low-gas pressure switches, gas relief valves, diaphragm valve gas bleed lines, "tell-tale" valve leakage indicator ports and pressure regulators. Control requirements are dictated by equipment standards and historically have also been dictated by insurance underwriters such as Factory Mutual (FM) and Industrial Risk Insurance (IRI).

Reliefs, vents and bleeds can convey raw fuel gas under normal or abnormal conditions and therefore must be piped to the outdoors to avoid a fire or explosion hazard in the building. This piping/tubing is usually joined to an appropriately sized manifold that is vented to the outdoors by one or more pipes.

Atmospheric vent piping is commonly terminated with a screened and turned down outlet or specially designed vent terminal fitting. Blockage by debris, water, ice or insects could cause dangerous control malfunctions of the equipment served by the vent. Where allowed by the listing of a device, such as a pressure regulator; for example, a flow-restrictor device, can be substituted for the vent to the outdoors. A vent-limiter device allows the device served to function as intended, and in the event of a diaphragm failure, the volume of gas escaping through the vent would be controlled by the vent-limiter device.

611.7 Relief opening. The design of the installation shall include provisions to permit nonrecirculating direct-fired industrial air heaters to operate at rated capacity without overpressurizing the space served by the heaters by taking into account the structure's designed infiltration rate, providing properly designed relief openings or an interlocked power exhaust system, or a combination of these methods. The structure's designed infiltration rate and the size of relief openings shall be determined by approved engineering methods. Relief openings shall be permitted to be louvers or counterbalanced gravity dampers. Motorized dampers or closable louvers shall be permitted to be used, provided they are verified to be in their full open position prior to main burner operation.

❖ Most direct-fired heater installations convey 100-percent outdoor air or a mixture of outdoor and recirculated air. Without a means of exhausting the outdoor air being introduced into a building, the building would pressurize, and the airflow through the heater would be retarded. This must not be allowed to happen because the heater is dependent upon a certain flow-through rate for proper burner operation. Exhaust systems, relief air openings, calculated exfiltration rates of the building, or a combination of these must be used to allow the heater to operate at its required airflow rate (see commentary, Section 611.5 for discussion of interlock requirements).

611.8 Access. Nonrecirculating direct-fired industrial air heaters shall be provided with access for removal of burners; replacement of motors, controls, filters and other working parts; and for adjustment and lubrication of parts requiring maintenance.

❖ See the commentary for Section 306.

611.9 Purging. Inlet ducting, where used, shall be purged by not less than four air changes prior to an ignition attempt.

❖ Purging will require that burner ignition be delayed for the time necessary to move a volume of outdoor air through the duct equal to 4 times the volume of the entire inlet duct. Purging will ensure that no flammable fuel gas/air mixtures are present in the combustion zone.

SECTION 612 (IFGC)
RECIRCULATING DIRECT-FIRED INDUSTRIAL AIR HEATERS

612.1 General. Recirculating direct-fired industrial air heaters shall be listed to ANSI Z83.18 and shall be installed in accordance with the manufacturer's installation instructions.

❖ Recirculating direct-fired industrial air heaters are usually used to condition very large volume buildings used as factories, mills, warehouses, etc. They most often consist of ductless blower and burner assemblies that turn over the air in the building through a vertical air discharge configuration (see commentary, Section 611.1). Figures 612.1(1) and 612.1(2) illustrate the basic difference between direct-fired and indirect-fired appliances. Direct-fired appliances do not use a heat exchanger to separate the combustion (flue) gases from the air being conditioned.

612.2 Location. Recirculating direct-fired industrial air heaters shall be installed only in industrial and commercial occupancies. Recirculating direct-fired air heaters shall not serve any area containing sleeping quarters. Recirculating direct-fired industrial air heaters shall not be installed in hazardous locations or in buildings that contain flammable solids, liquids or gases, explosive materials or substances that can become toxic when exposed to flame or heat.

❖ Industrial air heaters are large space heaters, commonly called air turnover units, used to temper the air within factory, industrial and storage facilities. The intent of this section is to prohibit the use of such equipment to serve sleeping quarters because direct-fired heaters are unvented and discharge all combustion by-products into the air being heated. There is a health concern because of the possible buildup of carbon monoxide, nitrogen oxides and other combustion gases.

Air recirculated across the burner of a direct-fired heater could be an ignition source for flammable or ex-

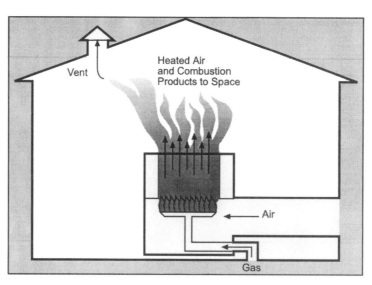

Figure courtesy of Reznor/Thomas & Betts Corporation

**Figure 612.1(1)
BASIC PRINCIPLE OF DIRECT-FIRED
SPACE HEATING APPLIANCE**

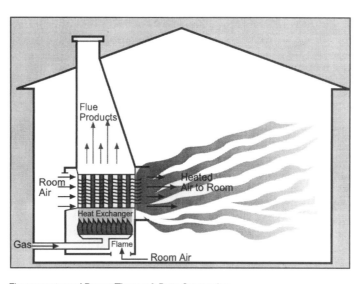

Figure courtesy of Reznor/Thomas & Betts Corporation

**Figure 612.1(2)
BASIC PRINCIPLE OF INDIRECT-FIRED
SPACE HEATING APPLIANCE**

plosive atmospheres or could create toxins by chemically altering substances that pass through the combustion zone.

Heaters used in hazardous areas and areas having substances present that are made toxic by exposure to flames must not recirculate air and therefore must condition only 100-percent outdoor air or transfer air from nonhazardous areas and areas with suitable atmospheres. For example, a direct-fired heater could serve as a 100-percent outdoor make-up air unit for a paint spray booth exhaust system but could not recirculate any air from within the booth. Passing flammable vapor-laden air across the burner would pose a serious hazard.

Because of their high thermal efficiency, direct-fired heaters are well suited for the heating of outdoor areas, make-up air heating and the heating of buildings with high ventilation rates.

612.3 Installation. Direct-fired industrial air heaters shall be permitted to be installed in accordance with their listing and the manufacturer's instructions. Direct-fired industrial air heaters shall be installed only in industrial or commercial occupancies. Direct-fired industrial air heaters shall be permitted to provide fresh air ventilation.

❖ The code does not include prescriptive installation requirements for industrial air heaters, but instead relies on the manufacturer's installation instructions and the listing requirements. This will ensure that the equipment is installed just as it was tested by the listing agency. The heaters are allowed only in industrial or commercial occupancies because of the products of combustion that are introduced into the occupied space (see commentary, Section 612.2). The code allows the outdoor air heated by the direct-fired heaters to supply some of the ventilation air required by Section 403 of the *International Mechanical Code®* (IMC®) (see commentary, Section 612.5).

612.4 Clearance from combustible materials. Direct-fired industrial air heaters shall be installed with a clearance from combustible material of not less than that shown on the label and in the manufacturer's instructions.

❖ The code relies on the listing and the manufacturer's installation instructions for the required clearance to combustibles, just as it does for other installation requirements in Section 612.3.

612.5 Air supply. Air to direct-fired industrial air heaters shall be taken from the building, ducted directly from outdoors, or a combination of both. Direct-fired industrial air heaters shall incorporate a means to supply outside ventilation air to the space at a rate of not less than 4 cubic feet per minute per 1,000 Btu per hour (0.38 m^3 per min per kW) of rated input of the heater. If a separate means is used to supply ventilation air, an interlock shall be provided so as to lock out the main burner operation until the mechanical means is verified. Where outside air dampers or closing louvers are used, they shall be verified to be in the open position prior to main burner operation.

❖ The required ventilation air controls the concentration of combustion by-products in the indoor air. The prescribed ventilation rate is intended to dilute the combustion gases introduced by the direct-fired equipment. If means other than the heater's air handler itself is used to supply the ventilation air required by this section, it must be interlocked with the heating equipment to ensure that the heater cannot operate if the ventilation air supply is inoperative.

612.6 Atmospheric vents, gas reliefs or bleeds. Direct-fired industrial air heaters with valve train components equipped with atmospheric vents, gas reliefs or bleeds shall have their atmospheric vent lines and gas reliefs or bleeds lead to the outdoors.

Means shall be employed on these lines to prevent water from entering and to prevent blockage by insects and foreign matter. An atmospheric vent line shall not be required to be provided on a valve train component equipped with a listed vent limiter.

❖ See the commentary for Section 611.6.

612.7 Relief opening. The design of the installation shall include adequate provision to permit direct-fired industrial air heaters to operate at rated capacity by taking into account the structure's designed infiltration rate, providing properly designed relief openings or an interlocked power exhaust system, or a combination of these methods. The structure's designed infiltration rate and the size of relief openings shall be determined by approved engineering methods. Relief openings shall be permitted to be louvers or counterbalanced gravity dampers. Motorized dampers or closable louvers shall be permitted to be used, provided they are verified to be in their full open position prior to main burner operation.

❖ Most direct-fired heater installations convey 100-percent outdoor air or a mixture of outdoor and recirculated air. Without a means of exhausting the outdoor air being introduced into a building, the building would pressurize, and the airflow through the heater would be retarded. This must not be allowed to happen because the heater is dependent upon a certain flow-through rate for proper burner operation. Exhaust systems, relief air openings, calculated exfiltration rates of the building or a combination of these must be used to allow the heater to operate at its required airflow rate (see commentary, Section 611.5 for a discussion of interlock requirements).

SECTION 613 (IFGC)
CLOTHES DRYERS

613.1 General. Clothes dryers shall be tested in accordance with ANSI Z21.5.1 or ANSI Z21.5.2 and shall be installed in accordance with the manufacturer's installation instructions.

❖ This section addresses clothes dryer appliances, Types 1 and 2. See the definition of "Clothes Dryer." Dryers are tested to the applicable safety standard for the appliance, and the manufacturer's installation instructions convey the information needed to duplicate the installation configuration that was tested and found to meet the requirements of the safety standard. The manufacturer's installation instructions are evaluated by the agency responsible for testing, listing and labeling the appliance, and will therefore prescribe an installation that is consistent with the appliance installation that was tested. Clothes dryers must be exhausted in compliance with Section 614 and the manufacturer's installation instructions (see commentary Figure 613.1).

SECTION 614 (IFGC)
CLOTHES DRYER EXHAUST

614.1 Installation. Clothes dryers shall be exhausted in accordance with the manufacturer's instructions. Dryer exhaust systems shall be independent of all other systems and shall convey the moisture and any products of combustion to the outside of the building.

❖ Clothes dryer exhaust systems must convey the moisture and any products of combustion directly to the exterior of the building (outdoors). The code does not use the term "dryer vent". Gas-fired clothes dryers are not "vented" as are other fuel-fired appliances, but rather are exhausted.

Clothes dryer exhaust ducts must be installed to comply with the dryer manufacturer's installation instructions and the requirements of this section. Dryers are designed and built to meet industry safety standards.

The clothes dryer manufacturer's installation instructions control the type of exhaust duct material allowed and the method of installation. For example, typical dryer installation instructions will require metallic duct materials and will impose more stringent length limitations for flexible metallic ducts than for rigid ducts because of the poorer flow characteristics of flexible duct materials.

Because clothes dryer exhaust contains high concentrations of combustible lint, debris and water vapor, dryer exhaust systems must be independent of all other systems. This requirement prevents the fire hazards associated with such an exhaust system from extending into or affecting other systems or other areas in the building. Additionally, this section intends to prevent products of combustion from entering the building through other systems.

Dryer exhaust ducts must be independent of other dryer exhaust ducts unless connected to an engineered exhaust system specifically designed to serve multiple dryers. Type I domestic dryers are not designed for connection to a common exhaust duct serving multiple dryers. For example, connecting multiple Type I dryers to a common duct riser would pressurize the riser and cause exhaust to back up into any dryer that was not operating. Also, the low-flow velocities and duct temperature losses could result in water vapor condensation and the buildup of lint and debris at low points.

Clothes dryer exhaust systems cannot terminate in or discharge to any enclosed space such as an attic or crawl space, regardless of whether the space is ventilated through openings to the outdoors. The high levels of moisture in the exhaust air can cause condensation to form on exposed surfaces or in insulation materials. Water vapor condensation can cause structural damage and deterioration of building materials and contribute to the growth of mold and fungus. Clothes dryer exhausts that discharge to enclosed spaces will also cause an accumulation of combustible lint and debris, creating a significant fire hazard. An improperly installed clothes dryer exhaust system not only reduces dryer efficiency and increases running time but can also cause a significant increase in exhaust temperature, in turn causing the dryer to cycle on its high limit control, which is an unsafe operating condition.

614.2 Duct penetrations. Ducts that exhaust clothes dryers shall not penetrate or be located within any fireblocking, draftstopping or any wall, floor/ceiling or other assembly required by the *International Building Code* to be fire-resistance rated, unless such duct is constructed of galvanized steel or aluminum of the thickness specified in Table 603.4 of the *International Mechanical Code* and the fire-resistance rating is maintained in accordance with the *International Building Code*. Fire dampers shall not be installed in clothes dryer exhaust duct systems.

❖ Rigid clothes dryer exhaust ducts are permitted to penetrate assemblies that are not fire-resistance-rated and building elements not used as fireblocking or draftstopping. In all other cases, ducts must be constructed of galvanized steel or aluminum of the thickness specified in Section 603.4 of the IMC, and the penetration must be protected to maintain the fire-resistance rating and integrity of the assembly or element being penetrated. Because of the strength and rigidity differences between steel and aluminum, aluminum ducts generally must be of a heavier (thicker) gage than steel ducts for a given application.

The metal thickness requirements of the IMC practically necessitate rigid pipe and all but rule out the use of flexible duct where the duct must penetrate fireblocking, draftstopping or a fire-resistance-rated assembly. Where penetrating fireblocking or draftstopping, the exhaust duct must be constructed of galvanized steel or aluminum, and the annular space around the duct must be fireblocked in accordance with the IBC.

Where penetrating a fire-resistance-rated assembly, the penetration must be protected. The requirements of this section, combined with Section 614.6.1, make a compelling case for always placing clothes dryers against outside walls to avoid long duct runs and penetrations of other than exterior walls.

Figure courtesy of Maytag Corporation

**Figure 613.1
TYPE 1 CLOTHES DRYER LABEL**

614.3 Cleaning access. Each vertical duct riser for dryers listed to ANSI Z21.5.2 shall be provided with a cleanout or other means for cleaning the interior of the duct.

❖ This section applies to Type 2 dryers. Lint and debris carried in the dryer exhaust will settle in the lowest point of any vertical riser in the system; therefore, an accessible means for removing accumulations in the system must be provided to prevent duct blockages and to eliminate the fire hazard of any combustible accumulations. Because of the difficulty of transporting suspended solids vertically against gravity, a vertical section of exhaust duct will eventually require maintenance. The exhaust duct connection to an individual dryer outlet is typically considered as a cleanout because the code requires a "means for cleaning" without specifying what that means can or cannot be. Where exhaust ducts can be accessed and/or readily disassembled, the intent of this section has been met. This section does not require that a tee and cap or similar arrangement be installed where ducts can otherwise be accessed for cleaning.

614.4 Exhaust installation. Exhaust ducts for clothes dryers shall terminate on the outside of the building and shall be equipped with a backdraft damper. Screens shall not be installed at the duct termination. Ducts shall not be connected or installed with sheet metal screws or other fasteners that will obstruct the flow. Clothes dryer exhaust ducts shall not be connected to a vent connector, vent or chimney. Clothes dryer exhaust ducts shall not extend into or through ducts or plenums.

❖ Exhaust ducts must connect directly to terminals that pass through the building envelope to the outdoor atmosphere. Attics and crawl spaces are not considered to be outdoors, and exhaust ducts cannot terminate in those spaces (see commentary, Section 614.1). Backdraft dampers must be installed in dryer exhaust ducts to avoid outdoor air infiltration during periods when the dryer is not operating and to prevent the entry of animals. These dampers should be designed and installed to provide an adequate seal when in a closed position so as to minimize air leakage (infiltration). A backdraft damper is usually of the gravity type that is opened by the energy of the exhaust discharge. Some dryer manufacturers prohibit the use of magnetic backdraft dampers because of the extra resistance that the exhaust flow must overcome.

Exhaust terminal opening size is also governed by the dryer manufacturer's instructions. Full-opening terminals present less resistance to flow and might be mandated by the dryer manufacturer. A "full opening" is considered to be an opening having no dimension less than the diameter of the exhaust duct. Dryer exhaust flow must not be restricted by screens or fastening devices such as sheet metal screws. Any type of screen would become completely blocked with fibers in a very short time. Consider that the filter screen integral with the appliance becomes restricted with lint in each cycle of operation. These restrictions and projections will promote the accumulation of combustible lint and debris in the exhaust duct, thereby creating a potential fire hazard and causing flow resistance. Pop-rivets and $^1/_4$-inch-long screws are commonly used to join dryer exhaust ducts because they do not catch lint and do not create an obstruction. Duct tape should not be relied upon as the sole means of joining dryer exhaust ducts because adhesives can deteriorate with age and when exposed to high temperatures, causing joints to separate. However, sealing joints is still desirable to limit leakage of lint, fibers, moisture vapor and combustion products into the occupied space.

614.5 Makeup air. Installations exhausting more than 200 cfm (0.09 m^3/s) shall be provided with makeup air. Where a closet is designed for the installation of a clothes dryer, an opening having an area of not less than 100 square inches (645 mm^2) for makeup air shall be provided in the closet enclosure, or makeup air shall be provided by other approved means.

❖ Make-up air must be supplied to compensate for the air exhausted by the dryer exhaust system. A typical domestic clothes dryer will exhaust approximately 200 cfm (0.09 m^3/s). For closet installations, the closet door or the closet enclosure must have an opening with a minimum area of 100 square inches (0.0645 m^2). Where louvers or grilles are used, the solid portion of the louver or grille should be evaluated in accordance with Section 304.10. Make-up air is necessary to prevent the room or space housing the dryer(s) from developing a negative pressure with respect to adjacent spaces or the outdoors, which could result in the improper and dangerous operation of the dryer and other fuel-burning appliances. The required amount of make-up air should be approximately equal to the amount exhausted and is normally supplied by infiltration of air from outdoors or through openings to the outdoors.

The make-up air not only supplies the air that is to be exhausted from the dryer, but also supplies combustion air for gas-fired appliances. Commercial dryer manufacturers' installation instructions will prescribe outdoor air opening requirements for supplying the necessary make-up and combustion air (see commentary, Section 304.4).

To simplify the installation, commercial and multiple dryer installations commonly have transfer or ducted openings to the outdoors that introduce outdoor air into the immediate vicinity of the appliances. Commercial units usually obtain make-up air from an enclosed accessible space (plenum) behind the units. This enclosure provides maintenance access; houses the unit's power, piping and duct connections and isolates the make-up air from the conditioned occupiable spaces. Make-up air is rarely supplied by the building's HVAC system because if it were, the HVAC system would have to be interlocked with the dryers to avoid starving any one dryer of make-up air if the HVAC system were not operating. Also, supplying make-up air through the HVAC system would waste energy because large volumes of conditioned air would be exhausted through the dryers. Additionally, cold outdoor air will normally have low moisture content and thus serves well as make-up air for drying operations. (The warmed outdoor air will have low relative humidity.)

614.6 Domestic clothes dryer ducts. Exhaust ducts for domestic clothes dryers shall be constructed of metal and shall have a smooth interior finish. The exhaust duct shall be a minimum nominal size of 4 inches (102 mm) in diameter. The entire exhaust system shall be supported and secured in place. The male end of the duct at overlapped duct joints shall extend in the direction of airflow. Clothes dryer transition ducts used to connect the appliance to the exhaust duct system shall be metal and limited to a single length not to exceed 8 feet (2438 mm) and shall be listed and labeled for the application. Transition ducts shall not be concealed within construction.

❖ Section 305.1 states that gas-fired equipment and appliances must be installed in accordance with the manufacturer's installation instructions for the listed and labeled equipment. An installation complying with the manufacturer's installation instructions is required, except where the code requirements are more stringent. In the case of this section and Section 614.6.1, the code specifically addresses dryer exhaust ducts, an installation requirement that is also addressed in the dryer manufacturer's installation instructions. The code requirements for the items addressed in these sections, such as maximum total developed length of dryer exhaust ducts, equivalent lengths of directional fittings and minimum nominal size to name a few, may parallel or exceed the applicable requirements in the manufacturer's installation instructions. The manufacturer's installation instructions could also contain requirements that exceed those in the code.

In all cases, the more restrictive requirements would apply. Therefore, the clothes dryer exhaust duct size must be determined by the clothes dryer manufacturer but must not be less than 4 inches (102 mm) in diameter.

Dryer exhaust ducts cannot be larger than 4 inches in diameter unless specifically allowed by the manufacturer. Note that enlarging a duct will cause a reduction in flow velocity, and dryer exhaust ducts must maintain sufficient flow velocity to transport lint and fibers through the duct to the discharge terminal.

The duct length, the number and degree of directional fittings, the smoothness of the duct interior wall and the type of exhaust outlet terminal all contribute to the overall friction loss of a clothes dryer exhaust system. When the friction loss is high enough to restrict the required exhaust flow, the duct system must be redesigned to allow the required flow rate. An exhaust duct may have to be shortened or rerouted to compensate for the flow resistance.

Excessive friction losses will also result in reduced flow velocities, which means that the exhaust ducts would be much more likely to collect debris. An improperly designed exhaust system will result in poor dryer performance and poor energy efficiency and can cause the appliance to cycle on its limit control, which can be hazardous.

Joints in the dryer exhaust system must be reasonably airtight, must have a smooth interior finish and must run in the direction of airflow. For example, the male end of each section of duct must point away from the dryer. This permits the dryer exhaust system to function as intended so that joints and connections do not serve as collection points for lint and debris.

Proper support is required and necessary for maintaining alignment of the dryer exhaust duct system and to prevent excessive stress on ducts and duct joints. A sagging duct will increase the internal resistance to airflow, reduce the efficiency of the system and cause the accumulation of lint and debris.

Section 614.6 specifically addresses transition duct connectors. Within the context of this section, a transition duct is a flexible connector used as a transition between the dryer outlet and the connection point to the exhaust duct system. Transition duct connectors must be listed and labeled as transition ducts for clothes dryer application. Transition ducts are currently listed to comply with UL 2158A and are not considered flexible air ducts or flexible air connectors subject to the material requirements of the IMC.

Transition ducts are flexible ducts constructed of a metalized (foil) fabric supported on a spiral wire frame. They are more fire resistant than the typical plastic spiral duct. Transition duct connectors are limited to 8 feet (1829 mm) in length and must be installed in compliance with their listing and the manufacturer's instructions. These duct connectors must not be concealed by any portion of the structure's permanent finish materials such as drywall, plaster, paneling, built-in furniture or cabinets or any other similar permanently affixed building component; they must remain entirely within the room in which the appliance is installed. Transition duct connectors cannot be joined together to extend beyond the 8-foot maximum length limit. Transition ducts are to be cut to length as needed to avoid excess duct and unnecessary bends.

Transition duct connectors are necessary for domestic dryers because of appliance movement, vibration and outlet location. In many cases, connecting a domestic clothes dryer directly to rigid duct would be difficult.

614.6.1 Maximum length. The maximum length of a clothes dryer exhaust duct shall not exceed 25 feet (7620 mm) from the dryer location to the outlet terminal. The maximum length of the duct shall be reduced 2.5 feet (762 mm) for each 45-degree (0.79 rad) bend and 5 feet (1524 mm) for each 90-degree (1.6 rad) bend.

Exception: Where the make and model of the clothes dryer to be installed is known and the manufacturer's installation instructions for such dryer are provided to the code official, the maximum length of the exhaust duct, including any transition duct, shall be permitted to be in accordance with the dryer manufacturer's installation instructions.

❖ The maximum length of 25 feet (7620 mm) and the associated reductions for changes in direction can be very restrictive and make it necessary in most cases to locate the clothes dryer close to an exterior wall. Locating the dryer in the middle of a house or apartment creates a design problem for the architect and the contractor. Exceeding the length requirement of this section will contribute to excessive pressure loss and potential fire hazards (see commentary, Section 614.6).

The length reduction for changes in direction apply to all changes in direction in the exhaust duct, including the first 90-degree (1.57 rad) turn required to run the duct inside a wall cavity. Any change in direction made with the 8-foot (1829 mm) transition duct connector does not require a reduction in length.

The exception allows the 25-foot duct length to be exceeded if the manufacturer's installation instructions specify a longer duct allowance. The manufacturer's instructions must be made available to the code official to allow verification of the allowable length and to allow inspection of the rough-in installation for compliance with those instructions. The code official will also have to verify that the specific clothes dryer is installed. The dryer chosen for application of the exception to this section must be installed before final approval can be given.

The exception could create a problem when the occupants move and take the dryer with them. The next occupant must either buy a dryer with an equivalent duct length allowance or install a duct that is compatible with the dryer to be used. If the new occupant is not aware of the duct requirements, he or she might install a standard dryer designed for only 25 feet of duct length. This could result in a buildup of lint in the duct and a potential fire hazard. The code official is usually not aware of the occupancy change and has no method of assuring that the proper dryer and duct are installed.

The code does not currently address the installation of booster fans that are marketed for the purpose of extending exhaust duct lengths. The concerns have been that dryers are not currently listed for use with external booster fans and there is no interlock to prevent dryer operation in the event of booster fan failure. Clothes dryer exhaust duct length problems could be eliminated if designers would give more thought to the location of laundry facilities.

614.6.2 Rough-in required. Where a compartment or space for a domestic clothes dryer is provided, an exhaust duct system shall be installed.

❖ If the design of a residence includes a laundry room or clothes dryer compartment, this section requires installation of the exhaust system prior to the final inspection. It should also be shown on the plans submitted for review. This will allow the plans examiner to advise the designer or builder of an exhaust duct problem prior to construction.

614.7 Commercial clothes dryers. The installation of dryer exhaust ducts serving Type 2 clothes dryers shall comply with the appliance manufacturer's installation instructions. Exhaust fan motors installed in exhaust systems shall be located outside of the airstream. In multiple installations, the fan shall operate continuously or be interlocked to operate when any individual unit is operating. Ducts shall have a minimum clearance of 6 inches (152 mm) to combustible materials.

❖ See the commentary for Sections 614.1, 614.4 and 614.5.

SECTION 615 (IFGC)
SAUNA HEATERS

615.1 General. Sauna heaters shall be installed in accordance with the manufacturer's installation instructions.

❖ Sauna heaters are used in steam baths and similar rooms to generate heat and steam. They must be installed in compliance with the manufacturer's instructions to ensure they are installed as designed and tested. The code relies on these installation instructions rather than stating prescriptive requirements that might contradict the manufacturer.

615.2 Location and protection. Sauna heaters shall be located so as to minimize the possibility of accidental contact by a person in the room.

❖ Sauna heaters produce steam by passing water over a heated surface. The choice of locations must consider the fact that a heat-producing appliance will be exposed within a small, limited-visibility room with unclothed occupants.

615.2.1 Guards. Sauna heaters shall be protected from accidental contact by an approved guard or barrier of material having a low coefficient of thermal conductivity. The guard shall not substantially affect the transfer of heat from the heater to the room.

❖ Guards are required to protect the occupants from being burned. The guards must be constructed of a material that is a poor conductor of heat (e.g., wood) so that the guard itself will not present a burn hazard. The design of the guard must protect the occupants but not have a negative impact on the flow of heat into the room from the heater.

615.3 Access. Panels, grilles and access doors that are required to be removed for normal servicing operations shall not be attached to the building.

❖ Access panels, covers and doors must not be made unusable by trim, woodwork or room enclosures that impede or interfere with access (see commentary, Section 306.1).

615.4 Combustion and dilution air intakes. Sauna heaters of other than the direct-vent type shall be installed with the draft hood and combustion air intake located outside the sauna room. Where the combustion air inlet and the draft hood are in a dressing room adjacent to the sauna room, there shall be provisions to prevent physically blocking the combustion air inlet and the draft hood inlet, and to prevent physical contact with the draft hood and vent assembly, or warning notices shall be posted to avoid such contact. Any warning notice shall be easily readable, shall contrast with its background and the wording shall be in letters not less than $^1/_4$ inch (6.4 mm) high.

❖ Combustion air and dilution air must not be taken from the sauna room because of the excessive water vapor in the air and the fact that the sauna room would probably be incapable of providing the required volume of combustion air. If a draft hood or combustion air inlet is obstructed, the appliance could malfunction, which could threaten the occupants. For example, a combustion air-starved heater would produce high levels of carbon monoxide that could enter the sauna room in the event of venting failure. In general, the code expresses concern for the use of fuel-fired appliances in small closed rooms, especially where the occupants are sleeping or would have impaired senses or a diminished ability to recognize danger.

615.5 Combustion and ventilation air. Combustion air shall not be taken from inside the sauna room. Combustion and ventilation air for a sauna heater not of the direct-vent type shall be provided to the area in which the combustion air inlet and draft hood are located in accordance with Section 304.

❖ This section would require the heater to be either a direct-vent type or a separated-combustion type in which the combustion chamber does not communicate with the sauna room (see commentary, Section 615.4).

615.6 Heat and time controls. Sauna heaters shall be equipped with a thermostat which will limit room temperature to 194°F (90°C). If the thermostat is not an integral part of the sauna heater, the heat-sensing element shall be located within 6 inches (152 mm) of the ceiling. If the heat-sensing element is a capillary tube and bulb, the assembly shall be attached to the wall or other support, and shall be protected against physical damage.

❖ A thermostat is required to limit the temperature in the sauna for fire safety. The control must sense the warmest air near the ceiling and must be protected from physical damage.

615.6.1 Timers. A timer, if provided to control main burner operation, shall have a maximum operating time of 1 hour. The control for the timer shall be located outside the sauna room.

❖ Timers are not required by the code but provide an extra level of protection when installed. To protect both the occupants and the building, timers must limit the heater operating time to 1 hour. Resetting the timer would require the occupant to exit the sauna, thus lessening the chances of overexposure.

615.7 Sauna room. A ventilation opening into the sauna room shall be provided. The opening shall be not less than 4 inches by 8 inches (102 mm by 203 mm) located near the top of the door into the sauna room.

❖ A ventilation opening is required to allow the escape of steam and heat and to supply ventilation for the occupants. The opening is required near the ceiling to take advantage of the natural tendency of heat and steam to rise, facilitating the exhaust.

615.7.1 Warning notice. The following permanent notice, constructed of approved material, shall be mechanically attached to the sauna room on the outside:

WARNING: DO NOT EXCEED 30 MINUTES IN SAUNA. EXCESSIVE EXPOSURE CAN BE HARMFUL TO HEALTH. ANY PERSON WITH POOR HEALTH SHOULD CONSULT A PHYSICIAN BEFORE USING SAUNA.

The words shall contrast with the background and the wording shall be in letters not less than $^1/_4$ inch (6.4 mm) high.

Exception: This section shall not apply to one- and two-family dwellings.

❖ A warning sign is required to alert users of the potential hazard associated with prolonged exposure to high temperature and humidity levels. The sign is not necessary in a one- and two-family dwelling where the user of the sauna is more familiar with the equipment.

SECTION 616 (IFGC)
ENGINE AND GAS TURBINE-POWERED EQUIPMENT

616.1 Powered equipment. Permanently installed equipment powered by internal combustion engines and turbines shall be installed in accordance with the manufacturer's installation instructions and NFPA 37.

❖ This section addresses gas-fired internal combustion engines and turbines. Engine- and turbine-driven electrical generators for private use are becoming more popular as are engine-driven cooling appliances and heat pumps. This equipment is used to power fire pumps, generators, water pumps, refrigeration machines and other stationary equipment.

NFPA 37 addresses the fire safety of this equipment, including requirements for enclosures, controls, fuel supplies, exhaust systems, cooling systems and combustion air.

SECTION 617 (IFGC)
POOL AND SPA HEATERS

617.1 General. Pool and spa heaters shall be tested in accordance with ANSI Z21.56 and shall be installed in accordance with the manufacturer's installation instructions.

❖ Pool and spa heaters are specialized water heaters very similar in design to hot water supply boilers and are used with swimming pools, recreational or therapeutic spas and hot tubs. These heaters are usually of the water-tube type and are designed for either indoor or outdoor installation.

SECTION 618 (IFGC)
FORCED-AIR WARM-AIR FURNACES

618.1 General. Forced-air warm-air furnaces shall be tested in accordance with ANSI Z21.47 or UL 795 and shall be installed in accordance with the manufacturer's installation instructions.

❖ Forced-air warm-air furnaces are considered to be central heating units and consist of burners or heating elements, heat exchangers, blowers and associated controls. Forced-air furnaces are made in many different configurations, including upflow, counterglow (down flow), horizontal flow, multi-position flow and indoor and outdoor units. Commentary Figure 502.1(2) is an example of a multi-position flow (i.e., up, down and horizontal flow) forced-air warm-air furnace.

618.2 Forced-air furnaces. The minimum unobstructed total area of the outside and return air ducts or openings to a forced-air warm-air furnace shall be not less than 2 square inches for each 1,000 Btu/h (4402 mm^2/W) output rating capacity of the furnace and not less than that specified in the furnace manufacturer's installation instructions. The minimum unobstructed total area of supply ducts from a forced-air warm-air furnace shall be not less than 2 square inches for each 1,000 Btu/h (4402 mm^2/W) output rating capacity of the furnace and not less than that specified in the furnace manufacturer's installation instructions.

Exception: The total area of the supply air ducts and outside and return air ducts shall not be required to be larger than the minimum size required by the furnace manufacturer's installation instructions.

❖ The aggregate area of all ducts or openings that convey supply air from the furnace or return air back to the furnace must be adequate to allow the required airflow through the furnace. A furnace that is "starved" for return air or is restricted by an inadequate supply air duct size will produce an abnormal temperature rise across the heat exchanger, which is both a fire hazard and detrimental to the furnace. The furnace output rating is not usually indicated on the label and would be determined as approximately the input rating in Btu/h (W) times the efficiency rating of the furnace. For example, 100,000 Btu/h (29 310 W) input times 0.80 (80-percent efficiency) is 80,000 Btu/h (23 448 W) output. Return or supply air openings required by this section must not be less than that specified by the furnace manufacturer's installation instructions.

The exception intends to clarify that if the furnace installation instructions specify a lesser return or supply area than this section, that lesser area is permitted. The code-specified minimum area applies where the furnace manufacturer does not specify a minimum area.

618.3 Dampers. Volume dampers shall not be placed in the air inlet to a furnace in a manner that will reduce the required air to the furnace.

❖ Dampers are usually avoided in return air ducts and openings because of the risk of starving the furnace for return air. If dampers are installed, the total unrestricted return air duct or opening area must be as required by the code with all of the dampers in the fully closed position.

618.4 Circulating air ducts for forced-air warm-air furnaces. Circulating air for fuel-burning, forced-air-type, warm-air furnaces shall be conducted into the blower housing from outside the furnace enclosure by continuous air-tight ducts.

❖ This section parallels the manufacturers' installation instructions that require return air ducts to extend from the furnace cabinet to the exterior of any closet, alcove or furnace room that encloses a furnace. If return air were drawn from within a furnace enclosure, negative pressures could be produced in the room, and negative pressure could cause combustion by-products to be drawn from the combustion chamber or venting system and introduced into the circulating airflow.

Negative and positive pressure differentials between the interior and exterior of a furnace enclosure could also cause hazardous burner and venting system malfunctions.

618.5 Prohibited sources. Outside or return air for a forced-air heating system shall not be taken from the following locations:

1. Closer than 10 feet (3048 mm) from an appliance vent outlet, a vent opening from a plumbing drainage system or the discharge outlet of an exhaust fan, unless the outlet is 3 feet (914 mm) above the outside air inlet.

2. Where there is the presence of objectionable odors, fumes or flammable vapors; or where located less than 10 feet (3048 mm) above the surface of any abutting public way or driveway; or where located at grade level by a sidewalk, street, alley or driveway.

3. A hazardous or insanitary location or a refrigeration machinery room as defined in the *International Mechanical Code*.

4. A room or space, the volume of which is less than 25 percent of the entire volume served by such system. Where connected by a permanent opening having an area sized in accordance with Section 618.2, adjoining rooms or spaces shall be considered as a single room or space for the purpose of determining the volume of such rooms or spaces.

Exception: The minimum volume requirement shall not apply where the amount of return air taken from a room or space is less than or equal to the amount of supply air delivered to such room or space.

5. A room or space containing an appliance where such a room or space serves as the sole source of return air.

Exception: This shall not apply where:

1. The appliance is a direct-vent appliance or an appliance not requiring a vent in accordance with Section 501.8.
2. The room or space complies with the following requirements:
    2.1. The return air shall be taken from a room or space having a volume exceeding 1

cubic foot for each 10 Btu/h (9.6 L/W) of combined input rating of all fuel-burning appliances therein.

2.2. The volume of supply air discharged back into the same space shall be approximately equal to the volume of return air taken from the space.

2.3. Return-air inlets shall not be located within 10 feet (3048 mm) of any appliance firebox or draft hood in the same room or space.

3. Rooms or spaces containing solid fuel-burning appliances, provided that return-air inlets are located not less than 10 feet (3048 mm) from the firebox of such appliances.

6. A closet, bathroom, toilet room, kitchen, garage, mechanical room, boiler room or furnace room.

❖ This section prohibits outdoor air and return air from being taken from locations that are potential sources of contamination, odor, flammable vapors or toxic substances and also from locations that would negatively affect the operation of the furnace itself or other fuel-burning appliances.

The intent of item 4 is to prevent the system from being starved for return air by the placement of the main or only return air intake in an area not meeting the volume requirements. Items 4 and 5 are airflow balance related, and items 1, 2, 3, 5 and 6 are contaminant related. It is not the intent of this section to prohibit the common practice of installing return air intakes in bedrooms and similarly sized rooms that typically have a volume that is far less than 25 percent of the total volume of the space served by the furnace. The return-air system must be able to convey the required air flow to the furnace regardless of the position of any doors to any rooms in the building served by the furnace.

It is the intent of this section to avoid arrangements that cause an air pressure imbalance, which can cause fuel-fired appliances to spill combustion products into the occupied space. Pressure imbalances can be avoided by making sure that the amount of supply air discharge to a room or space is approximately equal to the amount of return air taken from the room or space.

618.6 Screen. Required outdoor air inlets for residential portions of a building shall be covered with a screen having $^1/_4$-inch (6.4 mm) openings. Required outdoor air inlets serving a nonresidential portion of a building shall be covered with screen having openings larger than $^1/_4$ inch (6.4 mm) and not larger than 1 inch (25 mm).

❖ The inlet openings for outdoor air must be covered with screen or mesh material with a mesh opening size as specified. The screen openings must be small enough to keep out insects, rodents, birds, etc., while being large enough to prevent blockage of air flow by lint, debris and plant fibers.

618.7 Return-air limitation. Return air from one dwelling unit shall not be discharged into another dwelling unit.

❖ This section prohibits a forced-air heating/cooling system from serving more than one dwelling unit. Any arrangement in which dwelling units share all or part of an air distribution system would allow a communication of atmospheres between the units. This type of communication would spread odors, smoke, allergens, contaminants and disease-causing organisms from one dwelling unit to another and therefore must be avoided.

SECTION 619 (IFGC)
CONVERSION BURNERS

619.1 Conversion burners. The installation of conversion burners shall conform to ANSI Z21.8.

❖ The referenced standard is an installation standard that, in addition to the manufacturers' instructions, would govern the installation of conversion burners. Conversion burners are an assembly of components including burners, gas controls, blowers, safety devices and supporting means. These units are designed to convert an existing appliance from another fuel to gas, commonly from fuel-oil or coal (see commentary, Section 301.7). Conversion of an existing appliance to a different fuel can involve much more than installation of a conversion burner. It can also include the addition of safety controls and limits, combustion air and secondary air supplies, fuel gas piping installation, chimney or vent alterations and other modifications to the existing appliance and its control system.

SECTION 620 (IFGC)
UNIT HEATERS

620.1 General. Unit heaters shall be tested in accordance with ANSI Z83.8 and shall be installed in accordance with the manufacturer's installation instructions.

❖ Unit heaters are ductless warm-air space heaters that are self-contained and usually suspended from a ceiling or roof structure. Garages, workshops, warehouses, factories, gymnasiums, mercantile spaces and similar large, open spaces are the most common locations for unit heaters [see commentary Figures 620.1(1) through 620.1(5)].

620.2 Support. Suspended-type unit heaters shall be supported by elements that are designed and constructed to accommodate the weight and dynamic loads. Hangers and brackets shall be of noncombustible material.

❖ As with all suspended fuel-fired appliances, a support failure can result in a fire, explosion or injury to building occupants. The supports themselves must be properly designed. Equally important are the structural members to which the supports are attached, such as rafters, beams, joists and purlins. Brackets, pipes, rods, angle iron, structural members and fasteners must be designed for the dead and dynamic loads of the suspended appliance (see Commentary Figure 620.2).

SPECIFIC APPLIANCES FIGURE 620.1(1) - FIGURE 620.1(3)

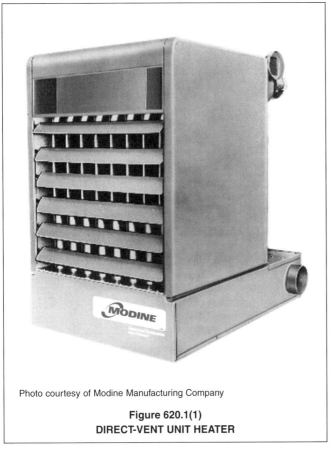

Photo courtesy of Modine Manufacturing Company

Figure 620.1(1)
DIRECT-VENT UNIT HEATER

Photo courtesy of Modine Manufacturing Company

Figure 620.1(2)
POWER-VENTED UNIT HEATER

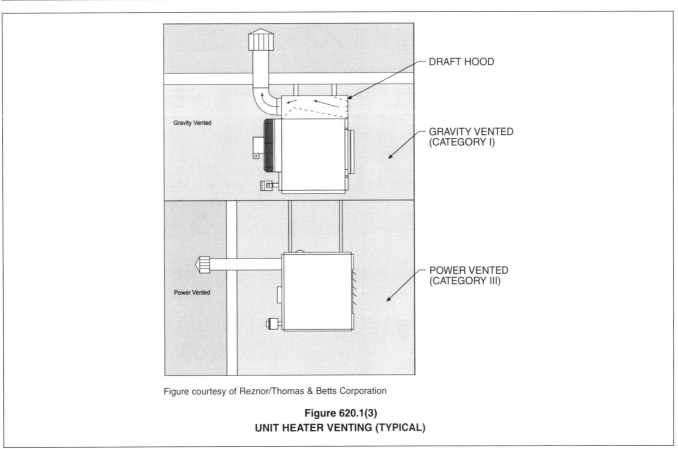

Figure courtesy of Reznor/Thomas & Betts Corporation

Figure 620.1(3)
UNIT HEATER VENTING (TYPICAL)

620.3 Ductwork. Ducts shall not be connected to a unit heater unless the heater is listed for such installation.

❖ Unit heaters are usually not designed to move air through ductwork. Unless specifically listed for the application, the fans or blowers on unit heaters are designed only for moving air across the heat exchanger without the added friction of ductwork. The addition of ductwork could create a hazard by restricting airflow through the heater.

620.4 Clearance. Suspended-type unit heaters shall be installed with clearances to combustible materials of not less than 18 inches (457 mm) at the sides, 12 inches (305 mm) at the bottom and 6 inches (152 mm) above the top where the unit heater has an internal draft hood or 1 inch (25 mm) above the top of the sloping side of the vertical draft hood.

Floor-mounted-type unit heaters shall be installed with clearances to combustible materials at the back and one side only of not less than 6 inches (152 mm). Where the flue gases are vented horizontally, the 6-inch (152 mm) clearance shall be measured from the draft hood or vent instead of the rear wall of the unit heater. Floor-mounted-type unit heaters shall not be installed on combustible floors unless listed for such installation.

Clearances for servicing all unit heaters shall be in accordance with the manufacturer's installation instructions.

Exception: Unit heaters listed for reduced clearance shall be permitted to be installed with such clearances in accordance with their listing and the manufacturer's instructions.

❖ The required clearances to combustible materials for suspended and floor mounted unit heaters are specified in this section and must be adhered to unless the

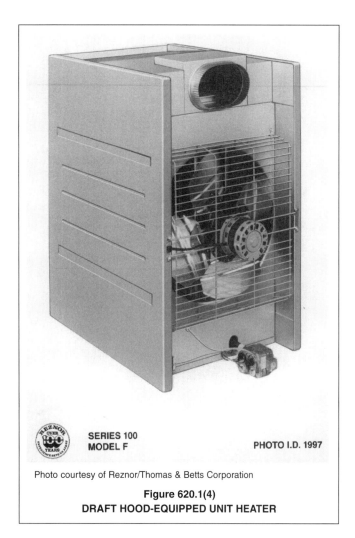

Photo courtesy of Reznor/Thomas & Betts Corporation

Figure 620.1(4)
DRAFT HOOD-EQUIPPED UNIT HEATER

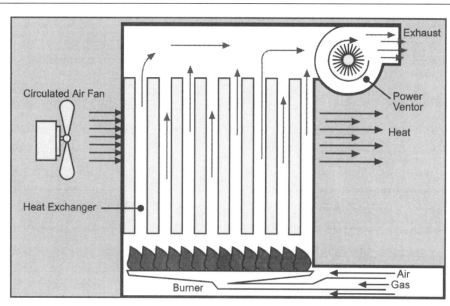

Figure courtesy of Reznor/Thomas & Betts Corporation

Figure 620.1(5)
FUNDAMENTAL COMPONENTS OF A UNIT HEATER

SPECIFIC APPLIANCES

Figure 620.2
UNIT HEATER SUPPORT

unit is listed for less clearance. The clearance reduction methods detailed in Section 308 may be applied where applicable. See the commentary for Section 308 for a complete discussion concerning clearance requirements and clearance reduction methods.

SECTION 621 (IFGC)
UNVENTED ROOM HEATERS

621.1 General. Unvented room heaters shall be tested in accordance with ANSI Z21.11.2 and shall be installed in accordance with the conditions of the listing and the manufacturer's installation instructions. Unvented room heaters utilizing fuels other than fuel gas shall be regulated by the *International Mechanical Code*.

❖ Unvented room heaters are limited-size gas-fired space heaters that discharge the combustion by-products into the space being heated. Like all appliances regulated by this code, unvented gas-fired room heaters must be listed and labeled, and their installation must comply with the manufacturer's installation instructions [see commentary Figures 621.1(1) and 621.1(2)].

621.2 Prohibited use. One or more unvented room heaters shall not be used as the sole source of comfort heating in a dwelling unit.

❖ Unvented room heaters are designed to supplement a central heating system to allow zone heating of particular rooms and spaces. Unvented room heaters are not intended for continuous use as would occur if they were the only source of heat in a building. One or more unvented room heaters used as the sole source of heat would not provide adequate heat distribution in most building arrangements and, depending on the heating load, could require continuous operation of the appliance. Unvented gas-log heaters are listed as room heaters and their installation in factory-built fireplaces is addressed in UL 127 (see commentary Figure 621.2). Fireplaces built to the current edition of UL 127 will have a label and installation instructions that will either allow or disallow the installation of an unvented gas log in the fireplace firebox (see Section 621.2).

Figure 621.1(1)
UNVENTED ROOM HEATER

Photo by D. F. Noyes Studio, courtesy of DESA International

Figure 621.1(2)
FIREBOX FOR UNVENTED ROOM HEATER

Photography by D. F. Noyes Studio, courtesy of DESA International

Figure 621.2
UNVENTED GAS LOG HEATER

621.3 Input rating. Unvented room heaters shall not have an input rating in excess of 40,000 Btu/h (11.7 Kw).

❖ The input rating limitation is consistent with the industry standard for such appliances and allows unvented room heaters to be categorized in the supplemental room heater classification.

621.4 Prohibited locations. Unvented room heaters shall not be installed within occupancies in Groups A, E and I. The location of unvented room heaters shall also comply with Section 303.3.

❖ In accordance with this section and Section 303.3, unvented gas-fired heaters are prohibited in assembly, educational and institutional use groups and in sleeping rooms, bathrooms and toilet rooms in all occupancies (see commentary, Section 303.3 for exceptions). These heaters are considered an unacceptable risk in such occupancies because of occupant density, occupant age, occupant physical condition and awareness and small room volumes.

621.5 Room or space volume. The aggregate input rating of all unvented appliances installed in a room or space shall not exceed 20 Btu/h per cubic foot (207 W/m^3) of volume of such room or space. Where the room or space in which the equipment is installed is directly connected to another room or space by a doorway, archway or other opening of comparable size that cannot be closed, the volume of such adjacent room or space shall be permitted to be included in the calculations.

❖ The Btu/h (W) input to room volume ratio limits the accumulation of combustion by-products in the building interior. Combustion by-products include CO_2, CO, NO_2 and H_2O. The required room volume would allow dilution of the combustion by-products by infiltration.

621.6 Oxygen-depletion safety system. Unvented room heaters shall be equipped with an oxygen-depletion-sensitive safety shutoff system. The system shall shut off the gas supply to the main and pilot burners when the oxygen in the surrounding atmosphere is depleted to the percent concentration specified by the manufacturer, but not lower than 18 percent. The system shall not incorporate field adjustment means capable of changing the set point at which the system acts to shut off the gas supply to the room heater.

❖ Because unvented heaters are not vented to the outdoors, they are required by the appliance standard to incorporate this extra safety feature. This safety system consists of a special pilot burner device that is incorporated with the appliance's flame safeguard device (see commentary, Section 602.2). The oxygen-depletion sensor is basically a pilot burner that is extremely sensitive to the oxygen content in the combustion air. See definition of oxygen depletion safety shutoff system. If the oxygen content (approximately 20 percent normal) drops to a predetermined level, the pilot flame will become "lazy," shifting from a stable horizontal flame to a less stable, more vertical flame, which is incapable of sufficiently heating the thermocouple or thermopile generator, resulting in main gas control valve shutdown and lockout. The predetermined oxygen level that activates burner shutdown is above the level at which incomplete combustion would start to occur.

The monitoring of oxygen levels in this manner is an indicator of approaching insufficient combustion air and is designed to prevent the formation of excess carbon monoxide. If the oxygen level reaches a low enough percentage of air, the level of oxygen will have an inverse relationship with the level of carbon monoxide. This means that as burners are increasingly starved for oxygen, the amount of carbon monoxide produced by the flames is increased. This relationship is valid for appliances in good working order. An improperly maintained, maladjusted or defective burner could produce abnormal (elevated) amounts of carbon monoxide regardless of the oxygen level in the combustion air, and, in such cases, the oxygen-depletion sensor device will react only to the oxygen level because it cannot directly detect burner malfunction or abnormal carbon monoxide production.

621.7 Unvented log heaters. An unvented log heater shall not be installed in a factory-built fireplace unless the fireplace system has been specifically tested, listed and labeled for such use in accordance with UL 127.

❖ Because the fireplace chimney damper can be closed while an unvented log heater (room heater) is operating, the firebox might reach surface temperatures higher than would occur with the damper open. The lack of chimney draft with the resultant dilution air introduction into the firebox would allow higher temperatures to occur in the fireplace components. To prevent surface temperatures from exceeding that allowed by UL 127, the standard now requires factory-built fireplace manufacturers to either test their units for use with unvented log heaters and meet the standard criteria or provide instructions and labels that prohibit such use.

If a factory-built fireplace has not been tested for use with an unvented appliance, the installation instructions and labels must state that the use is prohibited. Fireplaces manufactured before the UL standard added coverage for unvented log heaters could not have been tested for that use, and the manufacturers did not have the opportunity to include instructions and labels covering unvented log heaters for use with their units. The installation instructions for fireplace units that were built prior to the unvented heater coverage being added to the UL 127 standard stated that the fireplace was permitted for use only with solid fuel and decorative gas logs listed to Z21.60. Regardless of when the fireplace unit was manufactured, therefore, unvented log heaters can be installed only in factory-built fireplaces that specifically state that they are tested for use with unvented log heaters. The fireplace manufacturers will currently indicate whether or not their units are intended for use with unvented log heaters. If listed for the application, an unvented log heater can be used as a vented decorative appliance by opening the fireplace chimney damper.

SECTION 622 (IFGC)
VENTED ROOM HEATERS

622.1 General. Vented room heaters shall be tested in accordance with ANSI Z21.11.1 or ANSI Z21.86/CSA 2.32, shall be designed and equipped as specified in Section 602.2 and shall be installed in accordance with the manufacturer's installation instructions.

❖ This section addresses vented gas-fired space/room heaters, including direct-vent and vent- or chimney-connected appliances, which are limited-size room heaters designed to connect to a vent or chimney. As required by the appliance standard, these appliances are equipped with safety controls that will prevent gas flow to the burners in the event of ignition system failure [see commentary Figures 622.1(1) and 622.1(2)].

Photo courtesy of Quadra-Fire

**Figure 622.1(1)
VENTED GAS FIREPLACE HEATER**

SECTION 623 (IFGC)
COOKING APPLIANCES

623.1 Cooking appliances. Cooking appliances that are designed for permanent installation, including ranges, ovens, stoves, broilers, grills, fryers, griddles, hot plates and barbecues, shall be tested in accordance with ANSI Z21.1, ANSI Z21.58 or ANSI Z83.11 and shall be installed in accordance with the manufacturer's installation instructions.

❖ This section addresses cooking appliances in all occupancies that are designed for permanent installation, including, but not limited to, ranges, ovens, stoves, broilers, grills, fryers, griddles and barbecues. These appliances must be installed in accordance with the listing and the manufacturer's installation instructions [see commentary Figures 623.1(1) and 623.1(2)].

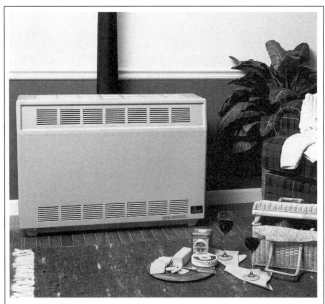

Photo courtesy of Empire Comfort Systems Inc.

**Figure 622.1(2)
VENTED ROOM HEATER**

The code intends to regulate the design, construction and installation of cooking appliances that are designed for permanent installation–heated counter-top appliances. Appliances that are not readily moveable to another location because of a gas-fuel supply connection would be considered permanently installed, even if they were on casters. Line equipment under a Type I hood, for example, is usually on casters and connected with quick-disconnect-type fuel supply lines to allow movement for routine cleaning. This kind of equipment would be considered permanently installed.

623.2 Prohibited location. Cooking appliances designed, tested, listed and labeled for use in commercial occupancies shall not be installed within dwelling units or within any area where domestic cooking operations occur.

❖ Commercial cooking appliances are tested and labeled to different standards than those listed for domestic use. Commercial cooking appliances generally are not insulated to the same level, have higher surface operating temperatures and require a much greater clearance to combustible material. The safety measures inherent to household cooking appliances, such as child-safe push-to-turn knobs and insulated oven doors, are not usually found in commercial cooking appliances. Commercial cooking appliances also have a greater ventilation air requirement for safe operation than household-type cooking appliances. For this reason, installation of commercial-type cooking appliances in dwellings is prohibited.

Photo courtesy of Blodgett Ovens

**Figure 623.1(1)
COMMERCIAL OVEN**

Photo courtesy of Blodgett Ovens

**Figure 623.1(2)
COMMERCIAL DECK OVEN**

SPECIFIC APPLIANCES

623.3 Domestic appliances. Cooking appliances installed within dwelling units and within areas where domestic cooking operations occur shall be listed and labeled as household-type appliances for domestic use.

❖ Cooking appliances used in dwelling units or in areas where domestic cooking operations occur require a greater degree of user protection and must be listed and labeled as household-type appliances for domestic use (see commentary, Section 623.2). To satisfy residential consumer demand for commercial appliances in the home, some manufacturers are producing listed household-type appliances that have the appearance of commercial application cooking appliances [see commentary Figures 623.3(1) and 623.3(2)].

623.4 Domestic range installation. Domestic ranges installed on combustible floors shall be set on their own bases or legs and shall be installed with clearances of not less than that shown on the label.

❖ This section requires the installation of a domestic range using the legs or base provided by the manufacturer where the range is resting on a combustible surface. These supports were part of the test set-up when the range was tested and must be included in the installation to ensure compliance with the listing. Clearances behind and to the sides of the range must also comply with the listing.

623.5 Open-top broiler unit hoods. A ventilating hood shall be provided above a domestic open-top broiler unit, unless otherwise listed for forced down draft ventilation.

❖ Open top broiler units generate smoke and grease-laden vapors as part of the cooking process. A ventilat-

Photo courtesy of Maytag Corporation

**Figure 623.3(1)
DOMESTIC (HOUSEHOLD TYPE) RANGE**

Photo courtesy of Maytag Corporation

**Figure 623.3(2)
HOUSEHOLD-TYPE COOKING APPLIANCE LABEL**

ing hood is required to evacuate the effluent from the kitchen. Broilers using forced downdraft ventilation are exempt from the hood requirement because they exhaust the effluent down through exhaust ducts installed in or under the floor.

623.5.1 Clearances. A minimum clearance of 24 inches (610 mm) shall be maintained between the cooking top and combustible material above the hood. The hood shall be at least as wide as the open-top broiler unit and be centered over the unit.

❖ The ventilation hood must be large enough and positioned in such a way that cooking effluent will be captured. The clearance to combustibles above the broiler surface must be maintained to avoid a potential fire hazard.

623.6 Commercial cooking appliance venting. Commercial cooking appliances, other than those exempted by Section 501.8, shall be vented by connecting the appliance to a vent or chimney in accordance with this code and the appliance manufacturer's instructions or the appliance shall be vented in accordance with Section 505.1.1.

❖ Unless exempted by Section 501.8, commercial cooking appliances must be vented by a chimney or venting system or by the ventilation exhaust hood (Type I or II) required by the IMC (see commentary, Section 505.1.1).

Because the hood will be installed in almost all cases, it is taken advantage of and used as the means of venting the appliances. In this case, the Type I or II hood required by the IMC serves a dual purpose as an exhaust system for cooking effluent and as a combustion product venting means.

SECTION 624 (IFGC)
WATER HEATERS

624.1 General. Water heaters shall be tested in accordance with ANSI Z21.10.1 and ANSI Z21.10.3 and shall be installed in accordance with the manufacturer's installation instructions Water heaters utilizing fuels other than fuel gas shall be regulated by the *International Mechanical Code*.

❖ Water heaters are recognized by the codes as both a plumbing and a mechanical appliance. Chapter 5 of the *International Plumbing Code*® (IPC®) and Chapter 10 of the IMC contain more detailed information concerning the testing and installation of water heaters than is given in this code. This section identifies those aspects of water heaters specifically related to fuel gas. The IMC addresses those aspects related to other fuel or power sources.

The standards listed are specific to gas-fired water heaters. ANSI Z21.10.1 is for water heaters with an input rating less than or equal to 75,000 Btu per hour. ANSI Z21.10.3 is for water heaters with an input rating greater than 75,000 Btu per hour and for circulating and instantaneous water heaters.

624.1.1 Installation requirements. The requirements for water heaters relative to sizing, relief valves, drain pans and scald protection shall be in accordance with the *International Plumbing Code*.

❖ This section refers the user to the IPC for generic requirements related to water heaters not related to the fuel source. These include connections to the potable water system, safety devices such as relief valves, drain pans for protection of the structure and sizing of the water heaters.

624.2 Water heaters utilized for space heating. Water heaters utilized both to supply potable hot water and provide hot water for space-heating applications shall be listed and labeled for such applications by the manufacturer and shall be installed in accordance with the manufacturer's installation instructions and the *International Plumbing Code*.

❖ Water heaters serving the dual purpose of supplying potable hot water and serving as a heat source for a space-heating system must be listed and labeled for that dual application. This section does not address water heaters used solely for space-heating applications, but rather addresses water heaters that serve a secondary purpose of space heating. The label will indicate whether the water heater is suitable for space heating [see commentary Figures 624.2(1) and (2)].

Chapter 5 of the IPC contains additional requirements for the proper installation of these appliances.

SECTION 625 (IFGC)
REFRIGERATORS

625.1 General. Refrigerators shall be tested in accordance with ANSI Z21.19 and shall be installed in accordance with the manufacturer's installation instructions.

Refrigerators shall be provided with adequate clearances for ventilation at the top and back, and shall be installed in accordance with the manufacturer's instructions. If such instructions are not available, at least 2 inches (51 mm) shall be provided between the back of the refrigerator and the wall and at least 12 inches (305 mm) above the top.

❖ Gas operated refrigerators, although still available, are rarely installed where they would be subject to the permit and inspection process. They usually are used in remote areas such as hunting cabins or in areas where electricity is unavailable or unreliable. At one time, they were used on many farms before electricity became readily available. Clearances are specified in the code for use where manufacturer's installation instructions are not available or do not address clearances (see commentary, Section 501.8).

SECTION 626 (IFGC)
GAS-FIRED TOILETS

626.1 General. Gas-fired toilets shall be tested in accordance with ANSI Z21.61 and installed in accordance with the manufacturer's installation instructions.

❖ Gas-fired toilets, similar to gas operated refrigerators, are generally located in remote locations such as hunting or fishing cabins, or in temporary locations such as construction sites. They provide an alternative where water or sewer service is unavailable or where it is not feasible to connect to the sewer for a temporary installation. The firing chamber, upon activation by the user, incinerates the wastes and exhausts the products of combustion through a vent.

626.2 Clearance. A gas-fired toilet shall be installed in accordance with its listing and the manufacturer's instructions, provided that the clearance shall in any case be sufficient to afford ready access for use, cleanout and necessary servicing.

❖ This section requires installation that complies with the manufacturer's instructions but emphasizes that, regardless of the instructions, room must be provided to allow cleaning, normal maintenance, and proper use.

SECTION 627 (IFGC)
AIR CONDITIONING EQUIPMENT

627.1 General. Gas-fired air-conditioning equipment shall be tested in accordance with ANSI Z21.40.1 or ANSI Z21.40.2 and shall be installed in accordance with the manufacturer's installation instructions.

FIGURE 624.2(1)
DIRECT-VENT WATER HEATER DESIGNED
FOR BOTH POTABLE WATER HEATING
AND SPACE HEATING APPLICATIONS

❖ Gas-fired air conditioning systems include absorption types and internal-combustion-engine-driven machines. The referenced ANSI standards contain requirements for gas-fired air conditioning equipment. ANSI Z21.40.1 regulates gas-fired absorption systems; ANSI Z21.40.2 regulates work-activated gas-fired air conditioning and heat pump systems. The code relies on the manufacturer's installation instructions and the referenced standards for installation requirements.

627.2 Independent piping. Gas piping serving heating equipment shall be permitted to also serve cooling equipment where such heating and cooling equipment cannot be operated simultaneously (see Section 402).

❖ Where gas supply piping serves both heating and cooling equipment, this section permits the piping to be sized based on the larger gas demand of the two systems rather than sizing based on the total demand of the two systems. This provision is not applicable unless the heating and cooling systems are incapable of operating simultaneously.

627.3 Connection of gas engine-powered air conditioners. To protect against the effects of normal vibration in service, gas engines shall not be rigidly connected to the gas supply piping.

❖ Because of the inherent vibration of a gas engine, a connector must be installed between the engine and the gas supply pipe so that the vibration does not place undue stress on the connector or have an adverse effect on the shutoff valve or the gas piping. Vibration transmitted to the connector and piping could eventually cause joint failure and/or connector failure. The connector must be designed and approved for this application.

627.4 Clearances for indoor installation. Air-conditioning equipment installed in rooms other than alcoves and closets shall be installed with clearances not less than those specified in Section 308.3 except that air-conditioning equipment listed for installation at lesser clearances than those specified in Section 308.3 shall be permitted to be installed in accordance with such listing and the manufacturer's instructions and air-conditioning equipment listed for installation at greater clearances than those specified in Section 308.3 shall be installed in accordance with such listing and the manufacturer's instructions.

Air-conditioning equipment installed in rooms other than alcoves and closets shall be permitted to be installed with reduced clearances to combustible material, provided that the combustible material is protected in accordance with Table 308.2.

❖ See the commentary for Section 308.3.

627.5 Alcove and closet installation. Air-conditioning equipment installed in spaces such as alcoves and closets shall be specifically listed for such installation and installed in accordance with the terms of such listing. The installation clearances for air-conditioning equipment in alcoves and closets shall not be reduced by the protection methods described in Table 308.2.

❖ Air conditioning equipment must be specifically listed for installation in rooms not large in comparison with the size of the equipment, such as alcoves and closets (see the definition of "Room large in comparison with size of equipment").

627.6 Installation. Air-conditioning equipment shall be installed in accordance with the manufacturer's instructions. Unless the equipment is listed for installation on a combustible surface such as a floor or roof, or unless the surface is protected in an approved manner, equipment shall be installed on a surface of noncombustible construction with noncombustible material and surface finish and with no combustible material against the underside thereof.

❖ The intent of this section is to prohibit the installation of gas-fired air-conditioning equipment on combustible surfaces unless specifically listed for such an installation or the surface is protected in a manner approved by the code official.

627.7 Plenums and air ducts. A plenum supplied as a part of the air-conditioning equipment shall be installed in accordance with the equipment manufacturer's instructions. Where a plenum is not supplied with the equipment, such plenum shall be installed in accordance with the fabrication and installation instructions provided by the plenum and equipment manufacturer. The method of connecting supply and return ducts shall facilitate proper circulation of air.

Where air-conditioning equipment is installed within a space separated from the spaces served by the equipment, the air circulated by the equipment shall be conveyed by ducts that are sealed to the casing of the equipment and that separate the circulating air from the combustion and ventilation air.

❖ This section requires installation of plenums and ducts that are part of an air-conditioning system to comply with the air-conditioning equipment manufacturer's requirements. Further guidance on plenums and ducts can be found in Chapter 6 of the IMC (see commentary, Section 618.4).

627.8 Refrigeration coils. A refrigeration coil shall not be installed in conjunction with a forced-air furnace where circulation of cooled air is provided by the furnace blower, unless the blower has sufficient capacity to overcome the external static resistance imposed by the duct system and cooling coil at the air throughput necessary for heating or cooling, whichever is greater. Furnaces shall not be located upstream from cooling units, unless the cooling unit is designed or equipped so as not to develop excessive temperature or pressure. Refrigeration coils shall be installed in parallel with or on the downstream side of central furnaces to avoid condensation in the heating element, unless the furnace has been specifically listed for downstream installation. With a parallel flow arrangement, the dampers or other means used to control flow of air shall be sufficiently tight to prevent any circulation of cooled air through the furnace.

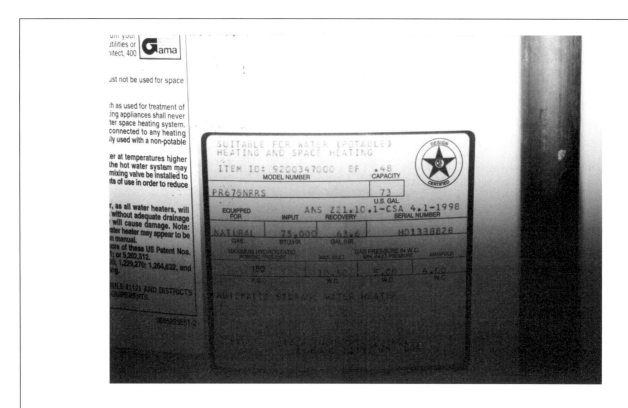

Figure 624.2(2)
WATER HEATER LABEL INDICATING SUITABILITY
FOR BOTH POTABLE WATER HEATING AND SPACE HEATING

Means shall be provided for disposal of condensate and to prevent dripping of condensate onto the heating element.

❖ A refrigeration coil adds static pressure loss to that of the furnace and duct system. This section requires that the furnace blower be sized to overcome the total resistance.

The location of a cooling coil downstream from a furnace will expose the coil to higher temperatures during the heating cycle. Most systems of this type will be listed, and the manufacturer will design the equipment to prevent the buildup of temperature and pressure in the cooling coils. This may be achieved using a method that allows the coolant to expand into the piping system or an expansion chamber.

Installing cooling coils upstream of the furnace might expose the burners and heat exchanger to condensation formed during a cooling cycle. Only furnaces listed for this application with corrosion-resistant elements are allowed. Refrigeration coils installed in a parallel arrangement with furnaces can also cause condensation problems if the cooled air is allowed to pass over the heat exchanger. Bypass dampers are usually installed in front of the heating unit to prevent this problem. A listed parallel system will have addressed the condensation problem in its design.

A drain pan or other approved method must be incorporated in the installation to capture the cooling coil condensation and prevent it from dripping onto furnace components such as the heat exchanger. Section 307 of the IMC addresses condensate disposal.

627.9 Cooling units used with heating boilers. Boilers, where used in conjunction with refrigeration systems, shall be installed so that the chilled medium is piped in parallel with the heating boiler with appropriate valves to prevent the chilled medium from entering the heating boiler. Where hot water heating boilers are connected to heating coils located in air-handling units where they might be exposed to refrigerated air circulation, such boiler piping systems shall be equipped with flow control valves or other automatic means to prevent gravity circulation of the boiler water during the cooling cycle.

❖ Where boilers are used in the same system with cooling units, the code requires that they be in parallel with controls to prevent chilled water from circulating into the boiler. Similarly, if heating coils are located where they could be exposed to cooled air, the code requires valves or other means to prevent chilled water from circulating through the boiler from the heating coils during the cooling cycle. In either case the efficiency of the system will be adversely affected, and boiler damage could result.

627.10 Switches in electrical supply line. Means for interrupting the electrical supply to the air-conditioning equipment and to its associated cooling tower (if supplied and installed in a location remote from the air conditioner) shall be provided within sight of and not over 50 feet (15 240 mm) from the air conditioner and cooling tower.

❖ This requirement is a personnel safety issue that is also required by the ICC *Electrical Code®* (ICC EC™).

SECTION 628 (IFGC)
ILLUMINATING APPLIANCES

628.1 General. Illuminating appliances shall be tested in accordance with ANSI Z21.42 and shall be installed in accordance with the manufacturer's installation instructions.

❖ Illuminating appliances must be listed and labeled as required for other appliances regulated by this code. These appliances include gas lamps designed for outdoor use (see commentary, Section 301.4).

628.2 Mounting on buildings. Illuminating appliances designed for wall or ceiling mounting shall be securely attached to substantial structures in such a manner that they are not dependent on the gas piping for support.

❖ Gas piping is not designed to act as a support for any appliance. All appliances must be independently supported to prevent stresses and strains on fuel supply connections.

628.3 Mounting on posts. Illuminating appliances designed for post mounting shall be securely and rigidly attached to a post. Posts shall be rigidly mounted. The strength and rigidity of posts greater than 3 feet (914 mm) in height shall be at least equivalent to that of a $2^1/_2$-inch-diameter (64 mm) post constructed of 0.064-inch-thick (1.6-mm) steel or a 1-inch (25.4 mm) Schedule 40 steel pipe. Posts 3 feet (914 mm) or less in height shall not be smaller than a $^3/_4$-inch (19.1 mm) Schedule 40 steel pipe. Drain openings shall be provided near the base of posts where there is a possibility of water collecting inside them.

❖ Regardless of the intended method of mounting and support of an appliance, the installation must be secure and must not transmit any loading to the fuel supply connection. A requirement is added to provide a drain at the base of the post to prevent collection of water that can deteriorate the post material.

628.4 Appliance pressure regulators. Where an appliance pressure regulator is not supplied with an illuminating appliance and the service line is not equipped with a service pressure regulator, an appliance pressure regulator shall be installed in the line to the illuminating appliance. For multiple installations, one regulator of adequate capacity shall be permitted to serve more than one illuminating appliance.

❖ Outdoor lighting appliances generally require low-pressure gas flow. Some of these lighting appliances are supplied by the manufacturer with a built-in pressure regulator set for the proper operating pressure. If, however, the regulator is not provided, this section requires the installation of one in the gas line serving the appliance. This regulator must be appropriate for the pressure required by the manufacturer. One pressure regulator with the required capacity may be installed to serve more than one lighting appliance if the piping pressure losses permit.

SECTION 629 (IFGC)
SMALL CERAMIC KILNS

629.1 General. Ceramic kilns with a maximum interior volume of 20 cubic feet (0.566 m³) and used for hobby and noncommercial purposes shall be installed in accordance with the manufacturer's installation instructions and the provisions of this code.

❖ Small ceramic kilns used for hobby and noncommercial purposes operate at high temperatures and therefore need to be regulated by the code as are other heat-producing appliances. The minimum level of safety required for installation of small kilns is compliance with the kiln manufacturer's installation instructions and any applicable requirements of this code such as clearance requirements.

SECTION 630 (IFGC)
INFRARED RADIANT HEATERS

630.1 General. Infrared radiant heaters shall be tested in accordance with ANSI Z83.6 and shall be installed in accordance with the manufacturer's installation instructions.

❖ This section addresses radiant heaters including ceramic element and steel tube-type designs. Infrared heaters are produced in both vented and unvented types and function by creating a very hot surface area from which heat energy is directly radiated. Infrared radiant heaters are usually suspended from ceilings or roofs. These heaters are typically used for "spot" heating in spaces that are otherwise unconditioned or that are not conditioned to human comfort levels. Radiant heaters have the advantage of being able to heat objects and personnel without having to heat the surrounding air [see commentary Figures 630.1(1) through 630.1(3)].

The code relies on the manufacturers' installation instructions for installation requirements. Maintaining the clearance to combustibles for radiant heaters is of paramount importance. Direct radiation is a very effective method of transferring heat energy, and improperly located combustible materials can be readily ignited [see commentary Figure 630.1(4)].

630.2 Support. Infrared radiant heaters shall be safely and adequately fixed in an approved position independent of gas and electric supply lines. Hanger and brackets shall be of noncombustible material.

❖ Supports must prevent radiant heaters from falling. A fall would mean losing the required clearance to combustibles; putting tension or pressure on electrical, fuel and vent connections; and dislocating that redirects the radiant output to where a fire hazard would result. Often, such heaters are improperly hung and restrained, allowing heater movement (swinging) to stress flexible (semi-rigid) gas connectors. Gas connectors must be designed and approved for applications in which the appliance moves because of expansion and contraction or lack of restraints. To accommodate expansion/contraction movement, some appliance manufacturers require flexible connectors to be coiled or shaped in a manner that allows limited movement without stressing the metal of the connector, similar to expansion loops and offsets in rigid piping systems.

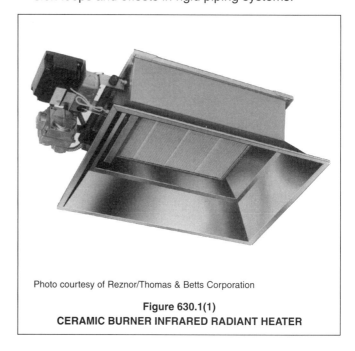

Photo courtesy of Reznor/Thomas & Betts Corporation

Figure 630.1(1)
CERAMIC BURNER INFRARED RADIANT HEATER

SECTION 631 (IFGC)
BOILERS

631.1 Standards. Boilers shall be listed in accordance with the requirements of ANSI Z21.13 or UL 795. If applicable, the boiler shall be designed and constructed in accordance with the requirements of ASME CSD-1 and as applicable, the ASME *Boiler and Pressure Vessel Code*, Sections I, II, IV, V and IX and NFPA 85.

❖ The scope of this section includes boilers in all occupancies including power plants; factories; industrial plants; schools; and institutional occupancies such as hospitals, commercial laundries, hotels and residential structures. Boilers are defined in Section 202 of this code [see commentary Figures 631.1(1) and 631.1(2)].

Boilers are potentially dangerous if not properly designed, constructed and operated, more so than many other appliances because of the potential explosion hazard associated with pressure vessels. In addition to this code, several industry standards are referenced. Manufactured gas-fired boilers must be listed to either ANSI Z21.13 or UL 795. Figure 202(3) shows a typical label with the required information for a listed boiler.

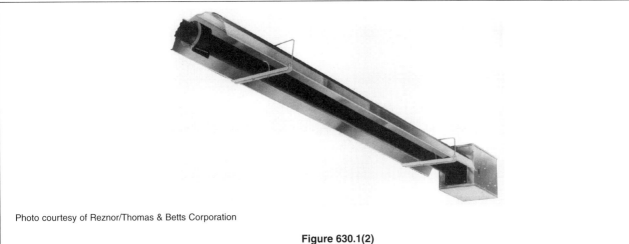

Figure 630.1(2)
TUBE-TYPE INFRARED RADIANT HEATER

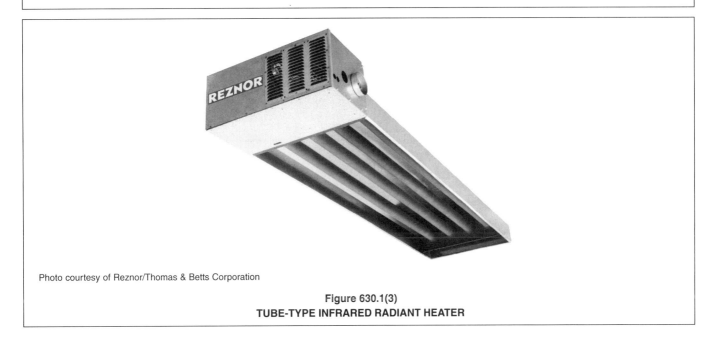

Figure 630.1(3)
TUBE-TYPE INFRARED RADIANT HEATER

Only the design and construction requirements of boilers are regulated by the referenced standards. The requirements for specific types of boilers can be found in the respective sections of the referenced standards.

631.2 Installation. In addition to the requirements of this code, the installation of boilers shall be in accordance with the manufacturer's instructions and the *International Mechanical Code*. Operating instructions of a permanent type shall be attached to the boiler. Boilers shall have all controls set, adjusted and tested by the installer. A complete control diagram together with complete boiler operating instructions shall be furnished by the installer. The manufacturer's rating data and the nameplate shall be attached to the boiler.

❖ This section governs the installation and commissioning of boilers and their control systems. The mechanical equipment requirements for approval, labeling, installation, maintenance, repair and alteration are regulated by the IMC and the manufacturer's instructions. The IMC contains installation requirements for boilers including provisions for shutoff valves, pressure relief valves, safety valves, electrical control wiring, blowoff valves, expansion tanks and low water cutoff controls [see commentary Figures 631.1(1) and (2) and 631.2(1) through 631.2(3)].

Complete operating instructions must be permanently affixed to the boiler upon completion of the installation. Boiler systems can be complex and generally require coordination of several pieces of equipment. The proper operating procedures, set points, etc., must be specified and included in these instructions. This is usually done in the control system design documentation, including such things as diagrams, system schematics and control sequence descriptions. Typically, an operating and maintenance (O&M) manual is given to the building owner/operator upon completion of the project. Along with operating and control procedures, the

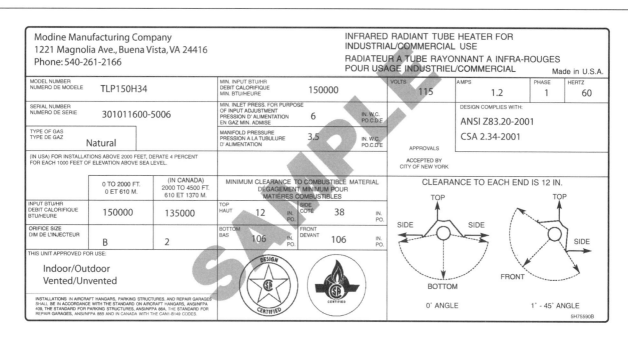

Figure courtesy of Modine Manufacturing Company

**Figure 630.1(4)
SAMPLE LABEL FOR INFRARED RADIANT TUBE HEATER**

O&M manual will clearly identify routine maintenance and calibration information. It is also typical for operation sequences and calibration information to be displayed under glass in a conspicuous place in the boiler room for use by operating and service personnel. The intent is to provide the owner/operator with all operating and control information necessary to properly operate and maintain the boiler and its associated controls and equipment.

631.3 Clearance to combustible materials. Clearances to combustible materials shall be in accordance with Section 308.4.

❖ Boilers, like all gas-fired equipment and appliances, operate at high temperatures that require adequate separation from combustibles. This section refers the user to Section 308 of this code for requirements for clearances to combustibles and allowable clearance reduction methods.

SECTION 632 (IFGC)
EQUIPMENT INSTALLED IN EXISTING UNLISTED BOILERS

632.1 General. Gas equipment installed in existing unlisted boilers shall comply with Section 631.1 and shall be installed in

SPECIFIC APPLIANCES

accordance with the manufacturer's instructions and the *International Mechanical Code*.

❖ This section addresses the replacement of equipment, such as burner assemblies, in existing unlisted boilers. The code relies on the manufacturer's installation instructions, the requirements of the IMC and Section 631 of this code.

Photo courtesy of Weil-McLain

Figure 631.1(1)
STEAM HEATING BOILER WITH GAUGE-GLASS,
LOW WATER CUTOFF CONTROL, PRESSURE GAUGE,
PRESSURE LIMIT CONTROL AND AUTOMATIC VENT DAMPER

Figure 631.1(2)
STEAM BOILER WITH POWER BURNER, LOW WATER
CUTOFF CONTROLS AND PRESSURE CONTROLS

Photo courtesy of McDonnell & Miller

Figure 631.2(1)
ELECTRONIC PROBE-TYPE LOW WATER
CUTOFF CONTROL WITH GAUGE GLASS
AND TRI-COCKS FOR STEAM BOILER APPLICATION

SECTION 633 (IFGC)
FUEL CELL POWER PLANTS

633.1 General. Stationary fuel-cell power plants having a power output not exceeding 1,000 kW shall be tested in accordance with ANSI Z21.83 and shall be installed in accordance with the manufacturer's installation instructions and NFPA 853.

❖ Fuel cell power plants convert hydrogen and oxygen into electrical and heat energy. Through a process called "reforming," natural gas can be converted to a hydrogen-rich gas that is commonly used as the fuel for stationary fuel cell plants. NFPA 853 is an installation standard that covers site and location requirements and piping, ventilation/exhaust and fire protection requirements (see Chapter 7).

Photo courtesy of McDonnell & Miller

Figure 631.2(2)
ELECTRONIC PROBE-TYPE LOW WATER CUTOFF
CONTROL FOR HOT WATER BOILER APPLICATION

Photo courtesy of McDonnell & Miller

Figure 631.2(3)
ELECTRONIC PROBE-TYPE LOW WATER CUTOFF
CONTROL FOR HOT WATER BOILER APPLICATION

SECTION 634 (IFGS)
CHIMNEY DAMPER OPENING AREA

634.1 Free opening area of chimney dampers. Where an unlisted decorative appliance for installation in a vented fireplace is installed, the fireplace damper shall have a permanent free opening equal to or greater than specified in Table 634.1.

❖ This section intends to ensure that an adequate fireplace damper opening is installed for all unlisted decorative appliances. The minimum damper opening size is part of the listing for listed appliances, but the installer and the code official might not have access to this information for an unlisted appliance. Section 301.3 requires all appliances to be listed.

The free opening size is a function of the height of the chimney and the appliance input rating. As stated in note "a," the opening areas in the table correspond to the cross-sectional area of round chimney flues having nominal sizes from 3 to 8 inches (76 to 203 mm), and the 64 square inches (41 290 mm^2) represents the cross-sectional area of a standard 8-inch by 8-inch chimney tile.

TABLE 634.1
FREE OPENING AREA OF CHIMNEY DAMPER FOR VENTING FLUE GASES
FROM UNLISTED DECORATIVE APPLIANCES FOR INSTALLATION IN VENTED FIREPLACES

CHIMNEY HEIGHT (feet)	MINIMUM PERMANENT FREE OPENING (square inches)[a]						
	8	13	20	29	39	51	64
	Appliance input rating (Btu per hour)						
6	7,800	14,000	23,200	34,000	46,400	62,400	8,000
8	8,400	15,200	25,200	37,000	50,400	68,000	86,000
10	9,000	16,800	27,600	40,400	55,800	74,400	96,400
15	9,800	18,200	30,200	44,600	62,400	84,000	108,800
20	10,600	20,200	32,600	50,400	68,400	94,000	122,200
30	11,200	21,600	36,600	55,200	76,800	105,800	138,600

For SI: 1 inch = 25.4 mm, 1 foot = 304.8 mm, 1 square inch = 645.16 m^2, 1 British thermal unit per hour = 0.2931 W.

a. The first six minimum permanent free openings (8 to 51 square inches) correspond approximately to the cross-sectional areas of chimneys having diameters of 3 through 8 inches, respectively. The 64-square-inch opening corresponds to the cross-sectional area of standard 8-inch by 8-inch chimney tile.

❖ See the commentary for Section 634.1.

Bibliography

The following resource materials are referenced in this chapter or are relevant to the subject matter addressed in this chapter.

ANSI Z21.60-00, *Decorative Gas Appliances for Installation in Solid-Fuel-Burning Fireplaces.* New York: American National Standards Institute, 2000.

NFPA 37-98, *Installation and Use of Stationary Combustion Engines and Gas Turbines.* Quincy, MA: National Fire Protection Association, 1998.

UL 103-98, *Factory-Built Chimneys, Residential Type and Building Heating Appliances.* Northbrook, IL: Underwriters Laboratories Inc., 1998.

UL 127-96, *Factory-Built Fireplaces.* Northbrook, IL: Underwriters Laboratories Inc., 1996.

CHAPTER 7
GASEOUS HYDROGEN SYSTEMS

General Comments

Chapter 7 is a newly created chapter that covers hydrogen generating systems, vehicle refueling systems, piping and storage systems and testing of piping systems. Fuel cell power plants are addressed in Section 633. The definition of "fuel gas" includes hydrogen in addition to the traditional fuel gases such as natural gas and LP-gases. Chapter 7 was created to segregate the hydrogen provisions from the provisions applicable to other fuel gases and thus avoid confusion regarding what applies to hydrogen and what does not. The hydrogen coverage is intended to be self-contained in this chapter, except that Chapters 1, 2, 8 and the applicable portions of Chapter 3 also apply in conjunction with Chapter 7.

Hydrogen, symbol H_2, is the lightest of all gases with a vapor density of 0.1, meaning that air is approximately 10 times heavier than hydrogen. At atmospheric pressure, it has a flammability range of 4.0 to 75.0 percent vapor in air by volume. Hydrogen is the most abundant element in the universe and is believed to be the element from which all other elements were created in the cosmos. Because of its small molecular size, hydrogen diffuses rapidly through porous materials. It also dissolves in and diffuses through metals slowly at ambient temperatures and more rapidly at elevated temperatures.

On earth, hydrogen is not found in uncombined form as free hydrogen except in minute traces. The majority of our hydrogen is locked up in water molecules, H_2O. Hydrogen is produced by several chemical processes, including the electrolysis of water and the reaction of natural gas or light hydrocarbons with steam in the presence of a catalyst (steam reforming). Most of the hydrogen now produced in the United States is on an industrial scale by the process of steam reforming, or as a byproduct of petroleum refining and chemicals production. However, there is growing interest in two different types of hydrogen generating appliances to produce hydrogen on-site at the customer's refueling facility or even at a private residence from either electricity or from natural gas, propane or other fuels.

Electrolysis of water breaks it down into its two building blocks, hydrogen and oxygen, by passing an electrical current through electrodes submerged in the water [see commentary Figures 700(1) and 702.1]. Adding an electrolyte such as a salt improves the conductivity of the water and increases the efficiency of the process. The hydrogen gas is collected and compressed for storage and the oxygen is typically released to the atmosphere.

Although hydrogen holds great promise as a clean, abundant and nonpolluting source of energy, there is no free ride in nature. Before we can use the energy of hydrogen, we must first expend an equal or greater amount of energy to produce, store and transport the gas. In other words, we must use energy to tear apart molecules to generate hydrogen before we can reap the energy obtained from recombining hydrogen and oxygen atoms in fuel cells. It is certainly worth the effort because fuel cells are more efficient and far less polluting than gasoline and diesel engines.

Purpose

Chapter 7 provides for the installation of hydrogen systems to mitigate the hazards that are associated with flammable gases and those that are unique to hydrogen. The use of hydrogen as a fuel for power generation and vehicles is expected to grow in the near future as the technology continues to advance and a hydrogen infrastructure is built. In fact, it is the lack of availability of hydrogen, not the lack of equipment that uses hydrogen that is preventing more widespread use of hydrogen fuel [see commentary Figures 700(2) and 700(3)].

Hydrogen is a colorless, odorless, tasteless, flammable, nontoxic gas that burns with an almost invisible bluish flame. A hydrogen flame is very pale and difficult to see in daylight. A major concern with hydrogen is the fact that odorants cannot be added because the chemicals involved would poison the fuel cells, thus hydrogen is not odorized and cannot be detected by any of the human senses.

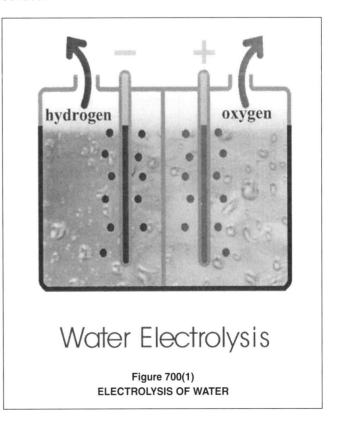

Figure 700(1)
ELECTROLYSIS OF WATER

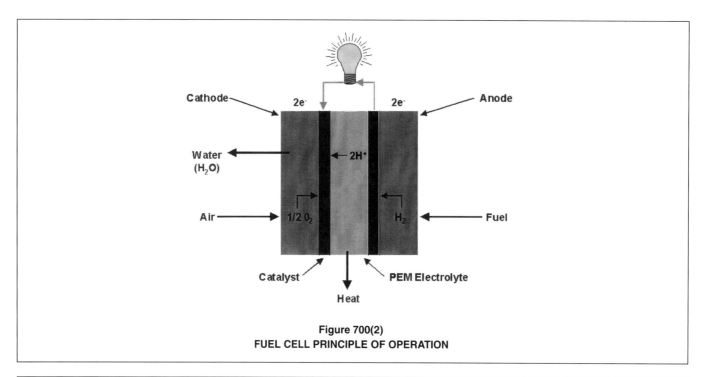

Figure 700(2)
FUEL CELL PRINCIPLE OF OPERATION

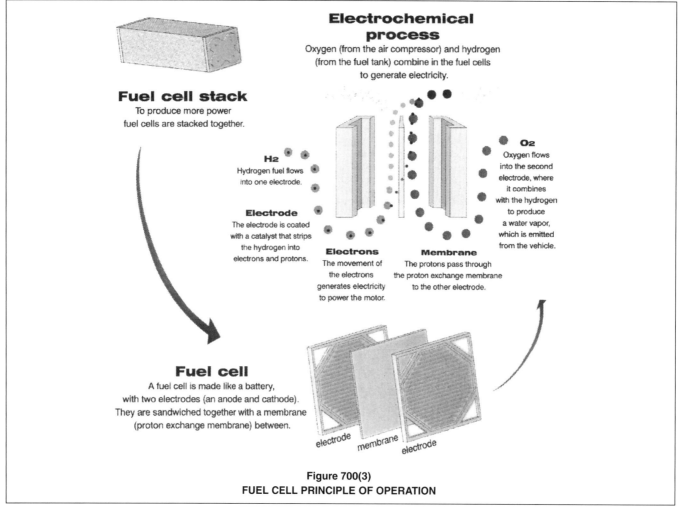

Figure 700(3)
FUEL CELL PRINCIPLE OF OPERATION

SECTION 701 (IFGC)
GENERAL

701.1 Scope. The installation of gaseous hydrogen systems shall comply with this chapter and Chapters 30 and 35 of the *International Fire Code*. Compressed gases shall also comply with Chapter 27 of the *International Fire Code* for general requirements. Containers provided with pyrophoric material shall also comply with Chapter 41 of the *International Fire Code*. Containers having residual gaseous hydrogen shall be considered as full for the purposes of the controls required.

❖ Chapter 30 of the *International Fire Code®* (IFC®) covers the use, handling and storage of compressed gases and Chapter 35 covers the use and storage of flammable gases. Chapters 30 and 35 of the IFC address the hazards of compressed hydrogen gas systems including stored hydrogen gas in pressure vessels. These referenced code sections are intended to reduce the risk posed by the inadvertent rupture of a pressure vessel and release of its hydrogen gas component or the leakage of the flammable gas associated with a piping rupture.

A reference to the commensurate IFC provisions appropriately serves to limit the quantities of compressed hydrogen gas present on both commercial and residential sites, and considers indoor and outdoor use aspects adequately. Reference to Chapter 41 covering pyrophoric materials is intended to address emerging technology (now being contemplated by the automotive industry) that allows for storage of higher densities of compressed hydrogen gas in containers with relatively lower applied pressures.

701.2 Permits. Permits shall be required as set forth in Section 106 and as required by the *International Fire Code*.

❖ This section refers to the applicable administrative rules in the IFC and *International Fuel Gas Code®* (IFGC®) governing the issuance, suspension, renewal or modification of a permit for work affecting gaseous hydrogen service systems.

SECTION 702 (IFGC)
GENERAL DEFINITIONS

702.1 Definitions. The following words and terms shall, for the purposes of this chapter and as used elsewhere in this code, have the meanings shown herein.

❖ Section 202 applies to this chapter. The terms that are unique to this chapter are defined herein for convenience.

HYDROGEN CUTOFF ROOM. A room or space which is intended exclusively to house a gaseous hydrogen system.

❖ An enclosed space used exclusively for a gaseous hydrogen system requires construction and protection that are unique to the hazards associated with these areas. The definition itself should not be interpreted to preclude H_2 piping systems from serving distributed hydrogen-using equipment and appliances located elsewhere on site or in the building. However, the appropriate material-specific quantity limitations of the *International Building Code®* (IBC®) and IFC must to be consulted.

HYDROGEN-GENERATING APPLIANCE. A self-contained package or factory-matched packages of integrated systems for generating gaseous hydrogen. Hydrogen-generating appliances utilize electrolysis, reformation, chemical or other processes to generate hydrogen.

❖ Hydrogen generators use water or hydrocarbon fuels as a feedstock for generation of pure hydrogen or a hydrogen-rich gas (see Figure 702.1). Hydrogen- generating appliances based on chemical reformers separate out the hydrogen from fossil fuels such as natural gas, propane, gasoline, etc. This is the same high-temperature chemical process used at large oil refineries to produce hydrogen. By generating the hydrogen on-site at the fueling station or customer's facility, these hydrogen-generating appliances avoid the high cost of either liquefying hydrogen and delivering it by cryogenic tanker truck, or installing a national hydrogen pipeline system that could cost many tens of billions of dollars. In effect, these on-site hydrogen-generating appliances take advantage of one of two existing energy infrastructures: either the natural gas distribution system or the electrical grid.

GASEOUS HYDROGEN SYSTEM. An assembly of piping, devices and apparatus designed to generate, store, contain, distribute or transport a nontoxic, gaseous hydrogen containing mixture having at least 95-percent hydrogen gas by volume and not more than 1-percent oxygen by volume. Gaseous hydrogen systems consist of items such as compressed gas containers, reactors and appurtenances, including pressure regulators, pressure relief devices, manifolds, pumps, compressors and interconnecting piping and tubing and controls.

❖ This term includes the source of hydrogen and all piping and devices between the source and the using equipment. The gas in a hydrogen system is above the upper flammability limit and is therefore, "too rich" to burn. Any leakage, however, will quickly form an explosive mixture in the ambient atmosphere.

SECTION 703 (IFGC)
GENERAL REQUIREMENTS

703.1 Hydrogen-generating and refueling operations. Ventilation shall be required in accordance with Section 703.1.1, 703.1.2 or 703.1.3 in public garages, private garages, repair garages, automotive service stations and parking garages which contain hydrogen-generating appliances or refueling systems. Such spaces shall be used for the storage of not more than three hydrogen-fueled passenger motor vehicles and have a floor area not exceeding 850 square feet (79 m^2). The maximum rated output capacity of hydrogen-generating appliances shall not exceed 4 standard cubic feet per minute (ft³/min) of hydrogen for each 250 square feet (23.2 m²) of floor area in such spaces. Such equipment and appliances shall not be installed in Group H occupancies except where the occupancy is specifically designed for hydrogen use, or in control areas where open use, handling or dispensing of combustible, flammable or explosive materials occurs. For the purpose of this section, rooms or spaces that are not part of the living space of a dwelling unit and that communicate directly with a private garage through openings shall be considered to be part of the private garage.

❖ This section intends to minimize the potential for explosions by limiting the source of hydrogen gas and by requiring sufficient ventilation to dissipate any leakage.

703.1.1 Natural ventilation. Indoor locations intended for hydrogen-generating or refueling operations shall communicate with the outdoors in accordance with Sections 703.1.1.1 through 703.1.1.2. The minimum cross-sectional dimension of air openings shall be 3 inches (76 mm). Where ducts are used, they shall be of the same cross-sectional area as the free area of the openings to which they connect. In such locations, equipment and appliances having an ignition source shall be located such that the source of ignition is not less than 12 inches (305 mm) below the ceiling.

❖ This section parallels the intent of Section 305.3, but, with respect to the ceiling instead of the floor. This section does not apply to spaces that only house vehicles and that do not contain hydrogen-generating or refueling operations. Because it is more buoyant, hydrogen will dissipate more quickly than natural gas, and much more quickly than either propane or gasoline, both of which have vapors that are heavier than air and will linger at an accident site. However, hydrogen and natural gas can both accumulate in unventilated pockets at the top of indoor structures and could represent a risk in such situations.

Similarly, gasoline fumes can accumulate at the floor level in unventilated spaces, posing a different risk. Thus, ignition sources must be avoided at the top of any unventilated spaces for hydrogen gas. Also, hydrogen is odorless, colorless and burns with a flame that is not generally visible to the human eye. This means that it is unlikely that people will be able to detect unsafe conditions without appropriate instrumentation [similar to carbon monoxide (CO) accumulation in a structure].

703.1.1.1 Two openings. Two permanent openings, one located entirely within 12 inches (305 mm) of the ceiling of the garage, and one located entirely within 12 inches (305 mm) of the floor of the garage, shall be provided in the same exterior wall. The openings shall communicate directly with the outdoors. Each opening shall directly communicate with the outdoors horizontally, and have a minimum free area of $^1/_2$ square foot per 1,000 cubic feet (1 m²/610 m³) of garage volume.

❖ The location requirement will prevent the openings from being more than 12 inches tall because the required openings must be entirely within the 12 inches of wall space measured down from the ceiling and up from the

Figure 702.1
HYDROGEN-GENERATING APPLIANCE

floor. The openings must be in the same wall to help create a gravity flow of gases driven by the natural buoyancy of hydrogen gas. The bottom opening is an air inlet and the top is an air outlet.

703.1.1.2 Louvers and grilles. In calculating the free area required by Section 703.1.1.1, the required size of openings shall be based on the net free area of each opening. If the free area through a design of louver or grille is known, it shall be used in calculating the size opening required to provide the free area specified. If the design and free area are not known, it shall be assumed that wood louvers will have 25-percent free area and metal louvers and grilles will have 75-percent free area. Louvers and grilles shall be fixed in the open position.

❖ This section recognizes that louvers and grilles are usually installed over air inlets and outlets to prevent rain, snow and animals from entering the building. When louvers or grilles are used, the solid portion of the louver or grille must be considered when determining the unobstructed (net clear) area of the opening.

Air openings are sized based on there being a free, unobstructed area for the passage of air into the space. Louvers or grilles placed over these openings reduce the area of the openings because of the area occupied by the solid portions of the grille or louver. The reduction in area must be considered because only the unobstructed area can be credited toward the required opening size.

The reduction in opening area caused by the presence of grilles or louvers will always require openings to be larger than determined from the sizing ratios of this chapter and larger than any duct of the minimum required size that might connect to these openings.

703.1.2 Mechanical ventilation. Indoor locations intended for hydrogen-generating or refueling operations shall be ventilated in accordance with Section 502.16 of the *International Mechanical Code*.

❖ Section 502.16 of the *International Mechanical Code®* (IMC®) provides criteria for the design and operation of a mechanical ventilation system in repair garages for natural-gas- and hydrogen-fueled vehicles. This section makes Section 502.16 of the IMC applicable to indoor spaces containing hydrogen-generating and/or refueling operations.

703.1.3 Specially engineered installations. As an alternative to the provisions of Section 703.1.1 and 703.1.2, the necessary supply of air for ventilation and dilution of flammable gases shall be provided by an approved engineered system.

❖ The code is not intended to inhibit innovative ideas or technological advances. A comprehensive regulatory document such as a fuel gas code cannot envision and then address all future innovations in the industry. As a result, a performance code must be applicable to and provide a basis for the approval of an increasing number of newly developed, innovative materials, systems and methods for which no code text or referenced standards yet exist. The fact that a material, product or method of construction is not addressed in the code is not an indication that prohibition of the material, product or method is intended. The code official is expected to apply sound technical judgement in accepting materials, systems or methods that, while not anticipated by the drafters of the current code text, can be demonstrated to offer equivalent performance. By virtue of its text, the code regulates new and innovative construction practices while addressing the relative safety of building occupants. The code official is responsible for determining whether a requested alternative provides a level of protection of the public health, safety and welfare as required by the code.

703.2 Containers, cylinders and tanks. Compressed gas containers, cylinders and tanks shall comply with Chapters 30 and 35 of the *International Fire Code*.

❖ The design and construction of containers must meet the requirements listed in Chapters 30 and 35 of the IFC and the standards referenced there.

703.2.1 Limitations for indoor storage and use. Flammable gas cylinders in occupancies regulated by the *International Residential Code* shall not exceed 250 cubic feet (7.1 m^3)at normal temperature and pressure (NTP).

❖ Flammable gas must not be stored or used where an accident could cause a loss of life. They must be stored and used under preventive safety guidelines. Cylinders with a capacity of 250 cubic feet (7.1 m^3) or less are allowed for maintenance of buildings, taking care of patients and equipment operation.

703.2.2 Design and construction. Compressed gas containers, cylinders and tanks shall be designed, constructed and tested in accordance with the Chapter 27 of the *International Fire Code*, ASME *Boiler and Pressure Vessel Code* (Section VIII) or DOTn 49 CFR, Parts 100-180.

❖ Chapter 27 of the IFC and the referenced standards establish rules of safety governing the design, fabrication and testing of storage vessels, including those for gaseous and liquid hydrogen.

703.3 Pressure relief devices. Pressure relief devices shall be provided in accordance with Sections 703.3.1 through 703.3.8. Pressure relief devices shall be sized and selected in accordance with CGA S-1.1, CGA S-1.2 and CGA S-1.3.

❖ Pressure relief devices are essential for cryogenics because of the high pressures and low temperatures at which cryogenics are maintained. Although storage tanks, other containers and transfer piping are normally well insulated, some heating of the contents will occur with time, causing internal pressures to increase. Pressure relief mechanisms are one method of relieving these overpressures and avoiding a hazardous situation. Three Compressed Gas Association standards that cover the full range of container types from portable to stationary are referenced.

Sections 703.3.1 through 703.3.8 contain requirements related to accessibility for maintenance, general sizing, installation requirements and device integrity.

703.3.1 Valves between pressure relief devices and containers. Valves including shutoffs, check valves and other mechanical restrictions shall not be installed between the pressure relief device and container being protected by the relief device.

> **Exception:** A locked-open shutoff valve on containers equipped with multiple pressure-relief device installations where the arrangement of the valves provides the full required flow through the minimum number of required relief devices at all times.

❖ Any valve or restriction placed between a pressure relief device and the vessel it protects could render the device useless if the valve was totally or partially closed. Restrictions in the inlet to a pressure relief device are dangerous because closing the valve in that location would allow overpressure in the container to build without relief.

703.3.2 Installation. Valves and other mechanical restrictions shall not be located between the pressure relief device and the point of release to the atmosphere.

❖ Valves and other restrictions in the outlet (discharge) piping from a pressure relief device would have the same effect as valves and restrictions placed upstream of the device (see commentary, Section 703.3.1).

703.3.3 Containers. Containers shall be provided with pressure relief devices in accordance with the ASME *Boiler and Pressure Vessel Code* (Section VIII), DOTn 49 CFR, Parts 100-180 and Section 703.3.7.

❖ This section states very clearly that pressure relief devices are required for gaseous hydrogen containers. The container standards referred to cover a broad range of container types, portable and stationary. Pressure relief devices are required only as dictated by the specifications to which the container was fabricated (i.e., ASME or DOTn).

703.3.4 Vessels other than containers. Vessels other than containers shall be protected with pressure relief devices in accordance with the ASME *Boiler and Pressure Vessel Code* (Section VIII), or DOTn 49 CFR, Parts 100-180.

❖ Just as Section 703.3.3 requires pressure relief devices on containers, this section mentions other hydrogen-containing vessels or components in which an overpressure condition could occur.

703.3.5 Sizing. Pressure relief devices shall be sized in accordance with the specifications to which the container was fabricated. The relief device shall be sized to prevent the maximum design pressure of the container or system from being exceeded.

❖ This section contains only general language that ensures that the pressure relief device is properly designed to fit the needs of the particular container. In most cases the manufacturer will already have the devices installed on the container; however, there are cases when a cryogenic system or container may be constructed to a user specification for a specific purpose. In those cases the relief valves must be sized and installed by the user.

703.3.6 Protection. Pressure relief devices and any associated vent piping shall be designed, installed and located so that their operation will not be affected by water or other debris accumulating inside the vent or obstructing the vent.

❖ As with all such protective devices, the relief devices must be located and protected to prevent submersion and the entrance of water, debris, insects, etc., all of which could impair operation of the device.

703.3.7 Access. Pressure relief devices shall be located such that they are provided with ready access for inspection and repair.

❖ These devices are extremely important safety controls that require periodic inspections and, in some cases, testing.

703.3.8 Configuration. Pressure relief devices shall be arranged to discharge unobstructed in accordance with Section 2209 of the *International Fire Code*. Discharge shall be directed to the outdoors in such a manner as to prevent impingement of escaping gas on personnel, containers, equipment and adjacent structures and to prevent introduction of escaping gas into enclosed spaces. The discharge shall not terminate under eaves or canopies.

> **Exception:** This section shall not apply to DOTn-specified containers with an internal volume of 2 cubic feet (0.057 m^3) or less.

❖ This section refers to the IFC regarding requirements for the design and location of hydrogen vent piping. The vent piping must be arranged so that the discharge itself does not cause additional damage simply because of the location chosen for the discharge point.

703.4 Venting. Relief device vents shall be terminated in an approved location in accordance with Section 2209 of the *International Fire Code*.

❖ Because of container locations, vent piping is often used to extend the discharge location of the pressure relief vent. The vent system will prevent excessive buildup of pressure in the system by venting gas to the outdoors. Reference to the IFC, addresses such issues as material requirements and joining methods, general support, height and separation distances for the proposed vent outlet location. The location of the vent outlet is critical because the ignition of vented hydrogen has the potential for damage to materials and harm to people from radiant heat exposure.

703.5 Security. Compressed gas containers, cylinders, tanks and systems shall be secured against accidental dislodgement in accordance with Chapter 30 of the *International Fire Code*.

❖ Pressure vessels and their connection valves and devices can be damaged by falling, rolling and tipping over. Therefore, they must be secured in place. For example, a cylinder containing any gas pressurized to a few thousand psi will become a missile if the valve is knocked off during a fall. This is especially dangerous when the gas is flammable.

703.6 Electrical wiring and equipment. Electrical wiring and equipment shall comply with the ICC *Electrical Code*.

❖ Electrical power, control and signal wiring associated with hydrogen systems must comply with NFPA 70. Some areas containing hydrogen system components could be classified as hazardous locations in the ICC *Electrical Code®* (ICC EC™).

SECTION 704 (IFGC)
PIPING, USE AND HANDLING

704.1 Applicability. Use and handling of containers, cylinders, tanks and hydrogen gas systems shall comply with this section. Gaseous hydrogen systems, equipment and machinery shall be listed or approved.

❖ All hydrogen system components must be listed for the application unless specifically approved by the code official. All components must be carefully chosen as suitable for the application, especially because of the extreme flammability, small molecule size, lack of detectable odor and high pressure associated with hydrogen.

704.1.1 Controls. Compressed gas system controls shall be designed to prevent materials from entering or leaving process or reaction systems at other than the intended time, rate or path. Automatic controls shall be designed to be fail safe in accordance with accepted engineering practice.

❖ "Fail-safe" mode means that if and when a device fails, the resulting condition will not create a hazard. For example, if a compressor motor starter fails, the contacts would fail in the open position, thereby stopping the motor or preventing it from starting.

704.1.2 Piping systems. Piping, tubing, valves and fittings conveying gaseous hydrogen shall be designed and installed in accordance with Sections 704.1.2.1 through 704.1.2.5, Chapter 27 of the *International Fire Code,* and ASME B31.3. Cast-iron pipe, valves and fittings shall not be used.

❖ All apparatus connected and part of the system must comply with this section, Chapter 27 of the IFC and the requirements of the referenced standard applicable to gaseous H_2.

704.1.2.1 Sizing. Gaseous hydrogen piping shall be sized in accordance with approved engineering methods.

❖ Hydrogen piping and manifold systems should be designed and constructed by competent personnel who are thoroughly familiar with the requirements for piping of flammable gases. Sizing information can be obtained from a variety of applicable codes and regulations; however, hydrogen piping systems should be designed to comply with the principles of ASME B31.3, which includes special requirements for hydrogen service.

704.1.2.2 Identification. Piping used to convey gaseous hydrogen shall be identified and marked "HYDROGEN," at intervals not exceeding 10 feet (3048 mm). Letters of such marking shall be in a color other than the color of the piping. Piping shall be identified a minimum of one time in each room or space through which it extends.

❖ This section applies to all piping, regardless of material. It should be noted that ANSI A13.1 details the valve and piping color coding, painting and labeling methods used in the United States.

704.1.2.3 Piping design and construction. Piping systems shall be Type 304, Type 304L or Type 316 stainless steel tubing listed or approved for hydrogen service and the use intended through the full range of pressure and temperature to which they will be subjected. Piping systems shall be designed and constructed to provide allowance for expansion, contraction, vibration, settlement and fire exposure.

❖ Hydrogen piping systems can consist of structural members, vacuum jackets, valve bodies and valve seats, electrical and thermal insulation, gaskets, seals, lubricants and adhesives, and will involve a multitude of materials. Hydrogen embrittlement involves many variables and can cause significant deterioration in the mechanical properties of certain metals. One must thoroughly review material selection methods and bills-of-lading, quality control procedures and material test reports used during manufacture to verify that the materials are suitable for hydrogen service.

Austenitic (300 series) stainless steels meeting ASME B31.3 are generally satisfactory for gaseous hydrogen service. Even plastic is allowed by the standard under controlled conditions (other provisions apply, consult ASME B31.3). The code specifies Type 304, 304L or 316 stainless steel piping and tubing listed or approved for gaseous hydrogen service. Gray, ductile or malleable cast-iron pipe, valves and fittings must not be used.

The bottom line is that the material(s) must be approved for hydrogen service. ASME B31.3 contains the material-specific provisions to do so for the rated pressure, volume and temperature of the gas or liquid transported. Care must be taken to not overstress components in systems because some of these materials may lose ductility if stressed beyond yield.

704.1.2.3.1 Prohibited locations. Piping shall not be installed in or through a circulating air duct, clothes chute, chimney or gas vent, ventilating duct, dumbwaiter or elevator shaft.

❖ There are two reasons why gas piping is not to be installed in the locations listed in this section as well as in similar areas. Gas piping must not be located where the spread of leaked gas throughout the building would be accelerated. An air supply duct or areas that create a shaft that provide a path for the gas to travel are considered to be locations where the problem of a gas leak could be compounded by spreading the gas or any resultant fire throughout the building. Also, in some locations, the gas piping could be subject to deterioration because of temperature or corrosive conditions such as in air ducts or chimneys.

Locating gas piping in certain areas or atmospheres may cause corrosion of the pipe, which in turn may cause a leak in the piping system. Within a supply air duct, the conditioned air may cause moisture to condense on or within the pipe, thereby causing corrosion of the pipe. Gas piping located within a chimney or a vent will be subjected to high temperatures and the corrosive effects of vent gases.

Additionally, piping located in a dumbwaiter or elevator shaft may be subject to an additional hazard resulting in mechanical damage to the pipe. Damage from a dumbwaiter or elevator impacting the gas piping may not only cause a leak that would allow the gas to escape and travel up the shaft but may also put any occupants within an elevator in immediate danger.

Gas piping is not prohibited in concealed spaces used as air plenums above a suspended ceiling. It is assumed that the piping system will not leak in a ceiling air plenum or any other location, for that matter. In fact, when viewed logically, it would be more hazardous for a leak to occur in a location where the air is static and an explosive hydrogen/air mixture could be created.

704.1.2.3.2 Piping in solid partitions and walls. Concealed piping shall not be located in solid partitions and solid walls, except where installed in a ventilated chase or casing.

❖ This section allows installation of gas piping within solid walls or partitions only if the piping is installed within a chase or casing to protect the pipe from stress and from corrosive effects of the wall material, such as concrete.

704.1.2.3.3 Piping in concealed locations. Portions of a piping system installed in concealed locations shall not have unions, tubing fittings, right or left couplings, bushings, compression couplings and swing joints made by combinations of fittings.

Exceptions:
1. Tubing joined by brazing.
2. Fittings listed for use in concealed locations.

❖ A concealed location is a location that requires the removal of permanent construction to gain access (see the definition of "Concealed location"). The space above a dropped ceiling having readily removable lay-in panels or other locations that have a removable access panel are not considered concealed locations for the purposes of this section. Concealed locations include wall, floor and ceiling cavities bounded by permanent finish materials such as gypsum board, masonry or paneling. Unions and mechanical joint tubing fittings are not permitted in concealed locations because they are more likely to loosen and leak than other joining means and fittings. Tubing fittings include flare, compression and similar proprietary-type fittings, all of which are mechanical joints.

704.1.2.3.4 Piping through foundation wall. Underground piping shall not penetrate the outer foundation or basement wall of a building.

❖ This section does not allow piping to enter or exit a building at any point below ground level. The risk of piping failure from settlement or corrosion is too great to allow underground penetrations; therefore, piping must enter or exit the building above ground.

704.1.2.3.5 Protection against physical damage. In concealed locations, where piping other than stainless steel piping, stainless steel tubing or black steel is installed through holes or notches in wood studs, joists, rafters or similar members less than 1 inch (25 mm) from the nearest edge of the member, the pipe shall be protected by shield plates. Shield plates shall be a minimum of $^1/_{16}$-inch-thick (1.6 mm) steel, shall cover the area of the pipe where the member is notched or bored, and shall extend a minimum of 4 inches (102 mm) above sole plates, below top plates and to each side of a stud, joist or rafter.

❖ Note that while Section 704.1.2.3 prescriptively limits hydrogen piping to stainless steel, materials other than stainless steel and approved for hydrogen service, such as aluminum and its alloys, copper alloys and titanium and its alloys, are often encountered (see commentary, Section 704.1.2.3).

704.1.2.3.6 Piping in solid floors. Piping in solid floors shall be laid in channels in the floor and covered in a manner that will allow access to the piping with a minimum amount of damage to the building. Where such piping is subject to exposure to excessive moisture or corrosive substances, the piping shall be protected in an approved manner. As an alternative to installation in channels, the piping shall be installed in a casing of Schedule 40 steel, wrought-iron, PVC or ABS pipe with tightly sealed ends and joints and the casing shall be ventilated to the outdoors. Both ends of such casing shall extend not less than 2 inches (51 mm) beyond the point where the pipe emerges from the floor.

❖ Piping must not be installed in any solid concrete or masonry floor construction. The potential for pipe damage resulting from slab settlement, cracking or the corrosive action of the floor material makes it imperative that one of the installation methods in this section be used. This section does not intend to allow any direct encasement of gas piping in solid concrete or masonry floor systems.

Gas piping installed within a solid floor system must be safeguarded by installation in a sealed casing or in a floor channel with a removable cover for pipe access. Either of these methods should provide reasonable protection of the pipe from the effects of settling, cracking and being in contact with corrosive materials.

Because casings constructed of metal could corrode where installed within a concrete slab on grade, consideration should be given to corrosion protection for the steel casing, or an alternate material should be chosen.

704.1.2.3.7 Piping outdoors. Piping installed above ground, outdoors, shall be securely supported and located where it will be protected from physical damage. Piping passing through an exterior wall of a building shall be encased in a protective pipe sleeve. The annular space between the piping and the sleeve shall be sealed from the inside such that the sleeve is ventilated to the outdoors. Where passing through an exterior wall of a building, the piping shall also be protected against corrosion by coating or wrapping with an inert material. Below-ground piping shall be protected against corrosion.

❖ Extraordinary protection of piping is required because of the potential danger of leakage. Wall sleeves are sealed only on the interior end so that any leakage in the sleeve will vent to the outdoors.

704.1.2.3.8 Settlement. Piping passing through interior concrete or masonry walls shall be protected against differential settlement.

❖ Building movement and shifting can put enormous stress on piping; therefore, it must be protected from stress by sleeves, oversized openings, expansion loops and similar measures.

704.1.2.4 Joints. Joints on piping and tubing shall be listed for hydrogen service, inclusive of welded, brazed, flared, socket, slip or compression fittings. Gaskets and sealants shall be listed for hydrogen service. Threaded or flanged connections shall not be used in areas other than hydrogen cutoff rooms or outdoors.

❖ Because hydrogen gas is composed of small molecules (2 hydrogen atoms), it is more difficult to contain, especially under the higher pressures anticipated for fuel delivery to vehicles. All joining methods for piping and tubing must be listed for hydrogen service, including welded, brazed, flared, socket, slip or compression fittings. Soft solder joints are not permitted. Threaded or flanged connections must not be used in areas other than hydrogen cutoff rooms or outdoors. Mechanical joints will be required to maintain electrical continuity or otherwise be connected with a bonding strip to avoid static discharge. Any gaskets or sealants must be listed or approved for hydrogen applications.

704.1.2.5 Valves and piping components. Valves, regulators and piping components shall be listed or approved for hydrogen service, shall be provided with access and shall be designed and constructed to withstand the maximum pressure to which such components will be subjected.

❖ Valves, gauges, regulators and other piping components must be listed or approved for hydrogen service for the rated pressure and temperature of the application. The manufacturer or hydrogen supplier should be consulted for valves, regulators and other accessories.

704.1.2.5.1 Shutoff valves on storage containers and tanks. Shutoff valves shall be provided on all storage container and tank connections except for pressure relief devices. Shutoff valves shall be provided with ready access.

❖ Containers, tanks and the associated connecting piping must have an accessible manual shutoff valve. Valves must not be installed between a pressure relief device (PRD) and the container protected by the PRD because this would circumvent the purpose of overpressure relief. Shutoff valves on tanks and containers must be conspicuous, and must not be hidden behind panels or access doors because they may have to be accessed quickly in an emergency.

704.2 Upright use. Compressed gas containers, cylinders and tanks, except those with a water volume less than 1.3 gallons (5 L) and those designed for use in a horizontal position, shall be used in an upright position with the valve end up. An upright position shall include conditions where the container, cylinder or tank axis is inclined as much as 45 degrees (0.79 rad) from the vertical.

❖ Containers, cylinders and tanks are to be used in an upright position to protect and provide access to the shutoff valve and to make sure that gaseous hydrogen is taken from the vessel where liquid is also present.

704.3 Material-specific regulations. In addition to the requirements of this section, indoor and outdoor use of hydrogen compressed gas shall comply with the material-specific provisions of Chapters 30 and 35 of the *International Fire Code*.

❖ In addition to meeting the requirements of this section, the indoor and outdoor use and storage of gaseous hydrogen needs to comply with the material-specific provisions of Chapters 30 and 35 of the IFC. For example, the storage and use of a compressed flammable gas system must comply with Chapters 30 and 35 of the IFC in addition to this section.

704.4 Handling. The handling of compressed gas containers, cylinders and tanks shall comply with Chapter 27 of the *International Fire Code*.

❖ The requirements of this section refer to Chapter 27 and the applicable material-specific requirements of the IFC for handling gas containers, cylinders and tanks.

SECTION 705 (IFGC)
TESTING OF HYDROGEN PIPING SYSTEMS

705.1 General. Prior to acceptance and initial operation, all piping installations shall be inspected and pressure tested to determine that the materials, design fabrication and installation practices comply with the requirements of this code.

❖ After installation, field-erected piping, tubing, hose and hose assemblies must be tested and proved hydrogen gas-tight for the rated pressure and temperature of the gas conveyed in that portion of the system.

Inspection of piping installations is intended to be a visual observation of the system and the testing procedure. The designer may require that welded joints in, for example, very high pressure applications be examined by a method that is capable of discovering internal defects that are not detectable by visual observation.

705.2 Inspections. Inspection shall consist of a visual examination of the entire piping system installation and a pressure test, prior to system operation. Engineered systems shall be designed using approved engineering methods and the inspection procedures of ASME B31.3, and such inspections shall be verified by the code official.

❖ In addition to pressure testing, the entire system must be visually inspected to verify that it has been properly installed, supported and protected from damage and that all required components are present.

705.3 Pressure test. The test pressure shall be not less than $1^1/_2$ times the proposed maximum working pressure, but not less than 5 pounds per square inch gauge (psig) (34.5 kPa gauge), irrespective of the design pressure. Where the test pressure exceeds 125 psig (862 kPa gauge), the test pressure shall not exceed a value that produces hoop stress in the piping greater than 50 percent of the specified minimum yield strength of the pipe. Testing of engineered systems shall utilize the testing procedures of ASME B31.3 provided that test duration and gauge accuracy are included in the procedures as specified in Sections 705.3.1 and 705.3.2.

❖ A testing and purging procedure should be prepared and reviewed. Although methods for testing hydrogen piping vary, an approved method such as outlined in ASME B31.3 often incorporates procedures that can be characterized as follows:

- Perform a pressure test (a mix of at least 10% helium in inert gas preferred) at 1 ½ times maximum working pressure, 30 minutes per 500 cubic feet (14.2 m^3) of pipe volume.
- After the pressure test, check for pressure decay. If some leakage is detected, use soap/water to find the local leaks (bubbles).
- Pressurize the piping with hydrogen and check for local leaks with a "sonic tester" or a "sniffer" (hand-held combustible gas detector).
- If the system "fails" the above procedure, purge the system, fix the leak, and repeat the process until it "passes." Sometimes a "sonic test" is used as part of yearly preventive maintenance.

ASME B31.3 procedures are referenced for engineered systems. Pressure testing in ASME B31.3 may be either by hydraulic or pneumatic means. Hydraulic tests are performed at 1½ times the maximum allowable working pressure (MAWP), times a temperature correction factor. In some situations, particularly when the hydrogen system contains catalysts or other materials that could be damaged by water, a pneumatic test is preferable to the hydraulic test. The flexibility is intentional and left to the discretion of the designer. The pneumatic test is typically performed at 1¼ times the MAWP rather than 1½ (other provisions apply, consult ASME B31.3).

Considering that some parts of a system might operate at up to 15,000 psi (103 421 kPa), the test pressure could be very high. Extreme care must be used when conducting this testing.

705.3.1 Test duration. The test duration shall not be less than $^1/_2$ hour for each 500 cubic feet (14.2 m^3) of pipe volume. For piping systems having a volume of more than 24,000 cubic feet (680 m^3), the duration of the test shall not be required to exceed 24 hours.

❖ As the piping system becomes larger, the test duration must be longer so that any leak can be detected. For example, a leakage rate of 1 cubic foot per hour would produce a more easily detectable pressure drop in a small (10 cubic foot) volume system. It would require a longer test period to detect the same leak in a large volume system because the leakage of 1 cubic foot represents a much smaller fraction of the total gas volume in the system. If a system has a volume of 501 cubic feet, the minimum test duration would be 1 hour. To put this in perspective, 500 cubic feet of piping system internal volume would equate to 83,333 feet of 1-inch Schedule 40 steel pipe or 9,747 feet of 3-inch Schedule 40 steel pipe. At 1/2 hour per each 500 cubic feet (14.2 m^3), a system having 24,000 cubic feet (680 m^3) of volume would require a test duration of 24 hours.

705.3.2 Test gauges. Gauges used for testing shall be as follows:

1. Tests requiring a pressure of 10 pounds per square inch (psi) (68.95 kPa) or less shall utilize a testing gauge having increments of 0.10 psi (.6895 kPa) or less.
2. Tests requiring a pressure of greater than 10 psi (68.95 kPa) but less than or equal to 100 psi (689.5 kPa) shall utilize a testing gauge having increments of 1 psi (6.895 kPa) or less.
3. Tests requiring a pressure test greater than 100 psi (689.5 kPa) shall utilize a testing gauge having increments of 2 psi (13.79 kPa) or less.

Exception: Measuring devices having an equivalent level of accuracy shall be permitted where approved by the design engineer and the code official.

❖ The smaller the increments are (the more detailed), the easier it is to detect a leak. Considering that hydrogen is not odorized, a leak-free system is even more imperative. Leak-indicating manometers and electronic instruments using pressure transducers are available that can read pressure changes with great accuracy, thus allowing detection of even the smallest leaks.

705.4 Detection of leaks and defects. The piping system shall withstand the test pressure specified without showing any evidence of leakage or other defects.

❖ Any pressure drop, no matter how small, is considered a failure of the test, except where it can be demonstrated that the drop was caused by a change in temperature or other such cause. This would be practically impossible to demonstrate in the field, especially for short test durations.

705.4.1 Corrections. Where leakage or other defects are located, the affected portion of the piping system shall be repaired and retested.

❖ Where a leak is detected by a method addressed in Section 705.3.2 or other accurate testing instrument, the defect must be corrected. Corrections include tightening of fittings or threaded joints and replacing defective pipe, tubing or fittings.

SECTION 706 (IFGC)
LOCATION OF GASEOUS HYDROGEN SYSTEMS

706.1 General. This section shall govern the location and installation of gaseous hydrogen systems.

Exceptions:
1. Dispensing equipment need not be separated from canopies that are constructed in accordance with the *International Building Code* and in a manner that prevents the accumulation of hydrogen gas.
2. Gaseous hydrogen systems located in a separate building designed and constructed in accordance with the *International Building Code* and NFPA 50A.
3. Gaseous hydrogen systems located inside a building in a hydrogen cutoff room designed and constructed in accordance with Section 706.3 and the *International Building Code.*
4. Gaseous hydrogen systems located inside a building not in a hydrogen cutoff room where the gaseous hydrogen system is listed and labeled for indoor installation and installed in accordance with the manufacturer's installation instructions.
5. Stationary fuel-cell power plants in accordance with Section 633.

❖ The overriding theme of the location requirements is to never permit the maximum concentration of flammable contaminants in air to exceed 25 percent of the lower flammability limit (LFL) for hydrogen during the period that a credible leak exists. This can be accomplished using natural or mechanical means to always assure adequate ventilation to prevent a hazardous buildup of hydrogen gas in buildings or confined spaces.

Accordingly, the five exceptions propose several alternatives to minimize the risk of a hydrogen incident.

For exception 1, the proximity of dispensing equipment to canopies need not be considered where constructed to comply with these provisions. As an alternative, the exception 2 refers to the associated standard dealing with gaseous hydrogen. Exception 3 provides for the indoor generation, compression and storage of gaseous hydrogen if these systems are located in a hydrogen cutoff room. Although the requirements for cutoff rooms provide many compensating safety factors, this language would require the appropriate Group H occupancy where a maximum allowable quantity threshold is exceeded. The final two exceptions address the circumstance where an appliance or factory-matched package of equipment is specifically tested for the indoor application.

706.2 Location on property. Gaseous hydrogen systems shall be located in accordance with Chapter 22 of the *International Fire Code.*

❖ The requirements of this section refer to Chapter 22 of the IFC specifically for locating hydrogen systems on property. The IFC requirements are based on the hazards associated with the specific materials (for example, compressed flammable gas). The requirements consider the risks presented to personnel, buildings and facilities resulting from radiant exposure.

706.3 Hydrogen cutoff rooms. Hydrogen cutoff rooms shall be designed and constructed in accordance with Sections 706.3.1 through 706.3.8 and the *International Building Code.*

❖ See the commentary for Sections 706.3.1 through 706.3.8.

706.3.1 Design and construction. Interior building openings shall be equipped with self-closing devices. Interior openings shall be electronically interlocked with the gaseous hydrogen system to prevent operation of the system when such openings are ajar or the room shall be provided with a mechanical exhaust ventilation system designed with a capture velocity at the opening of not less than 60 feet per minute (0.3048 m/s). Operable windows are prohibited in interior walls.

❖ The requirements of this section intend to detail the design and construction criteria for the cut-off room and relate to exception 3 of Section 706.1. The intent to isolate the hydrogen system from all other spaces in a building is accomplished by a sophisticated supervised enclosure or by a combination of enclosure and mechanical exhaust. Any required mechanical exhaust system should be supervised (that is, interlocked with the hydrogen system).

To allow for broader applications (such as emergency generators using fuel cell technology), interior wall openings are allowed as necessary for easy access to the systems. Hydrogen generation must be terminated through the use of interlocks if an interior door is left open for an inordinate amount of time. This precaution is needed to prevent the migration of flammable gas to adjacent areas that may not be properly ventilated. In applications where installation of interlocks is not practical, significant ventilation is required to ensure that a flammable mixture is not attained within the cut-off room and adjacent spaces.

706.3.2 Ventilation. Cutoff rooms shall be provided with mechanical ventilation in accordance with the applicable provisions for repair garages in Chapter 5 of the *International Mechanical Code*.

Exception: This section shall not apply to rooms provided with ventilation systems meeting the requirements of Section 706.3.1.

❖ This section intends to prevent a dangerous accumulation of flammable gas in the room through the use of an exhaust ventilation system. The Source-Book for Hydrogen Applications recommends ventilation at the rate of 1 cfm per square foot [0.00508 m^3/(s · m^2)] of floor area, which is relative to the requirements in Chapter 5 of the *International Mechanical Code®* (IMC®). The exception in Section 706.3.2 allows exhaust systems that are designed with a capture velocity of 60 fpm (0.3048 m/s) at the door opening. This capture velocity exceeds the requirements found in Chapter 5 of the IMC. Because hydrogen is nontoxic, if a release can be kept below its flammable limit, there is minimal hazard. The ventilation requirements are designed to perform that function.

This section can be viewed as an option to the mechanical ventilation provision of Section 706.3.1; however, a hydrogen cutoff room must always have a mechanical ventilation system, as stated in either Section 706.3.1 or this section.

706.3.3 Gas detection system. Hydrogen cutoff rooms shall be provided with an approved flammable gas-detection system in accordance with Sections 706.3.3.1 through 706.3.3.3.

❖ Some gases contain additives that produce pungent odors for easy recognition. Systems using non-odorized gases, such as hydrogen and nonodorized liquid natural gas, must have an approved flammable gas detection system installed.

706.3.3.1 System design. The flammable gas-detection system shall be listed for use with hydrogen and any other flammable gases used in the room. The gas detection system shall be designed to activate when the level of flammable gas exceeds 25 percent of the lower flammability limit (LFL) for the gas or mixtures present at anticipated temperature and pressure.

❖ The detection system must initiate the operations specified in Section 706.3.3.2 at any time that the flammable gas concentration exceeds one-fourth of the concentration necessary to support combustion.

Early detection of the presence of a flammable gas will allow adequate safeguards to be taken. Hydrogen fires are not normally extinguished until the supply of hydrogen has been shut off because of the danger of re-ignition or explosion. A gas detection system in the room or space housing a gaseous hydrogen system results in early notification of a leak that is occurring before the escaping gas reaches hazardous exposure concentration levels.

706.3.3.2 Operation. Activation of the gas detection system shall result in all of the following:

1. Initiation of distinct audible and visual alarm signals both inside and outside of the cutoff room.
2. Activation of the mechanical ventilation system.

❖ The detection system must activate the mechanical ventilation system that is required by either Section 706.3.1 or Section 706.3.2, in addition to causing alarms to activate.

The required local alarm is intended to alert the occupants to an emerging hazardous condition in the vicinity. The monitor control equipment must also initiate operation of the mechanical ventilation system in the event of a leak or rupture in the gaseous hydrogen system to prevent an accumulation of flammable gas.

706.3.3.3 Failure of the gas detection system. Failure of the gas detection system shall result in activation of the mechanical ventilation system, cessation of hydrogen generation and the sounding of a trouble signal in an approved location.

❖ Systems must be designed to be self monitoring and "fail-safe" in that all safety systems are activated to alert any occupants that a problem exists and to prevent more hydrogen from being generated by any appliances in the room.

706.3.4 Ignition source control. Open flames, flame-producing devices and other sources of ignition shall be controlled in accordance with Chapter 35 of the *International Fire Code*.

❖ This section is intended to prevent flammable gas releases from finding an open flame or other source of ignition. Similar requirements are identified in the IFC in locations where flammable gases are stored or used. The energy required for ignition of hydrogen-air mixtures is extremely small, and every effort must be made to control ignition sources until the area can be properly ventilated, thus removing the hazard.

706.3.5 Explosion control. Explosion control shall be provided in accordance with Chapter 9 of the *International Fire Code*.

❖ The requirements of this section are intended to address the circumstance resulting from a catastrophic failure of the cut-off room. It is the final safeguard in case other prevention methods fail (ventilation, alarms). An ignited hydrogen mixture produces large quantities of heat, causing a rapid expansion of the surrounding air. This can cause a pressure increase in a confined space and a catastrophic failure. Explosion control methods are identified in the IFC to prevent such a catastrophic failure. The explosion control requirements for hydrogen are consistent with the requirements in NFPA 50A, the IFC and the Sourcebook for Hydrogen Applications.

706.3.6 Standby power. Mechanical ventilation and gas detection systems shall be connected to a standby power system in accordance with Chapter 27 of the *International Building Code*.

❖ The ventilation system and gas detection system are life-safety systems and therefore must be dependable. Both safety systems must remain active in the event of a power failure of the primary power supply. Hydrogen is a colorless, odorless gas; a release might go undetected if detection systems are not functioning. The accumulation of hydrogen in an unventilated area can lead to mixtures in the flammable range if safety systems and mechanical ventilation systems are not in operation. Chapter 27 of the IFC addresses emergency and standby power requirements for emergency systems. It also allows an exception to the requirement for systems that are fail safe (see IFC Section 2704.7 Exception 4). This exception may be used in cut-off rooms where hydrogen is generated, but not stored. Any storage of hydrogen within the cut-off room would not qualify for the exception because, in the event of a power failure, there will be no way to detect or ventilate a release from a storage vessel.

706.3.7 Smoking. Smoking shall be prohibited in hydrogen cut-off rooms. "No Smoking" signs shall be provided at all entrances to hydrogen cutoff rooms.

❖ Smoking is prohibited because it is considered an ignition source (see Section 706.3.4). Smoking in a space that could contain an odorless, highly flammable gas is certainly an unacceptable risk. NO SMOKING signs should be conspicuously located throughout adjacent areas.

706.3.8 Housekeeping. The hydrogen cutoff room shall be kept free from combustible debris and storage at all times.

❖ Housekeeping should be a daily practice. The level of fire safety is greatly improved when areas are kept clean and neat. All waste generated should be removed from the cut-off room and safely disposed of each day.

SECTION 707 (IFGC)
OPERATION AND MAINTENANCE OF
GASEOUS HYDROGEN SYSTEMS

707.1 Maintenance. Gaseous hydrogen systems and detection devices shall be maintained in accordance with the *International Fire Code* and the manufacturer's installation instructions.

❖ This section refers to the IFC for maintenance of hydrogen systems. More specifically, Section 2703.2.6 of the IFC details required maintenance activities for hazardous materials storage and use. This includes maintenance of alarms, cylinders, ventilation systems and other devices needed to ensure safety of the gaseous hydrogen system. Failure to properly maintain any of these systems increases the likelihood of an unanticipated hydrogen release that could compromise the safety of the operation. This section is also consistent with the general maintenance provision of the IFC identified in Section 107.1.

Section 107.1 of the IFC does not identify who is responsible for maintenance because that determination should comply with the legal documents created between owners and occupants, such as a lease. However, the owner of a structure or premises is usually the party primarily responsible for its maintenance because the owner stands to gain the most from a well-maintained property. One of the underlying assumptions is that maintaining a commercial property in good condition allows the owner to recoup a substantial portion of his or her investment in maintenance. There are three factors that may influence owners to comply with code requirements:

- Code compliance requires only a small additional investment in the property;
- The owner has a long-term interest in the property; and
- The owner expects profitability after incurring the additional expense of complying with the code.

Although all these factors represent economic incentives, code officials should be equally aware of potential disincentives to compliance, such as assessable value, expiring tax credits or historic, architectural or aesthetic criteria. The code official need not belabor the justifications for compliance but should be prepared to acknowledge the owner's rationalizations for failure to comply.

This section also emphasizes that any "otherwise installed" system that currently exists must be maintained. For example, an existing fire protection system cannot be removed from a building because it is not required in new or existing buildings by current codes.

707.2 Purging. Purging of gaseous hydrogen systems shall be in accordance with Section 2211.8 of the *International Fire Code*.

❖ After installation, field-erected piping, tubing, hose and hose assemblies must be tested and proven hydrogen gas-tight for the rated pressure and temperature of the gas transported in that portion of the system.

This section refers to the applicable requirements of the IFC for the purging of vehicles powered by gaseous fuels.

Reliable purging procedures are essential to the safe use of hydrogen gas systems. Although any purging method must be approved, several methodologies in common use in the industry are outlined in ASME B31.3 (see commentary, Section 705.3).

The continuous flow method uses a continuous flow of the inert purge gas to remove the hydrogen gas and prevent air or moisture from entering the system.

The dilution method uses a sequence of pressurization and venting. This sequence is repeated several times and is very effective in removing gas from dead end piping such as pressure gauge lines.

SECTION 708 (IFGC)
DESIGN OF LIQUEFIED HYDROGEN SYSTEMS ASSOCIATED WITH HYDROGEN VAPORIZATION OPERATIONS

708.1 General. The design of liquefied hydrogen systems shall comply with Chapter 32 of the *International Fire Code*.

❖ The requirements of this section refer to Chapter 32 of the IFC specifically for the design of liquified hydrogen systems. The standards referenced there dealing with flammable cryogenic fluids must be adhered to in addition to complying with IFC Chapter 32.

Bibliography

The following resource materials are referenced in this chapter or are relevant to the subject matter addressed in this chapter.

ANSI A13.1-96, *Scheme for the Identification of Piping Systems.* New York, NY, American National Standards Institute, 1996.

Bain, A., Barclay, J., Bose, T., Edeskuty, F., Farlie, M., Hansel, J., Hay, R., and Swain, M. (et al), *Source Book for Hydrogen Applications,* Golden, CO, Hydrogen Research Institute and National Renewable Energy Laboratory, 1998.

CGA S-1.2-95, *Pressure Relief Device Standards - Part 2 - Cargo and Portable Tanks for Compressed Gases.* Chantilly, VA, Compressed Gas Association, 1995.

CGA S-1.3-95, *Pressure Relief Device Standards - Part 3 - Compressed Gas Stationary Storage Containers.* Chantilly, VA, Compressed Gas Association, 1995.

CGA G-5-2002, *Hydrogen*, 5th ed., Chantilly, VA, Compressed Gas Association, 2002.

CGA G-5.4-2001, *Standard for Hydrogen Piping Systems at Consumer Locations,* 2nd ed., Chantilly, VA, Compressed Gas Association, 2001.

CGA G-5.5-1996, *Hydrogen Vent Systems,* 1st ed., Chantilly, VA, Compressed Gas Association 1996.

DOTn 49 CFR 100-178 and 179-199. *Specification for Transportation of Explosive and Other Dangerous Articles, Shipping Containers.* Washington, D.C., U.S. Department of Transportation, 1994.

Hansel, James Gl, *Safety Considerations for Handling Hydrogen*, (A seminar presentation to Ford Motor Company), Allentown, PA, Air Products and Chemicals, June 12, 1998.

IBC-2003, *International Building Code.* Falls Church, VA, International Code Council, 2003.

IFC-2003 *International Fire Code.* Falls Church, VA, International Code Council, 2003.

NFPA 50A-99, *Gaseous Hydrogen Systems at Consumer Sites.* Quincy, MA, National Fire Protection Association, 1999.

NFPA 50B-99, *Liquefied Hydrogen Systems at Consumer Sites.* Quincy, MA, National Fire Protection Association, 1999.

IFGC/IFGS CHAPTER 8
REFERENCED STANDARDS

This chapter lists the standards that are referenced in various sections of this document. The standards are listed herein by the promulgating agency of the standard, the standard identification, the effective date and title, and the section or sections of this document that reference the standard. The application of the referenced standards shall be as specified in Section 102.8.

General Comments

Chapter 8 contains a comprehensive list of standards that are referenced in the code. It is organized to make locating specific document references easy.

It is important to understand that not every document related to mechanical system design, installation and construction is qualified to be a referenced standard. The International Code Council® (ICC®) has adopted a criterion that standards referenced in the International Codes and standards intended for adoption in the International Codes must meet to qualify as referenced standards. The policy is summarized as follows:

- Code references: The scope and application of the standard must be clearly identified in the code text.
- Standard content: The standard must be written in mandatory language and be appropriate for the
- subject covered. The standard cannot have the effect of requiring proprietary materials or prescribing a proprietary testing agency.
- Standard promulgation: The standard must be readily available and developed and maintained in a consensus process such as those used by ASTM or ANSI.

The ICC Code Development Procedures, of which the standards policy is a part, are updated periodically. A copy of the latest version can be obtained from the ICC offices.

Once a standard is incorporated into the code through the code development process, it becomes an enforceable part of the code. When the code is adopted by a jurisdiction, the standard also is part of that jurisdiction's adopted code. It is for this reason that the criteria were developed. Compliance with this policy means that documents incorporated into the code are developed through the use of a consensus process, are written in mandatory language, and do not mandate the use of proprietary materials or agencies. The requirement for a standard is representative of the most current body of available knowledge on the subject as determined by a broad range of interested or affected parties without dominance by any single interest group. A true consensus process has many attributes, including but not limited to:

- An open process that has formal (published) procedures that allow for the consideration of all view points,

- A definitive review period that allows for the standard to be updated and/or revised,
- A process of notification to all interested parties, and
- An appeals process.

Many available documents related to mechanical system design, installation and construction, though useful, are not standards and are not appropriate for reference in the code. Often, these documents are not developed or written with the intention of being used for regulatory purposes and are unsuitable for use as a standard because of extensive use of recommendations, advisory comments and nonmandatory terms. Typical examples include installation guides and manuals, handbooks and design aids.

The objective of ICC's standards policy is to provide regulations that are clear, concise and enforceable—thus the requirement that standards be written in mandatory language. This requirement is not intended to mean that a standard cannot contain informational or explanatory material that will aid the user of the standard in its application. When the standard's promulgating agency wants such material to be included, however, the information must appear in a nonmandatory location, such as an annex or appendix, and be clearly identified as not being part of the standard.

Overall, standards referenced by the code must be authoritative, relevant, up-to-date and, most important, reasonable and enforceable. Standards that comply with ICC's standards policy fulfill these expectations.

Purpose

This code contains numerous references to documents that are used to regulate materials and methods of construction. The references to these documents within the code text consist of the promulgating agency's initialism and its publication designation (for example ASME B1.20.1) and a further indication that the document being referenced is the one that is listed in Chapter 8. Chapter 8 contains all of the information that is necessary to identify the specific referenced document. Included is the following information on a document's promulgating agency (see commentary Figure 8):

- The promulgating agency (the agency's title),
- The promulgating agency's initialism, and
- The promulgating agency's address

REFERENCED STANDARDS

For example, a reference to an ASME standard within the code indicates that the document is promulgated by the American Society of Mechanical Engineers (ASME), which is located in New York City, New York. Chapter 8 lists the standard agencies alphabetically for ease of identification. Chapter 8 also includes the following information on the referenced document itself (see com- mentary Figure 8):

- The document's publication designation,
- The document's edition year,
- The document's title,
- Any addenda or revisions to the document known at the time of the code's publication, and
- Every section of the code in which the document is referenced.

For example, a reference to ASME B1.20.1 indicates this document can be found in Chapter 8 under the heading ASME. The specific standard's designation is B1.20.1. For convenience, these designations are listed in alphanumeric order. Chapter 8 shows that ASME B1.20.1 is titled *Pipe Threads, General Purpose (Inch)*; the applicable edition (that is, its year of publication) is 1992 and it is referenced in one specifically identified section of the code. Chapter 8 also indicates when a document has been discontinued or replaced by its promulgating agency. When a document is replaced by a different one, a note appears to tell the user the designation and title of the new document. The key aspect of the manner in which standards are referenced by the code is that a specific edition of a specific standard is clearly identified. The requirements necessary for compliance are, therefore, established and available to the code official, the mechanical contractor, the designer and the owner.

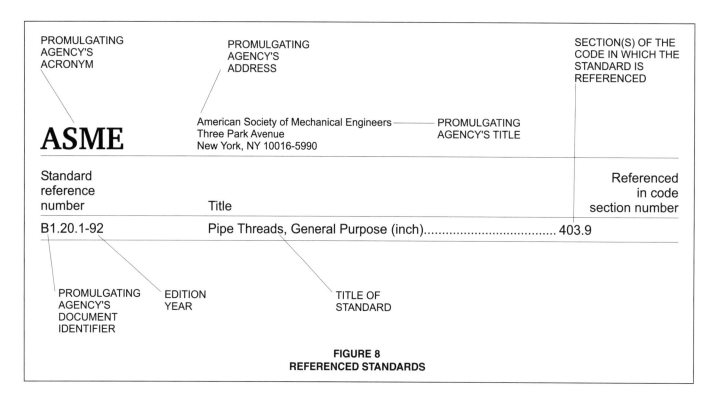

**FIGURE 8
REFERENCED STANDARDS**

REFERENCED STANDARDS

ANSI

American National Standards Institute
25 West 43rd Street
Fourth Floor
New York, NY 10036

Standard reference number	Title	Referenced in code section number
LC 1—97	Interior Gas Piping Systems Using Corrugated Stainless Steel Tubing	403.5.4
Z21.1—00	Household Cooking Gas Appliances	623.1
Z21.5.1—99	Gas Clothes Dryers - Volume I -Type 1 Clothes Dryers	613.1
Z21.5.2—99	Gas Clothes Dryers - Volume II- Type 2 Clothes Dryers with Z21.5.2a-99 and Z21.5.2b-99 Addenda	613.1, 614.3
Z21.8—94	Installation of Domestic Gas Conversion Burners	619.1
Z21.10.1—00	Gas Water Heaters - Volume I - Storage, Water Heaters with Input Ratings of 75,000 Btu per Hour or Less	624.1
Z21.10.3—98	Gas Water Heaters - Volume III - Storage, Water Heaters with Input Ratings Above 75,000 Btu per hour, Circulating and Instantaneous Water Heaters—with Z21.10.3a-99 Addendum	624.1
Z21.11.1—91	Gas-Fired Room Heaters -Volume I - Vented Room Heaters—with 1993 Addendum (Replaced by Z21.86-98/CSA 2.32 - M98, Vented Gas-Fired Space Heating Appliances)	622.1
Z21.11.2—96	Gas-Fired Room Heaters - Volume II - Unvented Room Heaters with Addendum A-97 and Addendum B-98	621.1
Z21.13—99	Gas-Fired Low-Pressure Steam and Hot Water Boilers—with Addenda Z21.13a-1993 and Z21.13b-1994	631.1
Z21.15—97	Manually Operated Gas Valves for Appliances, Appliance Connector Valves, and Hose End Valves	409.1.1
Z21.19—90	Refrigerators Using Gas (R 1999) Fuel—with Addenda Z721.19a-1992 (R1999) and Z21.19b-1995 (R1999)	625.1
Z21.40.1—96	Gas-Fired Heat Activated Air Conditioning and Heat Pump Appliances—with Z21.40.1a-98 Addendum	627.1
Z21.40.2—96	Gas-Fired Work Activated Air Conditioning and Heat Pump Appliances (Internal Combustion)—with Z21.40.2a-97 Addendum	627.1
Z21.42—93	Gas-Fired Illuminating Appliances	628.1
Z21.47—00	Gas-Fired Central Furnaces—with Addendum Z21.47a-00	618.1
Z21.48—92	Gas-Fired Gravity and Fan Type Floor Furnaces—with 1993 Addendum (Replaced by Z21.86-98/CSA 2.32-M98, Vented Gas-Fired Space Heating Appliances)	609.1
Z21.49—92	Gas-Fired Gravity and Fan-Type Vented Wall Furnaces—with 1993 Addendum B-94 (Replaced by Z21.86-98/CSA 2.32-M98, Vented Gas-Fired Space Heating Appliances)	608.1
Z21.50—98	Vented Decorative Gas Appliances	604.1
Z21.56—98	Gas-Fired Pool Heaters—with Addendum Z21.56a-99	616.1
Z21.58—95	Outdoor Cooking Gas Appliances—with Addendum Z21.58a-1998	623.1
Z21.60—00	Decorative Gas Appliances for Installation in Solid-Fuel Burning Fireplaces	602.1
Z21.61—83 (R 1996)	Toilets, Gas-Fired	625.1
Z21.69—97	Connectors for Movable Gas Appliances	411.1
Z21.83—98	Fuel Cell Power Plants	632.1
Z21.84—99	Manually-Lighted, Natural Gas Decorative Gas Appliances for Installation in Solid Fuel Burning Fireplaces	602.1, 602.2
Z21.86—98/CSA 2.32 M98	Gas-Fired Vented Space Heating Appliances	608.1, 609.1, 622.1
Z21.88—99	Vented Gas Fireplace Heaters	605.1
Z83.4—99	Non-Recirculating Direct-Gas-Fired Industrial Air Heaters with Addendum Z83.4a-2001	611.1
Z83.6—90 (R 1998)	Gas-Fired Infrared Heaters	630.1
Z83.8—96	Gas -Fired Unit Heaters—with Addendum Z83.8a-1997	620.1
Z83.9—96	Gas-Fired Duct Furnaces	610.1
Z83.11—00	Gas Food Service Equipment (Ranges and Unit Broilers), Baking and Roasting Ovens, Fat Fryers, Counter Appliances and Kettles, Steam Cookers, and Steam Generators	623.1
Z83.18—90	Recirculating Direct Gas-Fired Industrial Air Heaters with Addenda Z83a 2001 and Z83.18b 1992	612.1

ASME

American Society of Mechanical Engineers
Three Park Avenue
New York, NY 10016-5990

Standard reference number	Title	Referenced in code section number
B1.20.1—R92	Pipe Threads, General Purpose (inch)	403.9
B16.1—98	Cast Iron Pipe Flanges and Flanged Fittings, Class 25, 125 and 250	403.12
B16.20—98	Metallic Gaskets for Pipe Flanges Ring-Joint, Spiral-Wound, and Jacketed — with Addendum B16.20a-2000	403.12
B16.33—90	Manually Operated Metallic Gas Valves for Use in Gas Piping Systems up to 125 psig (Sizes 1/2 through 2)	409.1.1
B31.3—99	Process Piping	704.1.2, 705.2, 705.3

REFERENCED STANDARDS

B36.10M—00	Welded and Seamless Wrought-Steel Pipe	403.4.2
BPVC—01	ASME Boiler & Pressure Vessel Code (2001 Edition)	631.1, 703.2.2, 703.3.3, 703.3.4
CSD-1—98	Controls and Safety Devices for Automatically Fired Boilers with the ASME CSD-1a-1999 Addendum	631.1

ASTM

ASTM International
100 Barr Harbor Drive
West Conshohocken, PA 19428-2959

Standard reference number	Title	Referenced in code section number
A 53/A 53M—01	Specification for Pipe, Steel, Black and Hot Dipped Zinc-Coated Welded and Seamless	403.4.2
A 106—99	Specification for Seamless Carbon Steel Pipe for High-Temperature Service	403.4.2
A 254—97	Specification for Copper Brazed Steel Tubing	403.5.1
A 539—99	Specification for Electric Resistance-Welded Coiled Steel Tubing for Gas and Fuel Oil Lines	403.5.1
B 88—99	Specification for Seamless Copper Water Tube	403.5.2
B 210—00	Specification for Aluminum and Aluminum-Alloy Drawn Seamless Tubes	403.5.3
B 241/B 241M—00	Specification for Aluminum and Aluminum-Alloy, Seamless Pipe and Seamless Extruded Tube	403.4.4, 403.5.3
B 280—99	Specification for Seamless Copper Tube for Air Conditioning and Refrigeration Field Service	403.5.2
C 64—72 (1977)	Withdrawn No Replacement (Specification for Fireclay Brick Refractories for Heavy Duty Stationary Boiler Service)	503.10.2.5
C 315—00	Specification for Clay Flue Linings	501.12
D 2513—01A	Specification for Thermoplastic Gas Pressure Pipe, Tubing, and Fittings	403.6, 403.6.1, 403.11, 404.14.2

AWWA

American Water Works Association
6666 West Quincy Avenue
Denver, CO 80235

Standard reference number	Title	Referenced in code section number
C111—00	Rubber-Gasket Joints for Ductile-Iron Pressure Pipe and Fittings	403.12

CGA

Compressed Gas Association
1725 Jefferson Davis Highway, 5th Floor
Arlington, VA 22202-4102

Standard reference number	Title	Referenced in code section number
S-1.1—(1994)	Pressure Relief Device Standards—Part 1—Cylinders for Compressed Gases	703.3
S-1.2—(1995)	Pressure Relief Device Standards—Part 2—Cargo and Portable Tanks for Compressed Gases	703.3
S-1.3—(1995)	Pressure Relief Device Standards—Part 3—Stationary Storage Containers for Compressed Gases	703.3

CSA

CSA America Inc.
8501 E. Pleasant Valley Rd.
Cleveland, OH USA 44131-5575

Standard reference Partnumber	Title	Referenced in code section number
CSA 8—93 (Revision 1, 1999)	Requirements for Gas-Fired Log Lighters for Wood Burning Fireplaces	603.1

REFERENCED STANDARDS

DOTn

Department of Transportation
400 Seventh St. SW.
Washington, DC 20590

Standard reference number	Title	Referenced in code section number
49 CFR, Parts 192.281(e) & 192.283 (b)	Transportation of Natural and Other Gas by Pipeline: Minimum Federal Safety Standards	403.6.1
Parts 100-180	Hazardous Materials Regulations	703.2.2, 703.3.3, 703.3.4

ICC

International Code Council
5203 Leesburg Pike, Suite 600
Falls Church, VA 22041

Standard reference number	Title	Referenced in code section number
IBC—03	International Building Code®	102.2.1, 201.3, 301.10, 301.11, 301.12, 301.14, 302.1, 302.2, 305.6, 306.6, 401.1.1, 412.6, 413.3, 413.3.1, 501.1, 501.3, 501.12, 501.15.4, 609.3, 614.2, 706.1, 706.3
ICC EC—03	ICC Electrical Code™—Administrative Provisions	201.3, 306.3.1, 306.4.1, 306.5.2, 309.2, 413.8.2.4, 703.6
IEBC—03	International Existing Building Code™	101.2
IECC—03	International Energy Conservation Code®	301.2
IFC—03	International Fire Code®	201.3, 303.4, 401.2, 412.1, 412.6, 412.7, 412.7.3, 412.8, 413.1, 413.3, 413.3.1, 413.4, 413.8.2.5, 701.1, 701.2, 703.2, 703.2.2, 703.3.8, 703.4, 703.5, 704.1.2, 704.3, 704.4, 706.2, 706.3.4, 706.3.5, 706.3.6, 707.2, 707.1, 708.1
IMC—03	International Mechanical Code®	101.2.5, 201.3, 301.1.1, 301.13, 304.11, 501.1, 614.2, 618.5, 621.1, 624.1, 631.2, 632.1, 703.1.2, 706.3.2
IPC—03	International Plumbing Code®	201.3, 301.6, 624.1.1, 624.2
IRC—03	International Residential Code®.	703.2.1

MSS

Manufacturers Standardization Society of
the Valve and Fittings Industry
127 Park Street, Northeast
Vienna, VA 22180

Standard reference number	Title	Referenced in code section number
SP-6—96	Standard Finishes for Contact Faces of Pipe Flanges and Connecting-End Flanges of Valves and Fittings	403.12
SP-58—93	Pipe Hangers and Supports—Materials, Design and Manufacture	407.2

NFPA

National Fire Protection Association
1 Batterymarch Pike
P.O. Box 9101
Quincy, MA 02269-9101

Standard reference number	Title	Referenced in code section number
37—98	Installation and Use of Stationary Combustion Engines and Gas Turbines	616.1
50A—99	Gaseous Hydrogen Systems at Consumer Sites	706.1
51—97	Design and Installation of Oxygen-Fuel Gas Systems for Welding, Cutting, and Allied Processes	414.1
58—01	Liquefied Petroleum Gas Code	401.2, 402.6.1, 403.6.2, 403.11
82—99	Incinerators, Waste and Linen Handling Systems and Equipment	607.1
85—01	Boiler and Construction Systems Hazards Code	631.1
88B—97	Repair Garages	305.5
211—00	Chimneys, Fireplaces, Vents, and Solid Fuel-Burning Appliances	503.5.2, 503.5.3, 503.5.6.1, 503.5.6.3
853—00	Standard for the Installation of Stationary Fuel Cell Power Plants	633.1

REFERENCED STANDARDS

UL

Underwriters Laboratories Inc.
333 Pfingsten Road
Northbrook, IL 60062

Standard reference number	Title	Referenced in code section number
103—98	Factory-Built Chimneys, Residential Type and Building Heating Appliances—with Revisions thru March 1999	506.1
127—96	Factory-Built Fireplaces—with Revisions through November 1999	621.7
441—96	Gas Vents—with Revisions through April 1999	502.1
641—95	Type L Low-Temperature Venting Systems—with Revisions through April 1999	502.1
795—99	Commercial-Industrial Gas Heating Equipment	610.1, 618.1, 631.1
959—01	Medium Heat Appliance Factory-Built Chimneys	506.3
1738—93	Venting Systems for Gas Burning Appliances, Categories II, III and IV with Revisions through December 2000	502.1
1777—98	Chimney Liners—with Revisions through July 1998	501.12, 501.15.4

The appendices to the *International Fuel Gas Code®* (IFGC®) are explanatory commentaries of the code itself. Thus, there is no additional commentary for them.

APPENDIX A (IFGS)
SIZING AND CAPACITIES OF GAS PIPING

(This appendix is informative and is not part of the code.)

A.1 General. To determine the size of piping used in a gas piping system, the following factors must be considered:

(1) Allowable loss in pressure from point of delivery to equipment

(2) Maximum gas demand

(3) Length of piping and number of fittings

(4) Specific gravity of the gas

(5) Diversity factor

For any gas piping system, or special gas utilization equipment, or for conditions other than those covered by the tables provided in this code, such as longer runs, greater gas demands, or greater pressure drops, the size of each gas piping system should be determined by standard engineering practices acceptable to the code official.

A.2 Description of tables

A.2.1 General. The quantity of gas to be provided at each outlet should be determined, whenever possible, directly from the manufacturer's British thermal unit (Btu) input rating of the equipment that will be installed. In case the ratings of the equipment to be installed are not known, Table 402.2 shows the approximate consumption (in Btu per hour) of certain types of typical household appliances.

To obtain the cubic feet per hour of gas required, divide the total Btu input of all equipment by the average Btu heating value per cubic feet of the gas. The average Btu per cubic feet of the gas in the area of the installation can be obtained from the serving gas supplier.

A.2.2 Low pressure natural gas tables. Capacities for gas at low pressure [0.5 psig (3.5 kPa gauge) or less] in cubic feet per hour of 0.60 specific gravity gas for different sizes and lengths are shown in Tables 402.4(1) and 402.4(2) for iron pipe or equivalent rigid pipe, in Tables 402.4(7) through 402.4(9) for smooth wall semi-rigid tubing, and in Tables 402.4(14) through 402.4(16) for corrugated stainless steel tubing. Tables 402.4(1) and 402.4(7) are based upon a pressure drop of 0.3-inch water column (w.c.) (75 Pa), whereas Tables 402.4(2), 402.4(8) and 402.4(14) are based upon a pressure drop of 0.5-inch w.c. (125 Pa). Tables 402.4(9), 402.4(15) and 402.4(16) are special low-pressure applications based upon pressure drops greater than 0.5-inch w.c. (125 Pa). In using these tables, an allowance (in equivalent length of pipe) should be considered for any piping run with four or more fittings [see Table A.2.2].

A.2.3 Undiluted liquefied petroleum tables. Capacities in thousands of Btu per hour of undiluted liquefied petroleum gases based on a pressure drop of 0.5-inch w.c. (125 Pa) for different sizes and lengths are shown in Table 402.4(24) for iron pipe or equivalent rigid pipe, in Table 402.4(26) for smooth wall semi-rigid tubing, in Table 402.4(28) for corrugated stainless steel tubing, and in Tables 402.4(31) and 402.4(33) for polyethylene plastic pipe and tubing. Tables 402.4(29) and 402.4(30) for corrugated stainless steel tubing and Table 402.4(32) for polyethylene plastic pipe are based on operating pressures greater than 0.5 pounds per square inch (psi) (3.5 kPa) and pressure drops greater than 0.5-inch w.c. (125 Pa). In using these tables, an allowance (in equivalent length of pipe) should be considered for any piping run with four or more fittings [see Table A.2.2].

A.2.4 Natural gas specific gravity. Gas piping systems that are to be supplied with gas of a specific gravity of 0.70 or less can be sized directly from the tables provided in this code, unless the code official specifies that a gravity factor be applied. Where the specific gravity of the gas is greater than 0.70, the gravity factor should be applied.

Application of the gravity factor converts the figures given in the tables provided in this code to capacities for another gas of different specific gravity. Such application is accomplished by multiplying the capacities given in the tables by the multipliers shown in Table A.2.4. In case the exact specific gravity does not appear in the table, choose the next higher value specific gravity shown.

TABLE A.2.4
MULTIPLIERS TO BE USED WITH TABLES 402.4(1) THROUGH 402.4(21) WHERE THE SPECIFIC GRAVITY OF THE GAS IS OTHER THAN 0.60

SPECIFIC GRAVITY	MULTIPLIER	SPECIFIC GRAVITY	MULTIPLIER
.35	1.31	1.00	.78
.40	1.23	1.10	.74
.45	1.16	1.20	.71
.50	1.10	1.30	.68
.55	1.04	1.40	.66
.60	1.00	1.50	.63
.65	.96	1.60	.61
.70	.93	1.70	.59
.75	.90	1.80	.58
.80	.87	1.90	.56
.85	.84	2.00	.55
.90	.82	2.10	.54

A.2.5 Higher pressure natural gas tables. Capacities for gas at pressures greater than 0.5 psig (3.5 kPa gauge) in cubic feet per hour of 0.60 specific gravity gas for different sizes and lengths are shown in Tables 402.4(3) through 402.4(6) for iron pipe or

TABLE A.2.2
EQUIVALENT LENGTHS OF PIPE FITTINGS AND VALVES

		SCREWED FITTINGS[2]				90° WELDING ELBOWS AND SMOOTH BENDS[3]					
		45°/Ell	90°/Ell	180° close return bends	Tee	$R/d = 1$	$R/d = 1^{1}/_{3}$	$R/d = 2$	$R/d = 4$	$R/d = 6$	$R/d = 8$
k factor =		0.42	0.90	2.00	1.80	0.48	0.36	0.27	0.21	0.27	0.36
L/d ratio[4] *n* =		14	30	67	60	16	12	9	7	9	12
Nominal pipe size, inches	Inside diameter *d*, inches, Schedule 40[6]	*L* = Equivalent Length In Feet of Schedule 40 (Standard-Weight) Straight Pipe[6]									
$^1/_2$	0.622	0.73	1.55	3.47	3.10	0.83	0.62	0.47	0.36	0.47	0.62
$^3/_4$	0.824	0.96	2.06	4.60	4.12	1.10	0.82	0.62	0.48	0.62	0.82
1	1.049	1.22	2.62	5.82	5.24	1.40	1.05	0.79	0.61	0.79	1.05
$1^1/_4$	1.380	1.61	3.45	7.66	6.90	1.84	1.38	1.03	0.81	1.03	1.38
$1^1/_2$	1.610	1.88	4.02	8.95	8.04	2.14	1.61	1.21	0.94	1.21	1.61
2	2.067	2.41	5.17	11.5	10.3	2.76	2.07	1.55	1.21	1.55	2.07
$2^1/_2$	2.469	2.88	6.16	13.7	12.3	3.29	2.47	1.85	1.44	1.85	2.47
3	3.068	3.58	7.67	17.1	15.3	4.09	3.07	2.30	1.79	2.30	3.07
4	4.026	4.70	10.1	22.4	20.2	5.37	4.03	3.02	2.35	3.02	4.03
5	5.047	5.88	12.6	28.0	25.2	6.72	5.05	3.78	2.94	3.78	5.05
6	6.065	7.07	15.2	33.8	30.4	8.09	6.07	4.55	3.54	4.55	6.07
8	7.981	9.31	20.0	44.6	40.0	10.6	7.98	5.98	4.65	5.98	7.98
10	10.02	11.7	25.0	55.7	50.0	13.3	10.0	7.51	5.85	7.51	10.0
12	11.94	13.9	29.8	66.3	59.6	15.9	11.9	8.95	6.96	8.95	11.9
14	13.13	15.3	32.8	73.0	65.6	17.5	13.1	9.85	7.65	9.85	13.1
16	15.00	17.5	37.5	83.5	75.0	20.0	15.0	11.2	8.75	11.2	15.0
18	16.88	19.7	42.1	93.8	84.2	22.5	16.9	12.7	9.85	12.7	16.9
20	18.81	22.0	47.0	105.0	94.0	25.1	18.8	14.1	11.0	14.1	18.8
24	22.63	26.4	56.6	126.0	113.0	30.2	22.6	17.0	13.2	17.0	22.6

continued

TABLE A.2.2—continued
EQUIVALENT LENGTHS OF PIPE FITTINGS AND VALVES

		MITER ELBOWS[3] (No. of miters)					WELDING TEES		VALVES (screwed, flanged, or welded)			
		1-45°	1-60°	1-90°	2-90°[5]	3-90°[5]	Forged	Miter[3]	Gate	Globe	Angle	Swing Check
k factor =		0.45	0.90	1.80	0.60	0.45	1.35	1.80	0.21	10	5.0	2.5
L/d ratio[4] n =		15	30	60	20	15	45	60	7	333	167	83
Nominal pipe size, inches	Inside diameter d, inches, Schedule 40[6]	L = Equivalent Length In Feet of Schedule 40 (Standard-Weight) Straight Pipe[6]										
1/2	0.622	0.78	1.55	3.10	1.04	0.78	2.33	3.10	0.36	17.3	8.65	4.32
3/4	0.824	1.03	2.06	4.12	1.37	1.03	3.09	4.12	0.48	22.9	11.4	5.72
1	1.049	1.31	2.62	5.24	1.75	1.31	3.93	5.24	0.61	29.1	14.6	7.27
1 1/4	1.380	1.72	3.45	6.90	2.30	1.72	5.17	6.90	0.81	38.3	19.1	9.58
1 1/2	1.610	2.01	4.02	8.04	2.68	2.01	6.04	8.04	0.94	44.7	22.4	11.2
2	2.067	2.58	5.17	10.3	3.45	2.58	7.75	10.3	1.21	57.4	28.7	14.4
2 1/2	2.469	3.08	6.16	12.3	4.11	3.08	9.25	12.3	1.44	68.5	34.3	17.1
3	3.068	3.84	7.67	15.3	5.11	3.84	11.5	15.3	1.79	85.2	42.6	21.3
4	4.026	5.04	10.1	20.2	6.71	5.04	15.1	20.2	2.35	112.0	56.0	28.0
5	5.047	6.30	12.6	25.2	8.40	6.30	18.9	25.2	2.94	140.0	70.0	35.0
6	6.065	7.58	15.2	30.4	10.1	7.58	22.8	30.4	3.54	168.0	84.1	42.1
8	7.981	9.97	20.0	40.0	13.3	9.97	29.9	40.0	4.65	222.0	111.0	55.5
10	10.02	12.5	25.0	50.0	16.7	12.5	37.6	50.0	5.85	278.0	139.0	69.5
12	11.94	14.9	29.8	59.6	19.9	14.9	44.8	59.6	6.96	332.0	166.0	83.0
14	13.13	16.4	32.8	65.6	21.9	16.4	49.2	65.6	7.65	364.0	182.0	91.0
16	15.00	18.8	37.5	75.0	25.0	18.8	56.2	75.0	8.75	417.0	208.0	104.0
18	16.88	21.1	42.1	84.2	28.1	21.1	63.2	84.2	9.85	469.0	234.0	117.0
20	18.81	23.5	47.0	94.0	31.4	23.5	70.6	94.0	11.0	522.0	261.0	131.0
24	22.63	28.3	56.6	113.0	37.8	28.3	85.0	113.0	13.2	629.0	314.0	157.0

For SI: 1 foot = 305 mm, 1 degree = 0.01745 rad.

Note: Values for welded fittings are for conditions where bore is not obstructed by weld spatter or backing rings. If appreciably obstructed, use values for "Screwed Fittings."

1. Flanged fittings have three-fourths the resistance of screwed elbows and tees.
2. Tabular figures give the extra resistance due to curvature alone to which should be added the full length of travel.
3. Small size socket-welding fittings are equivalent to miter elbows and miter tees.
4. Equivalent resistance in number of diameters of straight pipe computed for a value of (f - 0.0075) from the *relation* (n - $k/4f$).
5. For condition of minimum resistance where the centerline length of each miter is between d and $2^1/_2 d$.
6. For pipe having other inside diameters, the equivalent resistance may be computed from the above n values.

Source: Crocker, S. *Piping Handbook*, 4th ed., Table XIV, pp. 100-101. Copyright 1945 by McGraw-Hill, Inc. Used by permission of McGraw-Hill Book Company.

equivalent rigid pipe, Tables 402.4(10) to 402.4(13) for semi-rigid tubing, Tables 402.4(17) and 402.4(18) for corrugated stainless steel tubing, and Tables 402.4(19) through 402.4(21) for polyethylene plastic pipe.

A.3 Use of capacity tables

A.3.1 Longest length method. This sizing method is conservative in its approach by applying the maximum operating conditions in the system as the norm for the system and by setting the length of pipe used to size any given part of the piping system to the maximum value.

To determine the size of each section of gas piping in a system within the range of the capacity tables, proceed as follows. (also see sample calculations included in this Appendix).

(1) Divide the piping system into appropriate segments consistent with the presence of tees, branch lines and main runs. For each segment, determine the gas load (assuming all appliances operate simultaneously) and its overall length. An allowance (in equivalent length of pipe) as determined from Table A.2.2 shall be considered for piping segments that include four or more fittings.

(2) Determine the gas demand of each appliance to be attached to the piping system. Where Tables 402.4(1) through 402.4(23) are to be used to select the piping size, calculate the gas demand in terms of cubic feet per hour for each piping system outlet. Where Tables 402.4(24) through 402.4(33) are to be used to select the piping size, calculate the gas demand in terms of thousands of Btu per hour for each piping system outlet.

(3) Where the piping system is for use with other than undiluted liquefied petroleum gases, determine the design system pressure, the allowable loss in pressure (pressure drop), and specific gravity of the gas to be used in the piping system.

(4) Determine the length of piping from the point of delivery to the most remote outlet in the building/piping system.

(5) In the appropriate capacity table, select the row showing the measured length or the next longer length if the table does not give the exact length. This is the only length used in determining the size of any section of gas piping. If the gravity factor is to be applied, the values in the selected row of the table are multiplied by the appropriate multiplier from Table A.2.4.

(6) Use this horizontal row to locate ALL gas demand figures for this particular system of piping.

(7) Starting at the most remote outlet, find the gas demand for that outlet in the horizontal row just selected. If the exact figure of demand is not shown, choose the next larger figure left in the row.

(8) Opposite this demand figure, in the first row at the top, the correct size of gas piping will be found.

(9) Proceed in a similar manner for each outlet and each section of gas piping. For each section of piping, determine the total gas demand supplied by that section.

When a large number of piping components (such as elbows, tees and valves) are installed in a pipe run, additional pressure loss can be accounted for by the use of equivalent lengths. Pressure loss across any piping component can be equated to the pressure drop through a length of pipe. The equivalent length of a combination of only four elbows/tees can result in a jump to the next larger length row, resulting in a significant reduction in capacity. The equivalent lengths in feet shown in Table A.2.2 have been computed on a basis that the inside diameter corresponds to that of Schedule 40 (standard-weight) steel pipe, which is close enough for most purposes involving other schedules of pipe. Where a more specific solution for equivalent length is desired, this may be made by multiplying the actual inside diameter of the pipe in inches by $n/12$, or the actual inside diameter in feet by n (n can be read from the table heading). The equivalent length values can be used with reasonable accuracy for copper or brass fittings and bends although the resistance per foot of copper or brass pipe is less than that of steel. For copper or brass valves, however, the equivalent length of pipe should be taken as 45 percent longer than the values in the table, which are for steel pipe.

A.3.2 Branch length method. This sizing method reduces the amount of conservatism built into the traditional Longest Length Method. The longest length as measured from the meter to the furthest remote appliance is only used to size the initial parts of the overall piping system. The Branch Length Method is applied in the following manner:

(1) Determine the gas load for each of the connected appliances.

(2) Starting from the meter, divide the piping system into a number of connected segments, and determine the length and amount of gas that each segment would carry assuming that all appliances were operated simultaneously. An allowance (in equivalent length of pipe) as determined from Table A.2.2 should be considered for piping segments that include four or more fittings.

(3) Determine the distance from the outlet of the gas meter to the appliance furthest removed from the meter.

(4) Using the longest distance (found in Step 3), size each piping segment from the meter to the most remote appliance outlet.

(5) For each of these piping segments, use the longest length and the calculated gas load for all of the connected appliances for the segment and begin the sizing process in Steps 6 through 8.

(6) Referring to the appropriate sizing table (based on operating conditions and piping material), find the longest length distance in the first column or the next larger distance if the exact distance is not listed. The use of alternative operating pressures and/or pressure drops will require the use of a different sizing table, but will not alter the sizing methodology. In many cases, the use of alternative operating pressures and/or pressure drops will require the approval of both the code official and the local gas serving utility.

(7) Trace across this row until the gas load is found or the closest larger capacity if the exact capacity is not listed.

(8) Read up the table column and select the appropriate pipe size in the top row. Repeat Steps 6, 7 and 8 for each pipe segment in the longest run.

(9) Size each remaining section of branch piping not previously sized by measuring the distance from the gas meter location to the most remote outlet in that branch, using the gas load of attached appliances and following the procedures of Steps 2 through 8.

A.3.3 Hybrid pressure method. The sizing of a 2 psi (13.8 kPa) gas piping system is performed using the traditional Longest Length Method but with modifications. The 2 psi (13.8 kPa) system consists of two independent pressure zones, and each zone is sized separately. The Hybrid Pressure Method is applied as follows.

The sizing of the 2 psi (13.8 kPa) section (from the meter to the line regulator) is as follows:

(1) Calculate the gas load (by adding up the name plate ratings) from all connected appliances. (In certain circumstances the installed gas load may be increased up to 50 percent to accommodate future addition of appliances.) Ensure that the line regulator capacity is adequate for the calculated gas load and that the required pressure drop (across the regulator) for that capacity does not exceed $^3/_4$ psi (5.2 kPa) for a 2 psi (13.8 kPa) system. If the pressure drop across the regulator is too high (for the connected gas load), select a larger regulator.

(2) Measure the distance from the meter to the line regulator located inside the building.

(3) If there are multiple line regulators, measure the distance from the meter to the regulator furthest removed from the meter.

(4) The maximum allowable pressure drop for the 2 psi (13.8 kPa) section is 1 psi (6.9 kPa).

(5) Referring to the appropriate sizing table (based on piping material) for 2 psi (13.8 kPa) systems with a 1 psi (6.9 kPa) pressure drop, find this distance in the first column, or the closest larger distance if the exact distance is not listed.

(6) Trace across this row until the gas load is found or the closest larger capacity if the exact capacity is not listed.

(7) Read up the table column to the top row and select the appropriate pipe size.

(8) If there are multiple regulators in this portion of the piping system, each line segment must be sized for its actual gas load, but using the longest length previously determined above.

The low pressure section (all piping downstream of the line regulator) is sized as follows:

(1) Determine the gas load for each of the connected appliances.

(2) Starting from the line regulator, divide the piping system into a number of connected segments and/or independent parallel piping segments, and determine the amount of gas that each segment would carry assuming that all appliances were operated simultaneously. An allowance (in equivalent length of pipe) as determined from Table A.2.2 should be considered for piping segments that include four or more fittings.

(3) For each piping segment, use the actual length or longest length (if there are sub-branchlines) and the calculated gas load for that segment and begin the sizing process as follows:

(a) Referring to the appropriate sizing table (based on operating pressure and piping material), find the longest length distance in the first column or the closest larger distance if the exact distance is not listed. The use of alternative operating pressures and/or pressure drops will require the use of a different sizing table, but will not alter the sizing methodology. In many cases, the use of alternative operating pressures and/or pressure drops may require the approval of the code official.

(b) Trace across this row until the appliance gas load is found or the closest larger capacity if the exact capacity is not listed.

(c) Read up the table column to the top row and select the appropriate pipe size.

(d) Repeat this process for each segment of the piping system.

A.4 Use of sizing equations. Capacities of smooth wall pipe or tubing can also be determined by using the following formulae:

(1) High Pressure [1.5 psi (10.3 kPa) and above]:

$$Q = 181.6 \sqrt{\frac{D^5 \cdot (P_1^2 - P_2^2) \cdot Y}{C_r \cdot fba \cdot L}}$$

$$= 2237 D^{2.623} \left[\frac{(P_1^2 - P_2^2) \cdot Y}{C_r \cdot L} \right]^{0.541}$$

SIZING AND CAPACITIES OF GAS PIPING

(2) Low Pressure [Less than 1.5 psi (10.3 kPa)]:

$$Q = 187.3 \sqrt{\frac{D^5 \cdot \Delta H}{C_r \cdot fba \cdot L}}$$

$$= 2313 \, D^{2.623} \left(\frac{\Delta H}{C_r \cdot L}\right)^{0.541}$$

where:

Q = Rate, cubic feet per hour at 60°F and 30-inch mercury column

D = Inside diameter of pipe, in.

P_1 = Upstream pressure, psia

P_2 = Downstream pressure, psia

Y = Superexpansibility factor = 1/supercompressibility factor

C_r = Factor for viscosity, density and temperature*

  = $0.00354 \, ST \left(\frac{Z}{S}\right)^{0.152}$

Note: See Table 402.4 for Y and C_r for natural gas and propane.

S = Specific gravity of gas at 60°F and 30-inch mercury column (0.60 for natural gas, 1.50 for propane), or = 1488μ

T = Absolute temperature, °F or = $t + 460$

t = Temperature, °F

Z = Viscosity of gas, centipoise (0.012 for natural gas, 0.008 for propane), or = 1488μ

fba = Base friction factor for air at 60°F (CF=1)

L = Length of pipe, ft

ΔH = Pressure drop, in. w.c. (27.7 in. H_2O = 1 psi)
(For SI, see Section 402.4)

A.5 Pipe and tube diameters. Where the internal diameter is determined by the formulas in Section 402.4, Tables A.5.1 and A.5.2 can be used to select the nominal or standard pipe size based on the calculated internal diameter.

TABLE A.5.1
SCHEDULE 40 STEEL PIPE STANDARD SIZES

NOMINAL SIZE (in.)	INTERNAL DIAMETER (in.)	NOMINAL SIZE (in.)	INTERNAL DIAMETER (in.)
1/4	0.364	1 1/2	1.610
3/8	0.493	2	2.067
1/2	0.622	2 1/2	2.469
3/4	0.824	3	3.068
1	1.049	3 1/2	3.548
1 1/4	1.380	4	4.026

TABLE A.5.2
COPPER TUBE STANDARD SIZES

TUBE TYPE	NOMINAL OR STANDARD SIZE inches	INTERNAL DIAMETER inches
K	1/4	0.305
L	1/4	0.315
ACR (D)	3/8	0.315
ACR (A)	3/8	0.311
K	3/8	0.402
L	3/8	0.430
ACR (D)	1/2	0.430
ACR (A)	1/2	0.436
K	1/2	0.527
L	1/2	0.545
ACR (D)	5/8	0.545
ACR (A)	5/8	0.555
K	5/8	0.652
L	5/8	0.666
ACR (D)	3/4	0.666
ACR (A)	3/4	0.680
K	3/4	0.745
L	3/4	0.785
ACR	7/8	0.785
K	1	0.995
L	1	1.025
ACR	1 1/8	1.025
K	1 1/4	1.245
L	1 1/4	1.265
ACR	1 3/8	1.265
K	1 1/2	1.481
L	1 1/2	1.505
ACR	1 5/8	1.505
K	2	1.959
L	2	1.985
ACR	2 1/8	1.985
K	2 1/2	2.435
L	2 1/2	2.465
ACR	2 5/8	2.465
K	3	2.907
L	3	2.945
ACR	3 1/8	2.945

A.6 Use of sizing charts. A third method of sizing gas piping is detailed below as an option that is useful when large quantities of piping are involved in a job (e.g., an apartment house) and material costs are of concern. If the user is not completely familiar with this method, the resulting pipe sizing should be checked by a knowledgeable gas engineer. The sizing charts are applied as follows.

(1) With the layout developed according to Section 106.3.1 of the code, indicate in each section the design gas flow under maximum operation conditions. For many layouts, the maximum design flow will be the sum of all connected loads. However, in some cases, certain combinations of utilization equipment will not occur simultaneously (e.g., gas heating and air conditioning). For these cases, the design flow is the greatest gas flow that can occur at any one time.

(2) Determine the inlet gas pressure for the system being designed. In most cases, the point of inlet will be the gas meter or service regulator, but in the case of a system addition, it could be the point of connection to the existing system.

(3) Determine the minimum pressure required at the inlet to the critical utilization equipment. Usually, the critical item will be the piece of equipment with the highest required pressure for satisfactory operation. If several items have the same required pressure, it will be the one with the greatest length of piping from the system inlet.

(4) The difference between the inlet pressure and critical item pressure is the allowable system pressure drop. Figures A.6(a) and A.6(b) show the relationship between gas flow, pipe size and pipe length for natural gas with 0.60 specific gravity.

(5) To use Figure A.6(a) (low pressure applications), calculate the piping length from the inlet to the critical utilization equipment. Increase this length by 50 percent to allow for fittings. Divide the allowable pressure drop by the equivalent length (in hundreds of feet) to determine the allowable pressure drop per hundred feet. Select the pipe size from Figure A.6(a) for the required volume of flow.

(6) To use Figure A.6(b) (high pressure applications), calculate the equivalent length as above. Calculate the index number for Figure A.6(b) by dividing the difference between the squares of the absolute values of inlet and outlet pressures by the equivalent length (in hundreds of feet). Select the pipe size from Figure A.6(b) for the gas volume required.

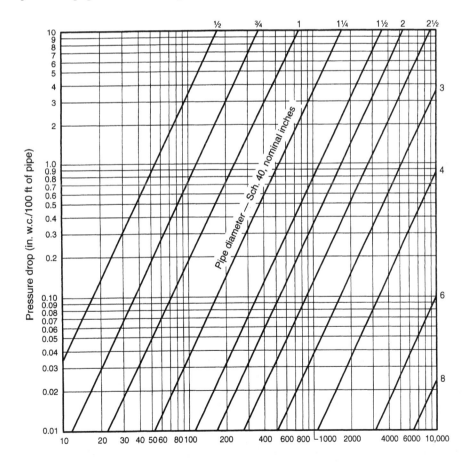

**FIGURE A.6 (a)
CAPACITY OF NATURAL GAS PIPING, LOW PRESSURE (0.60 WC)**

SIZING AND CAPACITIES OF GAS PIPING

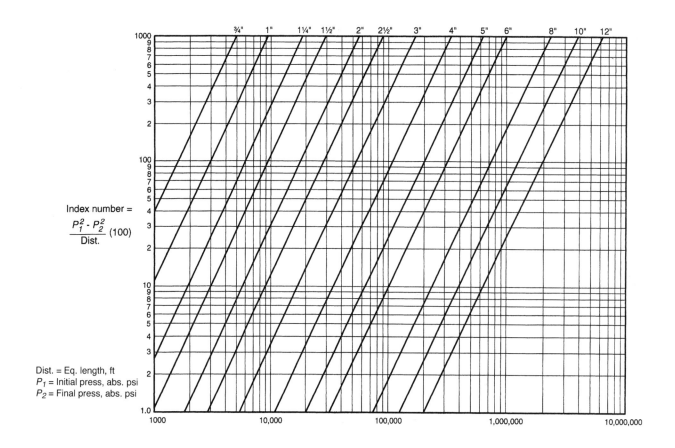

FIGURE A.6 (b)
CAPACITY OF NATURAL GAS PIPING, HIGH PRESSURE (1.5 psi and above)

A.7 Examples of piping system design and sizing

A.7.1 Example 1: Longest length method. Determine the required pipe size of each section and outlet of the piping system shown in Figure A.7.1, with a designated pressure drop of 0.5-inch w.c. (125 Pa) using the Longest Length Method. The gas to be used has 0.60 specific gravity and a heating value of 1,000 Btu/ft^3 (37.5 MJ/m^3).

Solution:

(1) Maximum gas demand for Outlet A:

$$\frac{\text{Consumption (rating plate input, or Table 402.2 if necessary)}}{\text{Btu of gas}} =$$

$$\frac{35,000 \text{ Btu per hour rating}}{1,000 \text{ Btu per cubic foot}} = 35 \text{ cubic feet per hour} = 35 \text{ cfh}$$

Maximum gas demand for Outlet B:

$$\frac{\text{Consumption}}{\text{Btu of gas}} = \frac{75,000}{1,000} = 75 \text{ cfh}$$

Maximum gas demand for Outlet C:

$$\frac{\text{Consumption}}{\text{Btu of gas}} = \frac{35,000}{1,000} = 35 \text{ cfh}$$

Maximum gas demand for Outlet D:

$$\frac{\text{Consumption}}{\text{Btu of gas}} = \frac{100,000}{1,000} = 100 \text{ cfh}$$

(2) The length of pipe from the point of delivery to the most remote outlet (A) is 60 feet (18 288 mm). This is the only distance used.

(3) Using the row marked 60 feet (18 288 mm) in Table 402.4(2):

  (a) Outlet A, supplying 35 cfh (0.99 m^3/hr), requires $^3/_8$-inch pipe.
  (b) Outlet B, supplying 75 cfh (2.12 m^3/hr), requires $^3/_4$-inch pipe.
  (c) Section 1, supplying Outlets A and B, or 110 cfh (3.11 m^3/hr), requires $^3/_4$-inch pipe.
  (d) Section 2, supplying Outlets C and D, or 135 cfh (3.82 m^3/hr), requires $^3/_4$-inch pipe.
  (e) Section 3, supplying Outlets A, B, C and D, or 245 cfh (6.94 m^3/hr), requires 1-inch pipe.

(4) If a different gravity factor is applied to this example, the values in the row marked 60 feet (18 288 mm) of Table 402.4(2) would be multiplied by the appropriate multiplier from Table A.2.4 and the resulting cubic feet per hour values would be used to size the piping.

SIZING AND CAPACITIES OF GAS PIPING

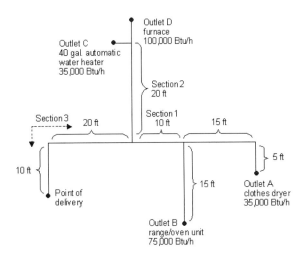

FIGURE A.7.1
PIPING PLAN SHOWING A STEEL PIPING SYSTEM

A.7.2 Example 2: Hybrid or dual pressure systems. Determine the required CSST size of each section of the piping system shown in Figure A.7.2, with a designated pressure drop of 1 psi (6.9 kPa) for the 2 psi (13.8 kPa) section and 3-inch w.c. (0.75 kPa) pressure drop for the 13-inch w.c. (2.49 kPa) section. The gas to be used has 0.60 specific gravity and a heating value of 1,000 Btu/ft^3 (37.5 MJ/m^3).

Solution

(1) Size 2 psi (13.8 kPa) line using Table 402.4(17).

(2) Size 10-inch w.c. (2.5 kPa) lines using Table 402.4(15).

(3) Using the following, determine if sizing tables can be used.

　(a) Total gas load shown in Figure A.7.2 equals 110 cfh (3.11 m^3/hr).

　(b) Determine pressure drop across regulator [see notes in Table 402.4 (17)].

　(c) If pressure drop across regulator exceeds $^3/_4$ psig (5.2 kPa), Table 402.4 (17) cannot be used. Note: If pressure drop exceeds $^3/_4$ psi (5.2 kPa), then a larger regulator must be selected or an alternative sizing method must be used.

　(d) Pressure drop across the line regulator [for 110 cfh (3.11 m^3/hr)] is 4-inch w.c. (0.99 kPa) based on manufacturer's performance data.

　(e) Assume the CSST manufacturer has tubing sizes or EHDs of 13, 18, 23 and 30.

(4) Section A [2 psi (13.8 kPa) zone]

　(a) Distance from meter to regulator = 100 feet (30 480 mm).

　(b) Total load supplied by A = 110 cfh (3.11 m^3/hr) (furnace + water heater + dryer).

　(c) Table 402.4 (17) shows that EHD size 18 should be used.

Note: It is not unusual to oversize the supply line by 25 to 50 percent of the as-installed load. EHD size 18 has a capacity of 189 cfh (5.35 m^3/hr).

(5) Section B (low pressure zone)

　(a) Distance from regulator to furnace is 15 feet (4572 mm).

　(b) Load is 60 cfh (1.70 m^3/hr).

　(c) Table 402.4 (15) shows that EHD size 13 should be used.

(6) Section C (low pressure zone)

　(a) Distance from regulator to water heater is 10 feet (3048 mm).

　(b) Load is 30 cfh (0.85 m^3/hr).

　(c) Table 402.4 (15) shows that EHD size 13 should be used.

(7) Section D (low pressure zone)

　(a) Distance from regulator to dryer is 25 feet (7620 mm).

　(b) Load is 20 cfh (0.57 m^3/hr).

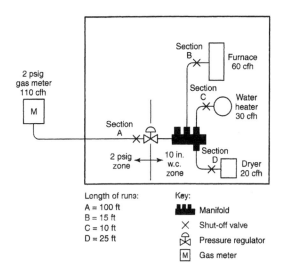

FIGURE A.7.2
PIPING PLAN SHOWING A CSST SYSTEM

　(c) Table 402.4(15) shows that EHD size 13 should be used.

A.7.3 Example 3: Branch length method. Determine the required semi-rigid copper tubing size of each section of the piping system shown in Figure A.7.3, with a designated pressure drop of 1-inch w.c. (250 Pa) (using the Branch Length Method). The gas to be used has 0.60 specific gravity and a heating value of 1,000 Btu/ft^3 (37.5 MJ/m^3).

Solution

(1) Section A

　(a) The length of tubing from the point of delivery to the most remote appliance is 50 feet (15 240 mm), A + C.

SIZING AND CAPACITIES OF GAS PIPING

(b) Use this longest length to size Sections A and C.

(c) Using the row marked 50 feet (15 240 mm) in Table 402.4(9), Section A, supplying 220 cfh (6.2 m^3/hr) for four appliances requires 1-inch tubing.

(2) Section B

  (a) The length of tubing from the point of delivery to the range/oven at the end of Section B is 30 feet (9144 mm), A + B.

  (b) Use this branch length to size Section B only.

  (c) Using the row marked 30 feet (9144 mm) in Table 402.4(9), Section B, supplying 75 cfh (2.12 m^3/hr) for the range/oven requires $^1/_2$-inch tubing.

(3) Section C

  (a) The length of tubing from the point of delivery to the dryer at the end of Section C is 50 feet (15 240 mm), A + C.

  (b) Use this branch length (which is also the longest length) to size Section C.

  (c) Using the row marked 50 feet (15 240 mm) in Table 402.4(9), Section C, supplying 30 cfh (0.85 m^3/hr) for the dryer requires $^3/_8$-inch tubing.

(4) Section D

  (a) The length of tubing from the point of delivery to the water heater at the end of Section D is 30 feet (9144 mm), A + D.

  (b) Use this branch length to size Section D only.

  (c) Using the row marked 30 feet (9144 mm) in Table 402.4(9), Section D, supplying 35 cfh (0.99 m^3/hr) for the water heater requires $^3/_8$-inch tubing.

(5) Section E

  (a) The length of tubing from the point of delivery to the furnace at the end of Section E is 30 feet (9144 mm), A + E.

  (b) Use this branch length to size Section E only.

  (c) Using the row marked 30 feet (9144 mm) in Table 402.4(9), Section E, supplying 80 cfh (2.26 m^3/hr) for the furnace requires $^1/_2$-inch tubing.

A.7.4 Example 4: Modification to existing piping system. Determine the required CSST size for Section G (retrofit application) of the piping system shown in Figure A.7.4, with a designated pressure drop of 0.5-inch w.c. (125 Pa) using the branch length method. The gas to be used has 0.60 specific gravity and a heating value of 1,000 Btu/ft^3 (37.5 MJ/m^3).

Solution

(1) The length of pipe and CSST from the point of delivery to the retrofit appliance (barbecue) at the end of Section G is 40 feet (12 192 mm), A + B + G.

(2) Use this branch length to size Section G.

(3) Assume the CSST manufacturer has tubing sizes or EHDs of 13, 18, 23 and 30.

(4) Using the row marked 40 feet (12 192 mm) in Table 402.4(14), Section G, supplying 40 cfh (1.13 m^3/hr) for the barbecue requires EHD 18 CSST.

(5) The sizing of Sections A, B, F and E must be checked to ensure adequate gas carrying capacity since an appliance has been added to the piping system (see A.7.1 for details).

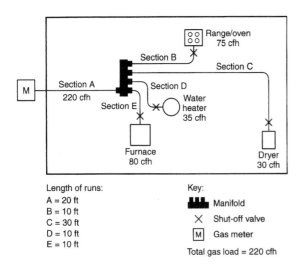

FIGURE A.7.3
PIPING PLAN SHOWING A COPPER TUBING SYSTEM

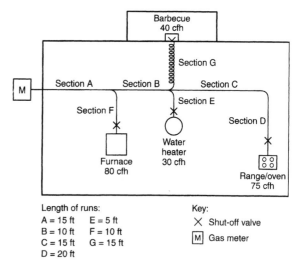

FIGURE A.7.4
PIPING PLAN SHOWING A MODIFICATION TO EXISTING PIPING SYSTEM

The appendices to the *International Fuel Gas Code®* (IFGC®) are explanatory commentaries of the code itself. Thus, there is no additional commentary for them.

APPENDIX B (IFGS)

SIZING OF VENTING SYSTEMS SERVING APPLIANCES EQUIPPED WITH DRAFT HOODS, CATEGORY I APPLIANCES, AND APPLIANCES LISTED FOR USE WITH TYPE B VENTS

(This appendix is informative and is not part of the code.)

EXAMPLES USING SINGLE APPLIANCE VENTING TABLES

Example 1: Single draft-hood-equipped appliance

An installer has a 120,000 British thermal unit (Btu) per hour input appliance with a 5-inch-diameter draft hood outlet that needs to be vented into a 10-foot-high Type B vent system. What size vent should be used assuming (a) a 5-foot lateral single-wall metal vent connector is used with two 90-degree elbows, or (b) a 5-foot lateral single-wall metal vent connector is used with three 90-degree elbows in the vent system?

Solution:

Table 504.2(2) should be used to solve this problem, because single-wall metal vent connectors are being used with a Type B vent.

(a) Read down the first column in Table 504.2(2) until the row associated with a 10-foot height and 5-foot lateral is found. Read across this row until a vent capacity greater than 120,000 Btu per hour is located in the shaded columns labeled "NAT Max" for draft-hood-equipped appliances. In this case, a 5-inch-diameter vent has a capacity of 122,000 Btu per hour and may be used for this application.

(b) If three 90-degree elbows are used in the vent system, then the maximum vent capacity listed in the tables must be reduced by 10 percent (see Section 504.2.3 for single appliance vents). This implies that the 5-inch-diameter vent has an adjusted capacity of only 110,000 Btu per hour. In this case, the vent system must be increased to 6 inches in diameter (see calculations below).

122,000 (.90) = 110,000 for 5-inch vent
From Table 504.2(2), Select 6-inch vent
186,000 (.90) = 167,000; This is greater than the required 120,000. Therefore, use a 6-inch vent and connector where three elbows are used.

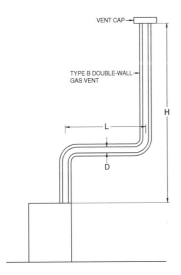

For SI: 1 foot = 304.8 mm, 1 British thermal unit per hour = 0.2931 W.

Table 504.2(1) is used when sizing Type B double-wall gas vent connected directly to the appliance.

Note: The appliance may be either Category I draft hood equipped or fan-assisted type.

**FIGURE B-1
TYPE B DOUBLE-WALL VENT SYSTEM SERVING A SINGLE APPLIANCE WITH A TYPE B DOUBLE-WALL VENT**

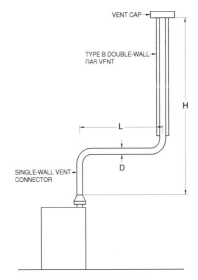

For SI: 1 foot = 304.8 mm, 1 British thermal unit per hour = 0.2931W.

Table 504.2(2) is used when sizing a single-wall metal vent connector attached to a Type B double-wall gas vent.

Note: The appliance may be either Category I draft hood equipped or fan-assisted type.

**FIGURE B-2
TYPE B DOUBLE-WALL VENT SYSTEM SERVING A SINGLE APPLIANCE WITH A SINGLE-WALL METAL VENT CONNECTOR**

SIZING OF VENTING SYSTEMS

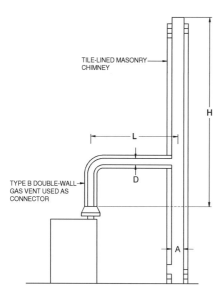

Table 504.2(3) is used when sizing a Type B double-wall gas vent connector attached to a tile-lined masonry chimney.

Note: "A" is the equivalent cross-sectional area of the tile liner.

Note: The appliance may be either Category I draft hood equipped or fan-assisted type.

**FIGURE B-3
VENT SYSTEM SERVING A SINGLE APPLIANCE
WITH A MASONRY CHIMNEY OF TYPE B
DOUBLE-WALL VENT CONNECTOR**

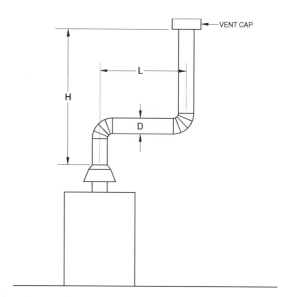

Asbestos cement Type B or single-wall metal vent serving a single draft-hood-equipped appliance [see Table 504.2(5)].

**FIGURE B-5
ASBESTOS CEMENT TYPE B OR SINGLE-WALL
METAL VENT SYSTEM SERVING A SINGLE
DRAFT-HOOD-EQUIPPED APPLIANCE**

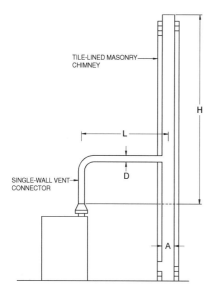

Table 504.2(4) is used when sizing a single-wall vent connector attached to a tile-lined masonry chimney.

Note: "A" is the equivalent cross-sectional area of the tile liner.

Note: The appliance may be either Category I draft hood equipped or fan-assisted type.

**FIGURE B-4
VENT SYSTEM SERVING A SINGLE APPLIANCE
USING A MASONRY CHIMNEY AND A
SINGLE-WALL METAL VENT CONNECTOR**

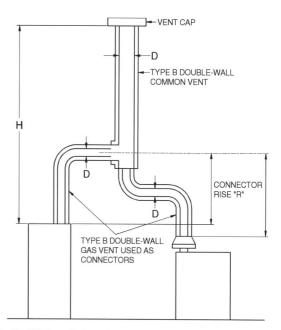

Table 504.3(1) is used when sizing Type B double-wall vent connectors attached to a Type B double-wall common vent.

Note: Each appliance may be either Category I draft hood equipped or fan-assisted type.

**FIGURE B-6
VENT SYSTEM SERVING TWO OR MORE APPLIANCES
WITH TYPE B DOUBLE-WALL VENT AND TYPE B
DOUBLE-WALL VENT CONNECTOR**

SIZING OF VENTING SYSTEMS

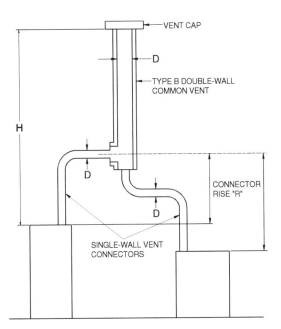

Table 504.3(2) is used when sizing single-wall vent connectors attached to a Type B double-wall common vent.

Note: Each appliance may be either Category I draft hood equipped or fan-assisted type.

**FIGURE B-7
VENT SYSTEM SERVING TWO OR MORE APPLIANCES
WITH TYPE B DOUBLE-WALL VENT AND
SINGLE-WALL METAL VENT CONNECTORS**

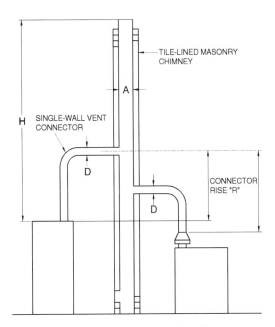

Table 504.3(4) is used when sizing single-wall metal vent connectors attached to a tile-lined masonry chimney.

Note: "A" is the equivalent cross-sectional area of the tile liner.

Note: Each appliance may be either Category I draft hood equipped or fan-assisted type.

**FIGURE B-9
MASONRY CHIMNEY SERVING TWO OR MORE APPLIANCES
WITH SINGLE-WALL METAL VENT CONNECTORS**

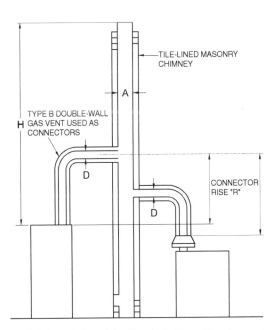

Table 504.3(3) is used when sizing Type B double-wall vent connectors attached to a tile-lined masonry chimney.

Note: "A" is the equivalent cross-sectional area of the tile liner.

Note: Each appliance may be either Category I draft hood equipped or fan-assisted type.

**FIGURE B-8
MASONRY CHIMNEY SERVING TWO OR MORE APPLIANCES
WITH TYPE B DOUBLE-WALL VENT CONNECTOR**

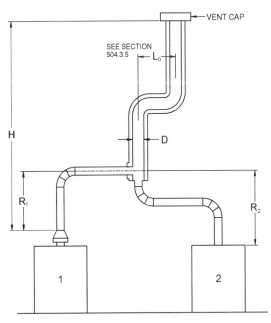

Asbestos cement Type B or single-wall metal pipe vent serving two or more draft-hood-equipped appliances [see Table 504.3(5)].

**FIGURE B-10
ASBESTOS CEMENT TYPE B OR SINGLE-WALL
METAL VENT SYSTEM SERVING TWO OR MORE
DRAFT-HOOD-EQUIPPED APPLIANCES**

SIZING OF VENTING SYSTEMS

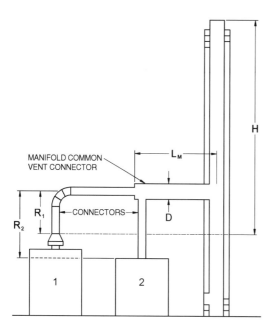

Example: Manifolded Common Vent Connector L_M shall be no greater than 18 times the common vent connector manifold inside diameter; i.e., a 4-inch (102 mm) inside diameter common vent connector manifold shall not exceed 72 inches (1829 mm) in length (see Section 504.3.4).

Note: This is an illustration of a typical manifolded vent connector. Different appliance, vent connector, or common vent types are possible. Consult Section 502.3.

FIGURE B-11
USE OF MANIFOLD COMMON VENT CONNECTOR

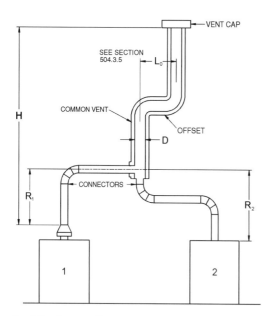

Example: Offset Common Vent

Note: This is an illustration of a typical offset vent. Different appliance, vent connector, or vent types are possible. Consult Sections 504.2 and 504.3.

FIGURE B-12
USE OF OFFSET COMMON VENT

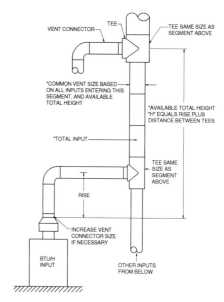

Vent connector size depends on:
- Input
- Rise
- Available total height "H"
- Table 504.3(1) connectors

Common vent size depends on:
- Combined inputs
- Available total height "H"
- Table 504.3(1) common vent

FIGURE B-13
MULTISTORY GAS VENT DESIGN PROCEDURE FOR EACH SEGMENT OF SYSTEM

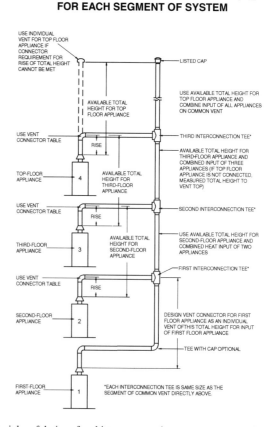

Principles of design of multistory vents using vent connector and common vent design tables (see Sections 504.3.11 through 504.3.17).

FIGURE B-14
MULTISTORY VENT SYSTEMS

SIZING OF VENTING SYSTEMS

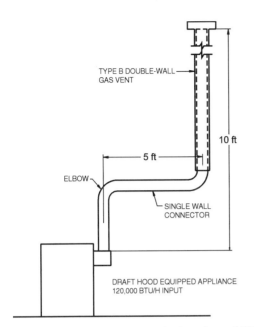

FIGURE B-15 (EXAMPLE 1)
SINGLE DRAFT-HOOD-EQUIPPED APPLIANCE

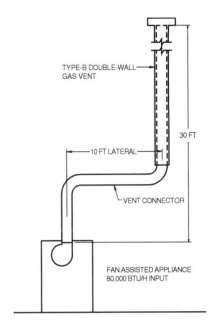

FIGURE B-16 (EXAMPLE 2)
SINGLE FAN-ASSISTED APPLIANCE

Example 2: Single fan-assisted appliance

An installer has an 80,000 Btu per hour input fan-assisted appliance that must be installed using 10 feet of lateral connector attached to a 30-foot-high Type B vent. Two 90-degree elbows are needed for the installation. Can a single-wall metal vent connector be used for this application?

Solution:

Table 504.2(2) refers to the use of single-wall metal vent connectors with Type B vent. In the first column find the row associated with a 30-foot height and a 10-foot lateral. Read across this row, looking at the FAN Min and FAN Max columns, to find that a 3-inch-diameter single-wall metal vent connector is not recommended. Moving to the next larger size single wall connector (4 inches), note that a 4-inch-diameter single-wall metal connector has a recommended minimum vent capacity of 91,000 Btu per hour and a recommended maximum vent capacity of 144,000 Btu per hour. The 80,000 Btu per hour fan-assisted appliance is outside this range, so the conclusion is that a single-wall metal vent connector cannot be used to vent this appliance using 10 feet of lateral for the connector.

However, if the 80,000 Btu per hour input appliance could be moved to within 5 feet of the vertical vent, then a 4-inch single-wall metal connector could be used to vent the appliance. Table 504.2(2) shows the acceptable range of vent capacities for a 4-inch vent with 5 feet of lateral to be between 72,000 Btu per hour and 157,000 Btu per hour.

If the appliance cannot be moved closer to the vertical vent, then Type B vent could be used as the connector material. In this case, Table 504.2(1) shows that for a 30-foot-high vent with 10 feet of lateral, the acceptable range of vent capacities for a 4-inch-diameter vent attached to a fan-assisted appliance is between 37,000 Btu per hour and 150,000 Btu per hour.

Example 3: Interpolating between table values

An installer has an 80,000 Btu per hour input appliance with a 4-inch-diameter draft hood outlet that needs to be vented into a 12-foot-high Type B vent. The vent connector has a 5-foot lateral length and is also Type B. Can this appliance be vented using a 4-inch-diameter vent?

Solution:

Table 504.2(1) is used in the case of an all Type B vent system. However, since there is no entry in Table 504.2(1) for a height of 12 feet, interpolation must be used. Read down the 4-inch diameter NAT Max column to the row associated with 10-foot height and 5-foot lateral to find the capacity value of 77,000 Btu per hour. Read further down to the 15-foot height, 5-foot lateral row to find the capacity value of 87,000 Btu per hour. The difference between the 15-foot height capacity value and the 10-foot height capacity value is 10,000 Btu per hour. The capacity for a vent system with a 12-foot height is equal to the capacity for a 10-foot height plus $2/5$ of the difference between the 10-foot and 15-foot height values, or $77,000 + 2/5 (10,000) = 81,000$ Btu per hour. Therefore, a 4-inch-diameter vent may be used in the installation.

EXAMPLES USING COMMON VENTING TABLES

Example 4: Common venting two draft-hood-equipped appliances

A 35,000 Btu per hour water heater is to be common vented with a 150,000 Btu per hour furnace using a common vent with a total height of 30 feet. The connector rise is 2 feet for the water heater with a horizontal length of 4 feet. The connector rise for the furnace is 3 feet with a horizontal length of 8 feet. Assume single-wall metal connectors will be used with Type B vent. What size connectors and combined vent should be used in this installation?

SIZING OF VENTING SYSTEMS

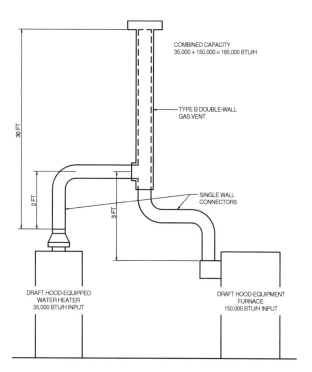

**FIGURE B-17 (EXAMPLE 4)
COMMON VENTING TWO DRAFT-
HOOD-EQUIPPED APPLIANCES**

Solution:

Table 504.3(2) should be used to size single-wall metal vent connectors attached to Type B vertical vents. In the vent connector capacity portion of Table 504.3(2), find the row associated with a 30-foot vent height. For a 2-foot rise on the vent connector for the water heater, read the shaded columns for draft-hood-equipped appliances to find that a 3-inch-diameter vent connector has a capacity of 37,000 Btu per hour. Therefore, a 3-inch single-wall metal vent connector may be used with the water heater. For a draft-hood-equipped furnace with a 3-foot rise, read across the appropriate row to find that a 5-inch-diameter vent connector has a maximum capacity of 120,000 Btu per hour (which is too small for the furnace) and a 6-inch-diameter vent connector has a maximum vent capacity of 172,000 Btu per hour. Therefore, a 6-inch-diameter vent connector should be used with the 150,000 Btu per hour furnace. Since both vent connector horizontal lengths are less than the maximum lengths listed in Section 504.3.2, the table values may be used without adjustments.

In the common vent capacity portion of Table 504.3(2), find the row associated with a 30-foot vent height and read over to the NAT + NAT portion of the 6-inch-diameter column to find a maximum combined capacity of 257,000 Btu per hour. Since the two appliances total only 185,000 Btu per hour, a 6-inch common vent may be used.

Example 5a: Common venting a draft-hood-equipped water heater with a fan-assisted furnace into a Type B vent

In this case, a 35,000 Btu per hour input draft-hood-equipped water heater with a 4-inch-diameter draft hood outlet, 2 feet of connector rise, and 4 feet of horizontal length is to be common vented with a 100,000 Btu per hour fan-assisted furnace with a 4-inch-diameter flue collar, 3 feet of connector rise, and 6 feet of horizontal length. The common vent consists of a 30-foot height of Type B vent. What are the recommended vent diameters for each connector and the common vent? The installer would like to use a single-wall metal vent connector.

Solution: - [Table 504.3(2)]

Water Heater Vent Connector Diameter. Since the water heater vent connector horizontal length of 4 feet is less than the maximum value listed in Section 504.3.2, the venting table values may be used without adjustments. Using the Vent Connector Capacity portion of Table 504.3(2), read down the Total Vent Height (H) column to 30 feet and read across the 2-foot Connector Rise (R) row to the first Btu per hour rating in the NAT Max column that is equal to or greater than the water heater input rating. The table shows that a 3-inch vent connector has a maximum input rating of 37,000 Btu per hour. Although this is greater than the water heater input rating, a 3-inch vent connector is prohibited by Section 504.3.21. A 4-inch vent connector has a maximum input rating of 67,000 Btu per hour and is equal to the draft hood outlet diameter. A 4-inch vent connector is selected. Since the water heater is equipped with a draft hood, there are no minimum input rating restrictions.

Furnace Vent Connector Diameter. Using the Vent Connector Capacity portion of Table 504.3(2), read down the Total Vent Height (H) column to 30 feet and across the 3-foot Connector Rise (R) row. Since the furnace has a fan-assisted combustion system, find the first FAN Max column with a Btu per hour rating greater than the furnace input rating. The 4-inch vent connector has a maximum input rating of 119,000 Btu per hour and a minimum input rating of 85,000 Btu per hour. The 100,000 Btu per hour furnace in this example falls within this range, so a 4-inch connector is adequate. Since the furnace vent connector horizontal length of 6 feet does not exceed the maximum value listed in Section 504.3.2, the venting table values may be used without adjustment. If the furnace had an input rating of 80,000 Btu per hour, then a Type B vent connector [see Table 504.3(1)] would be needed in order to meet the minimum capacity limit.

Common Vent Diameter. The total input to the common vent is 135,000 Btu per hour. Using the Common Vent Capacity portion of Table 504.3(2), read down the Total Vent Height (H) column to 30 feet and across this row to find the smallest vent diameter in the FAN + NAT column that has a Btu per hour rating equal to or greater than 135,000 Btu per hour. The 4-inch common vent has a capacity of 132,000 Btu per hour and the 5-inch common vent has a capacity of 202,000 Btu per hour. Therefore, the 5-inch common vent should be used in this example.

Summary. In this example, the installer may use a 4-inch-diameter, single-wall metal vent connector for the water heater and a 4-inch-diameter, single-wall metal vent connector for the furnace. The common vent should be a 5-inch-diameter Type B vent.

SIZING OF VENTING SYSTEMS

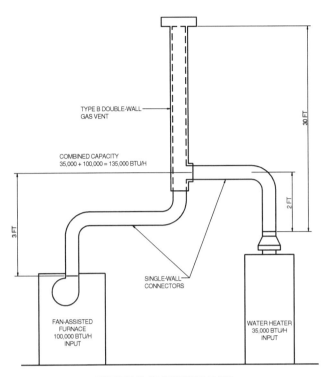

**FIGURE B-18 (EXAMPLE 5A)
COMMON VENTING A DRAFT HOOD WITH A FAN-ASSISTED FURNACE INTO A TYPE B DOUBLE-WALL COMMON VENT**

Example 5b: Common venting into a masonry chimney

In this case, the water heater and fan-assisted furnace of Example 5a are to be common vented into a clay tile-lined masonry chimney with a 30-foot height. The chimney is not exposed to the outdoors below the roof line. The internal dimensions of the clay tile liner are nominally 8 inches by 12 inches. Assuming the same vent connector heights, laterals, and materials found in Example 5a, what are the recommended vent connector diameters, and is this an acceptable installation?

Solution:

Table 504.3(4) is used to size common venting installations involving single-wall connectors into masonry chimneys.

Water Heater Vent Connector Diameter. Using Table 504.3(4), Vent Connector Capacity, read down the Total Vent Height (H) column to 30 feet, and read across the 2-foot Connector Rise (R) row to the first Btu per hour rating in the NAT Max column that is equal to or greater than the water heater input rating. The table shows that a 3-inch vent connector has a maximum input of only 31,000 Btu per hour while a 4-inch vent connector has a maximum input of 57,000 Btu per hour. A 4-inch vent connector must therefore be used.

Furnace Vent Connector Diameter. Using the Vent Connector Capacity portion of Table 504.3(4), read down the Total Vent Height (H) column to 30 feet and across the 3-foot Connector Rise (R) row. Since the furnace has a fan-assisted combustion system, find the first FAN Max column with a Btu per hour rating greater than the furnace input rating. The 4-inch vent connector has a maximum input rating of 127,000 Btu per hour and a minimum input rating of 95,000 Btu per hour. The 100,000 Btu per hour furnace in this example falls within this range, so a 4-inch connector is adequate.

Masonry Chimney. From Table B-1, the equivalent area for a nominal liner size of 8 inches by 12 inches is 63.6 square inches. Using Table 504.3(4), Common Vent Capacity, read down the FAN + NAT column under the Minimum Internal Area of Chimney value of 63 to the row for 30-foot height to find a capacity value of 739,000 Btu per hour. The combined input rating of the furnace and water heater, 135,000 Btu per hour, is less than the table value, so this is an acceptable installation.

Section 504.3.17 requires the common vent area to be no greater than seven times the smallest listed appliance categorized vent area, flue collar area, or draft hood outlet area. Both appliances in this installation have 4-inch-diameter outlets. From Table B-1, the equivalent area for an inside diameter of 4 inches is 12.2 square inches. Seven times 12.2 equals 85.4, which is greater than 63.6, so this configuration is acceptable.

Example 5c: Common venting into an exterior masonry chimney

In this case, the water heater and fan-assisted furnace of Examples 5a and 5b are to be common vented into an exterior masonry chimney. The chimney height, clay tile liner dimensions, and vent connector heights and laterals are the same as in Example 5b. This system is being installed in Charlotte, North Carolina. Does this exterior masonry chimney need to be relined? If so, what corrugated metallic liner size is recommended? What vent connector diameters are recommended?

Solution:

According to Section 504.3.20, Type B vent connectors are required to be used with exterior masonry chimneys. Use Table 504.3(8) to size FAN+NAT common venting installations involving Type-B double wall connectors into exterior masonry chimneys.

The local 99-percent winter design temperature needed to use Table 504.3(8) can be found in the ASHRAE *Handbook of Fundamentals*. For Charlotte, North Carolina, this design temperature is 19°F.

Chimney Liner Requirement. As in Example 5b, use the 63 square inch Internal Area columns for this size clay tile liner. Read down the 63 square inch column of Table 504.3(8a) to the 30-foot height row to find that the combined appliance maximum input is 747,000 Btu per hour. The combined input rating of the appliances in this installation, 135,000 Btu per hour, is less than the maximum value, so this criterion is satisfied. Table 504.3(8b), at a 19°F design temperature, and at the same vent height and internal area used above, shows that the minimum allowable input rating of a space-heating appliance is 470,000 Btu per hour. The furnace input rating of 100,000 Btu per hour is less than this minimum value. So this criterion is not satisfied, and an alternative venting design needs to be used, such as a Type B vent shown in Example 5a or a listed chimney liner system shown in the remainder of the example.

SIZING OF VENTING SYSTEMS

According to Section 504.3.19, Table 504.3(1) or 504.3(2) is used for sizing corrugated metallic liners in masonry chimneys, with the maximum common vent capacities reduced by 20 percent. This example will be continued assuming Type B vent connectors.

Water Heater Vent Connector Diameter. Using Table 504.3(1), Vent Connector Capacity, read down the Total Vent Height (H) column to 30 feet, and read across the 2-foot Connector Rise (R) row to the first Btu/h rating in the NAT Max column that is equal to or greater than the water heater input rating. The table shows that a 3-inch vent connector has a maximum capacity of 39,000 Btu/h. Although this rating is greater than the water heater input rating, a 3-inch vent connector is prohibited by Section 504.3.21. A 4-inch vent connector has a maximum input rating of 70,000 Btu/h and is equal to the draft hood outlet diameter. A 4-inch vent connector is selected.

Furnace Vent Connector Diameter. Using Table 504.3(1), Vent Connector Capacity, read down the Vent Height (H) column to 30 feet, and read across the 3-foot Connector Rise (R) row to the first Btu per hour rating in the FAN Max column that is equal to or greater than the furnace input rating. The 100,000 Btu per hour furnace in this example falls within this range, so a 4-inch connector is adequate.

Chimney Liner Diameter. The total input to the common vent is 135,000 Btu per hour. Using the Common Vent Capacity Portion of Table 504.3(1), read down the Vent Height (H) column to 30 feet and across this row to find the smallest vent diameter in the FAN+NAT column that has a Btu per hour rating greater than 135,000 Btu per hour. The 4-inch common vent has a capacity of 138,000 Btu per hour. Reducing the maximum capacity by 20 percent (Section 504.3.17) results in a maximum capacity for a 4-inch corrugated liner of 110,000 Btu per hour, less than the total input of 135,000 Btu per hour. So a larger liner is needed. The 5-inch common vent capacity listed in Table 504.3(1) is 210,000 Btu per hour, and after reducing by 20 percent is 168,000 Btu per hour. Therefore, a 5-inch corrugated metal liner should be used in this example.

Single-Wall Connectors. Once it has been established that relining the chimney is necessary, Type B double-wall vent connectors are not specifically required. This example could be redone using Table 504.3(2) for single-wall vent connectors. For this case, the vent connector and liner diameters would be the same as found above with Type B double-wall connectors.

TABLE B-1
MASONRY CHIMNEY LINER DIMENSIONS WITH CIRCULAR EQUIVALENTS[a]

NOMINAL LINER SIZE (inches)	INSIDE DIMENSIONS OF LINER (inches)	INSIDE DIAMETER OR EQUIVALENT DIAMETER (inches)	EQUIVALENT AREA (square inches)
4 × 8	2½ × 6½	4	12.2
		5	19.6
		6	28.3
		7	38.3
8 × 8	6¾ × 6¾	7.4	42.7
		8	50.3
8 × 12	6½ × 10½	9	63.6
		10	78.5
12 × 12	9¾ × 9¾	10.4	83.3
		11	95
12 × 16	9½ × 13½	11.8	107.5
		12	113.0
		14	153.9
16 × 16	13¼ × 13¼	14.5	162.9
		15	176.7
16 × 20	13 × 17	16.2	206.1
		18	254.4
20 × 20	16¾ × 16¾	18.2	260.2
		20	314.1
20 × 24	16½ × 20½	20.1	314.2
		22	380.1
24 × 24	20¼ × 20¼	22.1	380.1
		24	452.3
24 × 28	20¼ × 20¼	24.1	456.2
28 × 28	24¼ × 24¼	26.4	543.3
		27	572.5
30 × 30	25½ × 25½	27.9	607
		30	706.8
30 × 36	25½ × 31½	30.9	749.9
		33	855.3
36 × 36	31½ × 31½	34.4	929.4
		36	1017.9

For SI: 1 inch = 25.4 mm, 1 square inch = 645.16 m².

a. Where liner sizes differ dimensionally from those shown in Table B-1, equivalent diameters may be determined from published tables for square and rectangular ducts of equivalent carrying capacity or by other engineering methods.

SIZING OF VENTING SYSTEMS

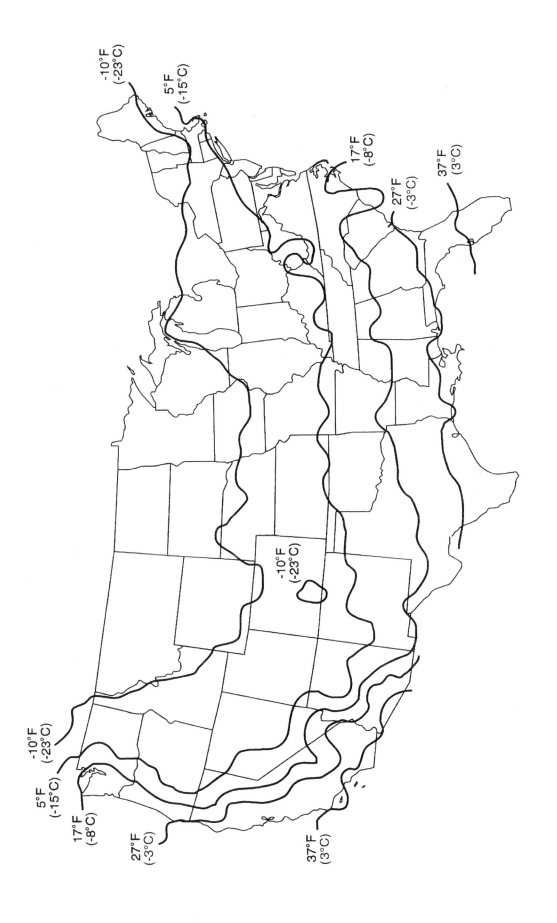

FIGURE B-19

The appendices to the *International Fuel Gas Code®* (IFGC®) are explanatory commentaries of the code itself. Thus, there is no additional commentary for them.

APPENDIX C (IFGS)

EXIT TERMINALS OF MECHANICAL DRAFT AND DIRECT-VENT VENTING SYSTEMS

(This appendix is informative and is not part of the code.)

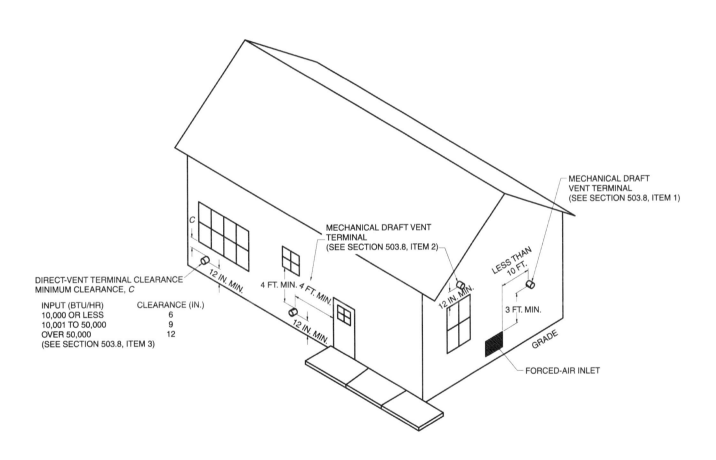

For SI: 1 inch = 25.4 mm, 1 foot = 304.8 mm, 1 British thermal unit per hour = 0.2931 W.

**APPENDIX C
EXIT TERMINALS OF MECHANICAL DRAFT AND DIRECT-VENT VENTING SYSTEMS**

The appendices to the *International Fuel Gas Code®* (IFGC®) are explanatory commentaries of the code itself. Thus, there is no additional commentary for them.

APPENDIX D (IFGS)

RECOMMENDED PROCEDURE FOR SAFETY INSPECTION OF AN EXISTING APPLIANCE INSTALLATION

(This appendix is informative and is not part of the code.)

The following procedure is intended as a guide to aid in determining that an appliance is properly installed and is in a safe condition for continuing use.

This procedure is predicated on central furnace and boiler installations, and it should be recognized that generalized procedures cannot anticipate all situations. Accordingly, in some cases, deviation from this procedure is necessary to determine safe operation of the equipment.

(a) This procedure should be performed prior to any attempt at modification of the appliance or of the installation.

(b) If it is determined there is a condition that could result in unsafe operation, the appliance should be shut off and the owner advised of the unsafe condition. The following steps should be followed in making the safety inspection:

1. Conduct a test for gas leakage. (See Section 406.6)

2. Visually inspect the venting system for proper size and horizontal pitch and determine there is no blockage or restriction, leakage, corrosion, and other deficiencies that could cause an unsafe condition.

3. Shut off all gas to the appliance and shut off any other fuel-gas-burning appliance within the same room. **Use the shutoff valve in the supply line to each appliance.**

4. Inspect burners and crossovers for blockage and corrosion.

5. **Applicable only to furnaces.** Inspect the heat exchanger for cracks, openings, or excessive corrosion.

6. **Applicable only to boilers.** Inspect for evidence of water or combustion product leaks.

7. Insofar as is practical, close all building doors and windows and all doors between the space in which the appliance is located and other spaces of the building. Turn on clothes dryers. Turn on any exhaust fans, such as range hoods and bathroom exhausts, so they will operate at maximum speed. Do not operate a summer exhaust fan. Close fireplace dampers. If, after completing Steps 8 through 13, it is believed sufficient combustion air is not available, refer to Section 304 of this code for guidance.

8. Place the appliance being inspected in operation. **Follow the lighting instructions.** Adjust the thermostat so appliance will operate continuously.

9. Determine that the pilot(s), where provided, is burning properly and that the main burner ignition is satisfactory by interrupting and reestablishing the electrical supply to the appliance in any convenient manner. If the appliance is equipped with a continuous pilot(s), test the pilot safety device(s) to determine if it is operating properly by extinguishing the pilot(s) when the main burner(s) is off and determining, after 3 minutes, that the main burner gas does not flow upon a call for heat. If the appliance is not provided with a pilot(s), test for proper operation of the ignition system in accordance with the appliance manufacturer's lighting and operating instructions.

10. Visually determine that the main burner gas is burning properly (i.e., no floating, lifting, or flashback). Adjust the primary air shutter(s) as required. If the appliance is equipped with high and low flame controlling or flame modulation, check for proper main burner operation at low flame.

11. Test for spillage at the draft hood relief opening after 5 minutes of main burner operation. Use a flame of a match or candle or smoke.

12. Turn on all other fuel-gas-burning appliances within the same room so they will operate at their full inputs. **Follow lighting instructions for each appliance.**

13. Repeat Steps 10 and 11 on the appliance being inspected.

14. Return doors, windows, exhaust fans, fireplace dampers, and any other fuel-gas-burning appliance to their previous conditions of use.

15. **Applicable only to furnaces.** Check both the limit control and the fan control for proper operation. Limit control operation can be checked by blocking the circulating air inlet or temporarily disconnecting the electrical supply to the blower motor and determining that the limit control acts to shut off the main burner gas.

16. **Applicable only to boilers.** Determine that the water pumps are in operating condition. Test low water cutoffs, automatic feed controls, pressure and temperature limit controls, and relief valves in accordance with the manufacturer's recommendations to determine that they are in operating condition.

INDEX

A

ACCESS, APPLIANCES
  Duct furnaces........................... 610.3
  General................................. 306
  Shutoff valves.......... 409.1.3, 409.3.1, 409.5
  Wall furnaces, vented................... 608.6
ADJUSTMENTS.................... 608.6, 621.6
ADMINISTRATION.................... Chapter 1
  Alternate materials and methods......... 105.2
  Alternate methods of sizing chimneys.... 503.5.5
  Appeals................................. 109
  Certificates............................ 104.8
  Duties and powers of code official...... 104
  Fees............................ 106.4, 106.5
  Inspections............. 104.4, 104.8, 107
  Liability............................... 103.4
  Permits................................. 106
  Plan review........................ 106.5.3(3)
  Severability............................ 101.5
  Scope................................... 101.2
  Title................................... 101.1
  Violations and penalties................ 108
AIR, COMBUSTION
  Defined................................. 202
  Requirements.................. 303.3(1), 304
AIR-CONDITIONING EQUIPMENT............ 627
  Clearances.............................. 308.3
**ALTERNATE MATERIALS AND
  METHODS**............................... 105.2
APPLIANCES
  Broilers for indoor use................. 623.5
  Connections to building piping.......... 411
  Cooking................................. 623
  Decorative.............................. 602
  Decorative vented.............. 202, 303.3(2),
                                   Table 503.4, 604
  Domestic ranges......................... 623.4
  Electrical.............................. 309
  Installation........................ Chapter 6
  Prohibited locations.................... 303.3
  Protection from damage.................. 303.4

B

BENDS, PIPE............................ 405
BOILERS
  Existing installations............. Appendix D
  Listed.................................. 631

  Prohibited locations.................... 303.3
  Unlisted................................ 632
  BUSHINGS............... 403.10.4(5), 404.3

C

CENTRAL FURNACES
  Defined................................. 202
  Existing installation.............. Appendix D
CERTIFICATES.......................... 104.8
CHIMNEYS........................... Chapter 5
  Alternate methods of sizing............. 503.5.5
  Clearance reduction..................... 308
  Damper opening area..................... 634
  Defined................................. 202
CLEARANCE REDUCTION................... 308
CLEARANCES
  Air-conditioning equipment.............. 627.4
  Boilers................................. 308.4
  Domestic ranges......................... 623.4
  Floor furnaces.................... 609.4, 609.6
  Open-top broiler units.................. 623.5.1
  Refrigerators........................... 625.1
  Unit heater............................. 620.4
CLOTHES DRYERS
  Defined................................. 202
  Exhaust................................. 614
  General................................. 613
CODE OFFICIAL
  Defined................................. 202
  Duties and powers....................... 104
COMBUSTION AIR
  Combination indoor and outdoor.......... 304.7
  Defined................................. 202
  Exhaust effect.......................... 304.4
  Free area........ 304.5.3.1, 304.5.3.2, 304.6.1,
                                   304.6.2, 304.10
  Horizontal ducts........................ 304.6.1
  Indoor.................................. 304.5
  Outdoor................................. 304.6
  Sources of (from)................... 304.11(4)
  Sauna heaters........................... 615.5
  Vertical ducts.......................... 304.6.1
COMPRESSED NATURAL GAS................ 413
CONCEALED PIPING...................... 404.2
CONDENSATE DISPOSAL................... 307
CONTROLS
  Boilers................................. 631.2

Gas pressure regulators 410.1, 628.4
CONVERSION BURNERS 503.12.1, 619
COOKING APPLIANCES 623
CORROSION PROTECTION 404.8
CREMATORIES 606
**CUTTING, NOTCHING, AND
   BORED HOLES** 302.3

D

DAMPERS, VENT 503.14
DECORATIVE APPLIANCES 602
DEFINITIONS . . . . Chapter 2
DIRECT VENT APPLIANCES
   Defined 202
   Installation 304.1
DITCH FOR PIPING . . . . 107.1(1)
DIVERSITY FACTOR . . . . 402.2, Appendix A
DRAFT HOODS 202, 503.12
DUCT FURNACES 202, 610

E

ELECTRICAL CONNECTIONS 309.2
EXHAUST SYSTEMS 202, 503.2.1, 503.3.4

F

FEES 104.8, 106.4, 106.5
FLOOD HAZARD 301.11
FLOOR FURNACES 609
FURNACES
   Central heating, clearance 308.3, 308.4
   Duct 610
   Floor 609
   Prohibited location 303.3
   Vented wall 608

G

GARAGE, INSTALLATION 305.3, 305.4, 305.5
GASEOUS HYDROGEN SYSTEMS . . . . Chapter 7
   General requirements 703
   Piping, use and handling 704
   Testing 705
GROUNDING, PIPE 309.1

H

HISTORIC BUILDINGS 102.6
**HOT PLATES AND LAUNDRY
   STOVES** . . . . 501.8(3), 623.1

I

ILLUMINATING APPLIANCES 628
INCINERATORS 606, 607

INFRARED RADIANT HEATERS 630
INSPECTIONS 104.4, 104.8, 107
INSTALLATION, APPLIANCES
   Garage 305
   Listed and unlisted appliances 301.3, 305.1
   General 301
   Specific appliances . . . . Chapter 6

K

KILNS 629

L

LIQUEFIED PETROLEUM GAS
   Defined 202
   Motor vehicle fuel-dispensing stations 412
   Storage 401.2
   Systems 402.5.1
   Piping material . . . . 403.6.2, 403.11(4)
   Thread compounds 403.9.3
   Size of pipe or tubing . . . . Appendix A
LOG LIGHTERS 603

M

MAKE-UP AIR HEATERS 611, 612
   Industrial 612
   Venting . . . . 501.8(9)
MATERIALS, DEFECTIVE
   Repair 301.9
   Workmanship and defects 403.7
METERS
   Interconnections 401.6
   Identification 401.7
   Multiple installations 401.6
**MINIMUM SAFE PERFORMANCE,
   VENT SYSTEMS** 503.3

O

OUTLET CLOSURES 404.12
   Location 404.13
OXYGEN DEPLETION SYSTEM
   Defined 202
   Unvented room heaters . . . . 303.3(3), 621.6

P

PIPE SIZING 402
PIPING
   Bends 405
   Changes in direction 405
   Drips and slopes 408
   Installation 404
   Inspection 406

INDEX

Materials 403
Purging............................ 406.7
Sizing............................... 402
Support 407, 415
Testing............................. 406
POOL HEATERS........................... 617
**POWERS AND DUTIES OF
THE CODE OFFICIAL** 104
PROHIBITED INSTALLATIONS
Floor furnaces 609.2
Fuel-burning appliances.............. 303.3
Piping in partitions 404.2
Plastic piping..................... 404.14.1
Unvented room heater 621.4
Vent connectors .... 503.3.3(4)
PURGING............................. 406.7

R

RADIANT HEATERS....................... 630
RANGES, DOMESTIC 623.4
REFRIGERATORS................. 501.8(6), 625
REGULATORS, PRESSURE.......... 410.1, 628.4
ROOFTOP INSTALLATIONS 306.5
ROOM HEATERS
Defined 202
Location 303.3
Unvented.......................... 621
Vented............................ 622

S

SAFETY SHUTOFF DEVICES
Flame safeguard device 602.2
Unvented room heaters 620.6
SAUNA HEATERS......................... 615
SCOPE............................... 101.2
SEISMIC RESISTANCE 301.12
SERVICE SPACE 306
SPA HEATERS 617
STANDARDS........................ Chapter 8
STRUCTURAL SAFETY 302.1
SUPPORTS, PIPING.................. 407, 415

T

TESTING 107
THIMBLE, VENT............. 503.6.3, 503.10.11,
503.10.15
THREADS
Damaged......................... 403.9.1
Specifications...................... 403.9
TOILETS 626

U

UNIT HEATERS......................... 620
UNLISTED BOILERS 632
UNVENTED ROOM HEATERS 621

V

VALIDITY........................... 106.4.2
**VALVES, MULTIPLE
HOUSE LINES**........................ 409.3
VALVES, SHUTOFF
Appliances 409.5
VENTILATING HOODS........... 503.2.1, 503.3.4
**VENTED DECORATIVE
APPLIANCES**......................... 604
VENTED ROOM HEATERS 622
VENTED WALL FURNACES 608
VENTS
Caps 504.3.13
Direct vent....................... 503.2.3
Equipment not requiring vents 501.8
Gas vent termination................ 503.6.7
General .... Chapter 5
Integral 505
Listed and labeled 502.1
Mechanical vent 505
VENT, SIZING
Category I appliances 502
Multiappliance 504.3
Multistory. 504.3.13, 504.3.14, 504.3.15, 504.3.16
Single appliance..................... 504.2
VIBRATION ISOLATION 301.8
VIOLATIONS AND PENALTIES 108

W

WALL FURNACES, VENTED 608
WARM AIR FURNACES 618
WATER HEATERS 624
WIND RESISTANCE.................... 301.10